Characterisation and Modelling of Manufacturing–Microstructure–Property–Mechanism Relationship for Advanced and Emerging Materials

Editors

Lihong Su
School of Mechanical, Materials, Mechatronic and Biomedical Engineering
University of Wollongong
Wollongong
Australia

Peitang Wei
State Key Laboratory of Mechanical Transmissions
Chongqing University
Chongqing
China

Xing Zhao
State Key Laboratory of High Performance Complex Manufacturing
Central South University
Changsha
China

Hui Wang
Institute of Industrial Science (IIS)
The University of Tokyo
Kashiwa
Japan

Editorial Office
MDPI
St. Alban-Anlage 66
4052 Basel, Switzerland

This is a reprint of articles from the Special Issue published online in the open access journal *Materials* (ISSN 1996-1944) (available at: www.mdpi.com/journal/materials/special_issues/manufacturing_Mmicrostructure_Pproperty_mechanism).

For citation purposes, cite each article independently as indicated on the article page online and as indicated below:

LastName, A.A.; LastName, B.B.; LastName, C.C. Article Title. *Journal Name* **Year**, *Volume Number*, Page Range.

ISBN 978-3-0365-7389-2 (Hbk)
ISBN 978-3-0365-7388-5 (PDF)

Characterisation and Modelling of Manufacturing–Microstructure–Property–Mechanism Relationship for Advanced and Emerging Materials

Editors

Lihong Su
Peitang Wei
Xing Zhao
Hui Wang

MDPI • Basel • Beijing • Wuhan • Barcelona • Belgrade • Manchester • Tokyo • Cluj • Tianjin

Contents

About the Editors

Lihong Su

Lihong Su received her Ph.D in Mechanical Engineering in 2013 from the University of Wollongong, Australia, where she is currently working as an Australian Research Council (ARC)-funded Discovery Early Career Researcher Award (DECRA) research fellow and lecture. She leads a research team on the 'Development and application of novel structural and functional materials with advanced manufacturing techniques'. Her major research interests include additive manufacturing, advanced rolling, severe plastic deformation, nano/ultrafine-grained materials, 2D materials, XRD and neutron diffraction, texture and residual stress analysis, advanced modelling, and characterisations. She has published around 100 high-quality journal and conference papers, including contributions of 11 invited or keynote talks in international conferences. She has been awarded 13 competitive research grants with USD 6.8 million of funding and 5 prestigious research awards, including the ARC DECRA Fellowship, Japan Society for the Promotion of Science (JSPS) Fellowship, France–Australia Science Innovation Collaboration (FASIC) Program Early Career Fellowship, and Australia Defence Materials Technology Centre (DMTC) Project Leadership Award.

Peitang Wei

Peitang Wei is currently an associate professor at Chongqing University, China. He is a famous young researcher in the research fields of micro-nano gear transmission, gear anti-fatigue design and manufacturing, and the mesoscopic mechanics modelling of advanced materials. He has already won several prestigious awards, including the second prize of the Science and Technology Progress Award of the Ministry of Education (China). He has been awarded more than 10 research or industrial projects from the Chinese government and industries. In addition, he has published 69 high-quality journal and conference papers and more than 10 patents. He has very strong collaborations with researchers from the UK and USA.

Xing Zhao

Xing Zhao received his Ph.D in mechanical engineering from the University of Wollongong, Australia, in 2015. He is currently a lecturer at Central South University, China. He is an internationally active researcher and expert in computational materials science and molecular dynamic simulation, particularly on the deformation of materials from the view of dislocation and twin activities. He has been invited to present his research in international conferences held in Australia, the USA, and China. He has published more than 30 high-quality papers, such as in *Acta Materialia*, the *Journal of Materials Science and Technology*, *Materials Science and Engineering*, and so on. He has successfully been awarded several competitive research funds based on his excellent research outputs.

Hui Wang

Hui Wang received his Ph.D in mechanical engineering in 2016 from the University of Wollongong, Australia. After working there as an associate research fellow for two years, he has joined the Institute of Industrial Science (IIS) at the University of Tokyo, Japan, as a project research fellow. His research interests include metal forming, materials characterisation, crystal plasticity finite element model, materials integration, machine learning, and deep learning. He has strong research collaborations with researchers from Japan, South Korea, Australia, and China. He has published around 30 high-quality journal papers, and he has often presented his research at international conferences and industrial workshops as an invited speaker. He has been awarded three grants from the Japanese government and industries.

materials

MDPI

Editorial

Characterization and Modelling of Manufacturing–Microstructure–Property–Mechanism Relationship for Advanced and Emerging Materials

Lihong Su [1,*], Peitang Wei [2], Xing Zhao [3] and Hui Wang [4,*]

1 School of Mechanical, Materials, Mechatronic and Biomedical Engineering, University of Wollongong, Wollongong, NSW 2522, Australia
2 State Key Laboratory of Mechanical Transmissions, Chongqing University, Chongqing 400044, China
3 State Key Laboratory of High Performance Complex Manufacturing, Central South University, Changsha 410083, China
4 Institute of Industrial Science (IIS), The University of Tokyo, Kashiwa 277-8574, Chiba, Japan
* Correspondence: lihongsu@uow.edu.au (L.S.); hui_wang@metall.t.u-tokyo.ac.jp (H.W.)

Depending on the state of its raw materials, final products, and processes, materials manufacturing can be classified into either top-down manufacturing and bottom-up manufacturing, or subtractive manufacturing (SM) and additive manufacturing (AM). Some important top-down manufacturing methods include casting [1], welding [2], hot rolling [3], cold rolling [4], cryo-forming [5], heat treatment [6], equal channel angular pressing [7], high pressure torsion, [8], and accumulative roll bonding [9], while some important bottom-up manufacturing techniques include powder metallurgy sintering [10–12] and filtered arc deposition [13]. The disadvantages of bottom-up manufacturing, in comparison to top-down manufacturing, are its limitations in fabricating large samples and its inability to avoid contamination and residual porosity in its final products. Recently, the AM approach has attracted increasing interest, and some important AM processes include wire-arc-based methods [14,15], laser-based AM methods [16], electron-beam-based AM methods [17], and cold spraying [18]. Regarding the development of the materials, both conventional materials with novel structures (such as nanograined microstructures or ultrafine-grained microstructures [19–22]) and novel materials (such as high entropy alloys with more than four principal alloying elements [23]) have become popular for material scientists and engineers, owing to their attractive physical and mechanical properties. In addition to experimental characterizations, advanced modelling also plays a critical role, particularly in revealing the phenomena that are hardly observed in experiments. Some important modelling tools that are related to this manufacturing include the finite element method (FEM) [24], crystal plasticity FEM [25], discrete element method (DEM) [26], molecular dynamic (MD) model [27], and atomistic-continuum coupled multiscale model [28]. With the quick development of computational techniques, these models are providing increasing contributions to our understanding of the manufacturing–structure–property–mechanism relationships between various materials under different conditions. This Special Issue has collected the recent advances and investigations within the research that has been filed on manufacturing, covering all of these above mentioned topics.

Hedhibi et al. [29] studied the influence of pseudo-ternary oxides on mechanical properties and microstructures, by comparing the activating tungsten inert gas (ATIG) weld with the conventional tungsten inert gas (TIG) weld. They optimized the composition of the flux and found an improvement in the ultimate tensile strength (UTS), from about 571 MPa for the conventional TIG weld to about 600 MPa for the optimal ATIG weld. Ahmed et al. [30] developed a novel refilling technique for the friction stir spot welding (FSSW) joints of AA6082-T6 sheets. The mechanical testing showed higher-bearing tensile shear loads in all the refilled FSSW joints than those that were given by the as-received

Citation: Su, L.; Wei, P.; Zhao, X.; Wang, H. Characterization and Modelling of Manufacturing–Microstructure–Property–Mechanism Relationship for Advanced and Emerging Materials. *Materials* **2023**, *16*, 2737. https://doi.org/10.3390/ma16072737

Received: 14 March 2023
Accepted: 28 March 2023
Published: 29 March 2023

FSSW joints. In order to improve the mechanical and tribological performances of an AZ91 Mg alloy, Ataya et al. [31] added short carbon fibers to the AZ91 matrix. They observed a large influence of the carbon fibers' orientation on both the compressive strength and the wear resistance of the Mg composite; however, they did not observe an obvious difference in the hardness. The paper published by Wan et al. [32] is about a laser-based AM of an Mg alloy. They successfully improved the wear and corrosion resistance of an AZ91D alloy by introducing an Al-Si alloy coating with an addition of Y_2O_3, using a laser cladding process. Yin et al. [33] studied the microstructures and textures of various automobile steels and their influences on quasistatic tensile deformation behavior, with a strain rate of 0.001 s^{-1} and dynamic tensile deformation behaviors with a wide strain rate range, varying from 33 to 600 s^{-1}. Zhang et al. [34] investigated the relationship between an ultrasonic vibration treatment and the microstructure evolution during the high-temperature forming process of 9310 steel. It was observed that the flow stress of the 9310 steel decreased with an increase in the deformation temperature or a decrease in the strain rate. Kubiak and Lesnikowski [35] investigated the influence of mechanical deformation on the characteristic impedance of sewed textile signal lines (TSLs). Regardless of the tensile forces, only the substrate weave was found not to affect the characteristic impedance change.

Chen et al. [36] studied the vibration characteristics of submarine-like structures with laminated materials that consisted of spherical, cylindrical, and cone shells with multiple built-in annular plates, based on a numerical model. Li et al. [37] developed a numerical discrete model that was based on a meshless Chebyshev-RPIM shape function, in order to study the vibration of a rotating cross-ply laminated combined conical–cylindrical shell in a thermal environment. In their study, Zhang et al. [38] focused on the vibration characteristics of a laminated composite double cylindrical shell system (LCDCSS), coupled with a variable number of annular plates. Dzwierzynska and Lechwar [39] proposed an algorithmic-aided approach for the design and optimization of the curvilinear steel bar structures of unit roofs, and their structural analysis was further verified by an FEM analysis, taking both the permanent and environmental loads into account. With the help of a computational fluid dynamics (CFD) model, Li et al. [40] simulated a high-pressure hydrogen flow through their newly proposed Tesla-type depressurization structure. They found that this pressure could be reduced by 237% if the standard orifice plate was replaced with a Tesla-type orifice structure. Furthermore, the subject of surface roughness in materials and manufacturing engineering has attracted increasing attention in recent years [41]. Lu et al. [42] studied the surface roughness of face gears, based on a non-contact measurement that was obtained via 3D optical scanning and FEM simulations. The paper by Yang et al. [43] is about the rough surface characterization parameter set (CPS) and redundant parameter set (RPS) that are used for surface modeling and performance. They successfully proposed a model for a performance evaluation of different workpiece surfaces, based on their capacity to fully cover the surface topography information. Li et al. [44] conducted MD simulations in order to study the hydrogen-induced dislocation nucleation (DN) and plastic deformation of <001> and <1–10> grain boundaries (GBs) in nickel bicrystals. Additionally, they also studied the influence of grain size on the hydrogen embrittlement (HE) of nanograined iron materials, which was also based on MD simulations [45]. Their study indicated that grain refinement could be an effective strategy for resisting H-induced brittle failure, owing to the fact that finer materials have a lower H concentration at the GBs, and an improved GB-mediated intergranular deformation, thus resulting in a lesser possibility of initiating cracks.

The variety and quality of all these papers are addressed to both academic and industrial researchers who are looking for new information that can contribute to the advancement of future research in these highly challenging fields. It is our hope, as guest editors, that you find this volume interesting. We would like to express our sincere gratitude to the authors for their contributions and cooperation during the editorial process. We are indebted to the reviewers for their constructive suggestions and comments. We thank the editorial team for their strong support throughout the entire process.

Author Contributions: The authors equally contributed to this work. All authors have read and agreed to the published version of the manuscript.

Acknowledgments: L.S. acknowledges the financial support from Australian Research Council (ARC) through the Discovery Early Career Researcher Award (DECRA) fellowship (No. DE180100124).

Conflicts of Interest: The authors declare no conflict of interest.

References

1. Su, L.; Li, H.; Lu, C.; Li, J.; Fletcher, L.; Simpson, I.; Barbaro, F.; Zheng, L.; Bai, M.; Shen, J.; et al. Transverse and z-Direction CVN Impact Tests of X65 Line Pipe Steels of Two Centerline Segregation Ratings. *Metall. Mater. Trans. A* **2016**, *47*, 3919–3932. [CrossRef]
2. Su, L.; Fei, Z.; Davis, B.; Li, H.; Bornstein, H. Digital Image Correlation Study on Tensile Properties of High Strength Quenched and Temped Steel Weld Joints Prepared by K-TIG and GMAW. *Mater. Sci. Eng. A* **2021**, *827*, 142033. [CrossRef]
3. Deng, G.Y.; Tieu, A.K.; Su, L.H.; Zhu, H.T.; Reid, M.; Zhu, Q.; Kong, C. Microstructural study and residual stress measurement of a hot rolling work roll material during isothermal oxidation. *Int. J. Adv. Manuf. Technol.* **2019**, *102*, 2107–2118. [CrossRef]
4. Deng, G.Y.; Tieu, A.K.; Si, L.Y.; Su, L.H.; Lu, C.; Wang, H.; Liu, M.; Zhu, H.T.; Liu, X.H. Influence of cold rolling reduction on the deformation behaviour and crystallographic orientation development. *Comput. Mater. Sci.* **2014**, *81*, 2–9. [CrossRef]
5. Xiong, H.; Su, L.; Kong, C.; Yu, H. Development of High Performance of Al Alloys via Cryo-Forming: A Review. *Adv. Eng. Mater.* **2021**, *23*, 2001533. [CrossRef]
6. Su, L.; Lu, C.; Tieu, K.; Deng, G. Annealing Behavior of Accumulative Roll Bonding Processed Aluminum Composites. *Steel Res. Int.* **2013**, *84*, 1241–1245. [CrossRef]
7. Deng, G.; Lu, C.; Su, L.; Tieu, A.K.; Li, J.; Liu, M.; Zhu, H.T.; Liu, X.H. Influence of outer corner angle (OCA) on the plastic deformation and texture evolution in equal channel angular pressing. *Comput. Mater. Sci.* **2014**, *81*, 79–88. [CrossRef]
8. Deng, G.; Bhattacharjee, T.; Chong, Y.; Zheng, R.; Bai, Y.; Shibata, A.; Tsuji, N. Influence of Fe addition in CP titanium on phase transformation, microstructure and mechanical properties during high pressure torsion. *J. Alloy. Compd.* **2020**, *822*, 153604. [CrossRef]
9. Su, L.; Lu, C.; Deng, G.; Tieu, K.; Sun, X. Microstructure and Mechanical Properties of 1050/6061 Laminated Composite Processed by Accumulative Roll Bonding. *Rev. Adv. Mater. Sci.* **2013**, *33*, 33–37.
10. Vo, T.D.; Tieu, A.K.; Wexler, D.; Su, L.; Nguyen, C.; Deng, G. Fabrication and Characterization of a Low-Cost Co-Free Al0.8CrFeNi2.2 Eutectic High Entropy Alloy Based Solid Self-Lubricating Composite; Microstructure, Mechanical and Wear Properties. *J. Alloys Compd.* **2022**, *928*, 167087. [CrossRef]
11. Deng, G.; Tieu, A.K.; Su, L.; Wang, P.; Wang, L.; Lan, X.; Cui, S.; Zhu, H. Investigation into reciprocating dry sliding friction and wear properties of bulk CoCrFeNiMo high entropy alloys fabricated by spark plasma sintering and subsequent cold rolling processes: Role of Mo element concentration. *Wear* **2020**, *460–461*, 203440. [CrossRef]
12. Deng, G.; Tieu, A.K.; Lan, X.; Su, L.; Wang, L.; Zhu, Q.; Zhu, H. Effects of normal load and velocity on the dry sliding tribological behaviour of CoCrFeNiMo0.2 high entropy alloy. *Tribol. Int.* **2020**, *144*, 106116. [CrossRef]
13. Sun, Y.; Lu, C.; Yu, H.; Tieu, A.K.; Su, L.; Zhao, Y.; Zhu, H.; Kong, C. Nanomechanical properties of TiCN and TiCN/Ti coatings on Ti prepared by Filtered Arc Deposition. *Mater. Sci. Eng. A* **2015**, *625*, 56–64. [CrossRef]
14. Wang, J.; Pan, Z.; Wang, L.; Su, L.; Carpenter, K.; Wang, J.; Wang, R.; Li, H. In-Situ Dual Wire Arc Additive Manufacturing of NiTi-Coating on Ti6Al4V Alloys: Microstructure Characterization and Mechanical Properties. *Surf. Coat. Technol.* **2020**, *386*, 125439. [CrossRef]
15. Wang, J.; Pan, Z.; Wang, Y.; Wang, L.; Su, L.; Cuiuri, D.; Zhao, Y.; Li, H. Evolution of crystallographic orientation, precipitation, phase transformation and mechanical properties realized by enhancing deposition current for dual-wire arc additive manufactured Ni-rich NiTi alloy. *Addit. Manuf.* **2020**, *34*, 101240. [CrossRef]
16. Yang, Y.; Hu, J.; Liu, X.Y.; Xu, W.; Li, B.; Ling, G.P.; Pang, X.Y.; Tian, Y.Z. Post treatment of an additively manufactured composite consisting of 304L stainless steel and CoCrFeMnNi high-entropy alloy. *Mater. Sci. Eng. A* **2022**, *831*, 142104. [CrossRef]
17. Zhao, X.; Li, S.; Zhang, M.; Liu, Y.; Sercombe, T.B.; Wang, S.; Hao, Y.; Yang, R.; Murr, L.E. Comparison of the microstructures and mechanical properties of Ti–6Al–4V fabricated by selective laser melting and electron beam melting. *Mater. Des.* **2016**, *95*, 21–31. [CrossRef]
18. Gao, P.; Zhang, C.; Wang, R.; Deng, G.; Li, J.; Su, L. Tamping effect during additive manufacturing of copper coating by cold spray: A comprehensive molecular dynamics study. *Addit. Manuf.* **2023**, *66*, 103448. [CrossRef]
19. Su, L.; Lu, C.; Deng, G.; Tieu, K. Microstructure and Mechanical Properties of AA5005/AA6061 Laminated Composite Processed by Accumulative Roll Bonding. *Metall. Mater. Trans. B* **2014**, *45*, 515–522. [CrossRef]
20. Su, L.H.; Lu, C.; Tieu, A.K.; He, L.Z.; Zhang, Y.; Wexler, D. Vacancy-assisted hardening in nanostructured metals. *Mater. Lett.* **2011**, *65*, 514–516. [CrossRef]
21. Deng, G.; Chong, Y.; Su, L.; Zhan, L.; Wei, P.; Zhao, X.; Zhang, L.; Tian, Y.; Zhu, H.; Tsuji, N. Mechanisms of remarkable wear reduction and evolutions of subsurface microstructure and nano-mechanical properties during dry sliding of nano-grained Ti6Al4V alloy: A comparative study. *Tribol. Int.* **2022**, *169*, 107464. [CrossRef]
22. Su, L.H.; Lu, C.; Deng, G.Y.; Tieu, K.; Zhang, L.C.; Guagliardo, P.; Samarin, S.N.; Williams, J.F. Vacancy-Type Defects Study on Ultra-Fine Grained Aluminum Processed by Severe Plastic Deformation. *Sci. Adv. Mater.* **2014**, *6*, 1338–1345. [CrossRef]

23. Tian, Y.-Z.; Peng, S.-Y.; Chen, S.-F.; Gu, Z.-J.; Yang, Y.; Shang, X.-L.; Deng, G.-Y.; Su, L.-H.; Sun, S.-J. Temperature-dependent tensile properties of ultrafine-grained C-doped CoCrFeMnNi high-entropy alloy. *Rare Met.* **2022**, *41*, 2877–2885. [CrossRef]
24. Deng, G.Y.; Zhu, H.T.; Tieu, A.K.; Zhu, Q.; Su, L.H.; Reid, M.; Wei, P.T.; Zhang, L.; Wang, H.; Zhang, J.; et al. Numerical Evaluation of a High Speed Steel Work Roll during Hot Strip Rolling Process. *Mater. Sci. Forum* **2017**, *904*, 55–60. [CrossRef]
25. Deng, G.Y.; Lu, C.; Tieu, A.K.; Su, L.H.; Huynh, N.N.; Liu, X.H. Crystal plasticity investigation of friction effect on texture evolution of Al single crystal during ECAP. *J. Mater. Sci.* **2010**, *45*, 4711–4717. [CrossRef]
26. Nguyen, V.D.X.; Tieu, A.K.; Andre, D.; Su, L.; Zhu, H. Discrete Element Method using Cohesive Platic Beam for Modeling Elasto-Plastic Deformation of Ductile Materials. *Comput. Part. Mech.* **2021**, *8*, 437–457. [CrossRef]
27. Zheng, X.; Su, L.; Deng, G.; Zhang, J.; Zhu, H.; Tieu, A.K. Study on Lubrication Characteristics of C4-Alkane and Nanoparticle during Boundary Friction by Molecular Dynamics Simulation. *Metals* **2021**, *11*, 1464. [CrossRef]
28. Zhang, J.; Zhang, L.; Tieu, A.K.; Michal, G.; Zhu, H.T.; Deng, G.Y. Finite-Temperature Multiscale Simulations for 3D Nanoscale Contacts. *Appl. Mech. Mater.* **2016**, *846*, 288–293. [CrossRef]
29. Hedhibi, A.C.; Touileb, K.; Djoudjou, R.; Ouis, A.; Alrobei, H.; Ahmed, M.M.Z. Mechanical Properties and Microstructure of TIG and ATIG Welded 316L Austenitic Stainless Steel with Multi-Components Flux Optimization Using Mixing Design Method and Particle Swarm Optimization (PSO). *Materials* **2021**, *14*, 7139. [CrossRef]
30. Ahmed, M.M.Z.; Seleman, M.M.E.-S.; Ahmed, E.; Reyad, H.A.; Alsaleh, N.A.; Albaijan, I. A Novel Friction Stir Deposition Technique to Refill Keyhole of Friction Stir Spot Welded AA6082-T6 Dissimilar Joints of Different Sheet Thicknesses. *Materials* **2022**, *15*, 6799. [CrossRef]
31. Ataya, S.; Seleman, M.M.E.-S.; Latief, F.H.; Ahmed, M.M.Z.; Hajlaoui, K.; Soliman, A.M.; Alsaleh, N.A.; Habba, M.I.A. Wear Characteristics of Mg Alloy AZ91 Reinforced with Oriented Short Carbon Fibers. *Materials* **2022**, *15*, 4841. [CrossRef] [PubMed]
32. Wan, X.; Tian, C.; Li, Y.; Zhou, J.; Qian, S.; Su, L.; Wang, L. Effect of Y_2O_3 Addition on Microstructure and Properties of Laser Cladded Al-Si Coatings on AZ91D Magnesium Alloy. *Materials* **2023**, *16*, 338. [CrossRef]
33. Yin, S.; Xue, Y.; Cui, H.; Pei, X.; Hu, C.; Wang, Y.; Tian, Q. Effect of Material Anisotropy on the Mechanical Response of Automotive Steel under High Strain Rates. *Materials* **2022**, *15*, 669. [CrossRef] [PubMed]
34. Zhang, Y.; Zhou, W.; Tang, J.; He, Y. Understanding Effects of Ultrasonic Vibration on Microstructure Evolution in Hot Forming Process via Cellular Automata Method. *Materials* **2022**, *15*, 7359. [CrossRef]
35. Kubiak, P.; Leśnikowski, J. Influence of Mechanical Deformations on the Characteristic Impedance of Sewed Textile Signal Lines. *Materials* **2022**, *15*, 1149. [CrossRef] [PubMed]
36. Chen, Z.; Zhong, R.; Hu, S.; Qin, B.; Zhao, X. Effect of Multiple Annular Plates on Vibration Characteristics of Laminated Submarine-Like Structures. *Materials* **2022**, *15*, 6357. [CrossRef]
37. Li, Z.; Hu, S.; Zhong, R.; Qin, B.; Zhao, X. Meshless Chebyshev RPIM Solution for Free Vibration of Rotating Cross-Ply Laminated Combined Cylindrical-Conical Shells in Thermal Environment. *Materials* **2022**, *15*, 6177. [CrossRef]
38. Zhang, Y.; Shi, D.; He, D. Vibration Characteristics of a Laminated Composite Double-Cylindrical Shell System Coupled with a Variable Number of Annular Plates. *Materials* **2022**, *15*, 4246. [CrossRef]
39. Dzwierzynska, J.; Lechwar, P. Algorithmic-Aided Approach for the Design and Evaluation of Curvilinear Steel Bar Structures of Unit Roofs. *Materials* **2022**, *15*, 3656. [CrossRef]
40. Li, B.; Liu, Y.; Li, J.; Liu, B.; Wang, X.; Deng, G. Investigation of a Novel Hydrogen Depressurization Structure Constituted by an Orifice Plate with Tesla-Type Channels. *Materials* **2022**, *15*, 4918. [CrossRef]
41. Nie, N.; Su, L.; Deng, G.; Li, H.; Yu, H.; Tieu, A.K. A review on plastic deformation induced surface/interface roughening of sheet metallic materials. *J. Mater. Res. Technol.* **2021**, *15*, 6574–6607. [CrossRef]
42. Lu, X.; Zhao, X.; Hu, B.; Zhou, Y.; Cao, Z.; Tang, J. A Measurement Solution of Face Gears with 3D Optical Scanning. *Materials* **2022**, *15*, 6069. [CrossRef]
43. Yang, D.; Tang, J.; Xia, F.; Zhou, W. Rough Surface Characterization Parameter Set and Redundant Parameter Set for Surface Modeling and Performance Research. *Materials* **2022**, *15*, 5971. [CrossRef] [PubMed]
44. Li, J.; Wu, Z.; Teng, L.; Deng, G.; Wang, R.; Lu, C.; Li, W.; Huang, X.; Liu, Y. Hydrogen-Induced Dislocation Nucleation and Plastic Deformation of <001> and <1–10> Grain Boundaries in Nickel. *Materials* **2022**, *15*, 6503. [PubMed]
45. Li, J.; Wu, Z.; Wang, F.; Zhang, L.; Zhou, C.; Lu, C.; Teng, L.; Lin, Q. Study on the Hydrogen Embrittlement of Nanograined Materials with Different Grain Sizes by Atomistic Simulation. *Materials* **2022**, *15*, 4589. [CrossRef] [PubMed]

materials

MDPI

Article

Mechanical Properties and Microstructure of TIG and ATIG Welded 316L Austenitic Stainless Steel with Multi-Components Flux Optimization Using Mixing Design Method and Particle Swarm Optimization (PSO)

Abdeljlil Chihaoui Hedhibi [1,2], Kamel Touileb [2], Rachid Djoudjou [2,*], Abousoufiane Ouis [2], Hussein Alrobei [2] and Mohamed M. Z. Ahmed [2,3,*]

1 Department of Mechanical Engineering, National Engineering School of Tunis (ENIT), El-Manar University, P.O. Box 37, Belvedere Tunis, Tunis 1002, Tunisia; a.hedhibi@psau.edu.sa
2 Department of Mechanical Engineering, College of Engineering in Al-Kharj, Prince Sattam bin Abdulaziz University, P.O. Box 655, Al-Kharj 16273, Saudi Arabia; k.touileb@psau.edu.sa (K.T.); a.ouis@psau.edu.sa (A.O.); h.alrobei@psau.edu.sa (H.A.)
3 Department of Metallurgical and Materials Engineering, Faculty of Petroleum and Mining Engineering, Suez University, Suez 43512, Egypt
* Correspondence: r.djoudjou@psau.edu.sa (R.D.); moh.ahmed@psau.edu.sa (M.M.Z.A.)

Citation: Hedhibi, A.C.; Touileb, K.; Djoudjou, R.; Ouis, A.; Alrobei, H.; Ahmed, M.M.Z. Mechanical Properties and Microstructure of TIG and ATIG Welded 316L Austenitic Stainless Steel with Multi-Components Flux Optimization Using Mixing Design Method and Particle Swarm Optimization (PSO). *Materials* **2021**, *14*, 7139. https://doi.org/10.3390/ma14237139

Academic Editors: Lihong Su and Andrea Di Schino

Received: 25 October 2021
Accepted: 22 November 2021
Published: 24 November 2021

Abstract: In this study, the effects of pseudo-ternary oxides on mechanical properties and microstructure of 316L stainless steel tungsten inert gas (TIG) and activating tungsten inert gas (ATIG) welded joints were investigated. The novelty in this work is introducing a metaheuristic technique called the particle swarm optimization (PSO) method to develop a mathematical model of the ultimate tensile strength (UTS) in terms of proportions of oxides flux. A constrained optimization algorithm available in Matlab 2020 optimization toolbox is used to find the optimal percentages of the selected powders that provide the maximum UTS. The study indicates that the optimal composition of flux was: 32% Cr_2O_3, 43% ZrO_2, 8% Si_2O, and 17% CaF_2. The UTS was 571 MPa for conventional TIG weld and rose to 600 MPa for the optimal ATIG flux. The obtained result of hardness for the optimal ATIG was 176 HV against 175 HV for conventional TIG weld. The energy absorbed in the weld zone during the impact test was 267 J/cm^2 for the optimal ATIG weld and slightly higher than that of conventional TIG weld 256 J/cm^2. Fracture surface examined by scanning electron microscope (SEM) shows ductile fracture for ATIG weld with small and multiple dimples in comparison for TIG weld. Moreover, the depth of optimized flux is greater than that of TIG weld by two times. The ratio D/W was improved by 3.13 times. Energy dispersive spectroscopy (EDS) analysis shows traces of the sulfur element in the TIG weld zone.

Keywords: ATIG welding; mixing design method; particle swarm optimization (PSO); pseudo-ternary flux; 316L SS; mathematical modeling; mechanical properties; microstructure

1. Introduction

Austenitic stainless steels are the most common stainless steel. They are used in many fields as oil, shipbuilding, machinery, and marine applications. They are characterized by good strength, high toughness, and excellent corrosion resistance. Fusion welding is the most popular method to join workpieces in many industrial applications.

Tungsten Inert Gas (TIG) welding is a widespread process in industries, but a limited thickness can be joined in a single pass with this process. TIG welding is also very sensitive to the chemical composition of the base metal. Moreover, materials thickness greater than 3 mm requires multiple passes to achieve full penetration weld; therefore, the productivity of the process is reduced. The activated tungsten inert gas (ATIG) technic is a variant of conventional TIG welding. In ATIG welding, a thin layer of flux is deposited on the workpiece before the welding operation. ATIG offers the possibility of increasing the penetration depth using the same equipment and parameters as conventional TIG. Comparatively to TIG, ATIG has many advantages. It eliminates the need for edge preparation, increases the penetration depth, and reduces the number of weld pass [1,2]. Three mechanisms have been proposed; the increase in ATIG weld penetration can be attributed to the reversal of Marangoni convection [3,4]. The second mechanism is the arc constriction proposed by D. S. Howse et al. [5,6]. The third mechanism, proposed by Sire, consists of a constriction of arc by using a flux characterized by high melting point and high electrical resistivity [7,8]. In recent years, considerable works were carried out using Taguchi method design to optimize TIG welding process parameters to improve both depth and weld aspects [9,10]. Some studies were oriented to study the influence of welding parameters on enhancing mechanical properties [11,12]. Other works were dedicated to optimizing the composition of paste to improve the depth and weld aspects [13,14] using the mixing method. Recently, other methods like genetic algorithms, simulated annealing, particle swarm optimization [15], etc., have been used to optimize solution in many industries [16,17]. In this work, the tensile strength of weld joints is optimized as an output parameter by varying the input parameters that are the combinations of oxides powder without varying the welding parameters.

The objective of this work is to elaborate an appropriate composition of flux to improve mechanical properties of the full penetrated 316L stainless steel ATIG weld. Mixing method simplex design degree four was combined with particle swarm optimization (PSO) method to optimize combination powders. The developed mathematical model and optimization methodology can help researchers and engineers working in developing fluxes for weld improvements. This study is a contribution to expanding research dedicated to ATIG welding of 316L.

2. Materials and Methods

2.1. Material

The material used in this study is the austenitic stainless steel grade 316L. The chemical composition is shown in Table 1. Rectangular pieces of 6 mm thickness were cut from the received plate to perform welding of 20 cm line.

Table 1. Chemical composition of 316L stainless steel.

Elements	C	Mn	Si	P	S	Cr	Ni	Mo	N	Cu	Fe
Weight %	0.026	1.47	0.42	0.034	0.0016	16.60	10.08	2.14	0.044	0.50	Balance

2.2. TIG-ATIG Welding Platform

First of all, the plates were cleaned with acetone. The oxide powders were dried in the furnace for 1 h at 180 °C to eliminate the humidity. Then, a thin layer of a mixed powder with methanol was applied using a brush to the surface subject to the welding as shown in Figure 1a. The mean coating density of flux was about 4–5 mg/cm^2.

The tungsten inert gas welding machine was used. The electrode used has a diameter of 3.2 mm and the torch was mounted on a motorized carriage as shown in Figure 1b. The experimental parameters selected for welding are presented in Table 2.

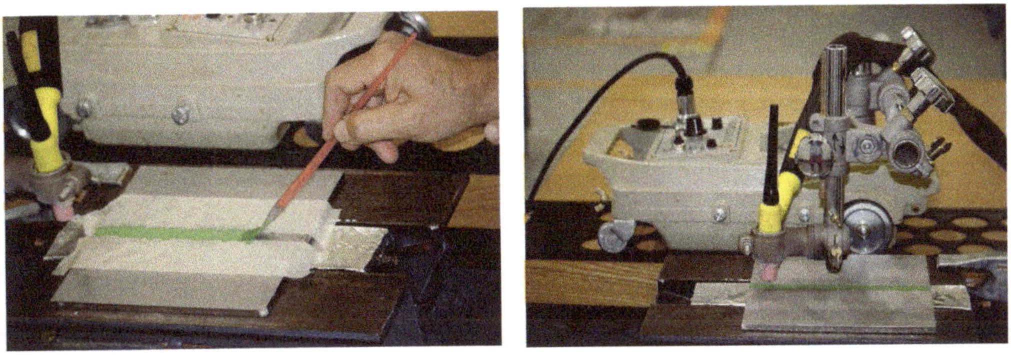

(a) (b)

Figure 1. The deposition of flux on the workpiece (**a**), motorized carriage (**b**).

Table 2. Welding parameters.

Parameters	Range
Welding speed	150 mm/min
Welding current	180 A
Arc length	2 mm
Electrode tip angle	45°
Shielding gas on the workpiece	Argon with flow rate 10 L/min
Shielding gas on the backside	Argon with flow rate 5 L/min
Polarity	DCEN

Figure 2a shows a good external aspect for TIG weld line, also surface ATIG weld line in Figure 2b is free of slugs, which indicates that the optimal flux is well consumed with small traces of residue.

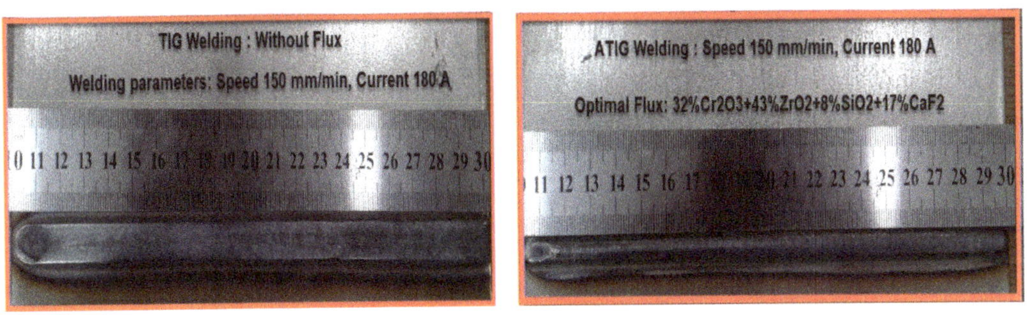

(a) (b)

Figure 2. The external weld line appearance for TIG (**a**) and ATIG (**b**) weld lines.

Pieces were cut from the welded plates for mechanical testing, weld morphology, and microstructure study according to the schematic drawing shown in Figure 3. The cutting machine used is Hydro-Jet Eco 0515 SL-Micro waterjet Cutting (KNUTH Germany).

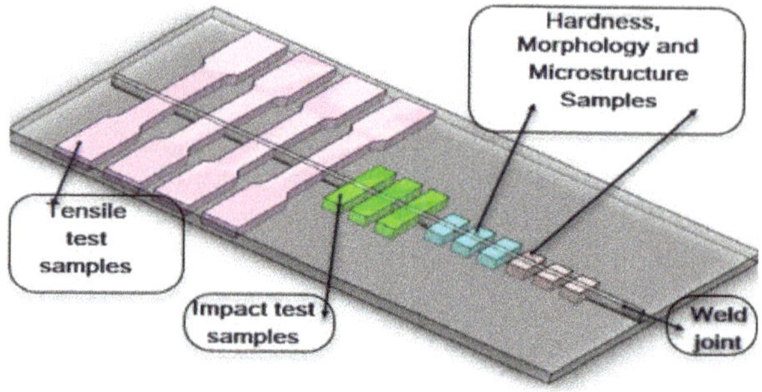

Figure 3. Schematic drawing showing test specimens.

2.3. Tensile Test

Design of experiment and mathematical modelling will be applied to the tensile strength property. Optimal flux, which will give the maximum UTS resulting from the tensile test, is used for the rest of the study.

The tensile tests were performed with a computer control electrohydraulic servo universal testing machine model WAW-300E (Jinan testing equipment IE, Jinan, China), at a test rate of 0.5 mm/min, at 0.5 kN/s load rate, and at low strain rate of 1.6×10^{-4} s^{-1}.

2.4. Weld Bead Aspect

The cross-sections of the weld beads for ATIG with optimal flux and conventional TIG were photographed using an optical microscope CAROLINA (CAROLINA, Burlington, NJ, USA). The morphology of the welds were checked using Motic Images plus version 2.0 software integrated with an optical microscope.

2.5. Microstructure Assessment

The microstructural characterization of the fusion zone of both TIG and ATIG welding has been analyzed. Micrographs were taken on JEOL JSM-7600F scanning electronic microscope, (SEM). The areas image processing software from Microvision Instruments (Microvision Instruments, Imager M2.m, Paris, France) was used to measure the ferrite volume proportions.

2.6. Hardness Test

Vickers hardness tests were performed by a digital hardness tester model HVS-50 (SCTMC, Shanghai, China) with a standard load of 98 N. Eight indentations in the weld bead on fusion zone (FZ) and at the heat-affected zone (HAZ) were performed on each sample, with about 0.5 mm between two indentations, as shown in Figure 4.

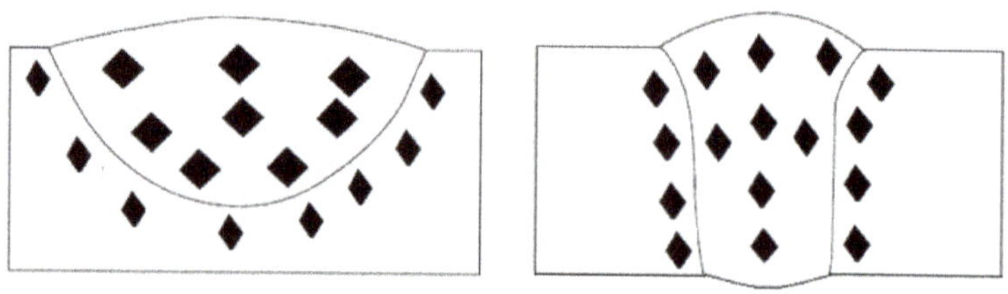

(a) (b)

Figure 4. Hardness test locations in the FZ and the HAZ of TIG (**a**) and ATIG (**b**) [13].

2.7. Impact Test

Impact testing was performed on three samples for TIG and three samples for ATIG weld only in the fusion zone (FZ) with the Charpy "V" notch impact testing machine model JBS-500 (Jinan testing equipment IE, Jinan, China), specimens are shown in Figure 5.

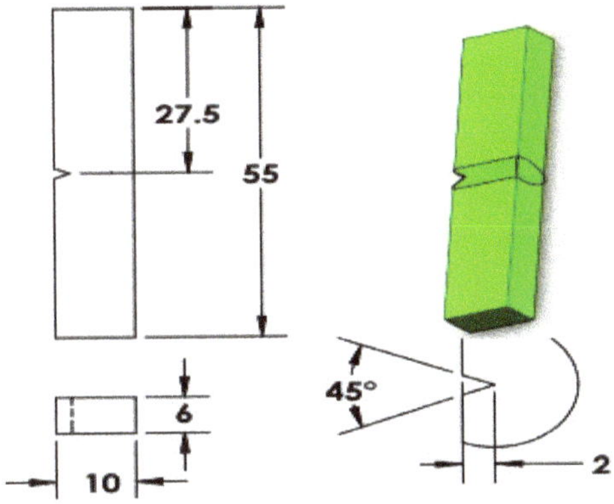

Figure 5. Sample for impact test "V" notch in the fusion zone (FZ) of TIG and ATIG Welded.

2.8. Design of Experiment Methodology

To get the maximum amount of useful information with the least amount of experimentation, the design of experiments (DOE) has been used; mixing method in Minitab 17 software was applied. First, to compare the effect of the mono-flux oxides on UTS of welds, thirteen oxides (SiO_2, TiO_2, Fe_2O_3, MnO_2, Cr_2O_3, ZrO_2, CaO, Mn_2O_3, V_2O_5, MoO_3, SrO, Co_2O_3, and MgO) were tested. Among these thirteen oxides, three oxides Cr_2O_3, TiO_2, and ZrO_2 that gave the best UTS were selected to be used in the mixing design method. Based on the simplex lattice degree four designs, nineteen combinations from the selected oxides have been prepared. For each combination, the three selected oxides Cr_2O_3, TiO_2, and ZrO_2 vary and 25% of (8% SiO_2 + 17% CaF_2) was added and kept fixed to get a pseudo ternary combination. Particle swarm optimization (PSO) method is used to establish an equation relating UTS to the proportions of the selected oxides. Finally, Matlab R2020 is carried out to obtain the optimal combination of oxides, hence, to maximize the UTS of weld through a constrained optimization algorithm.

Silica (SiO_2) has been added because it increases the current carrying capacity. Consequently, silicate (SiO_2) increases arc voltage and depth penetration of the weld bead [18]. On the other hand, the addition of calcium fluoride (CaF_2) had several advantages in that it reduces the dissolved silicon content of weld metal, prevents the deleterious effect of silicon on hot cracking, and lowers the melting range of the flux. Moreover, fluorine gases escaped from the weld pool interact with outer arc electrons leading to constrict arc [19]. The presence of fluorine in arc welding reduces the anode spot and tends to increase the energy density of the heat source and electromagnetic force in the weld pool. As a result, relatively narrow and deep weld morphology is formed [20,21]. Nineteen weld lines were executed and three samples were cut for the tensile test from each of the nineteen combinations.

2.9. Mathematical Modelling

For the first step of work, we developed a mathematical model where the UTS is written in terms of these selected oxides' percentages. The technic used to get a mathematical model will be presented later in this paper. In the second step, the optimal combination that maximizes UTS is determined.

Finally, based on the optimal formulation, ATIG and TIG weld lines have been carried out on a single plate.

3. Results and Discussion

3.1. Tensile Test

3.1.1. Selection of Candidate Oxides

Thirteen oxides were tested. The weld line was executed on the plain plate using welding parameters cited in Table 2. It can be seen in Table 3 that the highest value of UTS is 565 MPa, which was obtained for the sample welded with ZrO_2 flux, followed by the sample welded with Cr_2O_3 flux with 559 MPa and, then by the sample welded with TiO_2 flux with 542 MPa. Therefore, the selected oxides were Cr_2O_3, TiO_2, and ZrO_2.

Table 3. UTS (MPa) for different ATIG mono-oxide welds.

Oxide	SiO_2	TiO_2	Fe_2O_3	MnO_2	Cr_2O_3	ZrO_2	CaO	Mn_2O_3	V_2O_5	MoO_3	SrO	Co_2O_3	MgO
UTS	529	542	538	539	559	565	536	534	481	491	338	506	523

3.1.2. Mixture Design Combinations and UTS Response Values

Based on the simplex lattice degree four design, nineteen combinations have been prepared. As shown in Table 4, for each combination, the three selected oxides Cr_2O_3, TiO_2, and ZrO_2 vary and 25% of (8% SiO_2 + 17% CaF_2) is kept fixed.

Table 4. Different pseudo-ternary combinations with the three selected oxides.

Exp. No.	Cr_2O_3 [Weight %]	TiO_2 [Weight %]	ZrO_2 [Weight %]	$SiO_2 + CaF_2$ [Weight %]
1	56.25	18.75	0.00	25
2	56.25	0.00	18.75	25
3	37.50	18.75	18.75	25
4	37.50	0.00	37.50	25
5	37.50	37.50	0.00	25
6	18.75	56.25	0.00	25
7	18.75	37.50	18.75	25
8	18.75	18.75	37.50	25
9	0.00	56.25	18.75	25
10	0.00	37.50	37.50	25
11	0.00	18.75	56.25	25
12	25.00	25.00	25.00	25
13	50.00	12.50	12.50	25
14	12.50	50.00	12.50	25
15	12.50	12.50	50.00	25
16	75.00	0.00	0.00	25
17	0.00	75.00	0.00	25
18	0.00	0.00	75.00	25
19	18.75	0.00	56.25	25

Table 5 shows the Tensile test (UTS) results as well as the standards deviation (σ). The value of σ is less than 21 MPa for UTS results, which indicates that the disparities are acceptable.

Table 5. UTS of a set of experiments with the different combinations and the related standard deviation σ.

Exp. No.	Number of Tests	Max. UTS [MPa]	Min. UTS [MPa]	Average UTS [MPa]	σ Standard Deviation
1	3	577	552	564	12.58
2	3	592	587	590	2.51
3	3	595	575	578	16.16
4	3	590	557	579	19.05
5	3	570	538	551	16.82
6	3	577	541	558	18.14
7	3	594	560	582	19.34
8	3	590	556	579	19.62
9	3	585	557	575	15.88
10	3	600	563	587	20.55
11	3	594	561	581	17.78
12	3	594	558	582	20.78
13	3	592	558	579	18.35
14	3	596	560	582	19.50
15	3	593	562	582	17.61
16	3	570	558	563	6.02
17	3	555	542	550	6.80
18	3	585	565	580	11.54
19	3	593	591	592	1.15

3.1.3. Mathematical Model

In this study, the three selected oxides (Cr_2O_3, TiO_2, ZrO_2) percentages vary for the combined oxide with a fixed 25% of (SiO_2+ CaF_2) oxide to form the pseudo-ternary ATIG are used. The UTS is then expressed as a function of the percentages of the three oxides as follows:

UTS = f (X1, X2, X3, X4). X1, X2, X3 denote, respectively, the proportions (in terms of percentages) of the three oxides, whereas the proportion of the fourth oxide is set to be X4 = 25%.

The chosen mathematical model that describes the effect of these percentages on the UTS is the coupling of second-order model as proposed in [22] with conventional linear regression as in [23]. This model includes three components: (i) the linear effect of the proportions, (ii) their quadratic effects, and (iii) the interactions between those proportions. The general form of the mathematical model is given below in expression 1:

$$\text{UTS (predicted)} = (\alpha 1)*X(1) + (\alpha 2)*X(2) + (\alpha 3)*X(3) + (\alpha 4)*X(1)*X(1) + (\alpha 5)*X(2)*X(2) + (\alpha 6)*X(3)*X(3) + (\alpha 7)*X(1)*X(2) + (\alpha 8)*X(1)*X(3) + (\alpha 9)*X(2)*X(3) + (\alpha 10)*X(4) \quad (1)$$

With a fixed X(4) = 25%.

3.1.4. Optimization Process Details

The first step objective was to find the optimal parameters that minimize the quadratic error between the measured UTS and the UTS provided by the model. Thus, the modeling problem is converted into an optimization problem having the model coefficients as decision variables and the quadratic error as the objective function to be minimized. Since this problem may present many irregularities such as non-convexity of the criterion and many feasibility constraints, we used a metaheuristic technique called particle swarm optimization (PSO) to solve this problem. In PSO, a set of candidate solutions (coefficients of the model) are initialized randomly within the search-space limits and then "flown" progressively toward a sub-optimal solution while combining three components: (i) follow their current velocities, (ii) go back to their best positions visited so far, and (iii) go to the position of the best neighbor. Each particle in the group (called swarm) is assigned four vectors: the position including the model parameters, a velocity, the best personal position, and the global best position. The move equations of each particle at the kth iteration of the optimization process are provided below [24,25]:

$$V_{k+1}^{i} = w_k V_k^i + c_1 r_1 \left(P^i - \alpha^i\right) + c_2 r_2 \left(G^i - \alpha^i\right) \tag{2}$$

$$\alpha_{k+1}^{i} = \alpha_k^i + V_{k+1}^{i} \tag{3}$$

$$w_k = w_{max} - \frac{w_{max} - w_{min}}{\mathrm{k_{max}}} \times k \tag{4}$$

As provided in many papers such as [24,25], the inertia weight decreases linearly from 0.9 to 0.4. $c_1 = c_2 = 0.75$ are the cognitive and social factors. r_1 and r_2 are two random numbers generated randomly between 0 and 1. The model parameters' search limits are set, respectively, to -1 and +1 for the overall model stability [25]. The optimization process is stopped after a certain number of iterations (k_{max} = 5000). After this global search step, a local search is conducted to find better solutions around the global sub-optimal solution yielded by the PSO algorithm. After running the optimization process many times, the selected UTS model as expressed by X1, X2, and X3 is given below:

$$\begin{aligned} \text{UTS (predicted)} & \\ &= -0.7940(\%\mathrm{Cr_2O_3}) + 0.1706(\%\mathrm{TiO_2}) + 0.3442(\%\mathrm{ZrO_2}) \\ &+0.1094(\%\mathrm{Cr_2O_3})^2 + 0.0944(\%\mathrm{TiO_2})^2 + 0.0959(\%\mathrm{ZrO_2})^2 \\ &+0.2032(\%\mathrm{Cr_2O_3})(\%\mathrm{TiO_2}) + 0.2173(\%\mathrm{Cr_2O_3})(\%\mathrm{ZrO_2}) \\ &+0.2047(\%\mathrm{TiO_2})(\%\mathrm{ZrO_2}) + 0.4231 \end{aligned} \tag{5}$$

Three performance metrics are used to evaluate the effectiveness of the developed models as follows [25]:

Mean Absolute Percentage Error (MAPE (%)):

$$\mathrm{MAPE} = \frac{100}{\mathrm{N_2}} \sum_{t=1}^{N_2} \frac{|UTS(t) - UTSpredicted(t)|}{UTS\ mean} \tag{6}$$

- Coefficient of determination (R^2 (%))

$$\mathrm{R^2} = 100 \times \left(1 - \frac{\frac{1}{N_2}\sum_{t=1}^{N_2}(UTS(t) - UTSpredicted(t))^2}{\frac{1}{N_2}\sum_{t=1}^{N_2}(UTS(t) - UTS\ mean)^2}\right) \tag{7}$$

- Root Mean Square Error (RMSE (MPa))

$$\mathrm{RMSE} = \sqrt{\frac{1}{\mathrm{N_2}} \sum_{t=1}^{N_2} (UTS(t) - UTSpredicted(t))^2} \tag{8}$$

With N_2 = 19 (number of measurements)

The accuracy of this model as measured by the above cited measurement indicators are found to be as follows: MAPE = 0.7952%, R^2 = 72.04%, and RMSE = 5.5955 MPa.

The second step of the optimization process consists of finding the optimal percentages of the three oxides that provide the maximum UTS. To achieve this goal, a constrained optimization algorithm is used. This algorithm has provided the following optimal combination shown in Table 6. The predicted value is UTS = 588 MPa as shown in Table 7.

Table 6. The optimal composition of flux.

Variables	Cr_2O_3	TiO_2	ZrO_2	SiO_2	CaF_2
Single Percentages	32%	0%	43%	8%	17%
Combined Percentages		75% (Cr_2O_3 + ZrO_2)		25% (SiO_2 + CaF_2)	

Table 7. Predicted responses UTS of weld executed with optimal flux.

Response	Predicted Response (MPa)
UTS	588.27

3.1.5. Experimental Validation

The validation test was the last step in the experimental process. A confirmation test was performed according to the optimum flux composition. Table 8 shows that UTS for ATIG weld reached 600 MPa, which is higher than that of conventional TIG welding (571 MPa). Moreover, the UTS of ATIG weld is greater than the expected value calculated by the mathematical model (588 MPa). The UTS of the optimal flux (ATIG) weld (600 MPa) is close to that of base metal (624 MPa) shown in Table 9. The ATIG welding can reduce the heat input per unit length in welds and the residual stress of the weld can be reduced [26,27]. On the other hand, the increase of UTS about 29 MPA in favor of ATIG comparatively to conventional TIG weld can be attributed to the increase of the retained ferrite volume proportions ATIG weld. Moreover, high heat input in TIG weld exhibits a coarse ferrite distribution and lower strength than the lower heat input in ATIG weld, which has a relatively fine ferrite distribution [27]. The value of standard deviation (σ) is less than 5 MPa.

Table 8. Results of UTS (MPa) and standard deviation of TIG and ATIG.

Sample	Number of Tests	UTS Max.	UTS Min.	UTS Mean	Standards Deviation (σ)
TIG	4	578	568	571	4.57
ATIG	4	602	596	600	2.1

Table 9. UTS comparison between mathematical model, ATIG, TIG, and base metal.

Sample	UTS (MPa)
Expected value by mathematical model	588
Optimal combination (ATIG)	600
TIG	571
Base metal (SS316L)	624

3.1.6. Tensile Break Zone Investigation

Micrographs of the break zone in the tensile test were conducted with SEM. The images show the same profile with the formation of multiple dimples, which demonstrates that the fracture is in the ductile mode for both cases of TIG and ATIG weld as shown in Figure 6a,b, respectively.

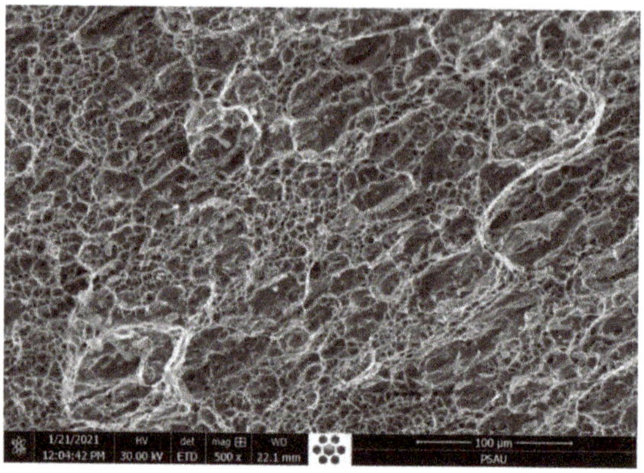

(**a**)

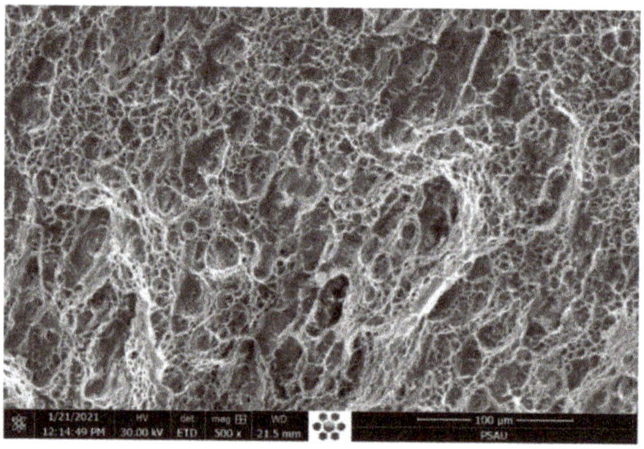

(**b**)

Figure 6. Fractograph of austenitic stainless steel 316L tensile test for TIG (**a**) and ATIG (**b**) welds (500×).

We notice that the fracture location for both welds occurs in base metal as shown in Figure 7. The results show the elongation percentages very close to 22.63% for ATIG specimen against 20.52% for TIG weld.

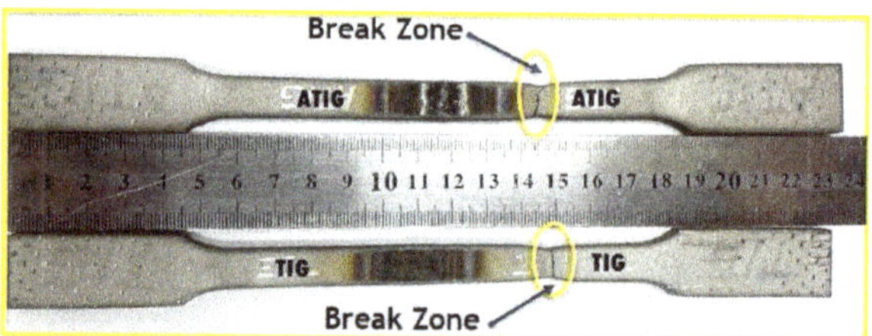

Figure 7. Failure zone in tensile test for ATIG and TIG welds.

3.2. Weld Bead Morphology

The values obtained for depth D and D/W ratio of the optimal flux ATIG weld are listed in Table 10. The results show that the depth of optimized flux is greater than TIG weld by two times. The ratio D/W was improved by 3.13 times. The weld of optimal flux is a full-penetrated weld (6.8 mm). The D/W value obtained for ATIG weld is 0.72.

This can be explained by the fact that Arc weld constriction and reversed Marangoni convection in ATIG occur when:

- Fluorine from flux migrates to the arc weld and contributes to enhancing the energy density according to the constriction arc mechanism explained before, as mentioned in several works [19,28].
- Oxygen liberated from oxides as surfactant element affects the surface tension of the molten metal resulting in a centripetal movement, the metal moves from the edges to the center as cited in related works [29,30]. A full penetration weld is performed in a single pass without edge preparation or the use of filler metal, which meets the needs of industries as shown in Figure 8b.

However, in TIG, Marangoni convection occurs. A molten metal moves from the center of the weld pool to the edges as pure metal leading to a wide and shallow weld bead as shown in Figure 8a.

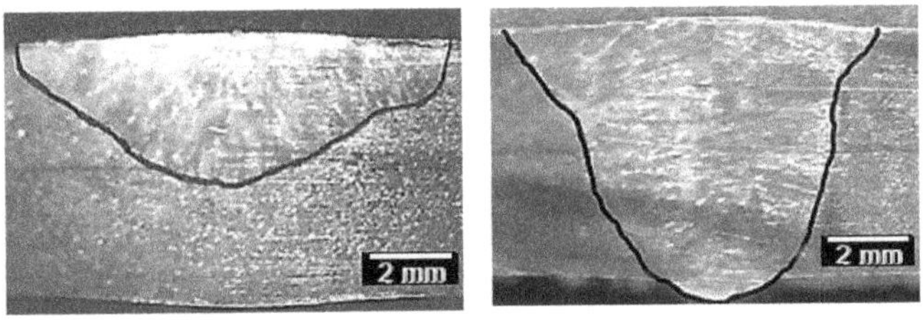

(**a**) (**b**)

Figure 8. Morphology of TIG (**a**) and ATIG (**b**) welds.

Table 10. Weldment bead profiles data of TIG and ATIG.

TIG			ATIG		
D (mm)	W (mm)	D/W	D (mm)	W (mm)	D/W
3.26	14.44	0.23	6.80	9.5	0.72

3.3. Microstructural Assessment

The differences in microstructure between TIG weld bead and ATIG weld are shown in Figure 9a,b and Figure 10a,b respectively. In both cases, the ferrite morphology has a discontinuous skeletal network of delta ferrite (δ) structures in a predominant austenite matrix. Alloy 316L with Creq/Nieq that is 1.67 solidifies in ferritic austenitic mode [31]. The delta ferrite is located mainly at the dendrite axes. The residual primary ferrite results from incomplete $\delta \rightarrow \gamma$ transformation during solidification.

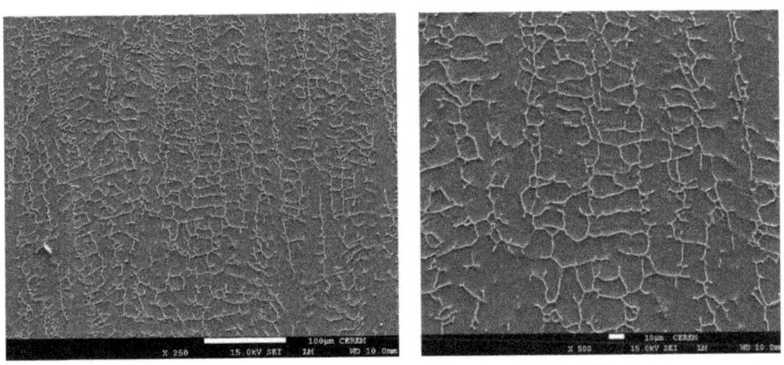

(**a**) (**b**)

Figure 9. SEM micrograph in TIG fusion zone of 316L (**a**) (250×), (**b**) (500×).

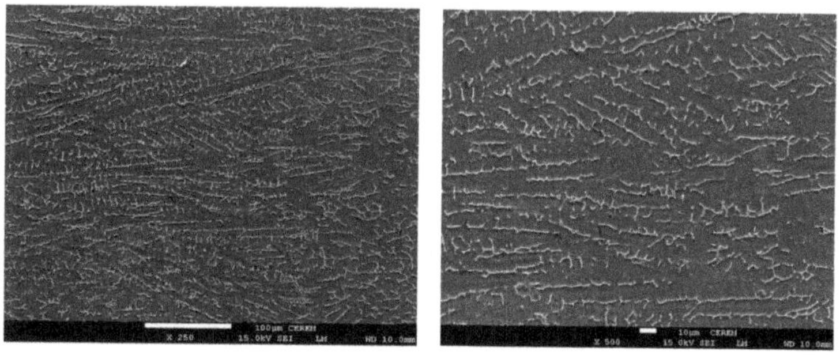
(a) (b)

Figure 10. SEM Micrograph in ATIG fusion zone of 316L (**a**) (250×), (**b**) (500×).

The delta ferrite (δ) in the TIG weld zone is coarser than that of the ATIG weld as shown in Figure 9b and Figure 10b, respectively. This difference in delta ferrite (δ) size is related to the heat provided by the weld bead, which is lesser in the case of ATIG weld.

In ATIG and TIG weld zone, the ferrite is the result of the incomplete primary delta ferrite- austenite transformation. The proportions of delta ferrite diminish from weld zone to base metal is as shown in Figure 11a,b.

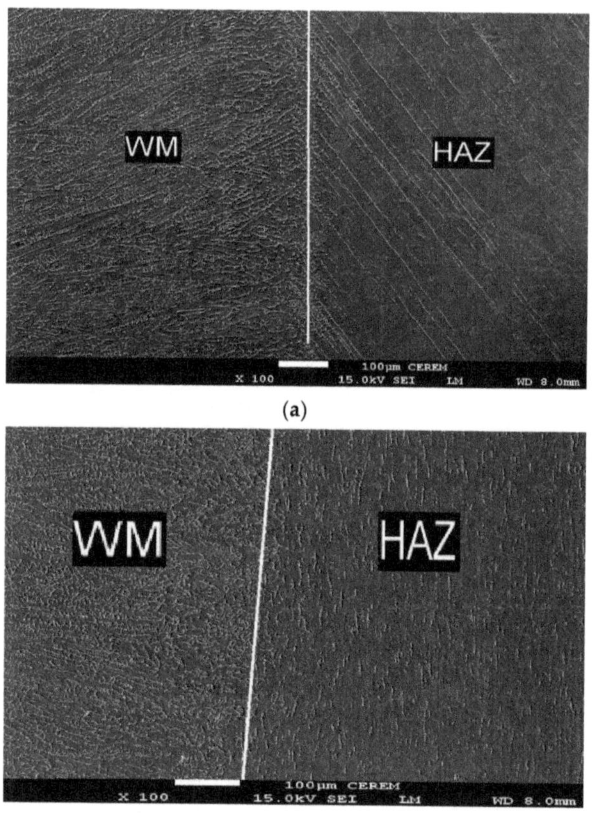

(a)
(b)

Figure 11. SEM Micrograph in transition zone (WZ/HAZ) in ATIG weld (**a**) and TIG weld (**b**) (100×).

In the ATIG heat-affected zone, austenitic grains crossed by parallel elongated inclined stringers of delta ferrite as shown in Figure 11a. In TIG heat-affected zone, the globular delta ferrite is randomly distributed and in the form of small slats of reduced size in the same rolling direction in a matrix of austenite as shown in Figure 11b.

Figure 12a,b represent the measurements of ferrite volume proportions in the austenite matrix. The advisable proper amount of δ-ferrite in austenitic stainless steel welds is less than 10% volume to ensure a better ductility, toughness, and corrosion resistance, and no less than 5% to avoid solidification cracking [32].

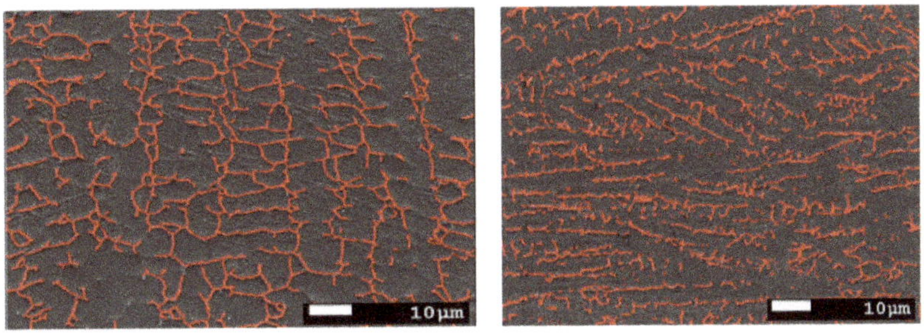

(**a**) (**b**)

Figure 12. Ferrite proportions measurements of 316L, (**a**) TIG Weld Zone and (**b**) ATIG Weld Zone (SEM. magnification 500×).

The results reported in Table 11 show the measurements taken in seven different locations in the weld zone and the mean value of ferrite volume. The proportions are up to 5.56% in the conventional TIG process and 8.63% for ATIG weld.

Table 11. Ferrite volume fraction measurement.

Sample	Number of Measurements	Max. δ-Ferrite %	Min. δ-Ferrite %	Mean δ-Ferrite %	Standards Deviation σ
TIG	7	6.2	4.8	5.56	0.35
ATIG	7	9	8.1	8.63	0.31

The higher retained ferrite may be attributed to the rate of cooling of the welds. Elements from optimal flux can constrict arc weld. The constriction of the arc will increase the temperature at the anode due to the increase in current density and arc voltage as reported in many studies [33,34].

The ATIG welding with optimal flux increased the energy density of the heat source that leads to low heat input. However, the arc heat of TIG welding without flux has a lower energy density; therefore, high heat input is provided to the workpiece. High heat input resulted in a slow cooling rate and a further transformation from ferrite phase to the austenite phase. Consequently, a ferrite volume proportion is reduced (5.56%). The ATIG welding was associated with a low heat input resulting in higher ferrite content (8.63%), which is in good agreement with several works [35,36]. On the other hand, the high heat input leads to coarse substructure during solidification in the case of TIG weld as shown in Figure 9a,b, which results in a more widely spaced ferrite network. But in the case of ATIG weld, heat provided is low; a fast cooling rate occurs and consequently a finer skeletal ferrite is formed as shown in Figure 10a,b [37].

3.4. Hardness Test

The hardness values are shown in Table 12. It is visible that the hardness of ATIG weld and TIG weld in both FZ and HAZ are very close. The delta-ferrite volume proportions in the weld metals are in both TIG and ATIG welds increased, and have a beneficial effect in increasing the hardness of as received 316L stainless steel welds (160 HV).

Table 12. Hardness values and standards deviation of TIG and ATIG.

Sample	Number of Tests	HV Max.	HV Min.	HV Mean	Standards Deviation σ
ATIG FZ	8	188	176	183	4.93
TIG FZ	8	199	175	185	8.98
ATIG HAZ	8	177	167	171	3.41
TIG HAZ	8	180	164	172	5.03

In ATIG weld, the property of hardness is not affected by the optimal flux used. On the other hand, the standard deviation is less than 9 HV, which attests to the small disparities in the obtained hardness values of the maximum and minimum. This result indicates good hardness homogeneities in the joints.

3.5. Impact Test

The impact tests were carried out on the fusion zone in ATIG and TIG welds. The experimental values obtained for the impact tests are shown in Table 13. The energy absorbed in the fusion zone in the case of ATIG weld (267 J/cm^2) is slightly higher than that of TIG weld (256 J/cm^2) by 11 J/cm^2. The standard deviation is less than 15 J/cm^2.

Table 13. Energy absorbed (J/cm^2) and standard deviation of TIG and ATIG at fusion zone.

Sample	Number of Tests	Absorbed Energy Min.	Absorbed Energy Max.	Absorbed Energy Mean	Standards Deviation (σ)
TIG	3	241	269	256	14.05
ATIG	3	254	281	267	13.58

Figure 12 represents the fractographs of the impact Charpy "V" notch test. The images show the formation of multiple dimples, which demonstrate that the fracture is a ductile mode in both cases of TIG and ATIG as shown in Figure 13a,b. However, in ATIG weld the dimples are finer with the presence of voids, which can explain the slightly high value of absorbed energy comparatively to TIG weld. The ductile fracture mode leads to good resistance to sudden impact loads. Multiple dimples attest for high impact energy withstand as reported by several authors [38,39].

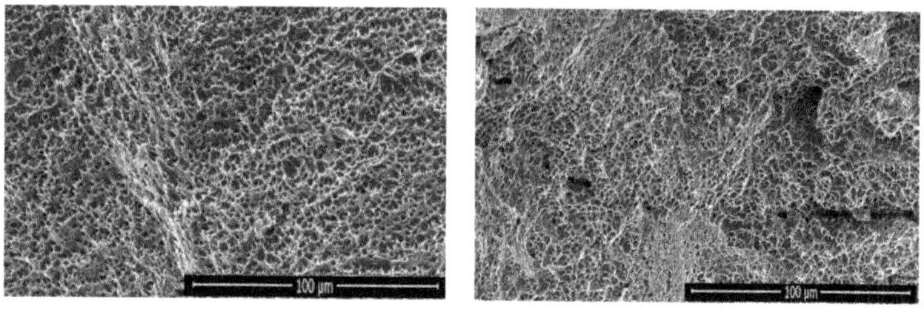

(a) (b)

Figure 13. Fractograph of Austenitic stainless steel 316L impact Charpy "V" notch for (**a**) TIG Welded Zone (500×) and (**b**) ATIG Welded Zone (500×).

The results of EDS/SEM analysis in Figure 14a and Table 14 are shown in the case of ATIG weld, with the same level of silicon as in base metal. There are no trace of unwanted elements in the case of ATIG welds. These results can explain the good resistance to impact test.

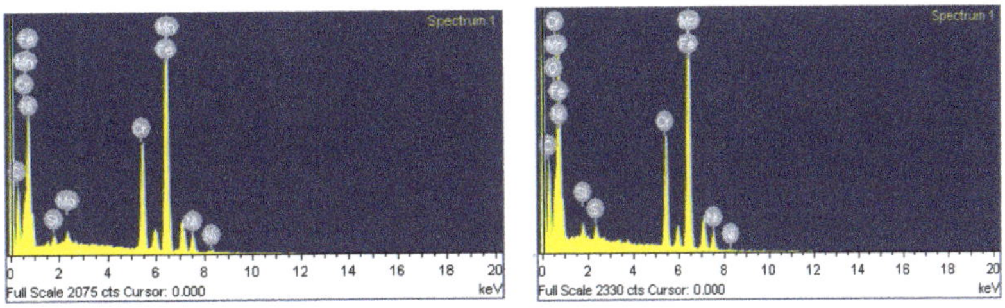

(a) (b)

Figure 14. EDS/SEM spectrum analysis of fracture face of ATIG welded zone (**a**) and TIG welded zone (**b**).

Table 14. Elements present in fracture face of ATIG welded zone and TIG welded zone.

Sample		C %	Si %	Cr %	Mn %	Fe %	Ni %	Mo %	O %	S %
ATIG weld	Weight %	16.26	0.43	14.80	1.75	58.10	7.09	1.57		
	Atomic %	47.29	0.54	9.94	1.11	36.34	4.22	0.54		
TIG weld	Weight %	17.898	0.68	14.86	1.39	54.67	6.73		3.20	0.58
	Atomic %	47.49	0.77	9.11	0.81	31.21	3.65		6.38	0.58

The results of EDS/SEM analysis in Figure 14b and in Table 14 show the presence of oxygen in TIG weld, which is probably ascribed to insufficient protection of the weld pool that affects mechanical properties, particularly the toughness. The presence of sulfur leads to the formation of low-melting eutectics with iron, chromium, and nickel, which can alter the mechanical properties of TIG weld beads [40]. The obtained results with EDS/SEM related to TIG weld may explain the decrease in resistance to impact test comparatively to the ATIG weld.

4. Conclusions

In the present work, ATIG weld has been investigated and compared to conventional TIG weld. The starting point of this work is elaborating the optimal flux to maximize the UTS ATIG weld. The mixing design of the experiment combined to the particle swarm optimization (PSO) method is used to minimize the number of trials, which reduces the cost of materials as well as time. Based on the obtained results in this investigation, the following conclusions can be drawn:

- Mixing design of the experiment combined to the particle swarm optimization (PSO) method is among the novelties of this work. Optimal flux was composed by 32% Cr_2O_3, 43% ZrO_2, 8% SiO_2, and 17% CaF_2. The optimal flux raised from this method improves the mechanical properties in comparison to conventional TIG weld bead. Transversal tensile testing of the produced ATIG weld has a UTS value (600 MPa) close to parent metal (624 MPa). On the other hand, the UTS of TIG weld fell to 571 MPa. ATIG welding is more resistant than conventional TIG welding to sudden impact loads. The ATIG weld hardness results were close to those of conventional TIG weld.
- The ATIG depth weld bead reached 6.80 mm and the weld aspect ratio increased 3.13. The ATIG depth weld was increased by two times in comparison to the conventional TIG weld bead. The fully penetrated ATIG weld is ascribed to two mechanisms cited earlier. The reversal Marangoni mechanism owing to oxygen liberated from flux and the constriction of ATIG arc weld related to the migration of fluorine from the flux to the arc.

- The microstructure of welds in both welds is composed of matrix austenite interspersed by skeleton δ-ferrite. The ferrite volume proportions in ATIG weld around 8.63% and decreased to 5.56% in conventional TIG welding. The δ-ferrite in TIG weld is coarser comparatively to that of ATIG δ-ferrite.

Author Contributions: Conceptualization, K.T. and A.C.H.; Methodology, K.T. and R.D.; Software, K.T. and A.C.H.; Validation, K.T., R.D., A.O. and A.C.H.; Formal analysis, K.T.; Investigation, K.T., R.D., A.O. and A.C.H.; Resources, H.A.; Data curation, K.T.; Writing—original draft preparation, K.T.; Writing—review and editing, K.T., R.D., A.O. and M.M.Z.A.; Visualization, K.T.; Supervision, K.T.; All authors have read and agreed to the published version of the manuscript.

Funding: This research received no external funding.

Institutional Review Board Statement: Not applicable.

Informed Consent Statement: Not applicable.

Data Availability Statement: The data used to support the findings of this study are included within the article.

Acknowledgments: The authors present their deep thanks and acknowledge the collaboration of El-Sayed M. Sherif and Hany S. Abdo from the Center of Excellence for Research in Engineering Materials (CEREM), King Saud University, Saudi Arabia, for their help in performing SEM images. The authors also express their sincere thanks to Sahbi Boubaker from the college of computer science and engineering, Jeddah University for his help in mathematical modeling.

Conflicts of Interest: The authors declare no conflict of interest.

References

1. Tathgir, S.; Bhattacharya, A. Activated-TIG Welding of Different Steels: Influence of Various Flux and Shielding Gas. *Mater. Manuf. Process.* **2016**, *31*, 235–342. [CrossRef]
2. Klobčar, D.; Tušek, J.; Bizjak, M.; Simončič, S.; Lešer, V. Active flux tungsten inert gas welding of austenitic stainless steel AISI 304. *Metalurgija* **2016**, *55*, 617–620.
3. Modenesi, P.J.; Neto, C.P.; Apolinario, E.R.; Dias, B.K. Effect of flux density and the presence of additives in ATIG welding of austenitic stainless steel. *Weld. Int.* **2015**, *29*, 425–432. [CrossRef]
4. Kumar, R.; Sundara Bharathi, S.R. A Review Study on A-TIG Welding of 316(L) Austenitic Stainless Steel. *Int. J. Emerg. Trends Sci. Technol. (IJETS)* **2015**, *2*, 2066–2072.
5. Howse, D.S.; Lucas, W. An investigation in to arc construction by active flux for TIG welding. *Sci. Technol. Weld. Join.* **2000**, *5*, 189–193. [CrossRef]
6. Kurtulmuş, M. Activated flux TIG welding of austenitic stainless steels. *Emerg. Mater. Res.* **2020**, *9*, 1041–1055. [CrossRef]
7. Sire, S.; Marya, S. On the development of a new flux bounded TIG process (FBTIG) to enhance weld penetrations in aluminium, 5086. *Int. J. Form. Process.* **2002**, *5*, 39–51. [CrossRef]
8. Jayakrishnan, S.; Chakravarthy, P.; Muhammed Rijas, A. Effect of Flux Gap and Particle Size on the Depth of Penetration in FBTIG Welding of Aluminium. *Trans. Indian Inst. Met.* **2017**, *70*, 1329–1335. [CrossRef]
9. Ambekar, S.D.; Wadhokar, S.R. Parametric Optimization of Gas metal arc welding process by using Taguchi method on stainless steel AISI 410. *Int. J. Res. Mod. Eng. Emerg. Technol.* **2015**, *3*, 1–9.
10. Srirangan, A.K.; Paulraj, S. Multi-response optimization of process parameters for TIG welding of Incoloy 800HT by Taguchi grey relational analysis. *Eng. Sci. Technol. Int. J.* **2016**, *19*, 811–817. [CrossRef]
11. Ramadan, N.; Boghdadi, A. Parametric Optimization of TIG Welding Influence On Tensile Strength of Dissimilar Metals SS-304 And Low Carbon Steel by Using Taguchi Approach. *Am. J. Eng. Res.* **2020**, *9*, 7–14.
12. Chaudhari, V.; Bodkhe, V.; Deokate, S.; Mali, B.; Mahale, R. Parametric optimization of TIG welding on SS 304 and MS using Taguchi approach. *Int. Res. J. Eng. Technol.* **2019**, *6*, 880–885.
13. Albaijan, I.; Hedhibi, A.C.; Touileb, K.; Djoudjou, R.; Ouis, A.; Alrobei, H. Effect of Binary Oxide Flux on Weld Shape, Mechanical Properties and Corrosion Resistance of 2205 Duplex Stainless Steel Welds. *Adv. Mater. Sci. Eng.* **2020**, *2020*, 5842741. [CrossRef]
14. Touileb, K.; Ouis, A.; Djoudjou, R.; Hedhibi, A.C.; Alrobei, H.; Albaijan, I.; Alzahrani, B.; Sherif, M.E.; Abdo, H.S. Effects of ATIG Welding on Weld Shape, Mechanical Properties, and Corrosion Resistance of 430 Ferritic Stainless Steel Alloy. *Metals* **2020**, *10*, 404. [CrossRef]
15. Ran, L.; Manshu, D.; Hongming, G. Prediction of Bead Geometry with Changing Welding Speed Using Artificial Neural Network. *Materials* **2021**, *14*, 1494.
16. Kshirsagar, R.; Jones, S.; Lawrence, J.; Tabor, J. Prediction of Bead Geometry Using a Two-Stage SVM–ANN Algorithm for Automated Tungsten Inert Gas (TIG) Welds. *J. Manuf. Mater. Process.* **2019**, *3*, 39. [CrossRef]

17. Las-Casas, M.S.; De Ávila, T.L.D.; Bracarense, A.Q.; Lima, E.J. Weld parameter prediction using artificial neural network: FN and geometric parameter prediction of austenitic stainless steel welds. *J. Braz. Soc. Mech. Sci. Eng.* **2018**, *40*, 26–35. [CrossRef]
18. Tseng, K.H.; Chen, K.L. Comparisons between TiO_2- and SiO_2-Flux Assisted TIG Welding Processes. *J. Nanosci. Nanotechnol.* **2012**, *12*, 6359–6367. [CrossRef]
19. Leconte, S.; Paillard, P.; Chapelle, P.; Henrion, G.; Saindrenan, J. Effects of flux containing fluorides on TIG welding process. *Sci. Technol. Weld. Join.* **2013**, *12*, 120–126. [CrossRef]
20. Neethu, N.; Togita, R.G.; Neelima, P.; Chakravarthy, P.; Narayana, M.S.V.S.; Nair, M.T. Effect of Nature of Flux and Flux Gap on the Depth-to-Width Ratio in Flux-Bounded TIG Welding of AA6061: Experiments and Numerical Simulations. *Trans. Indian Inst. Met.* **2019**, *72*, 1585–1588. [CrossRef]
21. Babbar, A.; Kumar, A.; Jain, V.; Gupta, D. Enhancement of Activated Tungsten Inert Gas (A-TIG) Welding Using Multi-component TiO2-SiO2-Al2O3 Hybrid Flux. *Measurement* **2019**, *148*, 106912. [CrossRef]
22. Ahmed, A.N.; Noor, C.W.M.; Allawi, M.F.; El-Shafie, A. RBF-NN-based model for prediction of weld bead geometry in Shielded Metal Arc Welding (SMAW). *Neural Comput. Appl.* **2018**, *29*, 889–899. [CrossRef]
23. Kumar, R.; Saurav, S.K. Modeling of TIG welding process by regression analysis and neural network technique. *Int. J. Mech. Eng. Technol. (IJMET)* **2015**, *6*, 10–27.
24. Kshirsagar, R.; Jones, S.; Lawrence, J.; Tabor, J. Optimization of TIG Welding Parameters Using a Hybrid Nelder Mead-Evolutionary Algorithms Method. *J. Manuf. Mater. Process.* **2020**, *4*, 10. [CrossRef]
25. Boubaker, S.; Kamel, S.; Kolsi, L.; Kahouli, O. Forecasting of One-Day-Ahead Global Horizontal Irradiation Using Block-Oriented Models Combined with a Swarm Intelligence Approach. *Nat. Resour. Res.* **2020**, *30*, 1–26. [CrossRef]
26. Patel, N.P.; Badheka, V.J.; Vora, J.J.; Upadhyay, G.H. Effect of Oxide Fluxes in Activated TIG Welding of Stainless Steel 316LN to Low Activation Ferritic/Martensitic Steel (LAFM) Dissimilar Combination. *Trans. Indian Inst. Met.* **2019**, *72*, 2753–2761. [CrossRef]
27. Goodwin, G.M.; Cole, N.C.; Slaughtera, D.G.M. Study of Ferrite Morphology in Austenitic Stainless Steel Weldments. *Weld. Res. Suppl.* **1972**, *12*, 425–429.
28. Ming, L.Q.; Hong, W.X.; Da, Z.Z.; Jun, W. Effect of activating flux on arc shape and arc voltage in tungsten inert gas welding. *Trans. Nonferrous Met. Soc. China* **2007**, *17*, 486–490.
29. Kulkarni, A.; Dwivedi, D.K.; Vasudevan, M. Study of Mechanism, Microstructure and Mechanical Properties of Activated Flux TIG Welded P91 Steel-P22 Steel Dissimilar Metal Joint. *Mater. Sci. Eng. A* **2018**, *50*, 309–323. [CrossRef]
30. Vora, J.J.; Badheka, V.J. Experimental Investigation on Mechanism and Weld Morphology of Activated TIG Welded Bead-on-plate Weldments of Reduced Activation Ferritic/martensitic Steel Using Oxide Fluxes. *J. Manuf. Process.* **2015**, *20*, 224–233. [CrossRef]
31. Kujanpaa, V.P.; Suutala, V.P.; Takalo, N.J.; Moisio, T.J.I. Solidification Cracking—Estimation of the Susceptibility of Austenitic and Austenitic-Ferritic Stainless Steel Welds. *Met. Constr.* **1980**, *12*, 282–285.
32. Kou, S. *Welding Metallurgy*, 2nd ed.; John Wiley & Sons: Hoboken, NJ, USA, 2003.
33. Dixit, P.; Suketu, J. Techniques to weld similar and dissimilar materials by ATIG welding—An overview. *Mater. Manuf. Process.* **2021**, *36*, 1–16.
34. Roy, S.; Samaddar, S.; Uddin, M.N.; Hoque, A.; Mishra, S.; Das, S. Effect of Activating Flux on Penetration in ATIG Welding of 316 Stainless Steel. *Indian Weld. J.* **2017**, *50*, 72–80. [CrossRef]
35. Vasudevan, M. Effect of A-TIG Welding Process on the Weld Attributes of Type 304LN and 316LN Stainless Steels. *J. Mater. Eng. Perform.* **2017**, *26*, 1325–1336. [CrossRef]
36. Suman, S.; Santanu, D. Effect of Polarity and Oxide Fluxes on Weld-bead Geometry in Activated Tungsten Inert Gas (A-TIG) Welding. *J. Weld. Join.* **2020**, *38*, 380–388.
37. Lippold, J.C.; Savage, W.F. Solidification of Austenitic Stainless Steel Weldments: Part 2-The Effect of Alloy Composition on Ferrite Morphology. *Weld. J.* **1980**, *59*, 48–58.
38. Jebaraj, A.V.; Kumar, T.S.; Manikandan, M. Investigation of Structure Property Relationship of the Dissimilar Weld Between Austenitic Stainless Steel 316L and Duplex Stainless Steel 2205. *Trans. Indian Inst. Met.* **2018**, *71*, 2593–2604. [CrossRef]
39. Charles, J. Composition and properties of duplex stainless steels. *Weld. World* **1995**, *36*, 43–54.
40. Harish, K.D.; Somi, R.A. Study of Mechanical Behavior in Austenitic Stainless Steel 316 LN Welded Joints. *Int. J. Mech. Eng. Rob. Res.* **2013**, *61*, 37–56.

materials

MDPI

Article

A Novel Friction Stir Deposition Technique to Refill Keyhole of Friction Stir Spot Welded AA6082-T6 Dissimilar Joints of Different Sheet Thicknesses

Mohamed M. Z. Ahmed [1,2,*], Mohamed M. El-Sayed Seleman [2], Essam Ahmed [2], Hagar A. Reyad [2], Naser A. Alsaleh [3] and Ibrahim Albaijan [1]

1 Mechanical Engineering Department, College of Engineering at Al Kharj, Prince Sattam Bin Abdulaziz University, Al Kharj 11942, Saudi Arabia
2 Department of Metallurgical and Materials Engineering, Faculty of Petroleum and Mining Engineering, Suez University, Suez 43512, Egypt
3 Department of Mechanical Engineering, College of Engineering, Imam Mohammad Ibn Saud Islamic University, Riyadh 11432, Saudi Arabia
* Correspondence: moh.ahmed@psau.edu.sa; Tel.: +966-115-888-273

Abstract: Joining dissimilar sheet thicknesses of AA6082-T6 alloys by friction stir spot welding (FSSW) provides many advantages in automotive and aerospace applications. The formed keyhole at the end of the FSSW process is one of the typical features after the welding process, which owns the same size as the rotating pin that remains at the joint center. This keyhole destroys the joint continuity and can stimulate serious stress concentration when the FSSW joint bears an external force. To solve this issue, a novel refilling technique was developed for the FSSW keyholes using a friction stir deposition (FSD) technique. The FSSW joints of AA6082-T6 sheets were welded at various rotation speeds from 400 to 1000 rpm and a constant dwell time of 3 s, where a 2 mm sheet thickness was an upper sheet, and a 1 mm sheet thickness was a lower sheet. All the keyhole refilling processes were achieved using a specially designed AA2011-T6 consumable rod to be used for friction stir deposition of continuous layers at a constant deposition parameter of 400 rpm consumable rod rotation speed and a 1 mm/min feed rate. The heat input energy for both the FSSW and refilled FSSW lap joints was calculated. In addition, the FSSW and the FSD temperatures were measured. Macrostructure, microstructure, and mechanical properties in terms of hardness and tensile shear maximum load were evaluated for both the friction stir spot welded (FSSWed) and the refilled FSSW lap joints. The obtained results showed that the keyhole could be successfully refilled with defect-free continuous multilayers after the refill friction stir spot welding (RFSSW) process. All the RFSSW lap joints showed higher tensile shear loads than that given by the FSSW (before refill) lap joints. The RFSSW joint (welded at 600 rpm/3 s and refilled at 400 rpm/1 mm/min) showed a higher tensile shear load of 5400 N ± 100 compared with that recorded by the unrefilled joint (4300 N ± 80). The fracture location and fracture surface of the FSSW and RFSSW were examined and discussed.

Keywords: aluminum alloys; AA6082-T6; AA2011-T6; friction stir spot welding; friction stir deposition; keyhole refilling; mechanical properties

Citation: Ahmed, M.M.Z.; El-Sayed Seleman, M.M.; Ahmed, E.; Reyad, H.A.; Alsaleh, N.A.; Albaijan, I. A Novel Friction Stir Deposition Technique to Refill Keyhole of Friction Stir Spot Welded AA6082-T6 Dissimilar Joints of Different Sheet Thicknesses. *Materials* **2022**, *15*, 6799. https://doi.org/10.3390/ma15196799

Academic Editors: Lihong Su, Xing Zhao, Peitang Wei and Hui Wang

Received: 24 August 2022
Accepted: 21 September 2022
Published: 30 September 2022

1. Introduction

Lightweight components are essential in the automotive and aerospace industries for the reduction of CO_2 emissions. The usage of aluminum alloys can reduce the vehicle structure's total weight [1,2]. Different spot joining techniques have been used in the automotive industry, such as riveting and resistance spot welding. Recently, to avoid the drawbacks of these spot welding techniques, alternative joining techniques in terms of friction stir welding (FSW) and its derivative friction stir spot welding (FSSW) processes are used for joining similar [3,4] and dissimilar structural materials [5–9] in engineering applications. The FSSW is a solid-state joining technique. It is achieved by applying a controlled thermomechanical process to produce spot lap joints based on the generated frictional heating and plastic deformation in the stir zone [10–12]. The quality of the friction stir spot welded (FSSWed) joints is controlled by welding process parameters, such as tool rotation speed [13,14], dwell time [1,15], plunge rate [16], axial downward force [17], and tool design [18,19]. The tool rotation speed plays a vital role in the frictional heat generation, whereas the downward force, plunging rate, dwell time, and tool design are credited to the material flow at the spot joint interface [20]. The most serious defect for the FSSWed joints is keyhole formation [21,22]. Nowadays, modified FSSW processes are developed to eliminate the keyhole defect, such as pinless friction stir spot welding [1,15], flat friction stir spot welding [23,24], and refill friction stir spot welding (RFSSW) [22,25,26]. For the pinless friction stir spot welding, the rotating tool has no pin, and its shoulder surface is recommended to be a scroll groove [1,15]. This process is only applicable to a weld aluminum sheet thickness of less than 2 mm [1]. The flat friction stir spot welding uses two different tools (a tool with a pin and a pinless tool) for producing the FSSW joint and refilling the formed keyhole [23,24]. In the first step, a specially designed back plate with a predrilled dent is utilized to yield a protuberance on the bottom side of the joint. In the second step, the protuberance and keyhole are removed by applying a pinless rotating tool and a smooth flat back plate [23]. This technique is complex and requires many precautions in design and implementation; in addition, it depends on the deformability of the welded materials. The RFSSW process contains four stages, which are friction heating, plunging, refilling, and joint forming; in addition, its tool rotation is complicated and mainly consists of three independent parts [22,25,26]. The first part is a clamping ring, which is used to retain the sheets and prevent the viscous plastic material from escaping. The other two parts are a sleeve and a pin; both are moved independently in opposite vertical directions to each other to realize the stirring and the refilling of the viscous plastic material. At the end of the process, the displaced material is initially pushed to the surface level, without the keyhole on the sheet surface. Furthermore, the process temperature may reach the melting point, and the formation of liquation cracks is difficult to avoid [27]. These complications have limited the use of this technique in many engineering applications.

Recently, friction stir deposition (FSD) as an additive manufacturing technology is recommended by many authors to build multilayers of different materials on various substrates [28–30]. This technology is a suitable technique to produce metallic parts for the automobile and aircraft industries. It is also a thermomechanical process based on the FSW principles, by adding a mechanism of material feeding to deposit alloys [29–31] and composites [28]. Perry et al. [32] concluded that the FSD path of AA2024 Al alloy at a 300 rpm tool rotational speed, 2 mm/s travel speed, and 0.85 mm/s feed rate reveals an almost fully recrystallized microstructure. Dilip et al. [33] investigated the microstructure of an AA2014-T6 friction stir deposited at an 800 rpm tool rotation speed, 1 mm/min feed rate, and 35 MPa axial pressure. The results showed that fine grains and refined second-phase precipitates are the microstructure features. Priedeman et al. [34] evaluated the microstructure features and hardness of FSD 110 Cu processed at a travel speed of 2.12 mm/s and rotational speed of 275 rpm. The results showed that the microstructure features are recrystallized fine grains throughout the deposited layers, with different degrees of grain refining; in addition, the hardness measurement showed that the deposited layers are softer than that of the base material.

Based on the literature review, although there is a great interest to publish many works in FSD, there is no attempt to use FSD in refilling the FSSW keyhole. Thus, the current study is considered the first try in the world to apply the FSD process for filling and repairing keyholes of the FSSW joints of AA6082-T6 with different sheet thicknesses produced at a constant dwell time of 3 s and different rotation speeds of 400, 600, 800, and 1000 rpm. A specially designed consumable rod of AA2011-T6 was suggested. The heat input energy at applied deposition process parameters of 400 rpm consumable rod rotation speed and 1 mm/min feed rate was studied, and a new equation to calculate the heat input energy was suggested. The current work aims also to conduct a comparative study of the FSSW and the RFSSW lap joints in terms of macrostructure, microstructure, hardness, maximum tensile shear, and fracture behaviors.

2. Materials and Methodology

2.1. Starting Materials

The starting materials to achieve dissimilar thickness friction stir spot welds were two thin sheets of AA6082-T6 with dimensions of 1 × 1000 × 1000 mm and 2 × 1000 × 1000 mm. Meanwhile, the keyhole filling material was AA2011-T6 rods with initial dimensions of 20 mm diameter and 100 mm length. Both the AA6082-T6 sheets and the AA2011-T6 rods were supplied by the Future Fond Company, Milano, Italy. The chemical compositions of the used AA6082-T6 sheets and the AA2011-T6 rods were performed using Foundry-Master Pro, Oxford Instruments, Abingdon, UK, and are listed in Table 1.

Table 1. The chemical composition (in wt.%) of AA6082-T6 sheets and AA2011-T6 rods.

Element	Si	Mg	Fe	Mn	Zn	Cr	Ti	Cu	Bi	Pb	Al
AA6082-T6	0.75	0.60	0.50	0.40	0.20	0.20	0.10	0.10	-	-	Bal
AA2011-T6	0.09	-	0.37	-	0.06	0.04	0.03	4.83	0.25	0.3	Bal

2.2. Friction Stir Spot Welding and Friction Stir Deposition Processes

The FSSW joints and the FSD refilled samples were carried out using the friction stir welding/processing machine (EG-FSW-M1) [35]. For the spot welding process, the two AA6082-T6 dissimilar sheets were cut to specimens with dimensions of 30 mm width and 100 mm length. The cut specimens were FSSWed in lap joints at a constant dwell time of 3 s and various rotational speeds of 400, 600, 800, and 1000 rpm, where the two different sheet thicknesses overlapped with a 30 mm length. The other FSSW parameters in terms of plunge rate and plunge depth were kept constant at 0.1 mm/s and 2.6 mm, respectively. The used FSSW tool was made of steel (AISI H13) with dimensions of 20 mm shoulder diameter, 2.6 mm pin length, and 5 mm pin diameter. The tool shoulder was flat, and the pin was cylindrical. The tool features and dimensions are given in Figure 1.

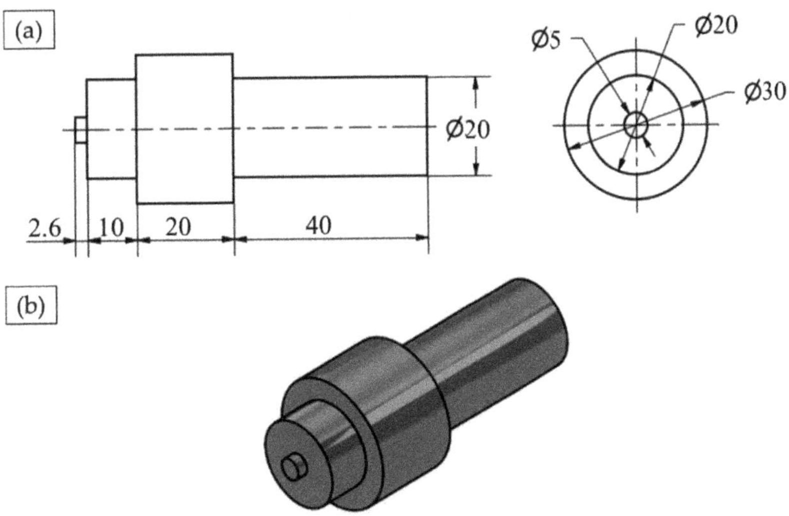

Figure 1. (**a**) Schematic of welding tool (all dimensions in mm) and (**b**) isometric view of designed FSSW tool.

Figure 2 illustrates the FSSW process stages, which involve three main steps: (1) fixing the lap joint specimens with a clamping system (Figure 2a), (2) the plunging and stirring stage with the rotating tool (Figure 2b), and (3) the retracting stage (Figure 2c). Finally, Figure 2d shows the representative spot-welded joint's top view with the formed keyhole. The temperature during the FSSW process at the applied different rotational speeds of 400 to 1000 rpm was recorded using a modern digital multimeter (type-UT61B, Zhejiang, China) with a thermocouple type "K".

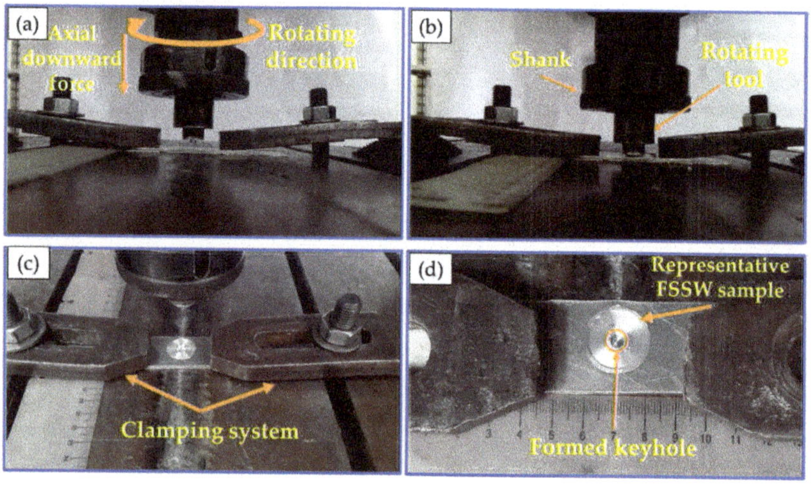

Figure 2. (**a–c**) summarizes the stages of the FSSW process of dissimilar sheet thickness AA6082-T welds. (**d**) shows the produced joint top view with the formed keyhole [36].

In order to explore the applicability of FSD for refilling the left keyhole after FSSW, an AA2011-T6 consumable rod was machined (according to the keyhole volumetric dimension) to a profile of a flat shoulder of 20 mm diameter and a tapered pin of 5 mm diameter and 3 mm pin length (as shown in Figure 3a) to be tried as a filling material in continuous layers via the FSD process. Figure 3 illustrates the FSD process stages, which involve three steps: fixing the AA 2011-T6 consumable rod in the spindle shank (Figure 3a) and rotating it at a constant rotation of speed 400 rpm while moving downward to reach the FSSW keyhole of the AA6082-T6 joint (Figure 3b). Finally, under a continuous feeding speed of 1 mm/min, the rod plastically deformed due to the high friction and the generated heat between the rod and the keyhole substrate, causing the material transfer from the consumable rod to the keyhole to build a material upward yielding refilling. Figure 3e shows the representative top view of the refilled FSSW joint.

The produced FSSWed and refilled friction stir spot welded (RFSSWed) joints were sectioned for macrostructure investigation, microstructure evaluation, and hardness test. The cross-sectional specimens were ground with SiC papers up to 2400 grit, then polished using a polishing vel-cloth in the presence of alumina paste suspension up to a surface finish of 0.05 μm. A chemical etcher consisting of 2.5 mL nitric acid (HNO_3), 1.5 mL hydrochloric acid (HCl), 95 mL distilled water, and 1 mL hydrofluoric (HF) was applied to all the polished specimens before microstructure investigation. The microstructure evaluation was performed using an Olympus optical microscope (OM) (BX41M-LED, Olympus, Tokyo, Japan). The hardness test was carried out along three lines across the transverse sections of the FSSWed and the refill friction stir spot welded (RFSSWed) joints for obtaining hardness profiles (Figure 4). The hardness measurements were carried out using an applied load of 5 N with a dwell time of 15 s using a Vickers hardness tester (model HWDV-75, TTS Unlimited, Osaka, Japan). The distance between every two indentations was set to 0.5 mm along the cross section of the produced FSSWed and RFSSWed joints. The tensile-shear test was performed at room temperature using a 30 ton universal tensile testing machine (Type-WDW-300D, Guangdong, China) at a constant loading rate of 0.1 mm/min. During the tensile-shear test, two backing sheets were used to ensure the application of the axial loading (Figure 5). After tensile shear testing, the fractured surfaces at the interface between the upper and lower sheets were examined using a scanning electron microscope (SEM-Quanta FEG 250-FEI Company, Hillsboro, OR, USA). For comparison purposes, the as-received materials were examined and tested using the previous described methods.

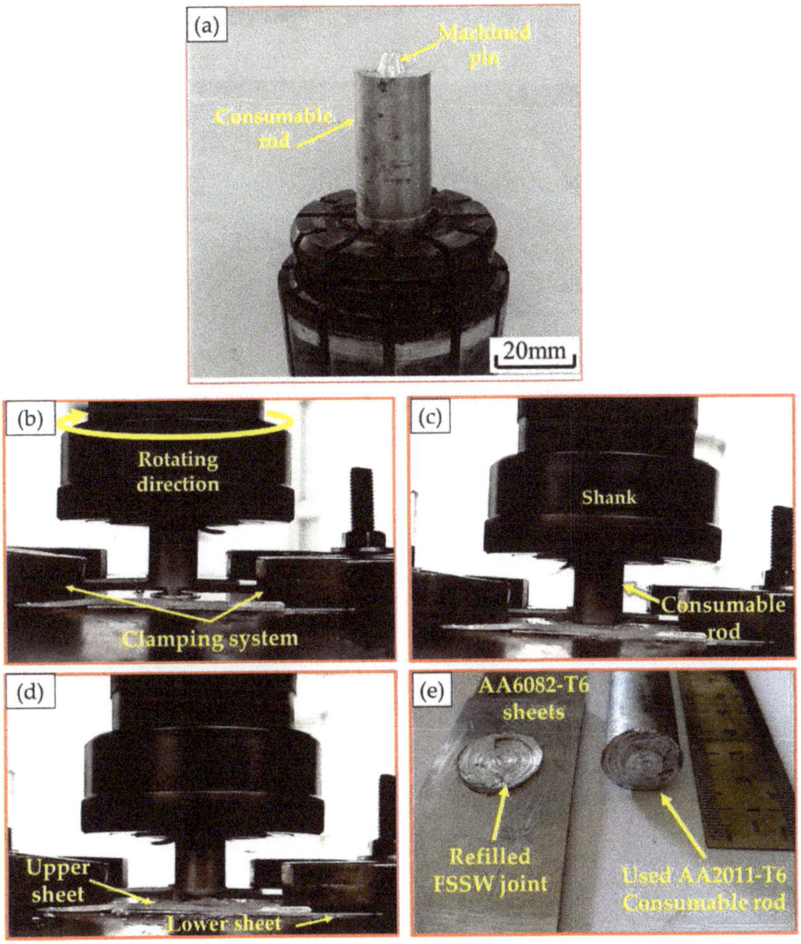

Figure 3. (**a**) Image shows the fixing of the machined AA 2011-T6 consumable rod for FSD, (**b–d**) summarizes the stages of the FSD process, and (**e**) shows the top view of the refilled keyhole FSSW dissimilar lap joint.

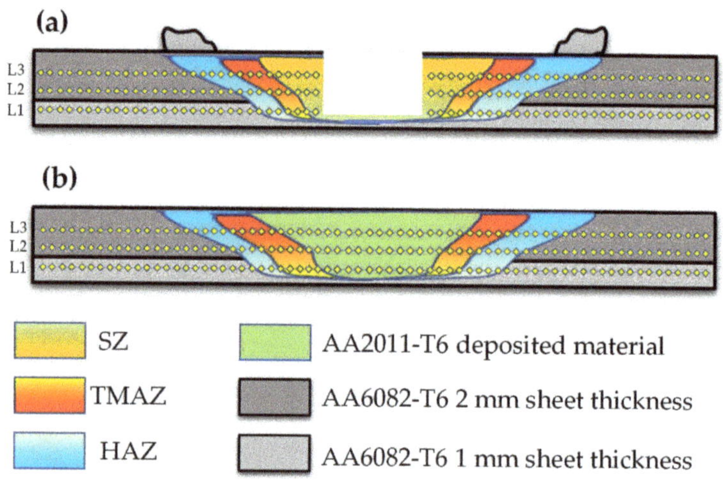

Figure 4. Schematics for the positions of the hardness test lines: (**a**) FSSWed AA6082-T6 dissimilar sheet thickness joint and (**b**) RFSSWed joint (all dimensions in mm).

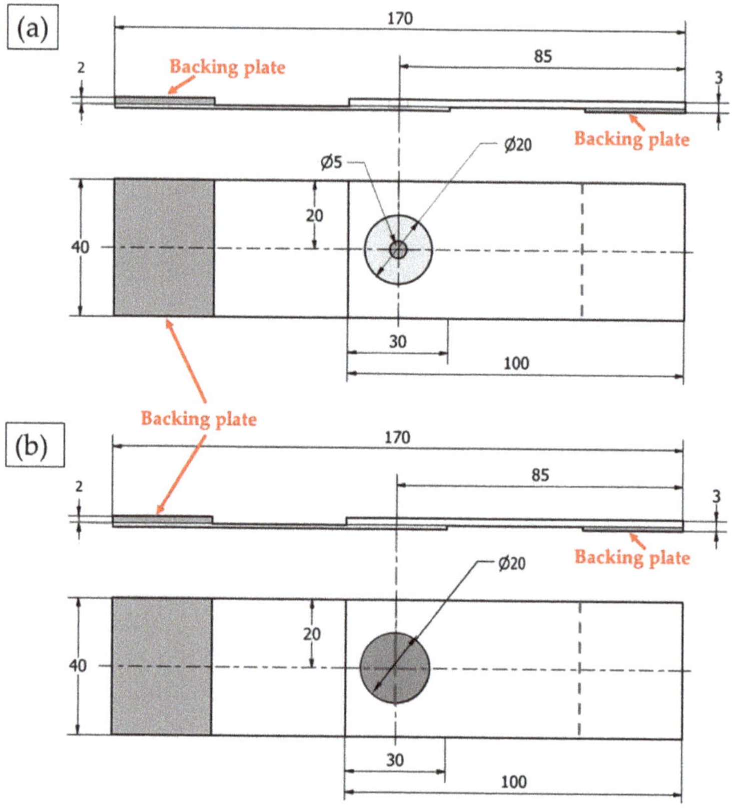

Figure 5. Schematic of tensile-shear test specimens: (**a**) the FSSWed AA6082-T6 dissimilar sheet thickness joint and (**b**) the RFSSWed joint (all dimensions in mm).

3. Results and Discussion

3.1. FSSWed and RFSSWed Joint Surfaces' Appearance

The top and bottom surfaces of the produced FSSWed AA6082-T6 dissimilar sheet thickness and their RFSSWed joints with AA2011-T6 deposited layers at the applied processing parameters (FSSW and refill processing parameters) are shown in Figure 6. The top and bottom surfaces of the FSSW lap joints welded at different rotation speeds of 400, 600, 800, and 1000 rpm and a constant dwell time of 3 s are represented in Figure 6a. Meanwhile, the top and the bottom surfaces of the RFSSW joints processed at a constant consumable rod rotation speed of 400 rpm and a constant feed rate of 1 mm/min are shown in Figure 6b. It can be remarked that the applied welding parameters for joining the dissimilar thicknesses of the AA6082-T6 thin sheet are appropriate to produce successful spot joints. The circular indentations are shown at the top surface views of the FSSW lap joints, indicating the shoulder projection for all the applied welding parameters. The sizes of the extruded flash materials are also the same (Figure 6a). Furthermore, the bottom surface views of the FSSW lap joints show the affected areas (dark areas) by thermal exposure due to the FSSW process (as shown in Figure 6a). These thermal affected areas become darker by increasing the rotation speeds from 400 to 1000 rpm at a constant dwell time of 3 s due to the increase in heat input generation with increasing rotation speed [6,36], as seen in Figure 6a.

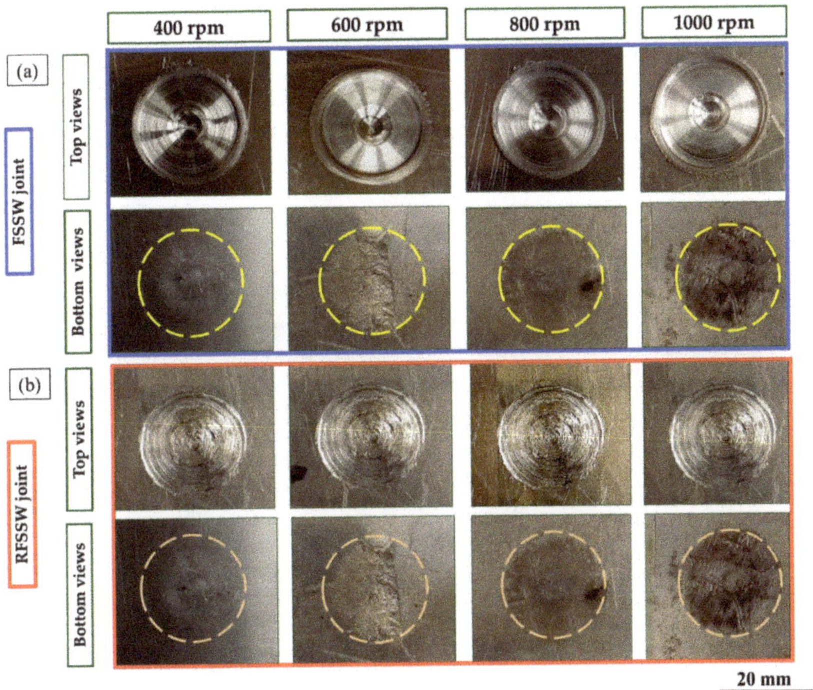

Figure 6. The visual appearance of the top and bottom surface views of (**a**) the AA6082-T6 FSSW joints FSSWed at a constant dwell time of 3 s and different rotation speeds of 400, 600, 800, and 1000 rpm and (**b**) the RFSSW joints after refilling the keyholes.

Figure 6b shows the top surface appearance of RFSSWed joints with FSD process parameters of 400 rpm consumable tool rotation speed and 1 mm/min constant feeding rate. Visually, it can be remarked that defect-free refilling using AA2011 material top surfaces was attained for all the processed joints. It can be concluded that the suggested FSD parameters succeeded to refill all the keyholes and the shoulder projections of the FSSW lap joints. In addition, there was no remarkable effect of the exposure thermal cycle during FSD on the bottom surface appearance of the RFSSW lap joints (Figure 6b) compared to that given by the unrefilled FSSW joints (Figure 6a).

3.2. Heat Input Energy Calculations and Peak Temperature Measurements

The total heat input energy is the summation of the generated heat input energy at the contact shoulder area and the pin surrounding contact area. The generated heat input energy during the FSSW process depends on the tool design, tool material, friction axial downward force, friction coefficient at the interface between the rotating pin and the surrounded stirring material, applied dwell time, and rotation speed [9,37]. The heat input is generated by converting the mechanical energy from the friction of the rotating tool through the joint thickness during the FSSW thermomechanical process. The heat input energy for the FSSW process (Q) can be calculated based on the following equations [9,38]:

$$Q = 1.083 \times \mu \times \frac{P}{K_A} \times \omega \times r \times t \quad (1)$$

where μ is the friction coefficient between the steel tool and the aluminum alloy sheet, which equals 0.4 [4]; P is the applied downward force (in N); K_A is the ratio of the shoulder contact area to cross-sectional area of the rotating tool; ω is rad/s and equals $\frac{\pi}{30}n$, where n is the applied rotational speed (in rpm); r is the pin radius (in m); and t is the dwell time (in s) during the FSSW process:

$$K_A = \frac{(\text{shoulder radius of the rotating tool})^2 - (\text{pin radius})^2}{(\text{shoulder radius of the rotating tool})^2} \quad (2)$$

where the shoulder radius of the rotating tool and pin radius is equal to 10 and 2.5 mm, respectively. Then $K_A = 0.9375$. From Equations (1) and (2), the calculated heat input energy is as follows:

$$Q = 1.1859 \times 10^{(-3)} \times P \times n \times t \text{ (J)} \quad (3)$$

The calculated Q of the FSSW joints is plotted against the applied rotation speed, as given in Figure 7. It can be noted that the increase of the rotation speeds from 400 to 1000 rpm leads to an increase in the generated heat input energy from 1434 to 3351 J during the FSSW process [39]. The FSSW thermal cycle exposure and the measured temperatures of the produced FSSW lap joints are shown in Figure 8.

The thermal cycles of the FSSW AA6082-T6 lap joint in terms of the measured working temperature against the consuming time to yield the spot joint at the applied rotation speeds of 400 to 1000 rpm are given in Figure 8. It can be observed that the experienced thermal cycle for all the processed joints gives the same trends at the welding parameters, with difference in the peak temperatures at each tool rotation speed. Each thermal cycle involves three stages, including rising temperature gradually by inserting the rotating tool pin at a constant feeding rate of 0.1 mm/s through the clamping of two different sheet thicknesses of AA6082-T6 (first stage). The second stage represents the dwell time required to perform the spot weld at near temperature stability. The last stage represents rotating tool retracting at the end of the spot-welding process. Table 2 illustrates the measured peak temperature in the stir zone (SZ) during the FSSW process. It can be noted that the peak temperature increases from to 225 ± 2 to 350 ± 2 with the increasing tool rotation speed from 400 to 1000 rpm, respectively, as a result of frictional heat during the stirring process [39].

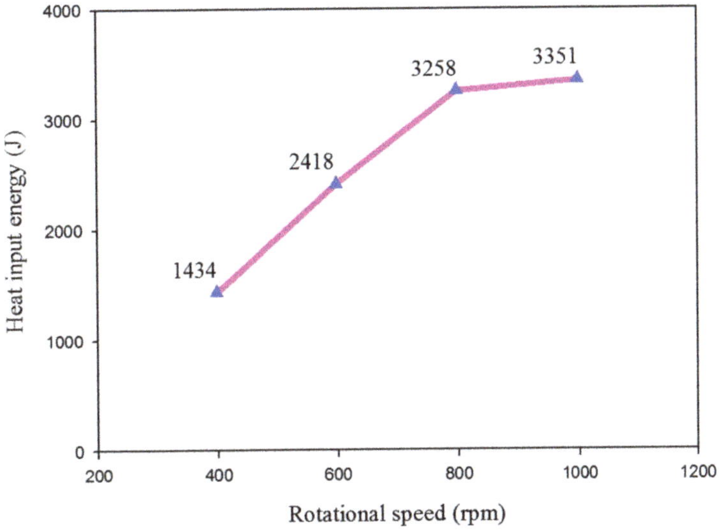

Figure 7. The calculated heat input energy against the applied rotational speeds of the FSSWed AA6082 joints.

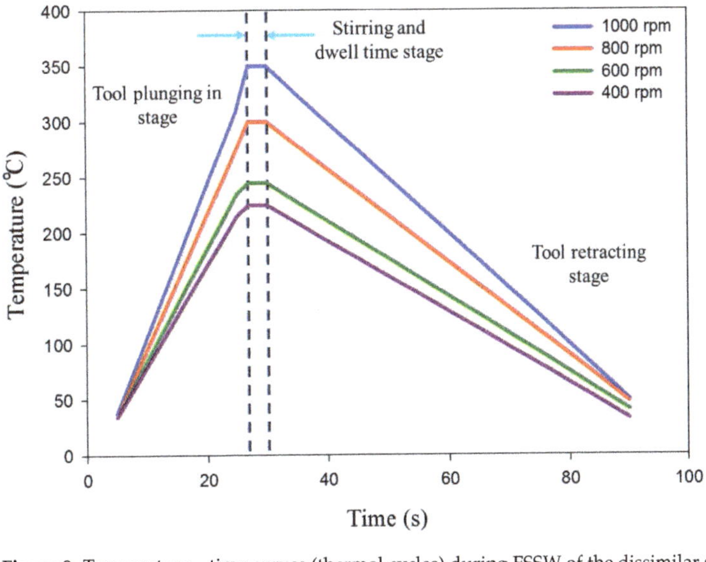

Figure 8. Temperature—time curves (thermal cycles) during FSSW of the dissimilar sheet thickness AA6082-T6 joints welded at different rotational speeds of 400, 600, 800, and 1000 rpm while keeping the dwell time constant (3 s).

Table 2. The peak measured temperature during the FSSW process.

Rotational speeds (rpm)	400	600	800	1000
Temperature (°C)	225 ± 2	245 ± 3	300 ± 5	350 ± 2

Based on the friction stir processing [40–42] and deposition [43,44] concepts, the total heat input (Q_t) to refill the FSSW keyhole via FSD is the summation of the frictional heat generated by the consumable tool shoulder between the rotating consumable pin and the surrounded material in the vicinity of the keyhole and at the consumable pin tip. In general, the heat generation is given as follows:

$$dQ = \omega \times dM = \omega \times r \times dF = \omega \times r \times \tau_{contact}\, dA \quad (4)$$

$$\tau_{contact} = \tau_{friction} = \mu \times P = \mu \times \frac{F}{A} \quad (5)$$

P is the contact pressure at friction interfaces, and it is equal to the axial downward force divided by the projected areas.

The heat input generation under the consumable rod shoulder (Q_1) is calculated by integration Equation (4).

$$Q_1 = \int_0^{2\pi}\int_{R_P}^{R_s} \omega \times r^2 \times \tau_{friction} \times d\theta \times dr = \frac{2}{3} \times \pi \times \omega \times \mu \times \frac{F}{A} \times \left(r_s^3 - r_p^3\right). \quad (6)$$

The heat input generation between the rotating consumable pin and the surrounded material in the vicinity of the keyhole (Q_2) is calculated by integration Equation (4).

$$Q_2 = \int_0^{2\pi}\int_0^{H_p} \omega \times r_p^2 \times \tau_{friction} \times d\theta \times dz = \frac{2}{3} \times \pi \times \omega \times \mu \times \frac{F}{A} \times r_p^2 \times H_p \quad (7)$$

The heat input generation at the consumable pin tip (Q_3) is calculated by integration Equation (4).

$$Q_3 = \int_0^{2\pi}\int_0^{R_p} \omega \times r_p^2 \times \tau_{friction} \times d\theta \times dr = \frac{2}{3} \times \pi \times \omega \times \mu \times \frac{F}{A} \times r_p^3 \quad (8)$$

The total heat input energy (Q_t) is calculated by summations of Equations (5)–(8):

$$Q_t = Q_1 + Q_2 + Q_3$$

$$Q_t = \frac{2}{3} \times \pi \times \omega \times \mu \times \frac{F}{A} \times \left(r_s^3 + r_p^2\, H_p\right) \quad (9)$$

$$Q_t = \frac{2}{3} \times \mu \times \omega \times F \times \left(r_s + 3\,\frac{H_p \times r_p^2}{r_s^2}\right) \quad (10)$$

where Q_t is in watt, μ is the friction coefficient between AA6082-T6 keyhole material and AA2011-T6 consumable rod material and equals 1.4 [45], ω is in rad/s and equals $\frac{\pi\, n}{30}$, n is the rotational speed (in rpm), F is the axial downward force (in N), r_s is the consumable rod shoulder radius (in m), r_p is the consumable rod pin radius (in m), H_p is the consumable rod pin height (in m), and t is the deposition time (in s).

$$Q_t = \frac{2\pi}{90} \times \mu \times n \times F \times \left(r_s + 3\,\frac{H_p \times r_p^2}{r_s^2}\right)$$

$$Q_t = 2250\ W$$

The refilling time of FSD = refill height $\times 60 = 180$ s.

The total heat input energy (QE) = $Q_t \times t$ (in J).

The knowledge of the thermal cycle in terms of the heat generation and the temperature history during the FSD process is very important in understanding the thermomechanical interaction occurring during the deposition process in the keyhole of the FSSW joint. The heat generates at the contact area between the consumable rod and the substrate (keyhole bottom) due to the friction during the stirring process. The heat to develop a bond between the keyhole bottom and the deposited material involves transferred heat to the FSD consumable tool, transferred heat to the keyhole bottom, and remaining heat for depositing a material [43]. When the generated peak temperature during the FSD process is below the melting point of the deposition material, the deformation is a solid-state process [43,46]. The thermal cycle to achieve refilling of the FSSW joints of AA6082-T6 is represented as temperature versus time, as given in Figure 9. It can be noted that the deposition peak temperature is 247 °C, which is below the melting point of the AA2011-T6 deposited material. This solid-state process promotes good bonding between the AA2011-T6 deposited material and the keyhole bottom of FSSW dissimilar thickness joints. A similar finding in terms of good bonding between the deposited materials and the substrates is reported by other researchers to build sound continuous layers of as-cast hypoeutectic A356 Al alloy on the AA2024 substrate [30] and also sound continuous layers of AA2011 in two temper conditions on the AA5083 substrate [29] without any defects at the interface.

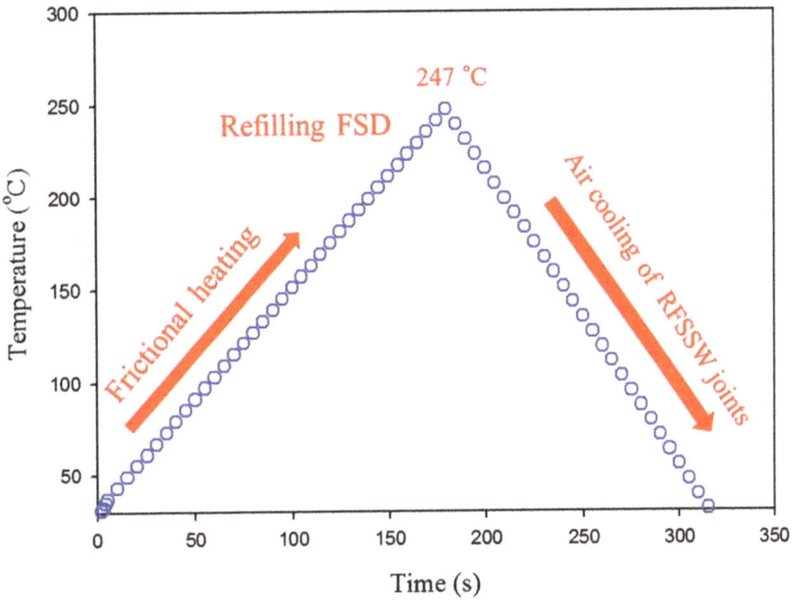

Figure 9. The temperature measurements against time during the FSD process.

3.3. *Macrographs of the FSSW and RFSSW Lap Joints*

Figure 10 illustrates the transverse macrographs of the FSSW AA6082-T6 dissimilar sheet thickness joints and their RFSSW counterparts. It can be seen that the applied FSSW parameters resulted in defect-free welds at the interface of the overlapped 1 mm and 2 mm AA6082-T6 sheets (Figure 10a–d). In addition, the excessive extruded flash material was eliminated as a result of the FSSW optimization welding parameters based on some experiments and published works [3,6]. The resulting spot welded joint has a characteristic hole in the middle of the joint; this hole is left by the pin after removal. This keyhole (material loss) is the main defect in the FSSW process, and it is a good representative of the pin dimensions and geometry (Figure 10a–d). Figure 10e–h represents the cross-section macrographs of the RFSSW joints. The keyhole refilling process was achieved at the deposition conditions of 400 rpm consumable rod rotation speed and 1 mm/min deposition rate. It can be remarked that the keyhole can be successfully refilled by the FSD process using the deposition material of AA2011-T6. Furthermore, the deposited materials bonded well with the keyhole inner surface without porosity and microcracks for all the refilled joints.

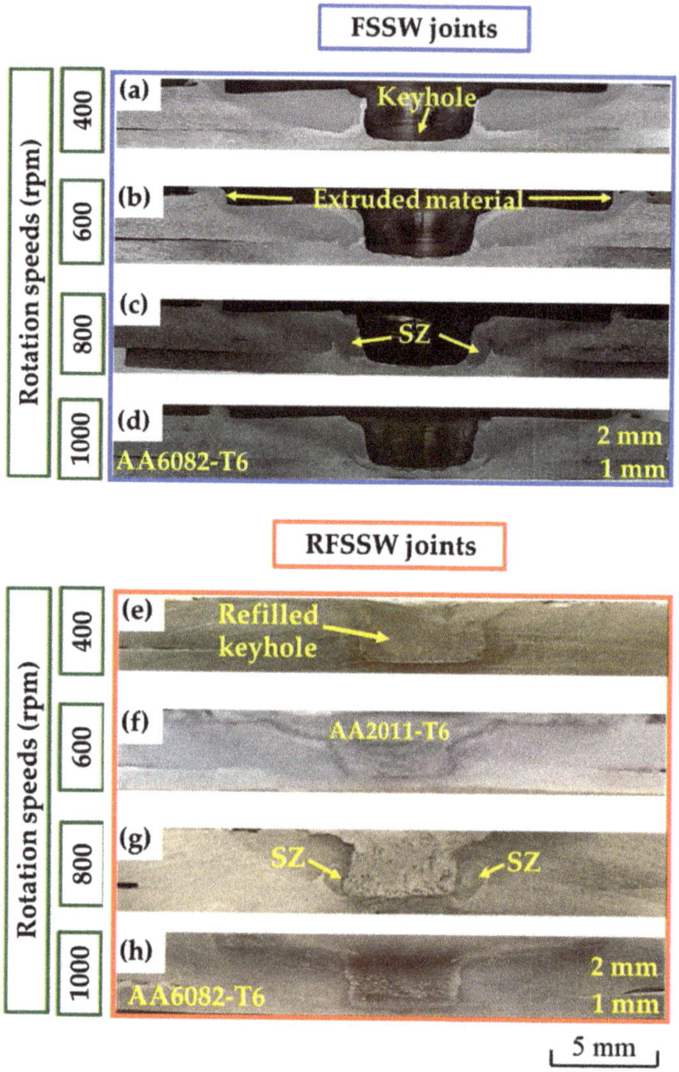

Figure 10. Transverse cross-section macrographs of (**a–d**) the FSSW lap joints of AA6082-T6 welded at a constant dwell time of 3 s and different rotational speeds of 400 to 1000 rpm [36] and (**e–h**) RFSSW deposited at a 400 rpm rotation speed and a 1 mm/min deposition rate.

3.4. Microstructure Investigation of the Base Materials

The microstructures of the base materials (BMs) for the two AA6082-T6 sheet thicknesses at 1 and 2 mm and the AA2011-T6 rod Al alloys are shown in Figure 11. It can be seen that the microstructures of the AA6082-T6 BMs show coarse and elongated grain structures, indicating the rolling directions. In addition, they contain comparatively coarse (Fe, Mn)$_3$SiAl$_{12}$ intermetallic and low-density fine dispersed Mg_2Si precipitates. These microstructure features are consistent with that reported in other works [2,36,47]. It can be observed also that the AA 6082-T6 sheets of 1 (Figure 11a) and 2 mm (Figure 11b) thickness have average grain sizes of 11 ± 2 and 17 ± 1.5 μm, respectively. The variation in the grain size values of the two sheets can be ascribed to imposing different rolling conditions to produce different sheet thicknesses. This reduction in thickness as a result of high accumulated plastic deformation promotes high internal strain, which increases the strength [48]. As reported in [28,29], the microstructure of the AA2011-T6 rod alloy shows coarse grains and the intermetallics of Al_7Cu_2Fe (rodlike shape) and Al_2Cu (almost spherical or irregular shapes), as shown in Figure 11c.

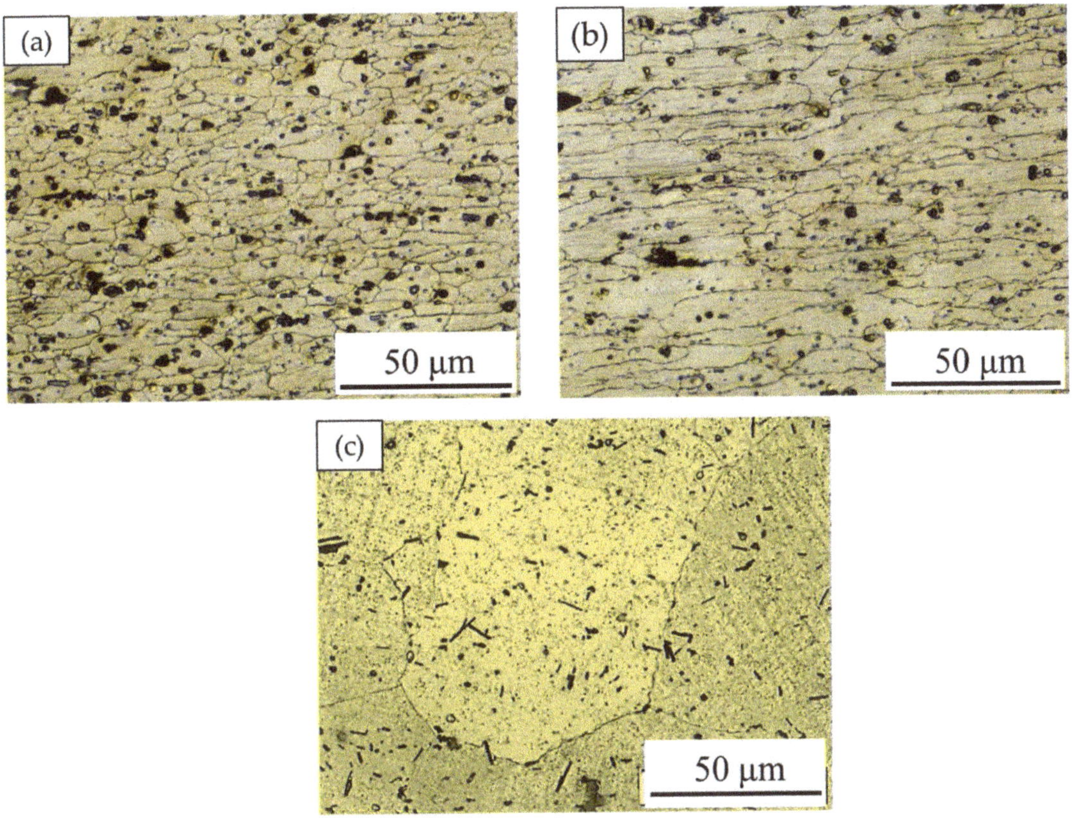

Figure 11. OM microstructure images of the starting BMs: (**a**) 1 mm, (**b**) 2 mm AA6082-T6, and (**c**) AA2011-T6 [29].

3.5. Hardness Results Evaluation

Hardness measurements were carried out on the transverse cross sections of both the AA6082-T6 FSSW joints of the dissimilar sheet thickness and RFSSW joints and plotted in a hardness profile. Figure 12 shows the hardness profile for the FSSW joints produced at a constant dwell time of 3 s and different rotation speeds of 400, 600, 800, and 1000 rpm. Meanwhile, Figure 13 represents the hardness profile of the RFSSW joints, refilled with AA2011-T6 via the FSD process at a 400 rpm rotation speed and a 1 mm/min feed rate.

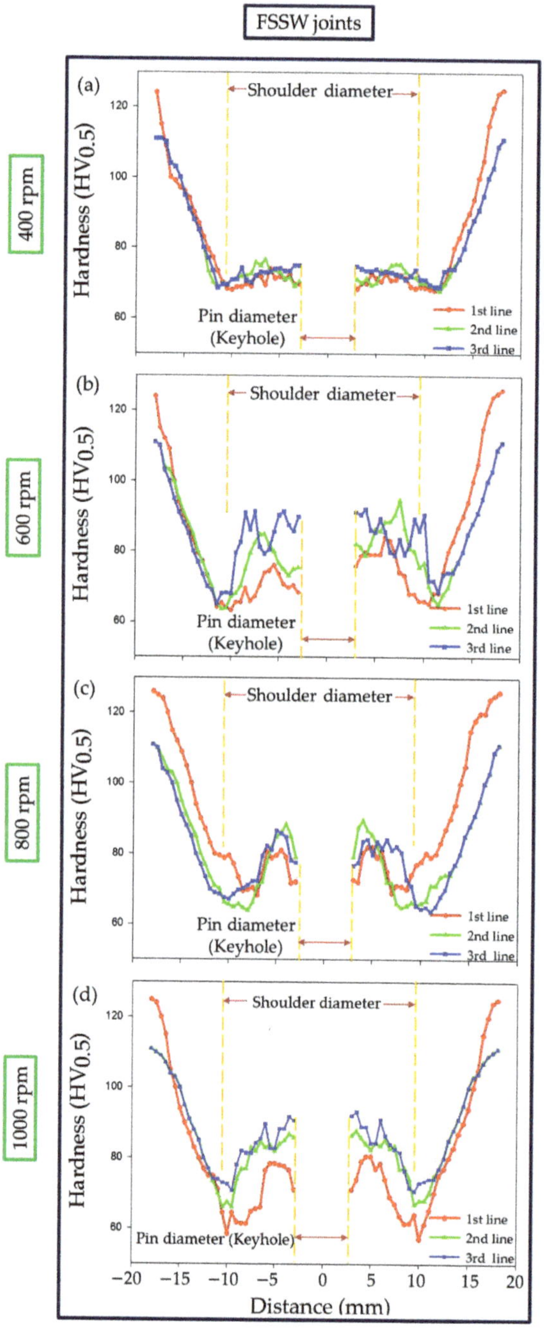

Figure 12. Hardness profile of the dissimilar sheet thickness AA6082-T6 FSSW joints produced at (**a**) 400 rpm, (**b**) 600 rpm, (**c**) 800 rpm, and (**d**) 1000 rpm rotation speeds and a constant dwell time of 3 s.

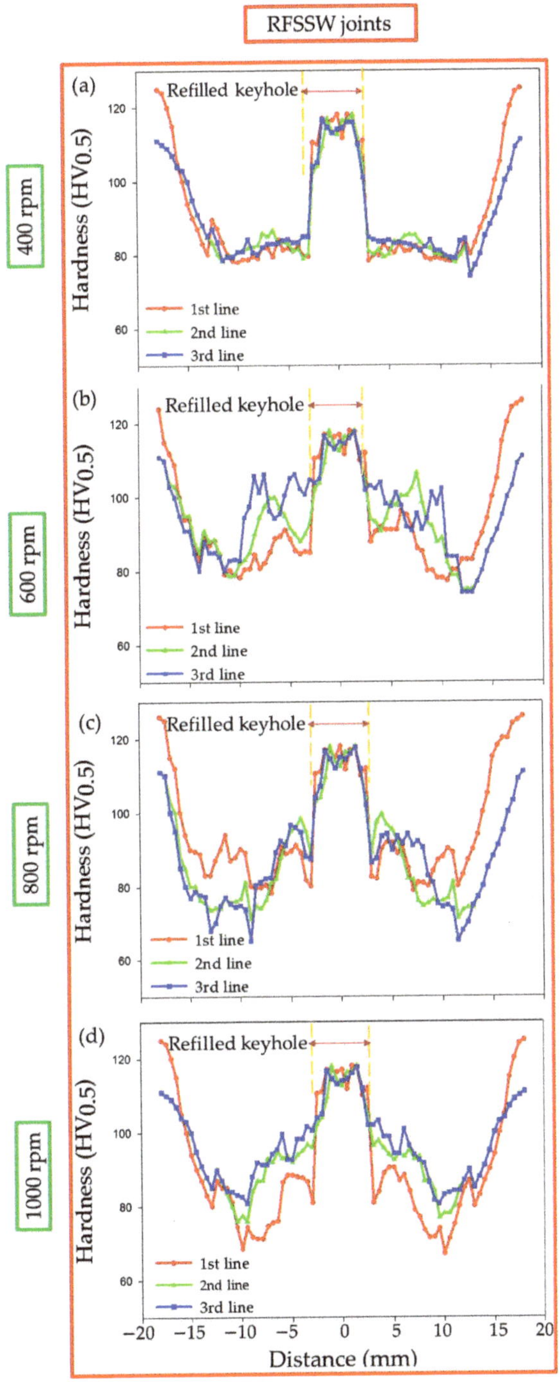

Figure 13. Hardness profile of the AA6082-T6 RFSSW joints spot welded at (**a**) 400 rpm, (**b**) 600 rpm, (**c**) 800 rpm, and (**d**) 1000 rpm rotation speeds with applying a constant dwell time of 3 s and refilled with AA2011-T6 via the FSD process at a 400 rpm rotation speed and a 1 mm/min feed rate.

Compared with the different sheet thickness AA6082-T6 BMs, the hardness of the weld zones, the SZ, thermomechanically affected zone (TMAZ), and the heat-affected zone (HAZ) of the FSSWed joints at the used rotation speeds of 400, 600, 800, and 1000 rpm significantly decreased, as shown in Figure 12a–d, respectively, due to annealing of the AA6082-T6 BMs during the FSSW process. It was reported that the hardness values across the weld zones of the joints of heat-treatable Al alloys are affected by thermal exposure during FSW [36,49]. Among the weld zones for each FSSWed joint, the minimum hardness values were observed in the HAZ due to the grain structure and overaging effect [36]. In contrast, the hardness improvement in the SZ is mainly due to the formation of dynamically recrystallized fine grains and the reprecipitation process that might take place during the cooling cycle. The TMAZ hardness gives lower values than the SZ and higher values than the HAZ. The increase in hardness of the TMAZ over the HAZ is likely due to the high dislocation density achieved by plastic deformation during the process of FSSW [14,47]. The highest hardness of the weld zones is achieved for the FSSWed joint at 600 rpm (Figure 12b). Thus, selected areas on the macrostructure of this joint (Figure 14a) were investigated using OM as a representative of the weld zone microstructure features. It can be seen that the welding zones on both sides of the keyhole center line look like a mirror and are arranged as SZ, TMAZ, and HAZ, as given in Figure 14b (left side) and Figure 14c (right side). The keyhole surrounding area is the SZ achieved during the FSSW process of the two AA6082-T6 sheets, and it is characterized by equiaxed refined grains (Figure 14e). Generally, optimizing the FSW [50,51] and the FSSW [3,36] parameters leads to grain refinement in the SZ due to dynamic recrystallization. These variations in microstructure features correspond to the variation in the hardness values in the weld zones (Figure 12b).

Instead of incomplete W-shape hardness profiles of the FSSWed joints (Figure 12), complete W-shape hardness profiles are achieved for the RFSSWed joints (Figure 13). As mentioned before in the experimental section, the FSD process was performed at the condition of 400 rpm AA2011-T6 consumable rod rotation speed and 1 mm/min feed rate. Thus, it is expected that the deposited material (keyhole refilling material) has the same microstructure feature and mechanical properties. The average measured hardness value of the refilled keyhole is 117 ± 5 HV for all the RFSSW joints, as shown in Figure 13a–d. During the deposition process, the material of the rotating consumable rod is subjected to severe plastic deformation at high homologous temperatures. As a result, the material deposited by the friction stirring process undergoes dynamic recrystallization and develops a very fine grain size [28,29]. El-Sayed Seleman et al. [28] studied utilizing the FSD process to manufacture continuous multilayer high-performance, metal-based AA2011/nano Al_2O_3 composites using AA2011 in two temper conditions of AA2011-T6 and AA2011-O as a matrix. The results showed that the microstructure of the starting, consumable rods, AA2011-T6 and AA2011-O, was changed from coarse grain to refined equiaxed grain throughout the deposited material. Additionally, the size of precipitates (Al_7Cu_2Fe and Al_2Cu) decreased with the stirring action during FSD for the deposited matrix and the produced composite. The dispersion and fragmentation of these intermetallics in the Al matrix have been detected in other works [29,33]. These microstructure features are gained for the deposited material in the keyholes of the FSSW joints. Figure 14f shows the microstructure of the AA6082-T6 RFSSWed joint welded at 600 rpm and 3 s with keyhole refilling by AA2011-T6 via the FSD conditions of 400 rpm and 1 mm/min. It can be seen that equiaxed refined grains (average grain size 2 ± 0.2 μm) due to the dynamic recrystallization compared with coarse grains AA2011-T6 rod BM (45 ± 8 μm) are detected. In addition, the intermetallics (Al_7Cu_2Fe and Al_2Cu) are highly dispersed and fragmented.

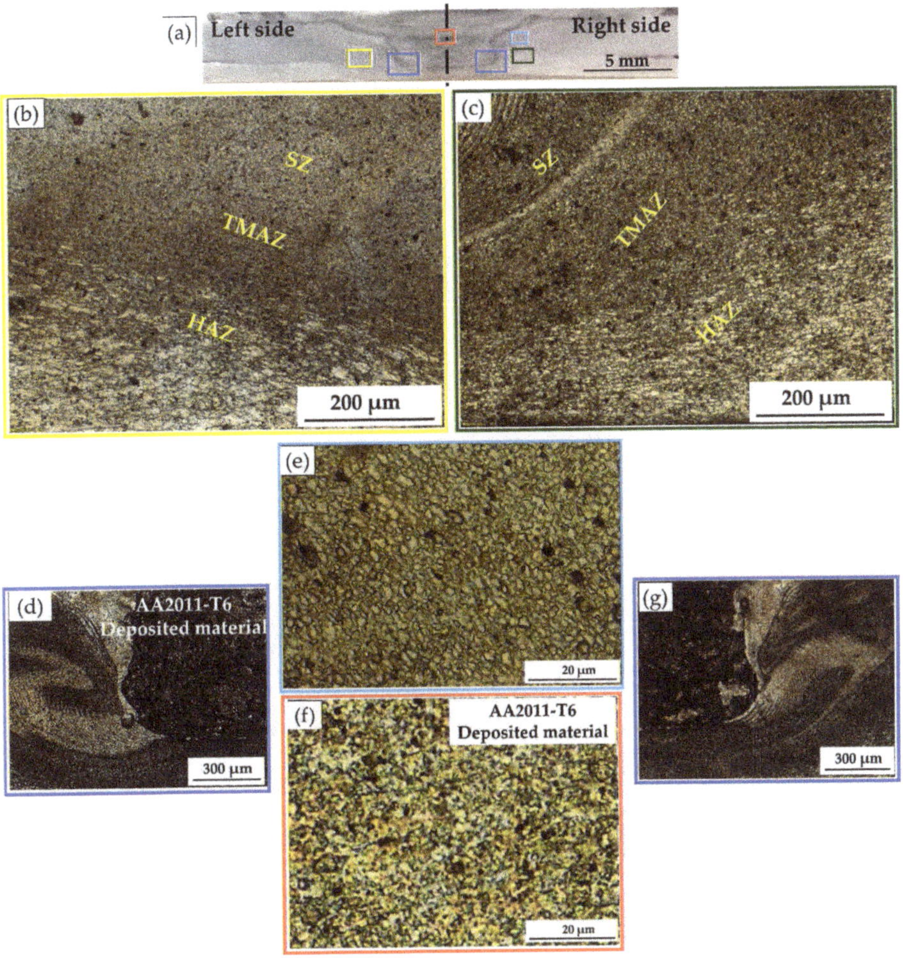

Figure 14. (**a**) Macrostructure image of the AA6082-T6 RFSSWed joint welded at 600 rpm and 3 s with keyhole refilling by AA2011-T6 via the FSD conditions of 400 rpm and 1 mm/min. And microstructure images of (**b**) the spot weld zones on the left side, (**c**) the spot weld zones on the right side, (**d**) weld zones/deposited material interface on the left side, (**e**) the SZ of the welded joint, (**f**) SZ of the deposited material and (**g**) weld zones/deposited material interface in the right side.

3.6. Tensile Shear Test and Fracture Surfaces

The tensile shear test is an important mechanical property for joint efficiency evaluation for vehicles and automotive structure designs. The tensile shear value of the spot welded joint is affected by FSSW process parameters [6,9,21,22]. The maximum tensile shear load of the FSSW and RFSSW lap joints of AA6082-T6 dissimilar sheet thickness is illustrated in Figure 15.

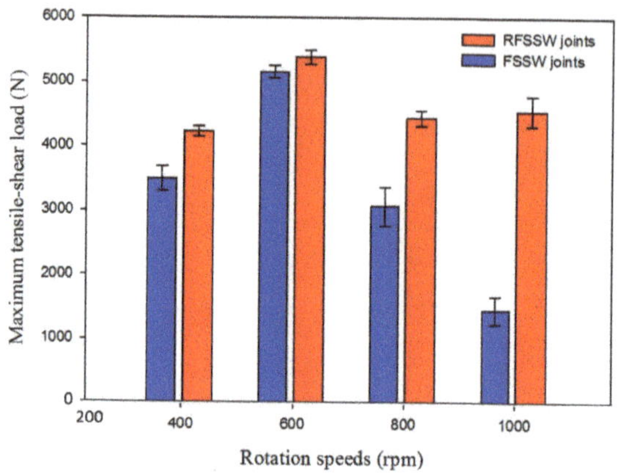

Figure 15. Maximum tensile-shear load of the AA6082-T6 FSSWed joints (before keyhole refilling) processed at a constant dwell time of 3 s and different rotation speeds of 400 to 1000 rpm and the RFSSW lap joints. All the keyholes of the FSSW joints refilled at the same FSD parameters of 400 rpm AA2011-T6 consumable rod rotation speed and 1 mm/min feed rate.

It can be noted that the FSSW joint welded at a constant dwell time of 3 s by applying a tool rotation speed of 600 rpm gives the highest tensile shear load of 4300 ± 80 N compared with that obtained by the other welded joints. The two AA6082-T6 sheets of this joint are still connected after the tensile shear testing, as given in Figure 16a–c. This indicates high joint efficiency and may be ascribed to a larger fully bonded section size and a lower hook height [36]. Besides, the enhancement in hardness in the SZ of this joint is over than that noted in the stir zones of the other spot-welded joints. In contrast, the FSSWed joints processed at the same dwell time (3 s) and different tool rotation speeds of 400, 800, and 1000 rpm are completely separated during the tensile shear testing, as given in Figure 17a. On the one hand, the failed joint FSSWed at a relatively lower rotation speed of 400 rpm may be attributed to insufficient heat input, which affects in mixing the dissimilar sheet thicknesses AA6082-T6 during the welding process. On the other hand, the tensile shear loads of the spot welds produced at higher tool rotation speeds of 800 and 1000 rpm illustrate a decrease in the maximum tensile shear load. This decrease in strength for these joints is ascribed to the increase in thermal softening in the SZ with the increasing the heat input and also to the reduced thickness of the upper sheet underneath the shoulder. Ohashi et al. [52] and Xie et al. [53] reported that the mechanical properties of the FSSW joints are mainly governed by both the weld bonded area and the upper sheet underneath the shoulder. In addition, the precipitate-free zones around the grain boundaries due to precipitate coarsening might have reduced the strength of the SZ [36]. From Figure 15, it can be also seen that all the RFSSW lap joints show higher tensile shear loads than that given by the FSSW (unrefilled FSSW) lap joints. The RFSSW joint (welded at 600 rpm/3 s and refilled at 400 rpm/1 mm/min) shows the highest tensile shear load of 5400 N ±100. Its enhancement in the tensile shear load is 25% over that recorded by the unrefilled spot joint. The two sheets of the RFSSW joint are still connected after the tensile shear testing, as given in Figure 16d–f. Furthermore, the other refilled joints show enhancements of 45%, 73%, and 273% over the unrefilled joints processed at 400, 800, and 1000 rpm, respectively. This enhancement for all the refilled joints is ascribed to the repairing of the keyhole defect by depositing high-strength AA 2011-T6 Al alloy via FSD and the strong bonding between the filling deposited material and the key hole interface, in addition to the microstructure features of the filling cavity of the keyhole. Both the deposition process and the good bonding between the keyhole interface and the FSD material filling the cavity are achieved in a solid-state process. Figure 17 displays the separated tested FSSW and RFSSW lap joint after the tensile shear test.

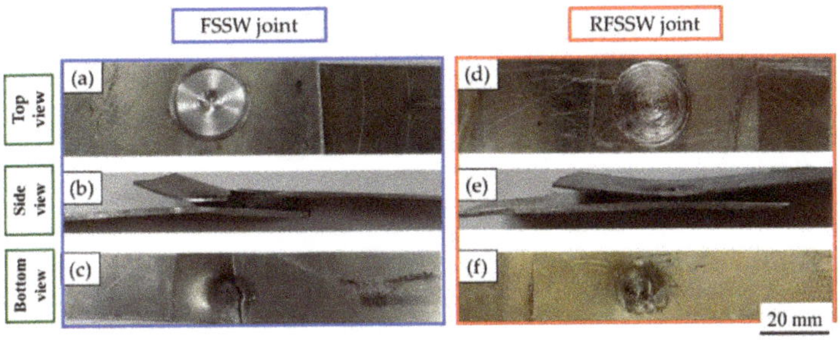

Figure 16. Top, side, and bottom photographs show the joint appearance of the AA6082-T6 dissimilar sheet thickness welded at 600 rpm and 3 s after tensile testing, where (**a**–**c**) the FSSW joint [36] and (**d**–**f**) the RFSSW joint are refilled with AA2011-T6 at 400 rpm and 1 mm/min.

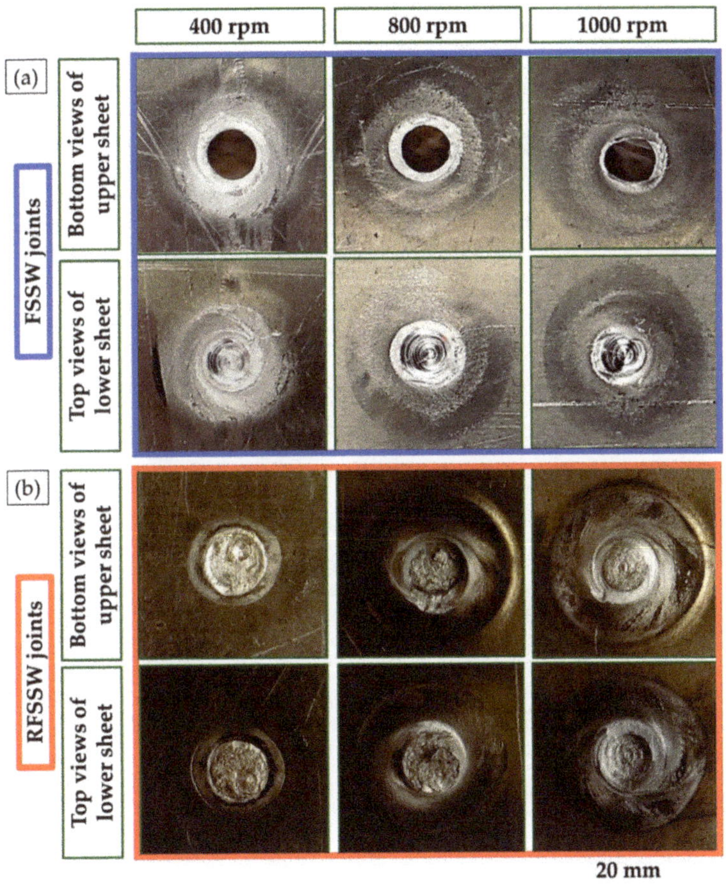

Figure 17. Photographs of the joints after tensile shear testing for (**a**) the FSSW lap joints welded at a dwell time of 3 s and rotation speeds of 400, 800, and 1000 rpm [36] and (**b**) the RFSSW joints welded at the same conditions and refilled at a 400 rpm AA2011-T6 consumable rod rotation speed and a 1 mm/min feed rate.

The fracture surfaces of the BMs after tensile testing are given in Figure 18. Two fracture modes in terms of ductile and brittle fractures are detected for both the 1 and 2 mm sheet thicknesses. The Al matrix shows a ductile fracture in terms of different shallow dimple sizes. While the (Fe, Mn)$_3$SiAl$_{12}$ and Mg$_2$Si coarse precipitates show a brittle fracture. Microvoids due to the pullout of the fragmented intermetallics are also noticed on the fracture surface of the two sheets. The dimple size is related to the initial grain structure of the two sheets' materials.

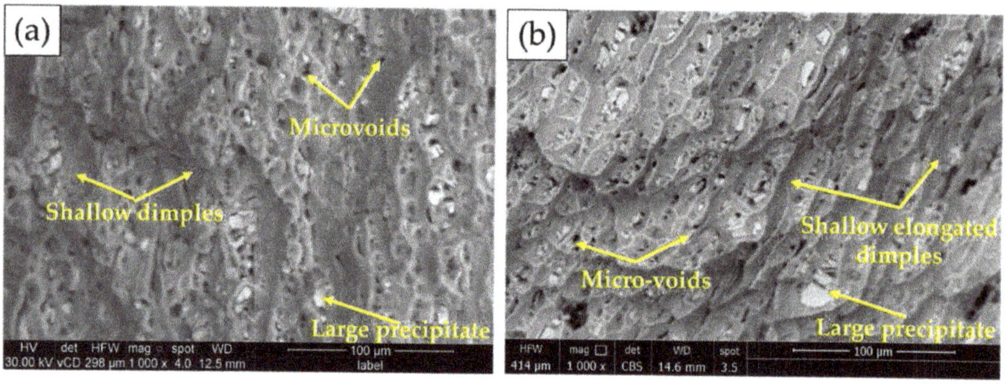

Figure 18. SEM images of the fracture surface of AA6082-T6 BMs: (**a**) 1 mm and (**b**) 2 mm [36].

Figure 19a illustrates a photograph of the top surface of the lower sheet of the AA6082-T6 FSSW joint welded at a constant dwell time of 3 s and a rotation speed of 400 rpm after tensile shear testing, and Figure 19b–d shows the SEM fracture surface at different magnifications. The spot weld joint is failed by a tensile shear mechanism. Under the tensile shear loading, the crack begins at the tip of the hook of the partially bonded region and propagates preferentially in a horizontal direction at the lap joint interface, shearing the SZ, causing failure. The fracture surface of the failed region shows a typically ductile mode fracture in terms of different sizes of equiaxed deep dimples (Figure 19d) compared with shallow elongated dimples of the fracture surface of both BMs (Figure 18), indicating the grain refining of the SZ during the FSSW process. Shear cleavage is observed, indicating the direction of the tensile shear mechanism.

Figure 20a shows a photograph of the top surface of the lower sheet of the AA6082-T6 RFSSW joint (spot welding conditions of 400 rpm and 3 s and refilled conditions of 400 rpm and 1 mm/min) after tensile shear testing. Figure 20b,c shows the SEM fracture surface at different magnifications. The refilled joint is failed also by the tensile shear mechanism. Under the tensile shear loading, the crack begins at the tip of the hook of the partially bonded region and propagates at the lap joint interface, shearing the SZ of the welded area and the deposited refilled material, causing failure. The fracture surface of the failed material shows two distinct regions (AA6082-T6 weld sheets and AA2011-T6 deposited material) having a well-bonded interface (Figure 20b). The refilled joint reveals ductile fractures in both regions as they have equiaxed fine grain structures due to the stirring action in both the weld region and the refilled keyhole (Figure 20b).

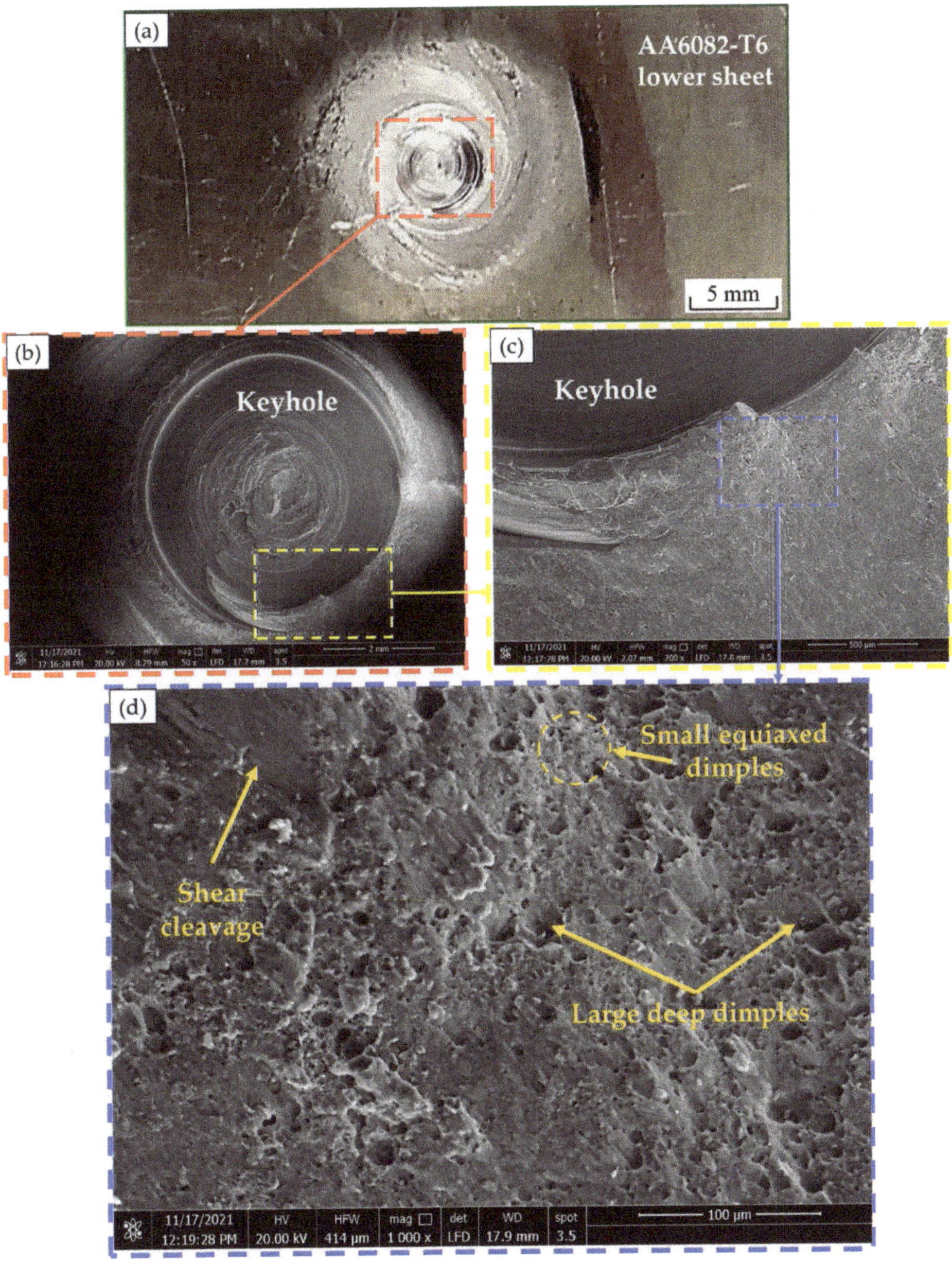

Figure 19. (**a**) Photograph of the top surface of the lower sheet of AA6082 -T6 FSSW (welded at 400 rpm and 3 s) after tensile shear testing and (**b–d**) SEM images of the fracture surface at different magnifications [36].

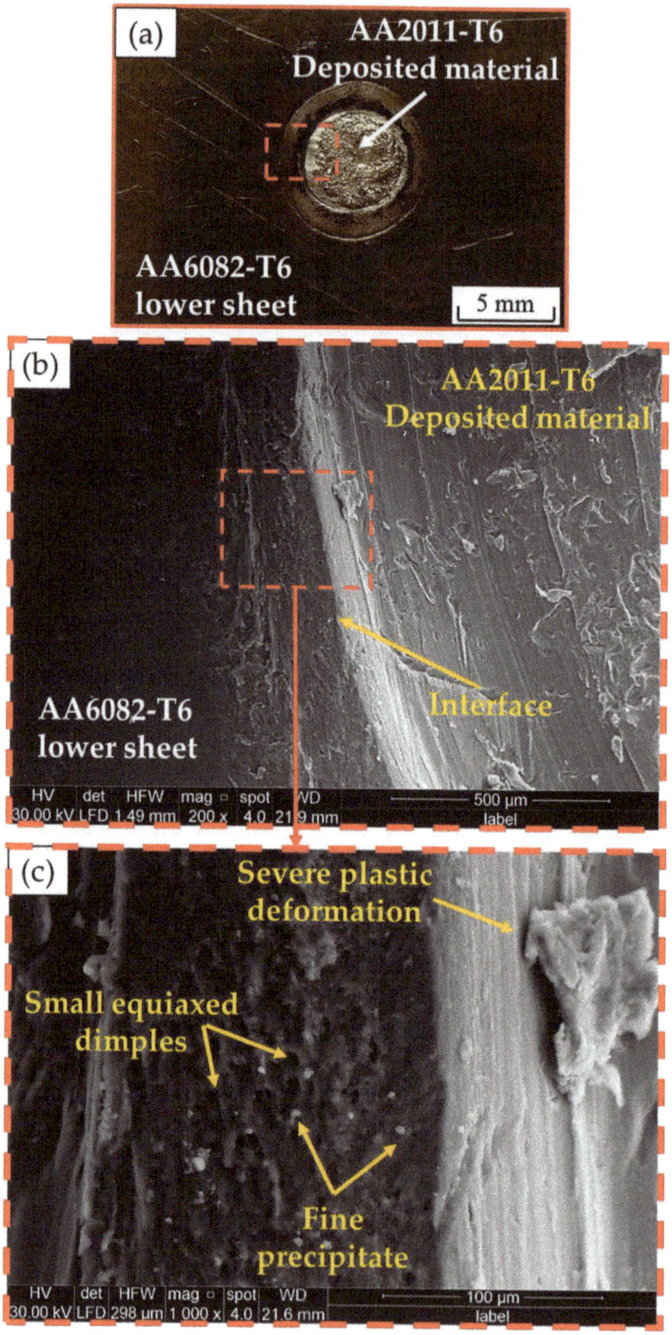

Figure 20. (**a**) Photograph of the top surface of the lower sheet of AA6082-T6 of the RFSSW joint (welded at 400 rpm and 3 s) after tensile shear testing and (**b**,**c**) SEM images of the fracture surface at different magnifications.

4. Conclusions

In this study, an innovative technique was developed based on the friction stir deposition and examined to refill the keyhole of the FSSWed aluminum samples. A comparative investigation and evaluation were carried out between the unrefilled FSSWed and RFSSWed samples. Based on that, the following conclusions can be outlined:

1. The applied FSSW parameters in terms of different rotational speeds of 400 to 1000 rpm and a constant dwell time of 3 s succeeded in spot-welding two different thin sheet thicknesses of AA6082-T6.
2. The applied FSD parameters in terms of a feed rate of 1 mm/min and an AA2011-T6 consumable rod rotational speed of 400 rpm succeeded in refilling FSSW keyholes and shoulder projections of all the produced AA6082-T6 FSSW lap joints with defect-free continuous multilayers.
3. All the RFSSW joints show higher bearing tensile shear loads than that given by the as-FSSWed joints.
4. Among all the RFSSW joints, the RFSSW joint (welded at 600 rpm/3 s and refilled at 400 rpm/1 mm/min) promotes the highest tensile shear load of 5400 N $\pm$ 100. Meanwhile, among all the FSSW joints, the FSSW joint (welded at 600 rpm/3 s) gives the highest tensile shear load of 4300 N $\pm$ 80.
5. The suggested FSD technique, including the consumable tool design and the FSD parameters, open new horizons for repairing the FSSW keyhole defect for different welded joints.

Author Contributions: Conceptualization, M.M.Z.A., N.A.A., H.A.R. and M.M.E.-S.S.; methodology, H.A.R.; software, H.A.R. and N.A.A.; validation, M.M.Z.A., N.A.A., H.A.R. and M.M.E.-S.S.; formal analysis, M.M.Z.A., N.A.A. and H.A.R.; investigation, H.A.R.; resources, H.A.R., I.A. and E.A.; data curation, H.A.R. and N.A.A.; writing—original draft preparation, H.A.R., M.M.E.-S.S. and M.M.Z.A.; writing—review and editing, M.M.Z.A., H.A.R., N.A.A., E.A. and M.M.E.-S.S.; visualization, I.A.; supervision, M.M.Z.A., E.A. and M.M.E.-S.S.; project administration, M.M.Z.A., N.A.A. and I.A.; funding acquisition, M.M.Z.A., N.A.A. and M.M.E.-S.S. All authors have read and agreed to the published version of the manuscript.

Funding: The authors extend their appreciation to the Deputyship for Research and Innovation, Ministry of Education, in Saudi Arabia for funding this research work through project number IF-PSAU-2021/01/18022.

Institutional Review Board Statement: Not applicable.

Informed Consent Statement: Not applicable.

Data Availability Statement: Data will be available upon request through the corresponding author.

Acknowledgments: The authors extend their appreciation to the Deputyship for Research and Innovation, Ministry of Education, in Saudi Arabia for funding this research work through project number IF-PSAU-2021/01/18022.

Conflicts of Interest: The authors declare no conflict of interest.

References

1. Suryanarayanan, R.; Sridhar, V.G. Effect of Process Parameters in Pinless Friction Stir Spot Welding of Al 5754-Al 6061 Alloys. *Metallogr. Microstruct. Anal.* **2020**, *9*, 261–272. [CrossRef]
2. Muhayat, N.; Priatmana Putra, B.; Triyono. Mechanical Properties and Microstructure of Friction Stir Spot Welded 6082-T6 Aluminium Alloy Joint. *MATEC Web Conf.* **2019**, *269*, 01005. [CrossRef]
3. Ahmed, M.M.Z.; Ahmed, E.; Hamada, A.S.; Khodir, S.A.; El-Sayed Seleman, M.M.; Wynne, B.P. Microstructure and Mechanical Properties Evolution of Friction Stir Spot Welded High-Mn Twinning-Induced Plasticity Steel. *Mater. Des.* **2016**, *91*, 378–387. [CrossRef]
4. Ahmed, M.M.Z.; Habba, M.I.A.; Jouini, N.; Alzahrani, B.; El-Sayed Seleman, M.M.; El-Nikhaily, A. Bobbin Tool Friction Stir Welding of Aluminum Using Different Tool Pin Geometries: Mathematical Models for the Heat Generation. *Metals* **2021**, *11*, 438. [CrossRef]

5. Ahmed, M.M.Z.; Ataya, S.; El-Sayed Seleman, M.M.; Ammar, H.R.; Ahmed, E. Friction Stir Welding of Similar and Dissimilar AA7075 and AA5083. *J. Mater. Process. Technol.* **2017**, *242*, 77–91. [CrossRef]
6. Ahmed, M.M.Z.; Abdul-Maksoud, M.A.A.; El-Sayed Seleman, M.M.; Mohamed, A.M.A. Effect of Dwelling Time and Plunge Depth on the Joint Properties of the Dissimilar Friction Stir Spot Welded Aluminum and Steel. *J. Eng. Res.* **2022**, *10*, 264–279. [CrossRef]
7. Ahmed, M.M.Z.; Ataya, S.; El-Sayed Seleman, M.M.; Mahdy, A.M.A.; Alsaleh, N.A.; Ahmed, E. Heat Input and Mechanical Properties Investigation of Friction Stir Welded Aa5083/Aa5754 and Aa5083/Aa7020. *Metals* **2021**, *11*, 68. [CrossRef]
8. Ahmed, M.M.Z.; El-Sayed Seleman, M.M.; Zidan, Z.A.; Ramadan, R.M.; Ataya, S.; Alsaleh, N.A. Microstructure and Mechanical Properties of Dissimilar Friction Stir Welded AA2024-T4/AA7075-T6 T-Butt Joints. *Metals* **2021**, *11*, 128. [CrossRef]
9. Ataya, S.; Ahmed, M.M.Z.; El-Sayed Seleman, M.M.; Hajlaoui, K.; Latief, F.H.; Soliman, A.M.; Elshaghoul, Y.G.Y.; Habba, M.I.A. Effective Range of FSSW Parameters for High Load-Carrying Capacity of Dissimilar Steel A283M-C/Brass CuZn40 Joints. *Materials* **2022**, *15*, 1394. [CrossRef]
10. Deng, L.; Li, S.; Ke, L.; Liu, J.; Kang, J. Microstructure and Fracture Behavior of Refill Friction Stir Spot Welded Joints of AA2024 Using a Novel Refill Technique. *Metals* **2019**, *9*, 286. [CrossRef]
11. Janga, V.S.R.; Awang, M.; Yamin, M.F.; Suhuddin, U.F.H.; Klusemann, B.; Dos Santos, J.F. Experimental and Numerical Analysis of Refill Friction Stir Spot Welding of Thin AA7075-T6 Sheets. *Materials* **2021**, *14*, 7485. [CrossRef]
12. Shi, Y.; Yue, Y.; Zhang, L.; Ji, S.; Wang, Y. Refill Friction Stir Spot Welding of 2198-T8 Aluminum Alloy. *Trans. Indian Inst. Met.* **2018**, *71*, 139–145. [CrossRef]
13. Fratini, L.; Barcellona, A.; Buffa, G.; Palmen, D. Friction Stir Spot Welding of AA6082-T6: Influence of the Most Relevant Process Parameters and Comparison with Classic Mechanical Fastening Techniques. *Proc. Inst. Mech. Eng. Part B J. Eng. Manuf.* **2007**, *221*, 1111–1118. [CrossRef]
14. Aydin, H.; Tuncel, O.; Umur, Y.; Tutar, M.; Bayram, A. Effect of Welding Parameters on Microstructure and Mechanical Properties of Aluminum Alloy AA6082-T6 Friction Stir Spot Welds. *Indian J. Eng. Mater. Sci.* **2017**, *24*, 215–227.
15. Tozaki, Y.; Uematsu, Y.; Tokaji, K. A Newly Developed Tool without Probe for Friction Stir Spot Welding and Its Performance. *J. Mater. Process. Technol.* **2010**, *210*, 844–851. [CrossRef]
16. Khosa, S.U.; Weinberger, T.; Enzinger, N. Thermo-Mechanical Investigations during Friction Stir Spot Welding (FSSW) of AA6082-T6. *Weld. World* **2010**, *54*, 134–146. [CrossRef]
17. Habeeb, S.S.; Katratwar, T. Analyzing Process of Friction Stir Spot Weld Joint. *Int. J. Sci. Technol. Eng.* **2016**, *3*, 308–313.
18. Pan, T.Y. Friction Stir Spot Welding (FSSW)—A Literature Review. *SAE Tech. Pap.* **2007**, *1*, 1702. [CrossRef]
19. Haneklaus, N.; Cionea, C.; Reuven, R.; Frazer, D.; Hosemann, P.; Peterson, P.F. Hybrid Friction Diffusion Bonding of 316L Stainless Steel Tube-to-Tube Sheet Joints for Coil-Wound Heat Exchangers. *J. Mech. Sci. Technol.* **2016**, *30*, 4925–4930. [CrossRef]
20. Suryanarayanan, R.; Sridhar, V.G. Studies on the Influence of Process Parameters in Friction Stir Spot Welded Joints—A Review. *Mater. Today Proc.* **2020**, *37*, 2695–2702. [CrossRef]
21. Silva, B.H.; Zepon, G.; Bolfarini, C.; dos Santos, J.F. Refill Friction Stir Spot Welding of AA6082-T6 Alloy: Hook Defect Formation and Its Influence on the Mechanical Properties and Fracture Behavior. *Mater. Sci. Eng. A* **2020**, *773*, 138724. [CrossRef]
22. Xu, Z.; Li, Z.; Ji, S.; Zhang, L. Refill Friction Stir Spot Welding of 5083-O Aluminum Alloy. *J. Mater. Sci Technol.* **2018**, *34*, 878–885. [CrossRef]
23. Singh, B.; Upadhyaya, R. Influence of Flat Friction Stir Spot Welding Process Parameters on Quality Characteristics of AA 6082 Weld. *J. Univ. Shanghai Sci. Technol.* **2021**, *23*, 123–133.
24. Jambhale, S.; Kumar, S.; Kumar, S. Characterization and Optimization of Flat Friction Stir Spot Welding of Triple Sheet Dissimilar Aluminium Alloy Joints. *Silicon* **2022**, *14*, 815–830. [CrossRef]
25. Feng, X.S.; Li, S.B.; Tang, L.N.; Wang, H.M. Refill Friction Stir Spot Welding of Similar and Dissimilar Alloys: A Review. *Acta Metall. Sin.* **2020**, *33*, 30–42. [CrossRef]
26. Tier, M.D.; Rosendo, T.S.; Dos Santos, J.F.; Huber, N.; Mazzaferro, J.A.; Mazzaferro, C.P.; Strohaecker, T.R. The Influence of Refill FSSW Parameters on the Microstructure and Shear Strength of 5042 Aluminium Welds. *J. Mater. Process. Technol.* **2013**, *213*, 997–1005. [CrossRef]
27. Zhao, Y.; Wang, C.; Li, J.; Tan, J.; Dong, C. Local Melting Mechanism and Its Effects on Mechanical Properties of Friction Spot Welded Joint for Al-Zn-Mg-Cu Alloy. *J. Mater. Sci. Technol.* **2018**, *34*, 185–191. [CrossRef]
28. El-Sayed Seleman, M.M.; Ataya, S.; Ahmed, M.M.Z.; Hassan, A.M.M.; Latief, F.H.; Hajlaoui, K.; El-Nikhaily, A.E.; Habba, M.I.A. The Additive Manufacturing of Aluminum Matrix Nano Al_2O_3 Composites Produced via Friction Stir Deposition Using Different Initial Material Conditions. *Materials* **2022**, *15*, 2926. [CrossRef]
29. Ahmed, M.M.Z.; El-Sayed Seleman, M.M.; Elfishawy, E.; Alzahrani, B.; Touileb, K.; Habba, M.I.A. The Effect of Temper Condition and Feeding Speed on the Additive Manufacturing of AA2011 Parts Using Friction Stir Deposition. *Materials* **2021**, *14*, 6396. [CrossRef]
30. Alzahrani, B.; Seleman, M.M.E.S.; Ahmed, M.M.Z.; Elfishawy, E.; Ahmed, A.M.Z.; Touileb, K.; Jouini, N.; Habba, M.I.A. The Applicability of Die Cast A356 Alloy to Additive Friction Stir Deposition at Various Feeding Speeds. *Materials* **2021**, *14*, 6018. [CrossRef]
31. Elfishawy, E.; Ahmed, M.M.Z.; El-Sayed Seleman, M.M. Additive Manufacturing of Aluminum Using Friction Stir Deposition. In *Minerals, Metals and Materials Series*; Springer: Berlin/Heidelberg, Germany, 2020; pp. 227–238.

32. Perry, M.E.J.; Griffiths, R.J.; Garcia, D.; Sietins, J.M.; Zhu, Y.; Yu, H.Z. Morphological and Microstructural Investigation of the Non-Planar Interface Formed in Solid-State Metal Additive Manufacturing by Additive Friction Stir Deposition. *Addit. Manuf.* **2020**, *35*, 101293. [CrossRef]
33. Dilip, J.J.S.; Janaki Ram, G.D. Microstructure Evolution in Aluminum Alloy AA 2014 during Multi-Layer Friction Deposition. *Mater. Charact.* **2013**, *86*, 146–151. [CrossRef]
34. Priedeman, J.L.; Phillips, B.J.; Lopez, J.J.; Tucker Roper, B.E.; Hornbuckle, B.C.; Darling, K.A.; Jordon, J.B.; Allison, P.G.; Thompson, G.B. Microstructure Development in Additive Friction Stir-Deposited Cu. *Metals* **2020**, *10*, 1538. [CrossRef]
35. Ahmed, M.M.Z.; El-Sayed Seleman, M.M.; Shazly, M.; Attallah, M.M.; Ahmed, E. Microstructural Development and Mechanical Properties of Friction Stir Welded Ferritic Stainless Steel AISI 409. *J. Mater. Eng. Perform.* **2019**, *28*, 6391–6406. [CrossRef]
36. Ahmed, M.M.Z.; Seleman, M.M.E.S.; Ahmed, E.; Reyad, H.A.; Touileb, K.; Albaijan, I. Friction Stir Spot Welding of Different Thickness Sheets of Aluminum Alloy AA6082-T6. *Materials* **2022**, *15*, 2971. [CrossRef]
37. Hirasawa, S.; Badarinarayan, H.; Okamotoc, K.; Tomimura, T.; Kawanami, T. Analysis of Effect of Tool Geometry on Plastic Flow During Friction Stir Spot Welding Using Particle Method. *J. Mater. Process. Technol.* **2010**, *210*, 1455–1463. [CrossRef]
38. Atak, A.; Şık, A.; Özdemir, V. Thermo-Mechanical Modeling of Friction Stir Spot Welding and Numerical Solution with the Finite Element Method. *Int. J. Eng. Appl. Sci.* **2018**, *5*, 70–75.
39. Kumar, R.R.; Kumar, A.; Kumar, S. Effect on Tool Design and Heat Input of Some Welding Parameters in Friction Stir Welded Interstitial Free Steels. *Int. J. Eng. Technol. Innov.* **2018**, *8*, 64–75.
40. John, J.; Shanmughanatan, S.P.; Kiran, M.B. ScienceDirect Effect of Tool Geometry on Microstructure and Mechanical Properties of Friction Stir Processed AA2024-T351 Aluminium Alloy. *Mater. Today Proc.* **2018**, *5*, 2965–2979. [CrossRef]
41. John, J.; Shanmuganatan, S.P.; Kiran, M.B.; Kumar, V.S.S. Investigation of Friction Stir Processing Effect on AA 2014-T6. *Mater. Manuf. Process.* **2018**, *34*, 159–176. [CrossRef]
42. Ma, Z.Y. Friction Stir Processing Technology: A Review. *Metall. Mater. Trans. A* **2008**, *39*, 642–658. [CrossRef]
43. Chaudhary, B.; Jain, N.K.; Murugesan, J. Development of Friction Stir Powder Deposition Process for Repairing of Aerospace-Grade Aluminum Alloys. *CIRP J. Manuf. Sci. Technol.* **2022**, *38*, 252–267. [CrossRef]
44. Zhang, Z.; Tan, Z.J.; Li, J.Y.; Zu, Y.F.; Liu, W.W.; Sha, J.J. Experimental and Numerical Studies of Re-Stirring and Re-Heating Effects on Mechanical Properties in Friction Stir Additive Manufacturing. *Int. J. Adv. Manuf. Technol.* **2019**, *104*, 767–784. [CrossRef]
45. Moran, J.; Sucharitakul, T. Variations in Dry Sliding Friction Coefficients with Velocity. *Recent Adv. Mech. Mater. Mech. Eng. Chem. Eng.* **2015**, 184–194.
46. Gandra, J.; Krohn, H.; Miranda, R.M.; Vilaça, P.; Quintino, L.; Dos Santos, J.F. Friction Surfacing—A Review. *J. Mater. Process. Technol.* **2014**, *214*, 1062–1093. [CrossRef]
47. Aydın, H.; Tunçel, O.; Tutar, M.; Bayram, A. Effect of Tool Pin Profile on the Hook Geometry and Mechanical Properties of a Friction Stir Spot Welded Aa6082-T6 Aluminum Alloy. *Trans. Can. Soc. Mech. Eng.* **2021**, *45*, 233–248. [CrossRef]
48. Ma, C.; Hou, L.; Zhang, J.; Zhuang, L. Influence of Thickness Reduction per Pass on Strain, Microstructures and Mechanical Properties of 7050 Al Alloy Sheet Processed by Asymmetric Rolling. *Mater. Sci. Eng. A* **2016**, *650*, 454–468. [CrossRef]
49. Ahmed, M.M.Z.; Wynne, B.P.; Rainforth, W.M.; Addison, A.; Martin, J.P.; Threadgill, P.L. Effect of Tool Geometry and Heat Input on the Hardness, Grain Structure, and Crystallographic Texture of Thick- Section Friction Stir-Welded Aluminium. *Metall. Mater. Trans. A* **2019**, *50*, 271–284. [CrossRef]
50. Ahmed, M.M.Z.; El-Sayed Seleman, M.M.; Touileb, K.; Albaijan, I.; Habba, M.I.A. Microstructure, Crystallographic Texture, and Mechanical Properties of Friction Stir Welded Mild Steel for Shipbuilding Applications. *Materials* **2022**, *15*, 2905. [CrossRef]
51. Seleman, M.M.E.S.; Ahmed, M.M.Z.; Ramadan, R.M.; Zaki, B.A. Effect of FSW Parameters on the Microstructure and Mechanical Properties of T-Joints between Dissimilar Al-Alloys. *Int. J. Integr. Eng.* **2022**, *14*, 1–12. [CrossRef]
52. Ohashi, R.; Fujimoto, M.; Mironov, S.; Sato, Y.S.; Kokawa, H. Effect of Contamination on Microstructure in Friction Stir Spot Welded DP590 Steel. *Sci. Technol. Weld. Join.* **2009**, *14*, 221–227. [CrossRef]
53. Xie, G.M.; Cui, H.B.; Luo, Z.A.; Yu, W.; Ma, J.; Wang, G.D. Effect of Rotation Rate on Microstructure and Mechanical Properties of Friction Stir Spot Welded DP780 Steel. *J. Mater. Sci. Technol.* **2016**, *32*, 326–332. [CrossRef]

materials

MDPI

Article

Wear Characteristics of Mg Alloy AZ91 Reinforced with Oriented Short Carbon Fibers

Sabbah Ataya [1], Mohamed M. El-Sayed Seleman [2], Fahamsyah H. Latief [1,3], Mohamed M. Z. Ahmed [4,*], Khalil Hajlaoui [1], Ahmed M. Soliman [5], Naser A. Alsaleh [1] and Mohamed I. A. Habba [6]

1 Department of Mechanical Engineering, College of Engineering, Imam Mohammad Ibn Saud Islamic University, Riyadh 11432, Saudi Arabia; smataya@imamu.edu.sa (S.A.); fhlatief@imamu.edu.sa (F.H.L.); kmhajlaoui@imamu.edu.sa (K.H.); naalsaleh@imamu.edu.sa (N.A.A.)
2 Department of Metallurgical and Materials Engineering, Faculty of Petroleum and Mining Engineering, Suez University, Suez 43512, Egypt; mohamed.elnagar@suezuniv.edu.eg
3 Department of Mechanical Engineering, Faculty of Engineering and Science, Universitas Nasional, Jakarta 12520, Indonesia
4 Mechanical Engineering Department, College of Engineering at Al Kharj, Prince Sattam bin Abdulaziz University, Al Kharj 16273, Saudi Arabia
5 Department of Mechanical Engineering, College of Engineering, Jouf University, Sakaka 72388, Saudi Arabia; amsoliman@ju.edu.sa
6 Mechanical Department, Faculty of Technology & Education, Suez University, Suez 43518, Egypt; mohamed.atia@suezuniv.edu.eg
* Correspondence: moh.ahmed@psau.edu.sa

Abstract: Light-weight metal matrix composites, especially magnesium-based composites, have recently become more widespread for high-efficiency applications, including aerospace, automobile, defense, and telecommunication industries. The squeeze cast AZ91 base material (AZ91-BM) and its composites having 23 vol.% short carbon fibers were fabricated and investigated. The composite specimens were machined normal to the reinforced plane (Composite-N) and parallel to the reinforced plane (Composite-P). All the as-casted materials were subjected to different tests, such as hardness, compression, and wear testing, evaluating the mechanical properties. Dry wear tests were performed using a pin-on-disk machine at room temperature under different applied wear loads (1–5 N) and different sliding distances (0.4461×10^4–3.12×10^4 m). The microstructures and worn surfaces of the fabricated AZ91-BM and the two composite specimens were investigated using a scanning electron microscope (SEM) equipped with an energy dispersive spectroscopy (EDS) advanced analysis system. The wear debris was collected and investigated also under the SEM. The results showed significant improvement in hardness, compressive strength, and wear resistance of the composite specimens (Composite-N and Composite-P) over the AZ91-BM. The compressive strength and wear resistance are more fibers orientation sensitive than the hardness results. When the fiber orientation is parallel to the sliding direction (Composite-N), the weight loss is somewhat lower than that of the fiber orientation perpendicular to the sliding direction (Composite-P) at a constant wear load of 2 N and the sliding distances of 0.4461×10^4, 1.34×10^4 , and 2.23×10^4 m. In contrast, the weight loss of Composite-P is lower than Composite-N, especially at the highest sliding distance of 3.12×10^4 m due to the continuous feeding of graphite lubricant film and the higher compressive strength. Plastic deformation, oxidation, and abrasive wear are the dominant wear mechanisms of AZ91-BM; in contrast, abrasive and delamination wear are mainly the wear mechanisms of the two composites under the applied testing conditions.

Keywords: magnesium alloy AZ91; short carbon fibers; microstructure; compressive strength; wear resistance; wear mechanisms; wear debris

Citation: Ataya, S.; El-Sayed Seleman, M.M.; Latief, F.H.; Ahmed, M.M.Z.; Hajlaoui, K.; Soliman, A.M.; Alsaleh, N.A.; Habba, M.I.A. Wear Characteristics of Mg Alloy AZ91 Reinforced with Oriented Short Carbon Fibers. *Materials* **2022**, *15*, 4841. https://doi.org/10.3390/ma15144841

Academic Editors: Lihong Su, Xing Zhao, Peitang Wei and Hui Wang

Received: 21 June 2022
Accepted: 10 July 2022
Published: 12 July 2022

1. Introduction

Nowadays, the demand for materials weight reduction for structural applications has significantly increased. For this purpose, magnesium (Mg) alloys are promising candidates to be introduced since they have outstanding physical and chemical properties, which are considered to have expectant prospects in the areas of aerospace, automotive, and electronics [1]. On the other hand, Mg alloys have low strength and poor wear resistance, which limits their applications for structural and anti-friction applications [2]. It is well known that Mg-based composites are superior in certain aspects compared with the monolithic magnesium alloy, such as high specific strength and excellent damping capacity, which have become one of the research focuses of many engineering applications.

Many scientists presented that the stability and mechanical properties of Mg alloys at high temperatures can be increased by adding the reinforcements into Mg alloys as a matrix to make Mg-based composites have better performance compared with its alloys [3,4]. Lately, many researchers put their concern on using discontinuous particles as reinforcement for metal-matrix composite (MMC) since they are low-cost and easy to fabricate through secondary processing, have better mechanical properties, and have good homogeneity [5,6]. Noted that one of the requirements for structural applications, the materials should have excellent wear resistance during services. El-Sayed Seleman et al. [7] reported that the incorporation of graphite powders up to 20 wt.% in the AA6016 aluminum matrix improved the wear resistance of aluminum/graphite composites compared to AA6016 aluminum alloy.

Liu et al. [8] stated that short carbon fiber lowered direct contact between the aluminum matrix and counterpart and enhanced the wear resistance. This composite system has been produced by the vacuum pressure infiltration technique.

Furthermore, the mechanical and wear behavior of boron carbide and graphite-reinforced AZ91 magnesium matrix hybrid composite was discussed by Aatthisugan et al. [9]. As their results, the incorporation of boron carbide into AZ91 alloy enhanced the wear behavior of Mg composite, whereas the addition of graphite into AZ91–B_4C composite as a hybrid reinforcement reduced the wear resistance of Mg composite. Lim et al. [10] investigated the sliding wear behavior of AZ91/SiCp composites at various loads ranging from 10−30 N. Compared with its matrix alloy, the wear behavior of Mg composite at 10 N was significantly improved by a 15–30% increment. A study on the dry sliding wear behavior of globular AZ91 alloy and AZ91/SiCp composites was carried out by Garcia-Rodriguez et al. [11]; they summarized that the wear mechanism of globular AZ91 alloy was varied at different loads and sliding speeds. An investigation on the wear resistance of AE42-based composite reinforced with 23% vol.% carbon short fibers was conducted by Ataya et al. [12], and they confirmed that the use of various loads at a constant sliding distance or vice versa promoted more weight loss of the AE42 matrix than its composite.

Herein, the carbon short fibers are incorporated in the AZ91 alloy to produce an Mg-based composite by the squeeze casting technique. To the best of our knowledge, there is still little published work found on the field of wear behavior of AZ91 alloy reinforced with oriented high volume fraction short carbon fibers. In fact, carbon short fibers are considered an important material for different engineering applications when used as reinforcement into Mg matrix to produce composites due to their isotropic and stable mechanical performance [13]. Thus, the combination of the AZ91 alloy and the short carbon fibers is interesting to be investigated to explore the wear behavior of the Mg-based composite containing an oriented high-volume fraction. Therefore, the present work aims to comprehend the wear behavior of the AZ91/23% vol.% short carbon fiber composite materials for the specimens machined normal and parallel to the reinforcing plane. The influence of different loads and sliding distances of the as-cast AZ91 base material (BM) and the produced composites were examined in the light of the oriented short carbon fiber. The microstructures of the produced materials are investigated; in addition, an extensive study of the worn surface under SEM is also undertaken to show the role of the reinforcing carbon fibers in the wear process. The hardness and compressive strength of the produced materials are also evaluated.

2. Methodology

Materials

The starting materials to produce Mg-based composite containing a high-volume fraction of short carbon fibers via an advanced squeeze casting technique were AZ91 and short carbon fibers. The used carbon fibers were coated with silicon to improve the wettability with the AZ91 matrix and to hinder carbide formation at the interface between the matrix and the reinforcement. Researchers reported using many coating materials such as pyrolytic carbon layer [14], copper and nickel [15], alumina [16], and silicon carbide [17]. The carbon fibers were supplied by SIGRAFIL (SGL Carbon GmbH), Germany. According to the supplier, the diameter and length were 5–6 μm and 80–120 μm, respectively, and the physicomechanical properties were 1.76 g/cm^3 density and 280 GPa elastic modulus. The composite was designed to have 23 vol.% short carbon fibers. Figure 1 summarizes the experimental procedure to produce and characterize the AZ91-BM and its composites. The as-cast-produced composite was machined normally and parallel to the reinforcing plane, as sketched in the flow chart (Figure 1). In the current study, the composite specimen machined normal to the reinforced plane is named Composite-N, and that machined parallel to the reinforced plane is named Composite-P.

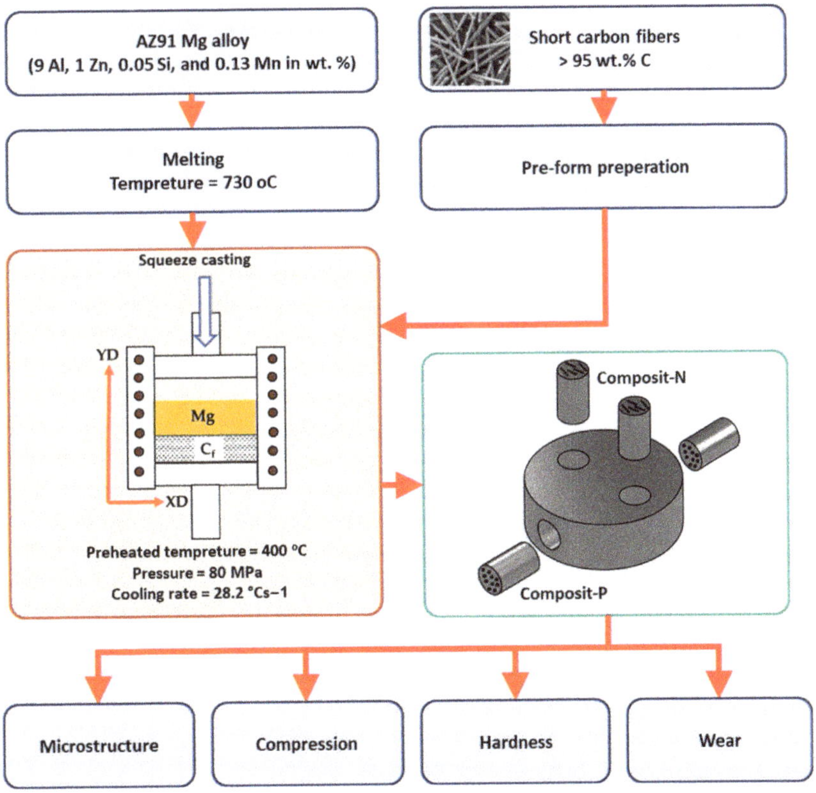

Figure 1. The flow chart summarizes the experimental procedure to produce and characterize the AZ91-BM and its composites.

Cylindrical specimens of the AZ91-BM, Composite-N, and Composite-P were machined in the dimensions of 6 mm in diameter and 9 mm in length for further testing and investigations. Hardness was measured using a Vickers hardness testing machine (Model-HWDV-7S) at 2 N load and 15 s dwell time loading conditions. Compression testing was applied at room temperature using a Universal Testing Machine (Schenck-Trebel RMC100, Deer Park, NY, USA) with a cross head speed of 0.01 mm/min. Microstructures were examined for the AZ91-BM and both composite specimens using a scanning electron microscope, Quanta FEG 250, Hillsboro, OR, USA, equipped with an energy-dispersive X-ray spectrometry (EDS) advanced system. Two SEM detectors, low k- Volt High-Contract Detector (vCD) and Everhart–Thornley Detector (ETD), were used to distinguish the microstructure features. Wear tests using a homemade pin-on-disc machine (WT-M1-SSMMR-CSE, Suez University, Suez, Egypt, Figure 2) were used to evaluate the wear behavior of the AZ91-BM and AZ91/23 vol.% short carbon fibers composites. The tested specimens were rubbed against a hard steel disk has a hardness of 64 HRC at different loads and sliding distances. The applied wear load ranged from 1 to 5 N, and the sliding distance ranged from 0.4461×10^4 to 3.12×10^4 m. After testing, the debris was collected for each wear condition. The worn surfaces and the collected debris were investigated using the SEM-EDS system.

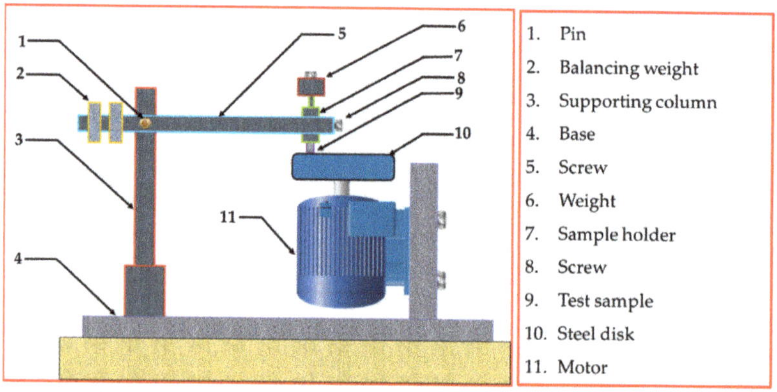

Figure 2. Schematic of homemade pin-on-disk wear testing machine.

3. Results and Discussion

3.1. Microstructures AZ91-BM and Composite Specimens

The morphologies of the intermetallic phases in Mg alloys are governed by some parameters, including the applied casting process and its variables, alloying elements, and cooling rate [18–20]. It is worth mentioning that the physical and mechanical properties of the as-cast material are related directly to the microstructure features. Thus, the SEM-EDS examinations were conducted to investigate the microstructure features of the as-cast AZ91-BM and the produced composites. Figure 3 shows SEM images of the AZ91-BM microstructure and the formed intermetallics. The microstructure involves a large α-Mg (Spot 2) dendritic structure (gray areas) with secondary arms, as given in Figure 3a,b. Two intermetallics were detected in the microstructure of AZ91-BM. The first appears as bright layers at the α-Mg grain boundaries (eutectic phase of β-$Mg_{17}Al_{12}$ intermetallic and α-Mg lamellar). It also appeared as a solid bright area on the α-Mg grain boundaries (spot 1 in Figure 3b and represented in Figure 3c). The different shapes of the β-$Mg_{17}Al_{12}$ intermetallics are attributed to the non-equilibrium solidification [21]. The second intermetallic is Al_4Mn (spot 3 in Figure 3b and represented in Figure 3e). These microstructure features in terms of intermetallic phases, morphologies, and their dispersion of the as-cast AZ91 matrix are typically agreed well with that examined and reported for the same alloy material in other works [22–24].

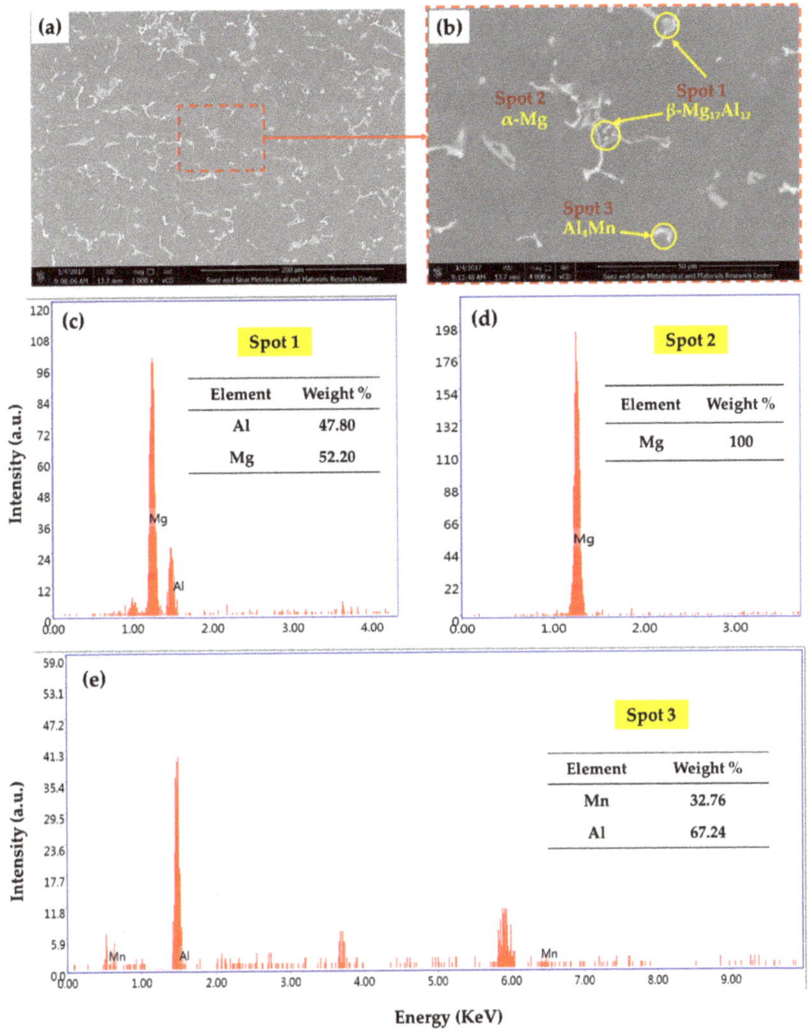

Figure 3. SEM images at (**a**) low and (**b**) high magnification of the AZ91-BM and the EDS analysis of the AZ91 intermetallics; (**c**) spot 1, (**d**) spot 2, and (**e**) spot 3 in subfigure (**b**).

Figure 4 shows the microstructures of the AZ91/23 vol.% short carbon fibers detected in Composite-N and Composite-P specimens. The carbon fiber dispersed and bonded well with the AZ91 matrix in both composite specimens. In addition, localized agglomeration of carbon fibers is also remarked in the microstructure investigation, as the carbon fibers tend to agglomerate [6] and as the result of the used pressure during the squeeze casting process. In composite-N, the carbon fibers mostly appeared in the longitudinal direction (Figure 4a,b). Some fibers are micro-cracked in the transverse direction with keeping their position in the Mg matrix, as shown in Figure 4b (ETD mode). This fiber fracture may be ascribed to the difference in the thermal expansion coefficient of the AZ91 Mg matrix and carbon fibers. In addition, interfacial debonding and sliding may also be expected in the Mg/carbon fiber composite system [25]. The axial thermal expansion coefficient of the carbon fibers has a more pronounced effect than that of the transverse direction on the interfacial bonding between the fibers and the Mg matrix composite during fast cooling of molten Mg to room temperature. This leads to partial debonding at the interface between the Mg matrix and the longitudinal fiber direction instead of the cross-section. In Composite-P, the circular cross-sections of the carbon fibers (Figure 4c,d) lie in the range of the as-received short carbon fiber diameters (5–6 μm). It can be remarked from Figure 4d (ETD mode) that the carbon fibers are bonded well with the Mg matrix without any cracking in the carbon cross-sections.

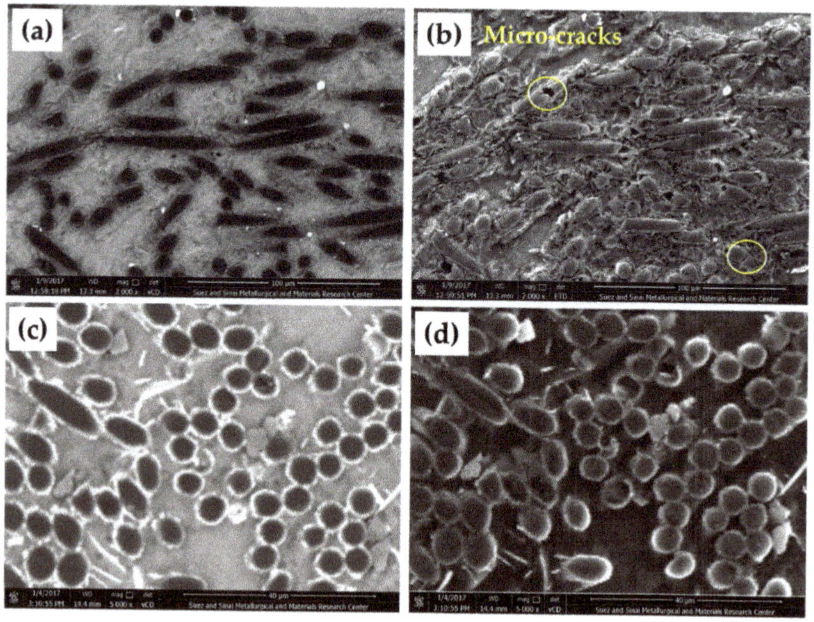

Figure 4. SEM images show the microstructures of Composite-N, (**a**) VCD, (**b**) EDT and Composite-P, (**c**) VCD, and (**d**) EDT.

3.2. Mechanical Properties

The AZ91-BM hardness value is 70 ± 3 HV. This value increased for the composite specimens containing 23 vol.% short carbon fibers to be 106 ± 4 HV and 111 ± 2 HV for the Composite-N and Composite-P, respectively. This improvement in hardness is ascribed to the action of adding a high-volume fraction (23 vol.%) of reinforcing carbon fiber. This result agreed well with that reported by many authors for different Mg-based composites containing different volume fractions of carbon fibers [12,22,26].

The compressive properties of the AZ91-BM and the two composites are presented in Figure 5, where Figure 5a illustrates the yield compressive strength (YCS) and ultimate compressive strength (UCS), and Figure 5b shows the influence of 23 vol.% short carbon fibers on the ductility of the compression tested specimens in terms of reduction in height (R%). It can be remarked that the two composites possess higher YCS and UCS than AZ91-BM (Figure 5a). The UYS of the Composite-N specimens improved by 38% over AZ91-BM, whereas no significant improvement in the UCS of the Composite-N compared to BM. For Composite-P, the enhancement in both YCS and UCS is remarkably observed over AZ91-BM and attained the improvement percentage of 124 and 20, respectively. This improvement in strength for the composite specimens can be attributed to the high modulus of elasticity of the short carbon fibers compared to the AZ91-BM [27,28]. The ductility loss for Composite-N and Composite-P are 48 and 72%, respectively, compared to AZ91-BM. The increase in hardness and strength is usually at the expense of ductility loss.

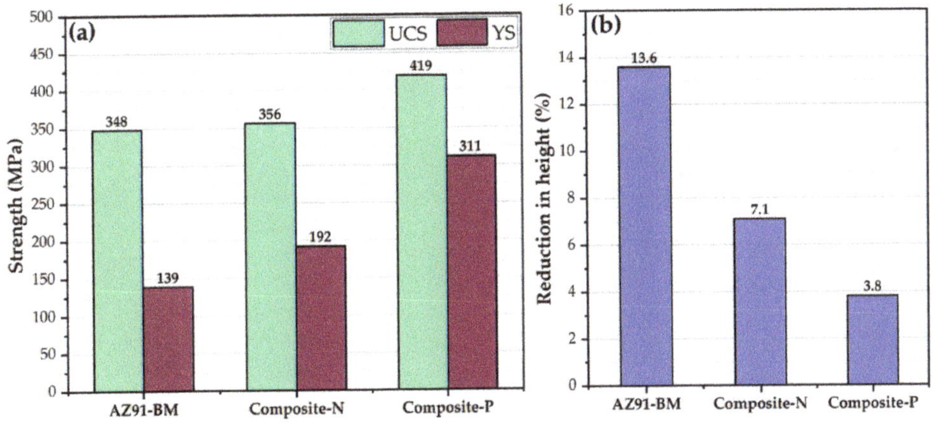

Figure 5. (**a**) Ultimate compressive strength and yield strength and (**b**) reduction in the height of the compression tested AZ91 matrix alloy and the composites.

AZ91-Mg alloy is widely used in automotive and aerospace industries due to its excellent combination of low density and high thermal conductivity. However, it suffers from poor wear resistance. To overcome this, high specific strength reinforcement phases (fibers, particulates, and whiskers) have been recommended to introduce to the Mg matrix to improve its wear resistance via producing AZ91-based composites [12,14]. Moreover, the wear behavior of the fiber-reinforced composite materials is governed by hardness, compressive strength, and fiber orientation with respect to sliding direction. The influence of oriented fibers on the wear properties is more complicated (especially for discontinuous fiber-reinforced composites) because of the random distribution of fibers. Thus, it is important to explore the wear properties of AZ91-BM and the produced composite specimens having oriented 23 vol.% short carbon fibers. The first group of the wear-tested specimens was conducted at a constant sliding distance of 1.34×10^4 m and different loads of 1, 2, 3, 4, and 5 N for the AZ91-BM, Composite-N, and Composite-P. Figure 6 shows the weight loss of the wear-tested materials (AZ91-BM and its composite specimens) against the applied wear loads. It can be observed that the weight loss of all the wear-tested materials increases with increasing the wear loads from 1 to 5 N for the BM and the two composite specimens. The weight loss of AZ91-BM is higher than that of both composites for all the applied wear loads from 1 to 3 N. After that, the Composite-N shows the highest weight loss at 5 N, whereas Composite-P displays the lowest weight loss at the higher wear loads of 4 and 5 N. The higher weight loss of AZ91-BM compared to the two composite specimens at a wide range of the applied wear loads (1–3 N) is ascribed to its lower hardness and compressive strength values compared to the AZ91-based composite specimens reinforced with the short carbon fibers. Carbon fibers have graphite-like layers, which are weakly bonded together. These layers have a very low coefficient of friction while sliding on another surface. During the wear testing, carbon fibers are fractured and spread on the worn surface of composites, forming a solid lubricant film. This lubricating film reduces the friction and dissipates the generated heat between the two rubbing surfaces of the wear-tested composite specimens and the steel disk counterpart. This leads to a reduced weight loss of both composites (Figure 6). Furthermore, at the mild wear loads of 2 and 3 N, the Composite-N shows an improvement in wear properties compared to the other tested specimens (AZ91-BM and Composite-P) because of the largest lubricant feeding area as the carbon fibers parallel to the hard rubbing surface. This promotes more graphite film than in the case of the fiber normal to the hard rubbing surface (Composite-P). At the same time, the applied shear force on the worn surface is lower than the cohesion force between the carbon fibers and the AZ91 matrix, which makes the carbon fibers achieve their role of lubrication. In contrast, at the wear loads of 4 and 5 N, the applied shear force on the worn surface is higher than the cohesion force between the carbon fibers and the AZ91 matrix. This causes carbon fibers to pull out due to the presence of the interfacial deboning as a result of thermal expansion mismatch between carbon fibers and the AZ91 matrix. The thermal expansion coefficient of AZ91 alloy is $26.8 \times 10^{-6}\ K^{-1}$, while the axial thermal expansion coefficient of carbon fiber is $1.48 \times 10^{-6}\ K^{-1}$ [25]. The pull-out of carbon fibers causes material loss without utilizing the inherent lubricity property. This phenomenon is remarked only with the Composite-N and combined with delamination layers of AZ91 matrix alloy due to the accumulated friction heat and severe plastic deformation. These cause deterioration in the wear properties of Composite-N compared to the AZ91-BM, and the Composite-P at the highest applied wear load of 5 N.

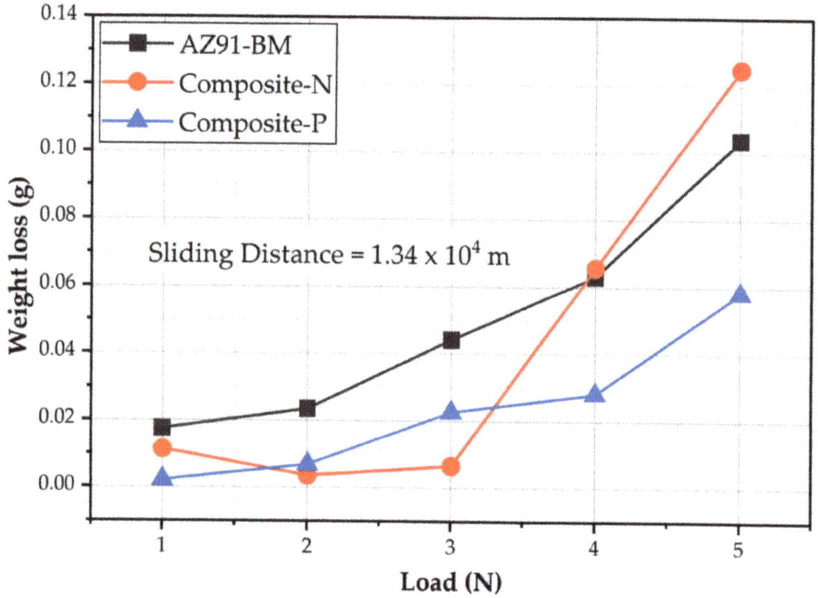

Figure 6. Weight loss against the applied wear loads for AZ91-BM and composite specimens wear-tested at a constant sliding distance of 1.34×10^4 m.

In order to examine the wear behavior in terms of weight loss of the AZ91-BM, Composite-N, and Composite-P at different sliding distances, the second group of the wear-tested specimens was conducted under a constant mild load of 2 N at different running sliding distances of 0.4461×10^4, 1.34×10^4, 2.23×10^4 and 3.12×10^4 m, as given in Figure 7a. It can be seen that both composites (Composite-N and Composite-P) display notable lower weight loss than AZ91-BM, indicating significant improvement in wear resistance in the presence of 23 vol.% oriented short carbon fibers. It is also observed that the weight loss values of Composite-N are lower than that given by Composite-p and AZ91-BM at the sliding distance range from 0.4461×10^4 to 2.23×10^4 m, whereas, at the highest sliding distance of 3.12×10^4 m, Composite-P shows the lowest weight loss of 0.01785 g compared to the AZ91-BM (0.0771 g) and the Composite-N (0.022 g). At a mild wear load of 2 N and applying the highest sliding distance of 3.13×10^4 m, the formed lubricating film is not enough to dissipate the generated friction heat, and then the accumulated heat weakens the bond strength between the carbon fibers and the surrounded Mg matrix as a result of the thermal expansion mismatch between the carbon fibers and the Mg matrix. This leads to the pull-out of the carbon fibers without complete utilizing of its inherent lubrication effect and increases the weight loss of composite-N. In contrast,

Composite-P does not suffer from the pull-out fiber phenomenon and utilizes completely the action of graphite fibers in the lubrication. In addition, the contentious lubricant film feeding when the fibers normal to the rotating wear machine desk, the Composite-P keeps its improvement in compressive strength.

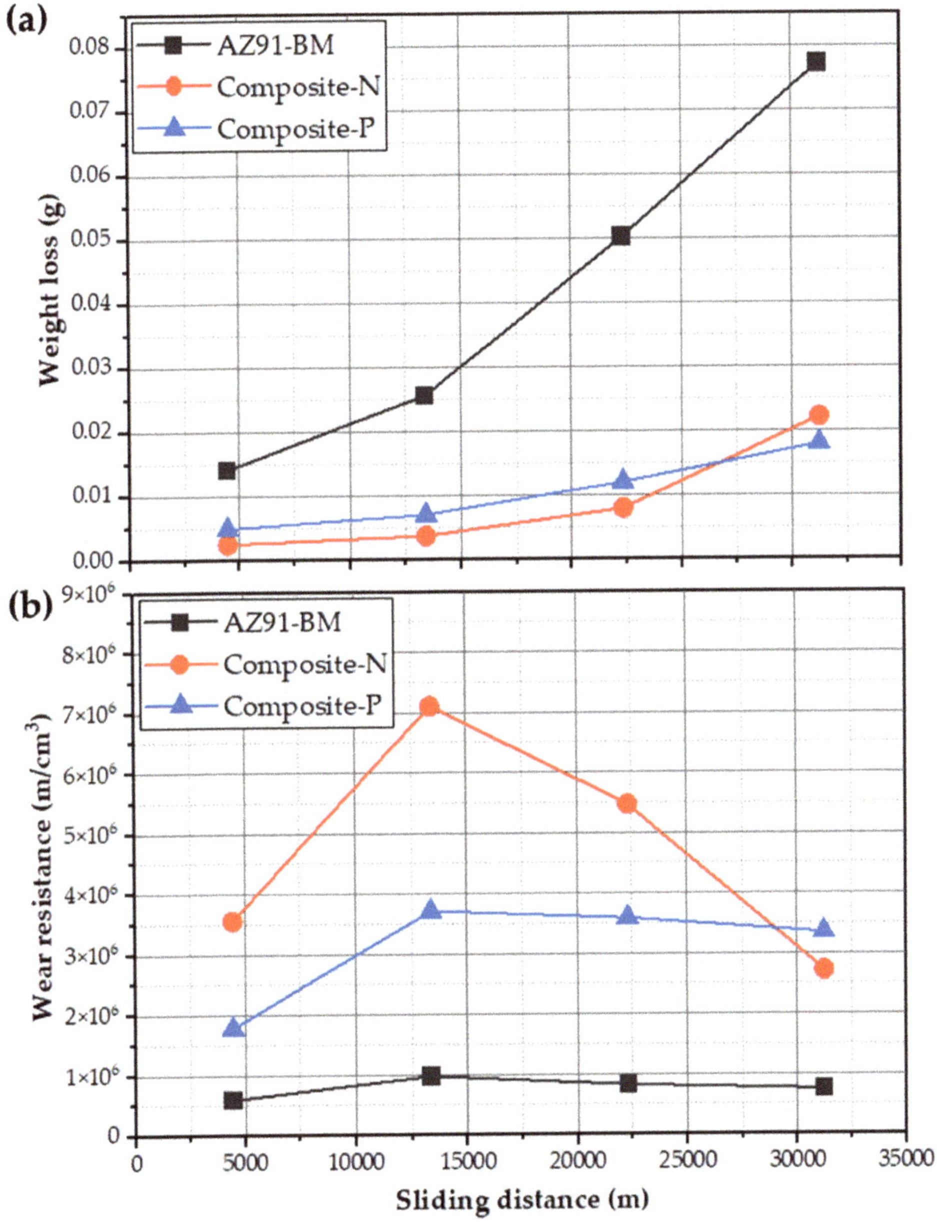

Figure 7. (**a**) Weight loss and (**b**) wear resistance against the applied sliding distances for AZ91-BM and composite specimens wear-tested at a constant wear load of 2 N.

Based on the collected data from the wear testing parameters and the density of the wear-tested specimens (AZ91-BM, Composite -N, and Composite-P), the wear properties are plotted as a wear resistance versus the applied sliding distances at a constant load of 2 N (Figure 7b). It can be seen that the incorporation of the oriented 23 vol.% short carbon fibers in the AZ91 matrix improves the wear resistance of the produced composite specimens over the AZ91-BM at all the applied sliding distances. In the beginning, there is an increase in the wear resistance in varying proportions for AZ91-BM, Composite-N, and Composite-P at the sliding distances from 0.4461×10^4 to 1.34×10^4 m. The slight increase in the wear resistance of the AZ91-BM may be ascribed to strain hardening, whereas a significant increase in the wear resistance of Composite-N and Composite-P is remarked at the same sliding distance range. This enhancement is likely due to the addition of a high-volume fraction of short carbon fibers. Above the sliding distance of 1.34×10^4 m, the wear resistance of Composite-N displays remarked to decrease with increasing the sliding distance. For AZ91-BM and Composite-P, the wear resistance curves show nearly a steady-state behavior with the increase in the sliding distance. Furthermore, Composite-P shows the highest wear resistance of 3.34×10^6 m/cm^3 compared to the wear resistance of AZ91-BM (0.749×10^6 m/cm^3) and Composite-N (2.71×10^6 m/cm^3) at the highest sliding distance of 3.12×10^4 m (Figure 7b). Finally, the Composite-N shows the lowest weight loss compared to the Composite-P and the AZ91-BM at a constant sliding distance of 1.34×10^4 m and the wear loads of 2 and 3 N. Additionally, it shows the lowest weight loss (the highest wear resistance) compared to the Composite-P and the AZ91-BM at a constant mild wear load of 2 N and the sliding distance range from 0.4461×10^4 to 3.12×10^4 m. In contrast, the Composite-P shows lower weight loss compared to the Composite-N and the AZ91-BM at a constant sliding distance of 1.34×10^4 m and the wear loads of 4 and 5 N. In addition, it shows the lowest weight loss (the highest wear resistance) compared to all the wear-tested specimens, especially at a wear load of 2 N and the highest sliding distance of 3.12×10^4 m.

To understand the wear mechanisms of AZ91-BM and the produced composite specimens containing oriented high-volume fraction of short carbon fibers, the worn surfaces and gathered debris of the wear-tested specimens were examined by SEM. Figure 8 displays the worn surface features of the wear-tested specimens of AZ91-BM, Composite-N, and Composite-P, and Figure 9 illustrates SEM images and EDS analysis of the gathered debris of the AZ91 matrix alloy and the composite specimens. The worn surface of the AZ91-BM (investigated using two SEM detectors, ETD and vCD modes) displays clear damage in the form of continuous scratches parallel to the sliding direction, plastic deformation, transverse microcracks, and delamination layers (Figure 8). There are also small smooth regions interspersed with wear scars. A notable feature in the SEM micro-images is inclined shear plates due to high surface stress [12]. The wear mechanisms are plastic deformation, delamination layer, and abrasion wear. The abrasion wear was caused by the free pull-out intermetallic particles and the formed magnesium oxide (Figure 9e) between the two rubbing surfaces, AZ91-BM and the hard steel disk. During the wear tests, the AZ91-BM is easily scraped due to its lower strength and hardness, which leads to material loss (Figure 9a,b). At the similar wear test condition, the worn surface of the two composites (Composite-N, Figure 8c; and Composite-P, Figure 8d) having 23 vol.% short carbon fibers displays other features. The scratches (wear track) of both composites become shallower than that in the AZ91-BM as a result of the short carbon fibers addition. Furthermore, the inclined Mg shear plates nearly disappeared from the worn surface of the composite specimens. In addition, smearing the worn surface with carbon film is generally remarked (Figure 8c,d). This carbon film lubricates the worn surface of the composites and decreases the degree of delamination [29] and the influence of abrasive wear [12]. This action generally reduces the wear loss of both composites as appears in their size of debris as given in Figure 9c for Composite-N and Figure 9d for Composite-P compared to the AZ91-matrix material's debris. It is worth mentioning that the AZ91 debris is very large and highly deformed with irregular shapes and dimensions, as shown in Figure 9a,b.

The formation of this type of debris can be ascribed to the delamination layers and an abrasive wear effect. However, the wear debris becomes smaller and lightly deformed for AZ91/23 vol.% carbon short fibers the Composite-N and the Composite-P as shown in Figure 9c,d, respectively. This decrease in the size of the debris is also detected in another work [7] and is mainly due to the role of carbon fibers in reducing the probabilities of direct contact between two worn surfaces and dissipating the frictional heat, which finally decreases the severity of micro-cutting effects. The mainly working wear mechanisms of the composites are likely to be abrasive and delamination wear [14]. It is also remarked that the worn surface of the Composite-N is smoother than that of the Composite-P. This may be ascribed to the intensity of the supplied lubricant film in the case of short carbon fibers being parallel to the sliding surface plane. The feeding of graphite film is discontinuous as the graphite fibers are separated by the Mg matrix. In the case of the short carbon fibers being perpendicular to the sliding surface plane, the feeding of graphite film is continuous. This leads to the higher wear resistance of Composite-P than Composite-N, especially at the highest distance of 3.12×10^4 m using an applied wear load of 2 N (Figure 7b).

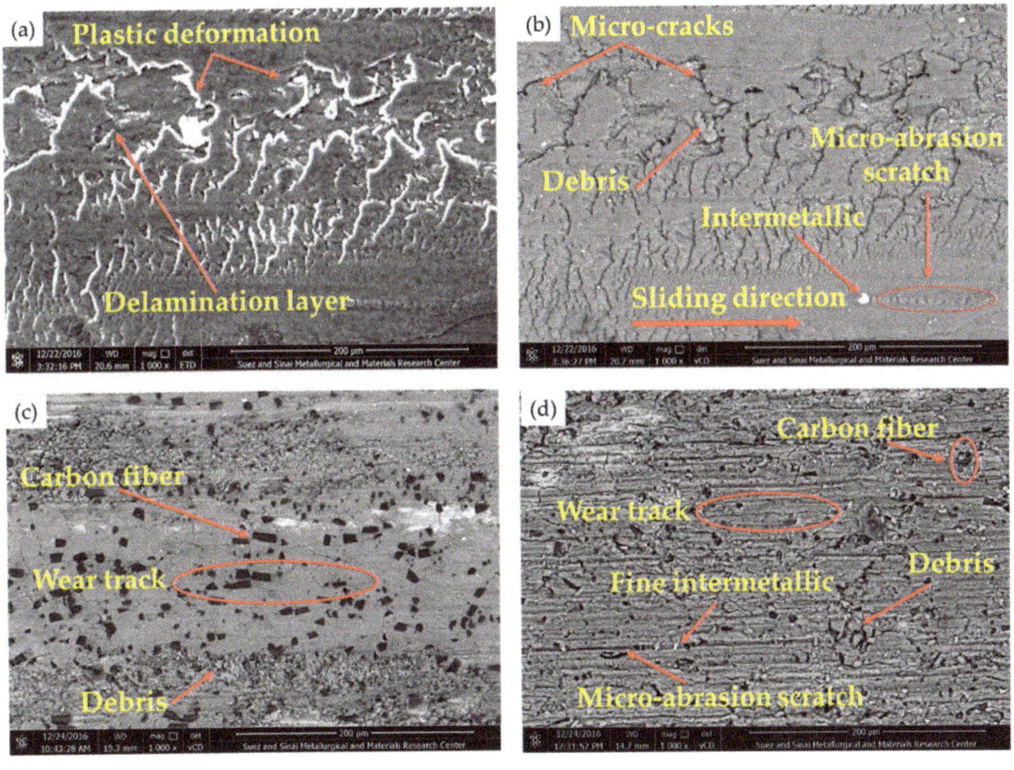

Figure 8. SEM images of the worn surfaces of the wear-tested specimens, where (**a**,**b**) AZ91-BM in ETD and vCD modes, respectively, (**c**) Composite-N and (**d**) Composite-P at the wear test conditions of wear load of 2 N and sliding distance of 1.34×10^4 m.

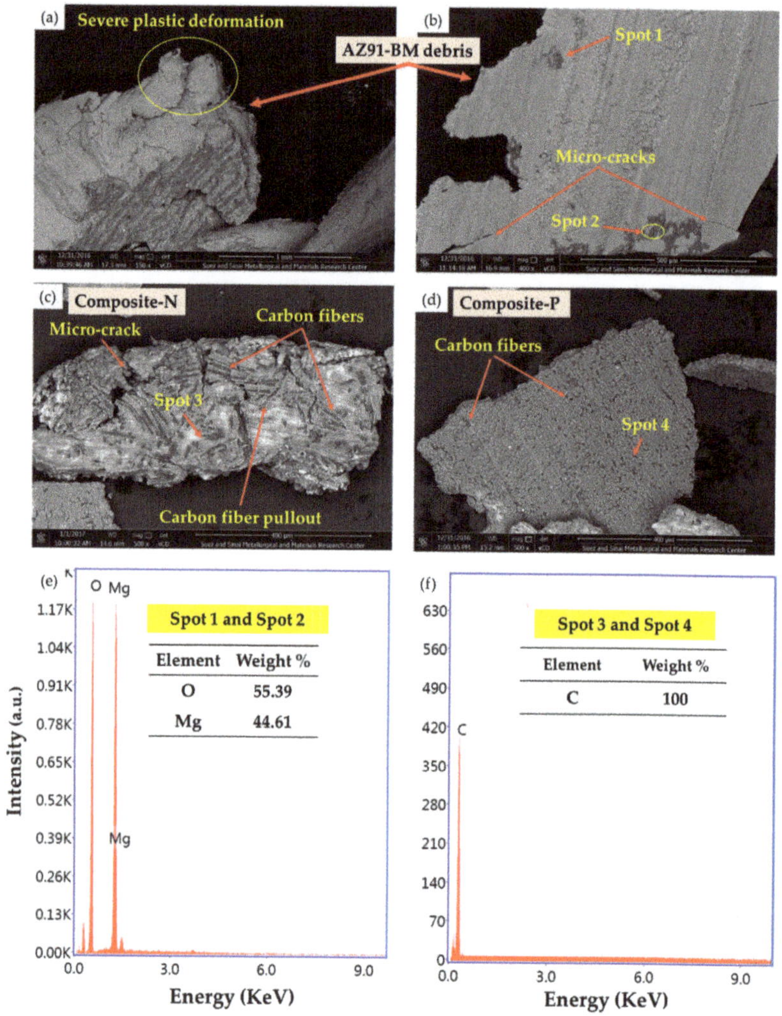

Figure 9. SEM images of the debris of the wear-tested specimens, where (**a**,**b**) AZ91-BM, (**c**) Composite-N (**d**) Composite-P, (**e**) EDS spot analysis of MgO, and (**f**) EDS spot analysis of Carbon fiber.

4. Conclusions

Based on the obtained results, the following conclusions can be outlined:

1. The reinforcing of AZ91 Mg alloy by 23 vol.% short carbon fibers improved the hardness of the two composites (Composite-N and Composite-P) by not less than 51%.
2. The YCS of the Composite-N and Composite-P were enhanced over AZ91-BM by 38% and 124%, respectively. In addition, Composite-P recorded the highest UCS compared to Composite-N and AZ91-BM.
3. The two composites display notable lower weight loss than AZ91-BM at a constant sliding distance of 1.34×10^4 m, and the applied wear loads from 1 to 3 N, indicating significant improvement in wear resistance in the presence of 23 vol.% oriented short carbon fibers.

4. Composite-P shows the lowest weight loss of 0.01785 g compared to the AZ91-BM (0.0771 g) and the Composite-N (0.022 g) at the wear conditions of 2 N applied load and the highest sliding distance of 3.12×10^4 m.
5. The two composites show higher wear resistance than AZ91-BM at a constant applied wear load of 2 N and various sliding distances from 0.4461×10^4 to 3.12×10^4 m. Furthermore, Composite-P possesses the highest wear resistance at a constant applied wear load of 2 N and the highest sliding distance of 3.12×10^4 m.
6. Plastic deformation, oxidation, and abrasive wear are the dominant wear mechanisms of AZ91-BM; in contrast, abrasive and delamination wear are mainly the wear mechanisms of the two composites under the applied testing conditions.

Author Contributions: Conceptualization, S.A., M.M.E.-S.S. and M.I.A.H.; Data curation, S.A., M.M.Z.A. and N.A.A.; Formal analysis, M.M.E.-S.S., S.A. and F.H.L.; Funding acquisition, K.H. and S.A.; Investigation, M.M.E.-S.S., N.A.A. and M.I.A.H.; Methodology, M.M.E.-S.S., N.A.A. and S.A.; Project administration, K.H. and S.A.; Resources, M.M.E.-S.S., K.H. and F.H.L.; Software, K.H., A.M.S., N.A.A. and F.H.L.; Supervision, S.A., M.M.E.-S.S. and M.I.A.H.; Validation, M.M.Z.A., K.H., N.A.A. and S.A.; Visualization, K.H., S.A. and F.H.L.; Writing—original draft, S.A. and F.H.L.; Writing—review and editing, M.M.Z.A., M.I.A.H., A.M.S., N.A.A. and M.M.E.-S.S. All authors have read and agreed to the published version of the manuscript.

Funding: Funded by the Deanship of Scientific Research at Imam Mohammad Ibn Saud Islamic University, KSA (Research Group no. RG-21-12-07).

Institutional Review Board Statement: Not applicable.

Informed Consent Statement: Not applicable.

Data Availability Statement: The data presented in this study are available on request from the corresponding author. The data are not publicly available due to the extremely large size.

Acknowledgments: The authors appreciate the Deanship of Scientific Research at Imam Mohammad Ibn Saud Islamic University for funding this work through Research Group no. RG-21-12-07.

Conflicts of Interest: The authors declare no conflict of interest.

References

1. Luo, K.; Zhang, L.; Wu, G.; Liu, W.; Ding, W. Effect of Y and Gd Content on the Microstructure and Mechanical Properties of Mg–Y–RE Alloys. *J. Magnes. Alloy* **2019**, *7*, 345–354. [CrossRef]
2. Lyu, J.; Kim, J.; Liao, H.; She, J.; Song, J.; Peng, J.; Pan, F.; Jiang, B. Effect of Substitution of Zn with Ni on Microstructure Evolution and Mechanical Properties of LPSO Dominant Mg–Y–Zn Alloys. *Mater. Sci. Eng. A* **2020**, *773*, 138735. [CrossRef]
3. Tokunaga, T.; Ohno, M.; Matsuura, K. Coatings on Mg Alloys and Their Mechanical Properties: A Review. *J. Mater. Sci. Technol.* **2018**, *34*, 1119–1126. [CrossRef]
4. Zhang, L.; Wang, Q.; Liao, W.; Guo, W.; Li, W.; Jiang, H.; Ding, W. Microstructure and Mechanical Properties of the Carbon Nanotubes Reinforced AZ91D Magnesium Matrix Composites Processed by Cyclic Extrusion and Compression. *Mater. Sci. Eng. A* **2017**, *689*, 427–434. [CrossRef]
5. Bakkar, A.; Ahmed, M.M.Z.; Alsaleh, N.A.; Seleman, M.M.E.S.; Ataya, S. Microstructure, Wear, and Corrosion Characterization of High TiC Content Inconel 625 Matrix Composites. *J. Mater. Res. Technol.* **2019**, *8*, 1102–1110. [CrossRef]
6. Shirvanimoghaddam, K.; Hamim, S.U.; Karbalaei Akbari, M.; Fakhrhoseini, S.M.; Khayyam, H.; Pakseresht, A.H.; Ghasali, E.; Zabet, M.; Munir, K.S.; Jia, S.; et al. Carbon Fiber Reinforced Metal Matrix Composites: Fabrication Processes and Properties. *Compos. Part A Appl. Sci. Manuf.* **2017**, *92*, 70–96. [CrossRef]
7. El-Sayed Seleman, M.M.; Ahmed, M.M.Z.; Ataya, S. Microstructure and Mechanical Properties of Hot Extruded 6016 Aluminum Alloy/Graphite Composites. *J. Mater. Sci. Technol.* **2018**, *34*, 1580–1591. [CrossRef]
8. Liu, L.; Li, W.; Tang, Y.; Shen, B.; Hu, W. Friction and Wear Properties of Short Carbon Fiber Reinforced Aluminum Matrix Composites. *Wear* **2009**, *266*, 733–738. [CrossRef]
9. Aatthisugan, I.; Razal Rose, A.; Selwyn Jebadurai, D. Mechanical and Wear Behaviour of AZ91D Magnesium Matrix Hybrid Composite Reinforced with Boron Carbide and Graphite. *J. Magnes. Alloy* **2017**, *5*, 20–25. [CrossRef]
10. Lim, C.Y.H.; Lim, S.C.; Gupta, M. Wear Behaviour of SiCp-Reinforced Magnesium Matrix Composites. *Wear* **2003**, *255*, 629–637. [CrossRef]
11. García-Rodríguez, S.; Torres, B.; Maroto, A.; López, A.J.; Otero, E.; Rams, J. Dry Sliding Wear Behavior of Globular AZ91 Magnesium Alloy and AZ91/SiCp Composites. *Wear* **2017**, *390–391*, 1–10. [CrossRef]

12. Ataya, S.; Naser Alsaleh, B.A.; Mohamed El-Sayed Seleman, B.M. Strength and Wear Behavior of Mg Alloy AE42 Reinforced with Carbon Short Fibers. *Acta Metall. Sin. (Engl. Lett.)* **2019**, *32*, 31–40. [CrossRef]
13. Yang, F.; Zhang, X.; Han, J.; Du, S. Mechanical Properties of Short Carbon Fiber Reinforced ZrB2-SiC Ceramic Matrix Composites. *Mater. Lett.* **2008**, *62*, 2925–2927. [CrossRef]
14. Qi, L.; Guan, J.; Liu, J.; Zhou, J.; Wei, X. Wear Behaviors of Cf/Mg Composites Fabricated by Extrusion Directly Following Vacuum Pressure Infiltration Technique. *Wear* **2013**, *307*, 127–133. [CrossRef]
15. Ureña, A.; Rams, J.; Escalera, M.D.; Sánchez, M. Characterization of Interfacial Mechanical Properties in Carbon Fiber/Aluminium Matrix Composites by the Nanoindentation Technique. *Compos. Sci. Technol.* **2005**, *65*, 2025–2038. [CrossRef]
16. Tang, Y.; Deng, Y.; Zhang, K.; Liu, L.; Wu, Y.; Hu, W. Improvement of Interface between Al and Short Carbon Fibers by α-Al2O3 Coatings Deposited by Sol-Gel Technology. *Ceram. Int.* **2008**, *34*, 1787–1790. [CrossRef]
17. Lee, C.W.; Kim, I.H.; Lee, W.; Ko, S.H.; Jang, J.M.; Lee, T.W.; Lim, S.H.; Park, J.P.; Kim, J.D. Formation and Analysis of SiC Coating Layer on Carbon Short Fiber. *Surf. Interface Anal.* **2010**, *42*, 1231–1234. [CrossRef]
18. Sarapure, S.; Satish, B.M.; Girish, B.M. Basawaraj Microstructure and Mechanical Behavior of Magnesium Alloy AZ91 Hybrid Composites. *IOP Conf. Ser. Mater. Sci. Eng.* **2018**, *310*, 012161. [CrossRef]
19. Tian, W.; Qi, L.; Zhou, J.; Guan, J. Effects of the Fiber Orientation and Fiber Aspect Ratio on the Tensile Strength of Csf/Mg Composites. *Comput. Mater. Sci.* **2014**, *89*, 6–11. [CrossRef]
20. Wang, X.J.; Hu, X.S.; Wu, K.; Deng, K.K.; Gan, W.M.; Wang, C.Y.; Zheng, M.Y. Hot Deformation Behavior of SiCp/AZ91 Magnesium Matrix Composite Fabricated by Stir Casting. *Mater. Sci. Eng. A* **2008**, *492*, 481–485. [CrossRef]
21. Ohno, M.; Mirkovic, D.; Schmid-Fetzer, R. Liquidus and Solidus Temperatures of Mg-Rich Mg-Al-Mn-Zn Alloys. *Acta Mater.* **2006**, *54*, 3883–3891. [CrossRef]
22. Anilan Ajukumar, K.; AjithKumar, K.K.; Kunjayyappan Ravikumar, K.; Deva Rajan, T.P.; Subramonia Pillai, U.T.; Chandrasekhara Pai, B. Fabrication and Characterization of Short Carbon Fiber Reinforced AZ91 Mg Alloy Composites. *Mater. Sci. Forum* **2012**, *710*, 347–352. [CrossRef]
23. Afsharnaderi, A.; Malekan, M.; Emamy, M.; Rasizadeh Ghani, J.; Lotfpour, M. Microstructure Evolution and Mechanical Properties of the AZ91 Magnesium Alloy with Sr and Ti Additions in the As-Cast and As-Aged Conditions. *J. Mater. Eng. Perform.* **2019**, *28*, 6853–6863. [CrossRef]
24. Liu, J.; Qi, L.H.; Guan, J.T.; Ma, Y.Q.; Zhou, J.M. Compressive Behavior of C Sf/AZ91D Composites by Liquid-Solid Extrusion Directly Following Vacuum Infiltration Technique. *Mater. Sci. Eng. A* **2012**, *531*, 164–170. [CrossRef]
25. Russell-Stevens, M.; Todd, R.; Papakyriacou, M. The Effect of Thermal Cycling on the Properties of a Carbon Fibre Reinforced Magnesium Composite. *Mater. Sci. Eng. A* **2005**, *397*, 249–256. [CrossRef]
26. Olszówka-Myalska, A.; Myalski, J. Magnesium Alloy AZ31—Short Carbon Fiber Composite Obtained by Pressure Die Casting. *Solid State Phenom.* **2015**, *229*, 115–122. [CrossRef]
27. Fu, S.Y.; Lauke, B.; Mäder, E.; Yue, C.Y.; Hu, X. Tensile Properties of Short-Glass-Fiber- and Short-Carbon-Fiber-Reinforced Polypropylene Composites. *Compos. Part A Appl. Sci. Manuf.* **2000**, *31*, 1117–1125. [CrossRef]
28. Kandemir, S.; Gavras, S.; Dieringa, H. High Temperature Tensile, Compression and Creep Behavior of Recycled Short Carbon Fibre Reinforced AZ91 Magnesium Alloy Fabricated by a High Shearing Dispersion Technique. *J. Magnes. Alloy* **2021**, *9*, 1753–1767. [CrossRef]
29. Daoud, A. Wear Performance of 2014 Al Alloy Reinforced with Continuous Carbon Fibers Manufactured by Gas Pressure Infiltration. *Mater. Lett.* **2004**, *58*, 3206–3213. [CrossRef]

materials

Article

Effect of Y_2O_3 Addition on Microstructure and Properties of Laser Cladded Al-Si Coatings on AZ91D Magnesium Alloy

Xiaofeng Wan [1], Chuang Tian [1], Yi Li [1], Jingling Zhou [1], Shuangqing Qian [1], Lihong Su [2,*] and Li Wang [1,*]

1 School of Mechanical and Manufacturing Engineering, Nantong University, Nantong 226019, China
2 School of Mechanical, Materials, Mechatronic and Biomedical Engineering, University of Wollongong, Wollongong, NSW 2522, Australia
* Correspondence: lihongsu@uow.edu.au (L.S.); wang1988@ntu.edu.cn (L.W.)

Abstract: The effect of Y_2O_3 addition on the microstructure and properties of the laser cladded Al-Si alloy coating on the surface of AZ91D magnesium alloy was investigated in this study. The experimental results showed that the Al-Si + Y_2O_3 cladding layers contained α-Mg, Mg_2Si, Al_4MgY and a small amount of $Al_{12}Mg_{17}$ phases. The coarse dendrites, reticulated eutectic structures and massive phases in the coatings tended to be refined and gradually uniformly distributed with the increased amount of Y_2O_3. The introduction of Y_2O_3 into the cladding layer favored the improvement of microhardness and wear resistance due to the grain refinement strengthening and dispersion strengthening. The addition of Y_2O_3 also promoted the reduction of localized corrosion sites and made the corrosion surface smoother, implying that the corrosion resistance of the Y_2O_3-modified coatings was better than that of the unmodified cladding layer.

Keywords: AZ91D magnesium alloy; laser cladding; Al-Si + Y_2O_3 coating; tribological property; wear resistance; corrosion resistance

Citation: Wan, X.; Tian, C.; Li, Y.; Zhou, J.; Qian, S.; Su, L.; Wang, L. Effect of Y_2O_3 Addition on Microstructure and Properties of Laser Cladded Al-Si Coatings on AZ91D Magnesium Alloy. *Materials* **2023**, *16*, 338. https://doi.org/10.3390/ma16010338

Academic Editor: Antonio Riveiro

Received: 28 November 2022
Revised: 26 December 2022
Accepted: 28 December 2022
Published: 29 December 2022

1. Introduction

Magnesium alloy has the advantages of high specific strength and low density, and is currently a common light metal material [1–3]. In recent years, it has been widely used in the fields of automobile, aviation, national defense, medical and electronics [4–6]. Other benefits of magnesium alloys include their high machinability, good castability, hot formability, and recyclable nature [7–9]. However, the surface performance of magnesium alloys is typically poor due to high chemical reactivity, weak hardness, poor corrosion resistance, and poor wear characteristics, which restricts their wide range of applications in various industries [10–13]. According to Abbas et al. [10], the wear rate of as-received magnesium alloy was greatly higher than the samples subjected to a laser remelting treatment. In the report by Liu and Guo [13], both microstructure and mechanical properties of the magnesium alloy fabricated by selective laser melting were dependent on the heat treatment conditions, such as the temperature and heat treatment duration.

In order to increase the wear and corrosion resistances of magnesium alloy, surface treatments such laser surface cladding [14–16], chemical plating [17], electroplating [18], and anodic oxidation are some effective methods [19,20], among which the laser surface cladding technique has garnered a significant amount of interest recently [21–23]. For example, a coating with TiC reinforce successfully enhanced the microhardness and wear resistance by about 5–6 and 4.5–5.8 times [16]. By applying Al-Si powders to three distinct magnesium alloys (AZ61, WE54 and ZK30) with a laser of Nd:YAG, Bernabe et al. [2] obtained coatings with no cracks, no pores, and strong metallurgical bonding. Lei et al. [24] produced dense, crack and porous-free Al-Si coatings with a saw-tooth form and strong metallurgical bonding on the surface of AZ91D magnesium alloy. Al + Al_2O_3 powders were laser-cladded onto a magnesium alloy surface by Hazra et al. [25], which resulted in decreased wear rate relative to the magnesium alloy substrate.

Laser cladding experiments on magnesium alloy surfaces to improve their properties have a mature technical foundation, and the addition of rare earth oxides plays an invaluable role in the modification of original powders in laser cladding. Zhu et al. [26] studied the impact of the addition of Y_2O_3 on the surface of magnesium alloy on Al-Cu cladding coating. Their findings demonstrated that Y_2O_3 considerably improved the mechanical characteristics and gave rise to a microstructure that was dendritic, striped, or dispersed granular. Yang and co-authors [27] prepared Al-TiC-Al_3Ti composites containing Y_2O_3 (0–2 wt.%) on AZ91D matrix by laser cladding. When an appropriate amount of Y_2O_3 was introduced into the coating, it is found that the TiC phase within the composite coating is fine and dispersed. The addition of Y_2O_3 enhances the mechanical properties and corrosion resistance of the cladding. By using laser surface cladding, Bu et al. [28] successfully created Al-TiC-Y_2O_3 composite coating on the surface of AZ63-Er alloy. The results showed that the hardness of the cladding layer containing Y_2O_3 is higher than that without adding Y_2O_3, and has the highest corrosion resistance as it contains 0.6% Y_2O_3. At present, there are very few studies on yttrium oxide modification of Al-Si coating. In this study, therefore, the effect of adding different amount of Y_2O_3 from 1 to 2 wt.% on the microstructure as well as properties of Al-Si laser cladded layer above AZ91D magnesium alloy surface has been analyzed. The properties of the coatings were systematically evaluated in terms of the microhardness measurement, wear and corrosion testing experiments.

2. Materials and Methods

2.1. Sample Preparation

The AZ91D magnesium alloy employed as the matrix material for this study has a geometry of $100 \times 50 \times 15$ mm^3. Table 1 displays the AZ91D chemical composition. The primary coating powder is Al-12Si (wt.%, the same as following). The Al-12Si powder has a size in the range from 100 to 150 μm. The rare earth oxide Y_2O_3 (99.99% purity and 75 μm mean powder size) with 1.0, 1.5 and 2.0 wt.%, respectively, has been incorporated into the primary powders to investigate how it affects the coatings microstructure and properties. Four groups of cladding powders, illustrated in Table 2, were created.

Table 1. Chemical composition (wt.%) of AZ91 magnesium alloy.

Al	Zn	Mn	Si	Fe	Cu	Ni	Be	Mg
9.3	0.63	0.32	0.05	0.003	0.021	0.001	0.001	Bal.

Table 2. Samples and composition (wt.%) of coatings.

Title 1	Title 2
AS	Al-12Si
AS1Y	Al-12Si + 1Y_2O_3
AS1.5Y	Al-12Si + 1.5Y_2O_3
AS2Y	Al-12Si + 2Y_2O_3

Before laser cladding, use 60 to 1500 mesh metallographic sandpaper to grind the surface of the substrate, wash away the abrasive debris with deionized water and dry it. The surface was then roughened by sandblasting with Al_2O_3, and finally ultrasonically cleaned with acetone, alcohol and deionized water in sequence. At least three specimens per group were made in order to evaluate performance and study microstructure and to acquire accurate data.

2.2. Laser Cladding Process

For the cladding treatment, a YAG fiber laser having the maximum power of 6 kW was employed. Laser cladding used synchronous powder feeding. Studies have shown that high output power is conducive to the formation of a more uniform microstructure, but too high power or too low scanning speed will lead to surface evaporation and pit formation, while too low power or too high scanning speed will lead to insufficient melting, inhomogeneous particle distribution and failure of the bonding interface. Following several experiments, the following technical settings were determined to be optimal: 1 kW laser power, 6 mm/s scanning velocity, 3 mm laser beam size, and 1.2 r/min powder feeding rate. Meanwhile, high-purity argon was employed as the shielding gas due to the easy oxidation of the laser cladding procedure. Figure 1 shows a schematic illustration of the laser cladding process.

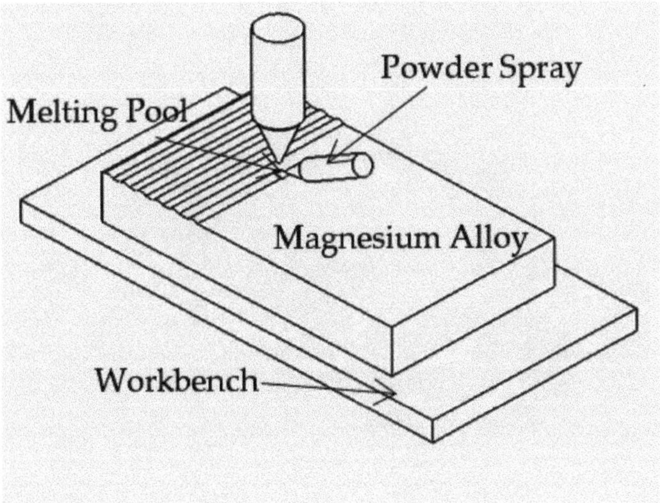

Figure 1. Schematic illustration of the laser cladding process.

2.3. Microstructure and Phase Analysis

The specimens were machined into size of 10 × 10 × 5 mm^3 by wire cutting for testing. From the cross-section of the laser-clad samples, microstructural investigations were performed using a GeminiSEM 300 scanning electron microscope (SEM). Phase constituents were analyzed using an Ultima IV multipurpose X-ray diffractometer (XRD), with the scanning 2θ in a range from 10 to 90°. Surface morphologies of the as-cladded coatings, as well as the coatings after wear test and corrosion test, were characterized by optical microscope (OM) and SEM, respectively.

2.4. Investigation of the Mechanical, Wear and Corrosion Properties

A TMVS-1 Vickers microhardness tester was used to measure the microhardness, with a force of 100 gf (0.98 N) and hold for 10 s during measurement. Each microhardness point was determined by performing five consecutive measurements on the coated cross section at the same depth and averaging the results. The sliding tribological test was performed in dry condition using an MPX-3G pin-disk friction wear tester. The grinding process lasted 15 min with a 50 N experimental load. The wear volume and worn surface morphologies have been analyzed following the previous studies [29,30]. Use epoxy resin to encapsulate the sample to be tested, and use CHI660e electrochemical workstation to test the Tafel polarization curve. The working electrode was the sample, the auxiliary electrode was a platinum sheet, and the reference electrode was a saturated calomel electrode. The electrolyte employed was a 3.5% NaCl solution.

3. Results and Discussion

3.1. Microstructural Study

The XRD patterns of the laser-clad Al-12Si coating and the Y_2O_3-modified coatings are shown in Figure 2. It can be seen that α-Mg and Mg_2Si phases have formed in the unmodified coating, and the diffraction peak intensities of these two phases are relatively high. Moreover, they are also accompanied by the production of a certain amount of $Al_{12}Mg_{17}$ and Al_3Mg_2 phases. In addition to Mg_2Si and $Al_{12}Mg_{17}$ phases, Al_4MgY is newly formed phase in the Y_2O_3 modified coating. Furthermore, no other contamination phases were observed within the coatings, which might be because rare earth elements have a purifying impact on the molten pool during process.

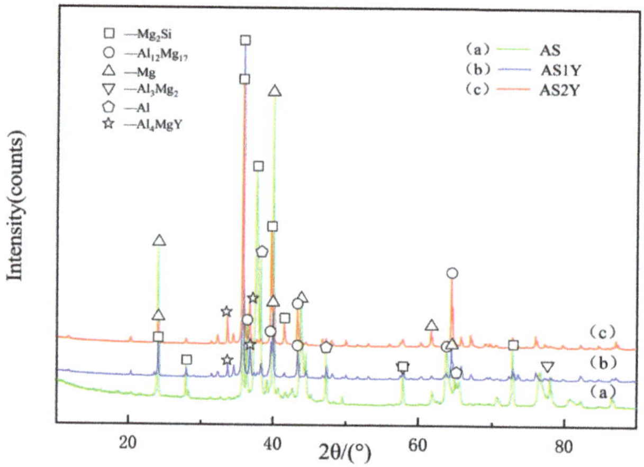

Figure 2. XRD patterns of the laser cladded coatings on AZ91D substrate: (**a**), (**b**) and (**c**).

Figure 3 displays the optical micrographs of the Al-Si powder coated samples with and without Y_2O_3. By comparing the morphologies of these four samples, it can be clearly observed that the eutectic structure and particle phase of the coating without adding rare earth oxides are coarse and massive, and the distribution of the microstructure is relatively concentrated. After adding Y_2O_3, the thick eutectic structure and massive phase structure become small and distributed uniformly, and the dendritic structure is significantly refined. The results show that rare earth has an obvious effect on the microstructure refinement of the laser cladded layer.

Figure 4 displays the cross-sectional morphology of the coating as seen by scanning electron microscopy. From the figure, it can be seen that the coating is divided into the cladding layer, the bonding zone and substrate. The results show that there is a strong metallurgical connection at the contact between the molten cladding layer and the substrate. There are no noticeable pores or flaws in any of the cladding layers. From each cladding layer area, the dendrite structure obviously shows a tendency to grow towards the top of the cladding layer. The cladding layer has a large number of dendrites and part of the cellular structure as shown in Figure 4a. Figure 4b,c show that there are clear columnar dendrites close to the substrate at the bottom of the cladding layer, which should be due to the large heat dissipation rate at the substrate. The temperature gradient becomes smaller, and the nucleus-forming rate is improved, thus resulting in the grain growth in a very short time with relatively strong directionality [31]. Furthermore, the cladding layer microstructure changes from coarse directed dendrites to fine equiaxed dendrites. The cladding layer structures are changed due to the addition of Y_2O_3, which acts as the nucleation center of dendrites [8], preventing dendrite coarsening and decreasing the dendrite arm spacing underneath the molten pool's non-equilibrium solidification and the influence of convective disturbance. As shown in Figure 4b, some dendritic and rod-like structures are distributed within the cladding coating layer, and the dendrite distribution is aggregated. Figure 4c shows that the rod-like structure in the cladding layer becomes short and thick with a little petal-like structure, and the structure distribution is more uniform. Figure 4d displays the AS2Y cladding layer. The directionality of the dendritic structure in the coating is no longer obvious from Figure 4d, and the eutectic structure tends to be more refined. The microstructure of the Al-Si + Y_2O_3 coating exhibits a dendritic eutectic structure with varying degrees of refinement in comparison to the coating without Y_2O_3, and the dendritic eutectic tends to distribute uniformly, which would be beneficial to a high microhardness, low wear rate and great corrosion performance of the coatings [26].

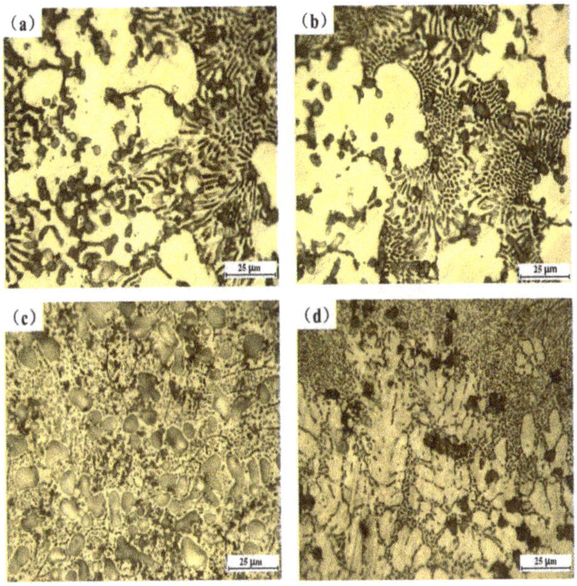

Figure 3. Optical micromorphology of cladded samples: (**a**) AS, (**b**) AS1Y, (**c**) AS1.5Y, and (**d**) AS2Y.

As shown in Figure 4e, three typical microstructural areas of A, B, and C are selected on the surface of the AS coating for SEM-EDS chemical composition analysis. Among them, zones A and B are blocky and strip-shaped with lighter color, respectively. The microstructures measured in area C are small particles with darker color. Based on the test results in Table 3, it is found that the structure measured in the strip region B is rich in Mg and Al elements, with an atomic ratio of nearly 1.4:1, so it can be inferred that the intermediate phase in the region is mainly $Al_{12}Mg_{17}$. It can be seen from the measurement results of C that the main elements in this area are Mg and Si. Combined with XRD analysis, it can be concluded that the substances in this area are mainly Mg_2Si. The high magnification SEM image of AS2Y coating presents that the microstructure morphologies are fine stripes, and the structures are more dense and uniform, which have been shown in Figure 4f. Combined with the SEM-EDS measurement results in Table 3, it can be seen that most of the Y_2O_3 particles have dissolved and decomposed into Y due to the strong metallurgical reaction. The Y atoms in the cladding coating are primarily distributed at the grain boundaries due to the greater atomic radius of Y of 0.18 nm [32]. Rare earth element Y acts as a non-homogeneous nucleus to inhibit grain growth by dragging effect on grain boundaries, thus inhibiting the growth of dendritic grains [33]. In addition, the high melting point of Y_2O_3, because of the increased content, it requires more energy to disintegrate and reduces the molten pool lifespan, thus the crystal does not have enough time to grow [16].

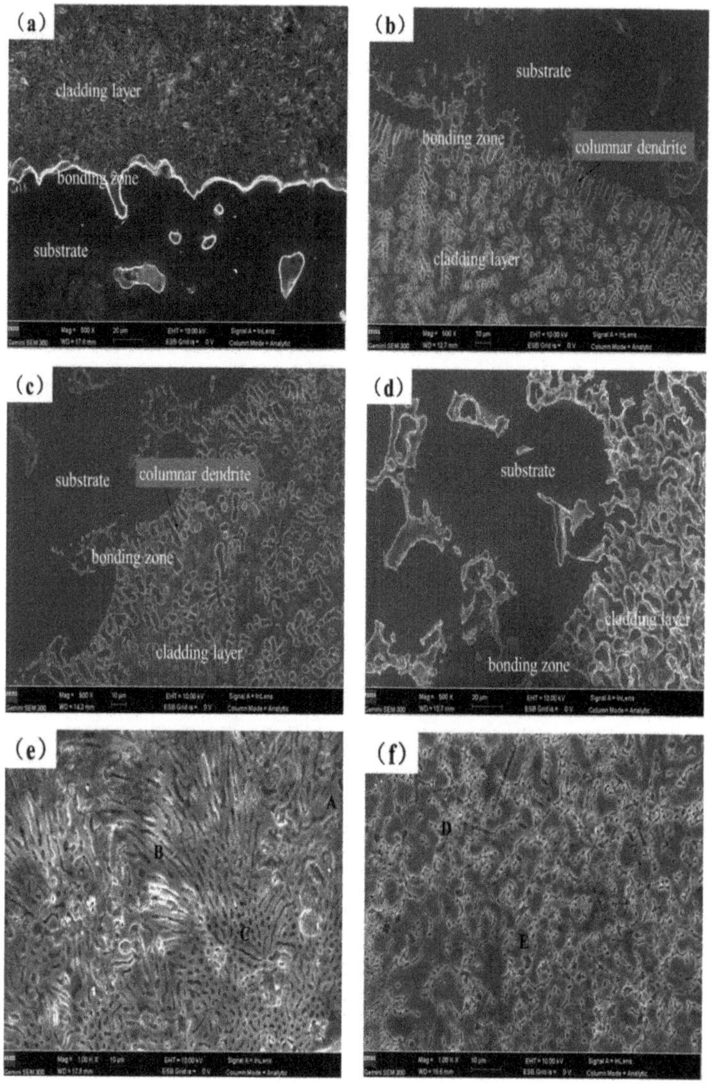

Figure 4. SEM microstructures of laser cladded coatings in this study: (**a**) AS, (**b**) AS1Y, (**c**) AS1.5Y, (**d**) AS2Y, (**e**) AS under magnification (Positions A, B and C are the EDS analysis area), and (**f**) AS2Y under magnification (Positions D and E are the EDS analysis area).

Table 3. Chemical compositions of the microstructure analyzed by SEM-EDS (wt.%).

Position	Al	Mg	Si	Y	O
A	41.0	58.1	0.9	0	0
B	42.1	51.3	6.6	0	0
C	14.7	45.8	39.6	0	0
D	74.7	3.6	19.6	2.0	0.1
E	65.6	4.6	28.7	1.2	0

3.2. Microhardness Analysis

Figure 5 shows the microhardness measurement results and indentation morphologies of the laser cladded sample sections. The curves in Figure 5 show that all the cladded samples have a similar tendency in terms of the microhardness distribution over the coating depth, which is consistent with the variation of the indentation dimension. The test results show that the test indentation steadily increases from the upper surface of the cladding to the substrate, indicating that the cladding surface has the highest hardness. Measurements of the hardness of coatings containing Y_2O_3 in comparison to the AS cladding layer show a significant improvement. The microhardness of all cladded coatings is higher than that of the substrate, which is about 75 HV, and the average microhardness of the cladding layer of AS2Y coating reaches the highest (about 270 HV), and the average microhardness of the AS cladding layer is only 220 HV. The addition of Y_2O_3 can refine the grain and introduce more grain boundaries. The grain boundary reinforcement and dislocation strengthening are conducive to the improvement in the coating microhardness. Additionally, the production of intermetallic compounds Mg_2Si and Al_4MgY has a favorable impact on enhancing the hardness of the laser-clad layer, as evidenced by the examination of microstructure, XRD, and SEM-EDS data [4].

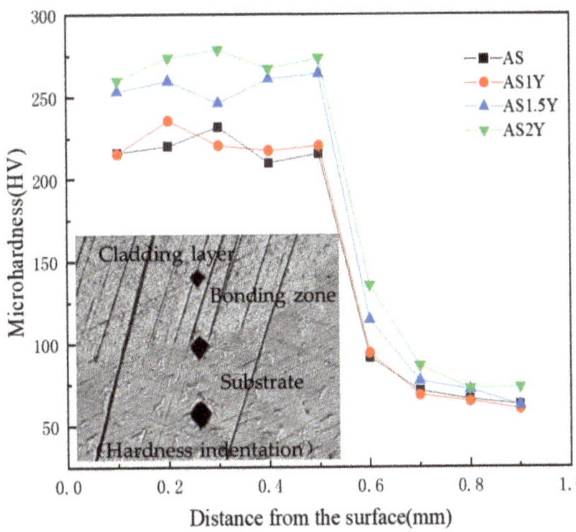

Figure 5. Microhardness distribution across the cross-section of coatings.

3.3. Wear Analysis

A key indicator of abrasion resistance is the friction coefficient. Figure 6a shows the average friction coefficient of various coatings investigated in the current study. The average friction coefficient of the coating shows a lower trend with increasing Y_2O_3 content, which may indicate a steady increase in the wear resistance of the coating. The transient friction coefficients versus time are shown in Figure 6b. As can be seen from the figure, the friction coefficient rises rapidly with increasing friction time, and then tends to stabilize. The rise in abrasive dust at the surface, which alters the dimension of the practical contact area and the localized contact load, may be the cause of the rapid increase in friction coefficient [34]. Because floating slag and gas can inevitably generate impurities and pores in the surface layer, AS1Y coating's average friction coefficient is a little bit higher than coatings without Y_2O_3 addition. However, the overall dynamic friction coefficient variation is smoother for the AS1Y coatings compared to the coatings without Y_2O_3. Rare earth Y elements could present wetting effect within the microstructure on account of its surface activity and mobility in the molten pool, which makes the microstructure distribution more uniform. According to the results, the friction coefficient of AS2Y coating displays the lowest value, which is about 60% lower than that of AS coating, and the fluctuation is the most stable, showing good wear resistance. It is worth to note that the friction coefficients obtained in this study in Figure 6 are much lower than the generally reported dry sliding friction coefficient in a range of about 0.5–0.8 for most of metallic materials [35–38]. According to the results displayed in Figure 6, it has been found that addition of Y_2O_3 acts as a solid lubricant during dry sliding and is effective into reducing the friction coefficient. The friction coefficient of the iron-based alloy coating prepared by Wang et al. [39] was reduced by about 0.1 due to the addition of La_2O_3. He et al. [40] fabricated Al-TiC-CeO_2 coatings with a minimum friction coefficient of about 0.5. The friction coefficient of the ASY coatings prepared in this work is significantly reduced.

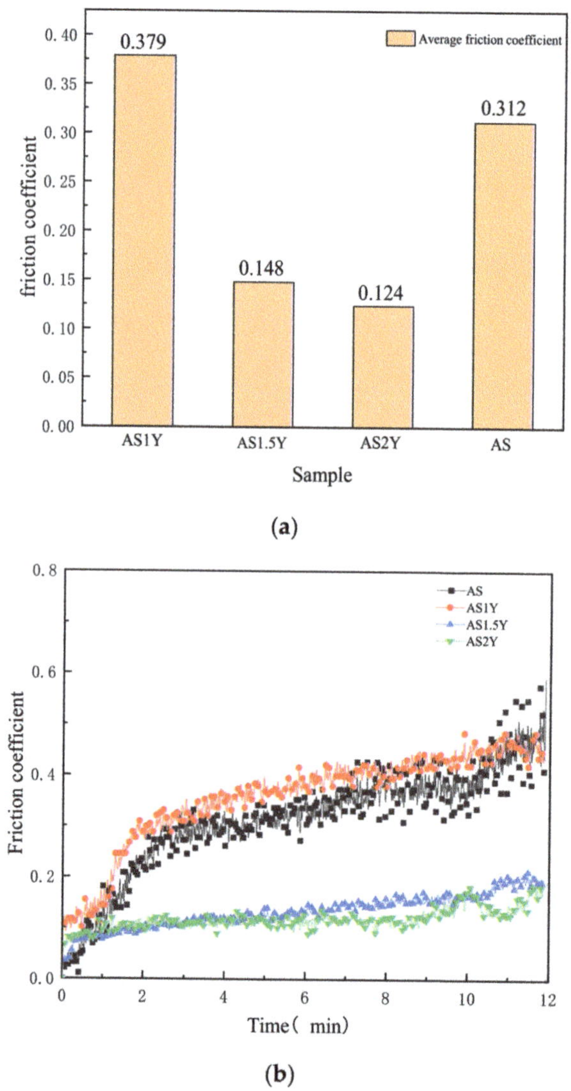

Figure 6. Comparison of the friction coefficient between the unmodified and Y_2O_3-modified coatings: (**a**) average friction coefficient, and (**b**) friction coefficient evolution history.

Figure 7 shows the wear volume loss of the cladding layer after 15 min of anti-grinding. The figure clearly shows that the AS2Y cladding layer's wear volume loss is substantially smaller than that of the AS coating, demonstrating that the addition of Y_2O_3 can somewhat reduce the coating's wear loss and increase its wear resistance. In fact, the evolution tendency of the wear volume loss in Figure 7 is consistent with the microhardness results in Figure 5. According to Deng et al. [29,30,38], a higher wear resistance or lower wear volume loss is generally obtained by increasing the mechanical strength and microhardness of a metallic material.

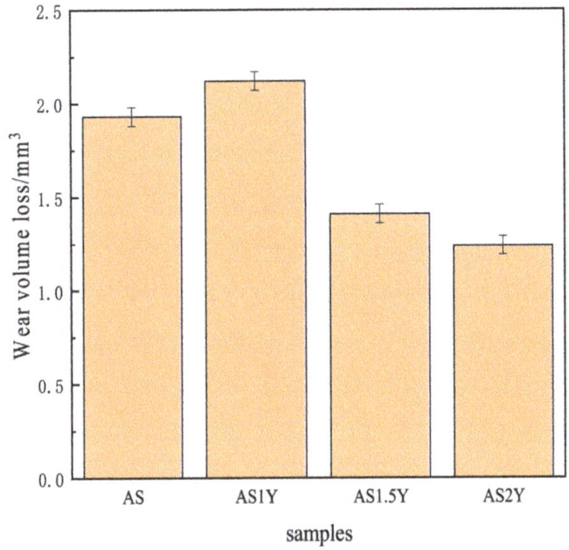

Figure 7. Influence of Y_2O_3 addition on the wear volume losses of the coatings.

The worn surface morphologies of various coatings as seen under SEM are displayed in Figure 8. Comparing the wear scars of the four samples, it is observed that under the same magnification, the width of the wear scars gradually decreases with the increase of Y_2O_3 content. The wear scar width of the coating falls from approximately 500 to 330 μm, as illustrated in Figure 8b–d, as the Y_2O_3 content rises from 1 to 2 wt%. The wear mechanism of the coatings investigated is a combination of abrasive and adhesive wear [2], and the adhesive zones are clearly visible. As seen in Figure 8b–d, the degree of wear of coatings containing Y_2O_3 gradually declines, with minor wear and micro-cuts serving as the primary wear features. The improvement in wear resistance is due to the formation of a refined Mg2Si phase in the microstructure, which acts as a pressure-bearing agent on the friction surface. Meanwhile, the addition of Y_2O_3 may improve the microstructure of the coating and make the hard phase dispersed distribution, thus improving the uniformity of the microstructure. As mentioned above, the microstructure of the coating is altered by the addition of Y_2O_3, going from columnar dendrites to a fine-grained cellular network structure, greatly increasing the hardness of the composite coating. Therefore, it is more difficult for the friction pair to press into the coating surface, and the wear resistance of the coating can be significantly improved. In addition, Y_2O_3 may hasten the convection of the molten pool and encourage a uniform distribution of the hard phase, which may in part contribute to the wear resistance of the cladding layer due to the reinforcement of the dispersion.

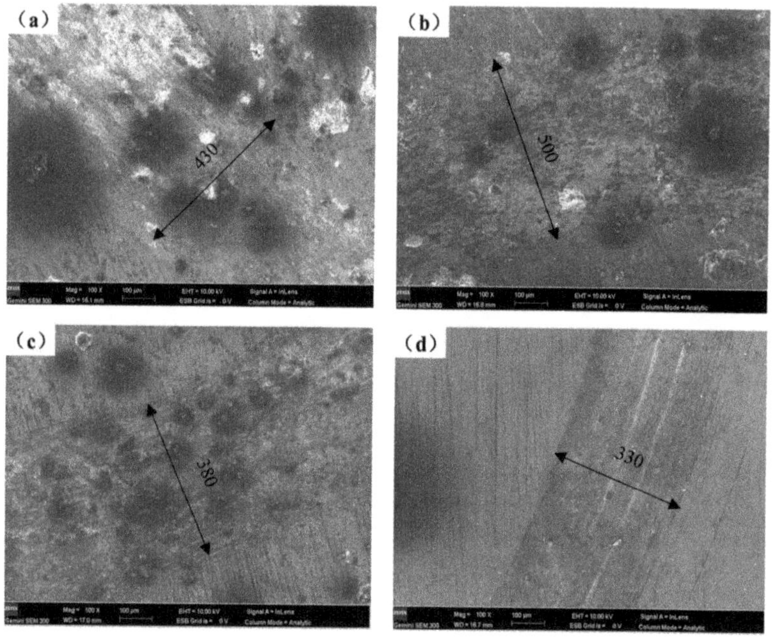

Figure 8. Worn surface morphologies of different coatings: (**a**) AS, (**b**) AS1Y, (**c**) AS1.5Y, and (**d**) AS2Y.

3.4. Corrosion Resistance Analysis

Figure 9 depicts the potentiodynamic polarization curves of the test specimens in a 3.5 weight percent solution of NaCl at room temperature, and Table 4 lists the electrochemical parameters that were obtained from Figure 9. Lower corrosion current values correspond to lower corrosion rates and greater corrosion resistance of the coating in terms of corrosion dynamics [33]. The data in Table 4 shows that the corrosion resistance of the coating created by laser cladding on the AZ91D magnesium alloy substrate is greatly improved compared to the uncoated AZ91D, and with the content of Y_2O_3 increasing the corrosion current density of the coating generally increases first and then decreases. When the Y_2O_3 content in the coating reaches 2%, the corrosion potential of AS2Y sample is the highest (−1.284 V), and the corrosion current density reaches the lowest (7.027×10^{-5} A/cm^2) among all the samples tested, which indicates the best corrosion resistance. Figure 10 shows the surface corrosion morphologies of AZ91D, AS and AS2Y samples after salt-immersion tests in 3.5 wt.% NaCl solution for 72 h. It can be observed that the corrosion surface of AZ91D in Figure 10a tends to develop vertically accompanied by deep and large corroded pores, which indicates the poor corrosion resistance. In the case of AS sample in Figure 10b, the occurrence of homogeneously uniform features with few pitting corrosion sites is evidenced at the specimen surface. Figure 10c shows no obvious pitting, with some micro-cracking on the surface. Figure 10d corrosion surface smooth without cracks produced, a very small part of the surface shows micro-pitting. In contrast, the surface of AS2Y sample in Figure 10e presents quite flat characteristics without pitting corrosion sites, indicating a superior corrosion resistance. Thus, adding Y_2O_3 to coatings is a viable technique to improve the corrosion behavior of AZ91D magnesium alloy, much like adding calcium to AZ31 magnesium alloy [41].

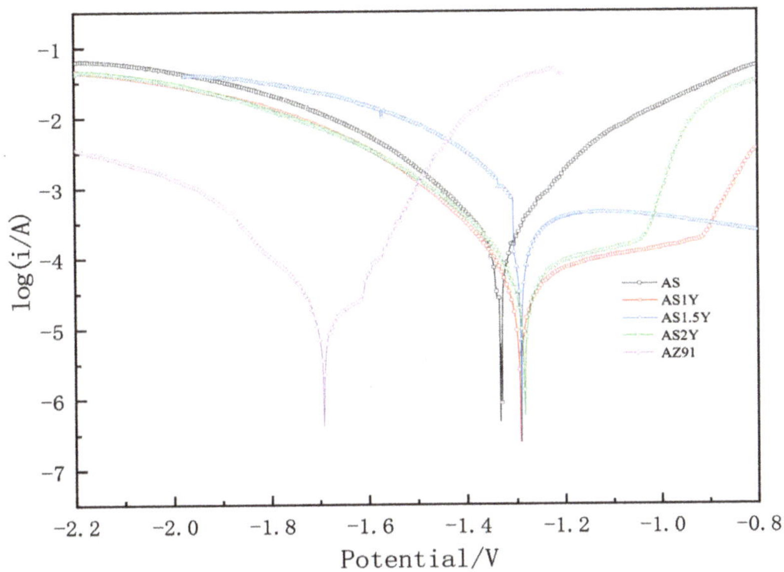

Figure 9. Potentiodynamic polarization curve.

Table 4. Electrochemical parameters calculated from the results in Figure 9.

Sample	E_{corr}/V	$I_{corr}/(A{\cdot}cm^{-2})$	β_a/V	β_c/V	R_p/Ω
AZ91D	−1.692	1.22×10^{-3}	0.376	−0.164	40.696
AS	−1.333	7.466×10^{-5}	0.127	−0.143	391.706
AS1Y	−1.293	9.85×10^{-5}	0.330	−0.122	393.162
AS1.5Y	−1.287	8.737×10^{-5}	0.296	−0.151	497.988
AS2Y	−1.284	7.027×10^{-5}	0.391	−0.129	600.158

The results show that the improved corrosion resistance of the Y_2O_3-modified coatings is mostly the result of structural changes within the coatings, suggesting that the corrosion of the Y_2O_3-modified specimens is in the passivation phase [26]. Shao et al. [42] recently reported the link between the microstructure and corrosion behavior of the magnesium alloy AZ91D. The introduction of rare earth elements into coatings has the effect of purifying the melt, purifying grain boundaries and inhibiting the growth of crystal columns for the molten pool metal in the laser cladding process [43], which has also been observed in the cross-sectional morphologies of the coatings mentioned above. As the Y_2O_3 content increases, the content of non-metallic inclusions and impurities in the coating structure are significantly reduced, and the metallurgical structure of the coating tends to denser, especially the existence of porosity and inhomogeneity would be reduced. Similar phenomenon has also been found by Qi et al. [44]. Meanwhile, the coarse dendrites, reticulated eutectic structures and bulk grains in the Al-Si + Y_2O_3 cladding layer tend to be refined. Thus, in general, refined grain size conduces to suppression of corrosion rate. In addition, the dispersion distribution of compound phases such as Mg_2Si and Al_4MgY plays a positive influence in improving the average corrosion potential and delaying the galvanic corrosion of the cladding layer, and the formation of these phases greatly reduces the amount of α-Mg used as the anode, which reduces the occurrence of intergranular corrosion, thereby lowering the rate of corrosion and increasing the corrosion resistance of samples.

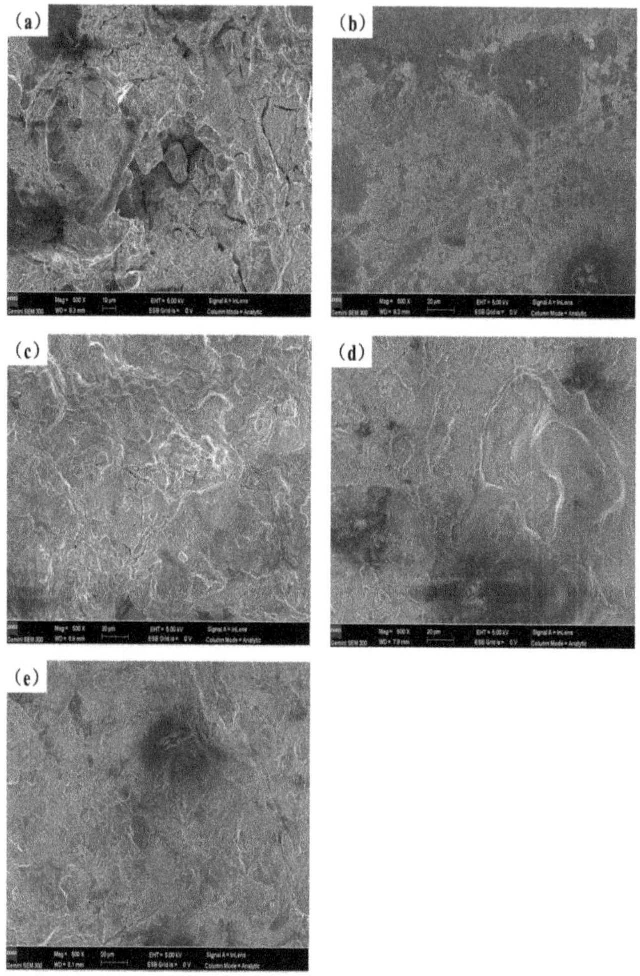

Figure 10. Surface morphology of different coatings after corrosion test in 3.5 wt.% NaCl solution: (**a**) AZ91D, (**b**) AS, (**c**) AS1Y, (**d**) AS1.5Y, (**e**) AS2Y.

4. Conclusions

By using laser cladding technology, Al-Si + Y_2O_3 coatings were created on the surface of AZ91D magnesium alloys, and the impact of Y_2O_3 content on the microstructure and characteristics of the Al-Si coatings was investigated. The outcomes of the trial revealed that:

(1) The microstructures of Al-Si + Y_2O_3 cladding layer include α-Mg, Mg_2Si, Al_4MgY phases and a small amount of $Al_{12}Mg_{17}$ phase. With the increase of the amount of Y_2O_3 added, the coarse dendrites, reticulated eutectic structures and massive phases in the Al-Si + Y_2O_3 cladding layer tend to be refined and gradually uniform distribution.

(2) With an increase in Y_2O_3 content, the microhardness of the Al-Si coating increases. When the content of Y_2O_3 reaches about 2 wt.%, the microhardness of the coating reaches the greatest with an value of about 270 HV.

(3) Friction experiments show that increasing the Y_2O_3 content, the average friction coefficient and the wear scar width of the coating have decreased. In addition, the

Y_2O_3 addition promotes the reduction of wear volume loss of the cladding layer, which effectively decreases the wear rate of the magnesium substrate.

(4) The Y_2O_3-modified coatings have greater corrosion resistance than the untreated coatings. The addition of Y_2O_3 promotes the reduction of localized corrosion sites and makes the corrosion surface smoother, implying that the corrosion resistance increases significantly.

Author Contributions: Conceptualization, X.W.; methodology, X.W., C.T. and Y.L.; validation, C.T., Y.L. and S.Q.; formal analysis, C.T. and Y.L.; investigation, X.W., C.T., L.S. and L.W.; data curation, C.T. and J.Z.; writing—original draft preparation, C.T.; writing—review and editing, X.W., L.S. and L.W.; supervision, L.S. and L.W. All authors have read and agreed to the published version of the manuscript.

Funding: Technological Production Translation of Jiangsu Province (BA2015129); and Australian Research Council (DE180100124).

Institutional Review Board Statement: Not applicable.

Informed Consent Statement: Not applicable.

Data Availability Statement: Not applicable.

Acknowledgments: The authors are grateful to the support of Jiangsu Overseas Visiting Scholar Program for University Prominent Young & Middle-aged Teachers and Presidents (2017), the "Qinglan" Project Foundation of Jiangsu Province (2016), and the Technological Production Translation of Jiangsu Province (BA2015129). The authors thank Nantong University Analysis & Testing Center for the technical support. The author L.S. would like to acknowledge the support from Australian Research Council (ARC) through Discovery Early Career Researcher Award (DECRA) fellowship.

Conflicts of Interest: The authors declare no conflict of interest.

References

1. Jiang, J.H.; Zhou, Q.; Yu, J.S.; Ma, A.B.; Song, D.; Lu, F.M.; Zhang, L.Y.; Yang, D.H.; Chen, J.Q. Comparative analysis for corrosion resistance of micro-arc oxidation coatings on coarse-grained and ultra-fine grained AZ91D Mg alloy. *Surf. Coat. Technol.* **2013**, *216*, 259–266. [CrossRef]
2. Carcel, B.; Sampedro, J.; Ruescas, A.; Toneu, X. Corrosion and wear resistance improvement of magnesium alloys by laser cladding with Al-Si. *Phys. Procedia* **2011**, *12*, 353–363. [CrossRef]
3. Liu, F.J.; Ji, Y.; Meng, Q.S.; Li, Z.S. Microstructure and corrosion resistance of laser cladding and friction stir processing hybrid modification Al-Si coatings on AZ31B. *Vacuum* **2016**, *133*, 31–37. [CrossRef]
4. Yang, Y.; Wu, H. Improving the wear resistance of AZ91D magnesium alloys by laser cladding with Al–Si powders. *Mater. Lett.* **2009**, *63*, 19–21. [CrossRef]
5. Cao, X.; Jahazi, M.; Immarigeon, J.P.; Wallace, W. A review of laser welding techniques for magnesium alloys. *J. Mater. Process. Technol.* **2006**, *171*, 188–204. [CrossRef]
6. Asghar, O.; Li-Yan, L.; Yasir, M.; Chang-Jiu, L.; Cheng-Xin, L. Enhanced Tribological Properties of LA43M Magnesium Alloy by Ni60 Coating via Ultra-High-Speed Laser Cladding. *Coatings* **2020**, *10*, 638. [CrossRef]
7. Zhang, X.L.; Zhang, K.M.; Zou, J.X. Microstructures and properties in surface layers of Mg-6Zn-1Ca magnesium alloy laser-clad with Al-Si powders. *Trans. Nonferrous Met. Soc. China* **2018**, *28*, 96–102. [CrossRef]
8. Liu, J.L.; Yu, H.J.; Chen, C.Z.; Weng, F.; Dai, J.J. Research and development status of laser cladding on magnesium alloys: A review. *Opt. Lasers Eng.* **2017**, *93*, 195–210. [CrossRef]
9. Mo, N.; Tan, Q.Y.; Bermingham, M.; Huang, Y.D.; Dieringa, H.; Hort, N.; Zhang, M.X. Current development of creep-resistant magnesium cast alloys: A review. *Mater. Des.* **2018**, *155*, 422–442. [CrossRef]
10. Abbas, G.; Li, L.; Ghazanfar, U.; Liu, Z. Effect of high power diode laser surface melting on wear resistance of magnesium alloys. *Wear* **2006**, *260*, 175–180. [CrossRef]
11. Ren, Z.; Zhao, Y.; Han, G.; Wang, W.; Zhou, K.; He, T.; Sun, Y. Laser-Arc Hybrid Cladding of Al-Mg Alloy Coating on AZ80 Mg Alloy: Effect of Laser Beam Oscillations Amplitude. *Materials* **2022**, *15*, 7272. [CrossRef] [PubMed]
12. Riquelme, A.; Rodrigo, P. An Introduction on the Laser Cladding Coatings on Magnesium Alloys. *Metals* **2021**, *11*, 1993. [CrossRef]
13. Liu, S.; Guo, H. Influence of Heat Treatment on Microstructure and Mechanical Properties of AZ61 Magnesium Alloy Prepared by Selective Laser Melting (SLM). *Materials* **2022**, *15*, 7067. [CrossRef] [PubMed]
14. Taltavull, C.; Torres, B.; López, A.J.; Rodrigo, P.; Rams, J. Novel laser surface treatments on AZ91 magnesium alloy. *Surf. Coat. Technol.* **2013**, *222*, 118–127. [CrossRef]

15. Wang, A.H.; Yue, T.M. YAG laser cladding of an Al-Si alloy onto an Mg/SiC composite for the improvement of corrosion resistance. *Compos. Sci. Technol.* **2001**, *61*, 1549–1554. [CrossRef]
16. Zhang, H.X.; Dai, J.J.; Ma, Z.W.; Wang, X.Y.; Zhang, N.L. Effect of Y_2O_3 on microstructures and wear resistance of TiC reinforced Ti-Al-Si coating by laser cladding on TC4 alloy. *Surf. Rev. Lett.* **2019**, *26*, 1950077. [CrossRef]
17. Thakur, A.; Gharde, S.; Kandasubramanian, B. Electroless nickel fabrication on surface modified magnesium substrates. *Def. Technol.* **2019**, *15*, 636–644. [CrossRef]
18. Kang, Z.X.; Li, W. Facile and fast fabrication of superhydrophobic surface on magnesium alloy by one-step electrodeposition method. *J. Ind. Eng. Chem.* **2017**, *50*, 50–56. [CrossRef]
19. Saranya, K.; Bhuvaneswari, S.; Chatterjee, S.; Rajendran, N. Titanate incorporated anodized coating on magnesium alloy for corrosion protection, antibacterial responses and osteogenic enhancement. *J. Magnes. Alloy.* **2022**, *10*, 1109–1123. [CrossRef]
20. Pan, S.; Tu, X.H.; Yu, J.X.; Zhang, Y.; Miao, C.P.; Xu, Y.L.; Fu, R.; Li, J.Y. Optimization of AZ31B Magnesium Alloy Anodizing Process in NaOH-Na2SiO3-Na2B4O7 Environmental-Friendly Electrolyte. *Coatings* **2022**, *12*, 578. [CrossRef]
21. Tan, C.; Zhu, H.; Kuang, T.; Shi, J.; Liu, H.; Liu, Z. Laser cladding Al-based amorphous-nanocrystalline composite coatings on AZ80 magnesium alloy under water cooling condition. *J. Alloys Compd.* **2017**, *690*, 108–115. [CrossRef]
22. Riquelme, A.; Rodrigo, P.; Escalera-Rodriguez, M.D.; Rams, J. Evaluation of the Wear Resistance and Cor-rosion Behavior of Laser Cladding Al/SiC Metal Matrix Composite Coatings on ZE41 Magnesium Alloy. *Coatings* **2021**, *11*, 639. [CrossRef]
23. Tokunaga, T.; Ohno, M.; Matsuura, K. Coatings on Mg alloys and their mechanical properties: A review. *J. Mater. Sci. Technol.* **2018**, *34*, 1119–1126. [CrossRef]
24. Lei, Y.W.; Sun, R.L.; Tang, Y.; Niu, W. Experimental and thermodynamic investigations into the micro-structure of laser clad Al–Si coatings on AZ91D alloys. *Surf. Coat. Technol.* **2012**, *207*, 400–405. [CrossRef]
25. Hazra, M.; Mondal, A.K.; Kumar, S.; Blawert, C.; Dahotre, N.B. Laser surface cladding of MRI 153M mag-nesium alloy with (Al + Al_2O_3). *Surf. Coat. Technol.* **2009**, *203*, 2292–2299. [CrossRef]
26. Zhu, R.D.; Li, Z.Y.; Li, X.X.; Sun, Q. Microstructure and properties of the low-power-laser clad coatings on magnesium alloy with different amount of rare earth addition. *Appl. Surf. Sci.* **2015**, *353*, 405–413. [CrossRef]
27. Yang, L.Q.; Li, Z.Y.; Zhang, Y.Q.; Wei, S.Z.; Wang, Y.J.; Kang, Y. In-situ TiC-Al_3Ti reinforced Al-Mg com-posites with Y_2O_3 addition formed by laser cladding on AZ91D. *Surf. Coat. Technol.* **2020**, *383*, 125249. [CrossRef]
28. Bu, R.; Jin, A.X.; Sun, Q.; Zan, W.; He, R.L. Study on laser cladding and properties of AZ63-Er alloy for automobile engine. *J. Mater. Res. Technol.* **2020**, *9*, 5154–5160. [CrossRef]
29. Deng, G.; Chong, Y.; Su, L.; Zhan, L.; Wei, P.; Zhao, X.; Zhang, L.; Tian, Y.; Zhu, H.; Tsuji, N. Mechanisms of remarkable wear reduction and evolutions of subsurface microstructure and nano-mechanical properties during dry sliding of nano-grained Ti6Al4V alloy: A comparative study. *Tribol. Int.* **2022**, *169*, 107464. [CrossRef]
30. Deng, G.; Zhao, X.; Su, L.; Wei, P.; Zhang, L.; Zhan, L.; Chong, Y.; Zhu, H.; Tsuji, N. Effect of high pressure tor-sion process on the microhardness, microstructure and tribological property of Ti6Al4V alloy. *J. Mater. Sci. Technol.* **2021**, *94*, 183–195. [CrossRef]
31. Wang, X.; Pan, X.D.; Niu, Q.; He, X.Q. Microstructure and properties of laser clad Al-Si coating on AZ33M magnesium alloy. *Heat Treat. Met.* **2021**, *46*, 202–206. [CrossRef]
32. Li, H.C.; Wang, D.G.; Chen, C.Z.; Weng, F. Effect of CeO_2 and Y_2O_3 on microstructure, bioactivity and degradability of laser cladding CaO-SiO_2 coating on titanium alloy. *Colloids Surf. B Biointerfaces* **2015**, *127*, 15–21. [CrossRef] [PubMed]
33. Quazi, M.M.; Fazal, M.A.; Haseeb, A.S.M.A.; Yusof, F.; Masjuki, H.H.; Arslan, A. Effect of rare earth elements and their oxides on tribo-mechanical performance of laser claddings: A review. *J. Rare Earths* **2016**, *34*, 549–564. [CrossRef]
34. Deng, G.; Tieu, A.K.; Lan, X.; Su, L.; Wang, L.; Zhu, Q.; Zhu, H. Effects of normal load and velocity on the dry sliding tribological behaviour of CoCrFeNi$Mo_{0.2}$ high entropy alloy. *Tribol. Int.* **2020**, *144*, 106116. [CrossRef]
35. Saffarzade, P.; Amadeh, A.A.; Agahi, N. Study of tribological and friction behavior of magnesium phos-phate coating and comparison with traditional zinc phosphate coating under dry and lubricated conditions. *Tribol. Int.* **2020**, *144*, 106122. [CrossRef]
36. Ataya, S.; Seleman, M.M.E.; Latief, F.H.; Ahmed, M.M.Z.; Hajlaoui, K.; Soliman, A.M.; Alsaleh, N.A.; Habba, M.I.A. Wear characteristics of Mg alloy AZ91 reinforced with oriented short carbon fibers. *Materials* **2022**, *15*, 4841. [CrossRef]
37. Fabre, A.; Masse, J.E. Friction behavior of laser cladding magnesium alloy against AISI 52100 steel. *Tribol. Int.* **2012**, *46*, 247–253. [CrossRef]
38. Deng, G.; Tieu, A.K.; Su, L.; Wang, P.; Wang, L.; Lan, X.; Cui, S.; Zhu, H. Investigation into reciprocating dry sliding friction and wear properties of bulk CoCrFeNiMo high entropy alloys fabricated by spark plasma sintering and subsequent cold rolling processes: Role of Mo element concentration. *Wear* **2020**, *460–461*, 203440. [CrossRef]
39. Wang, Q.; Yang, J.; Niu, W.; Li, Y.; Mao, X.; Wang, Y.; Zhang, K. Effect of La_2O_3 on microstructure and properties of Fe-based alloy coatings by laser cladding. *Optik* **2021**, *245*, 167653. [CrossRef]
40. He, X.; Kong, D.; Song, R. Microstructures and Properties of Laser Cladding Al-TiC-CeO_2 Composite Coatings. *Materials* **2018**, *11*, 198. [CrossRef]
41. Chaudry, U.M.; Farooq, A.; Tayyab, K.B.; Malik, A.; Kamran, M.; Kim, J.; Li, C.; Hamad, K.; Jun, T.S. Corrosion behavior of AZ31 magnesium alloy with calcium addition. *Corros. Sci.* **2022**, *199*, 110205. [CrossRef]
42. Cui, C.; Wu, M.P.; Miao, X.J.; Gong, Y.L.; Zhao, Z.S. The effect of laser energy density on the geometric characteristics, microstructure and corrosion resistance of Co-based coatings by laser cladding. *J. Mater. Res. Technol.* **2021**, *15*, 2405–2418. [CrossRef]

43. Shao, Z.; Nishimoto, M.; Muto, I.; Sugawara, Y. Real-time in situ observation of the corrosion process of die-cast AZ91D magnesium alloy in NaCl solutions under galvanostatic polarization. *Corros. Sci.* **2021**, *192*, 109834. [CrossRef]
44. Qi, J.; Ye, Z.; Gong, N.; Qu, X.; Mercier, D.; Swiatowska, J.; Skeldon, P.; Marcus, P. Formation of a trivalent chromium conversion coating on AZ91D magnesium alloy. *Corros. Sci.* **2021**, *186*, 109459. [CrossRef]

materials

Article

Effect of Material Anisotropy on the Mechanical Response of Automotive Steel under High Strain Rates

Sheng Yin [1], Yi Xue [2], Haotian Cui [2], Xinhua Pei [1], Chundong Hu [2], Yangxin Wang [2] and Qingchao Tian [2,*]

1 Technology Institute, Meishan Iron & Steel Co., Ltd., Nanjing 210000, China; yinsheng@baosteel.com (S.Y.); peixinhua@baosteel.com (X.P.)
2 State Key Laboratory of Advanced Special Steel, Shanghai University, Shanghai 200434, China; xueyi18@shu.edu.cn (Y.X.); cuihaotian@shu.edu.cn (H.C.); huchundong@shu.edu.cn (C.H.); wangyangxin@shu.edu.cn (Y.W.)
* Correspondence: tctian@shu.edu.cn

Abstract: A constitutive model for automobile steel with high elongation needs to be established to predict the dynamic deformation behavior under hydroforming applications. In order to clarify the confusing discrepancy in the essential parameters of the classical Cowper-Symonds (C-S) model, a series of automobile structural steels have been employed to investigate the strain rate response by conducting tensile dynamic deformation. Metallographic microscopy and orientation distribution functions were used to characterize the microstructure and texture components of the steels. The microstructure observation discloses that the matrix of all steels is mainly of ferrite and the texture constituent provides a framework for steel to withstand external deformation. The C-S model can be applied to simulate the dynamic deformation with satisfied expectations. It is concluded that the essential parameters D and p in the model show a specific relationship with the steel grade, and the parameter D is proportional to the steel grade and related to material anisotropy, while the parameter p is inversely proportional to the steel grade and has close links with the grain boundary characteristics.

Keywords: dynamic deformation behavior; Cowper-Symonds model; texture; Hall–Petch relationship; microstructure

Citation: Yin, S.; Xue, Y.; Cui, H.; Pei, X.; Hu, C.; Wang, Y.; Tian, Q. Effect of Material Anisotropy on the Mechanical Response of Automotive Steel under High Strain Rates. *Materials* **2022**, *15*, 669. https://doi.org/10.3390/ma15020669

Academic Editor: Andrea Di Schino

Received: 5 December 2021
Accepted: 10 January 2022
Published: 17 January 2022

1. Introduction

The mechanical behavior of automotive structural steels ensures the safety of automobile components during the occurrence of extreme collisions by the absorption of impact energy; therefore, the dependent variables of not only the quasi-static deformation but also the dynamic response to collision must be taken into account in order to predict the failure process under varied strain rate conditions [1,2]. A high-speed servohydraulic testing machine and modified Hopkinson bar system can be employed to record the mechanical behavior of materials over a wide range of strain rates [3,4]. Furthermore, researchers have designed different Hopkinson-like rods or employed resistance strain gauges to reduce the influence of stress fluctuation at high strain rates, and these attempts can reduce the fluctuation amplitude to a certain degree [5–9]; standard testing method has been established to obtain valid stress–strain data in the dynamic test with the consideration of stress equilibrium to avoid system ringing [10].

Since steel structure is very sensitive to the change of strain rate, diverse models have been established to predict the dynamic deformation behavior of structural steels [11–15]. The Johnson-Cook (J-C) model defines the constitutive relation of the effective yield stress with strain, strain rate and temperature [11]. In order to obtain an accurate rate-prediction model, Chen et al. [12,13] studied the mechanical behavior of Q420 and Q345 steels with strain rates from 0.001 s^{-1} to 288 s^{-1} and 0.001 s^{-1} to 330 s^{-1}, respectively. Compared with the J-C models, a modified Hollomon-Voce model is proposed to predict the rate-dependent response. According to the dynamic response of S235, S690QL and S960QL, Alabi et al. [14,15] proposed the strain hardening exponent and strain sensitivity to explain the mechanical performance of these materials even in the presence of flaws. Nevertheless, the classical Cowper-Symonds (C-S) constitutive model has been widely applied to describe the performance of materials under dynamic deformation situations with good reliability [16,17].

Automobile steel plate with higher total elongation and moderate strength is required for some hydroforming applications [18,19], in which the deformation rate of different parts varies. Therefore, a constitutive model for such steels needs to be established to predict the dynamic deformation behavior under hydroforming conditions. However, confusing significant discrepancy has been found in the essential parameters D and p of the C-S model after conducting dynamic deformation of the steels in comparison with former research [20–24], and the influence factors seems to be in close relation with the material microstructural character, in which limited research studies have mentioned before. Whereas this paper firstly builds the C-S model of a series of structural steels by using a Zwick/Roell HTM5020 testing machine to minute the dynamic response, and then discloses the inherent regularity of the observed microstructure to the parameters of the C-S model.

2. Experimental

All test steels were prepared subsequently by vacuum melting, casting and hot rolling. The cast ingots were reheated to 1210–1235 °C, followed by hot-rolling to the thickness of about 2 mm with the finish temperature (FT) of 840 °C and the coiling temperature (CT) of about 550–570 °C. The chemical composition and rolling temperature of the steels are listed in Table 1; it can be seen that the steel sheets were characterized by low carbon content of 0.07 wt.% and were prepared from the viewpoint of economical use of alloying elements by gradually increasing the contents of Nb and Ti for heightened steel grade matching with controlled rolling and controlled cooling process. The varied steel grades in Table 1 are expected to help understand the effect of orientation on the mechanical response of such type of automotive steels.

Table 1. Chemical composition and rolling temperature of the prepared steels.

Steel Grade	Chemical Composition/wt. %				Roll Temperature/°C	
	C	Si + Mn + Al	Nb	Ti	FT	CT
Q380	0.07	≤1.0	0.015	0.013	840	570
Q420	0.07	≤1.0	0.03	0.013	840	570
Q460	0.07	≤1.5	0.04	0.02	840	560
Q500	0.07	≤1.5	0.05	0.02	840	550

All tensile specimens used in this study were spark cut from the prepared steel sheet along the rolling and the transvers directions, respectively, and the geometry of quasi-static (EN ISO 6892-1 2009) and dynamic specimens (ISO 26203-2 2011) are shown in Figure 1, guaranteeing that the round-trip number of the stress wave is higher than 10 times to assure the validity of the stress–strain data [5,20–22]. It is noted that the different sampling norms are considered from engineering practice, and the elongation of small-size specimen is slightly higher than that of the large-size one, while yield strength and tensile strength are identical.

A standard tensile test was conducted under a strain rate of 0.001 s^{-1} using a universal electromechanical testing machine (ZwickRoell, Ulm, Germany), and dynamic tests were carried out at 1, 2, 3, 4, 5, 10 and 18 m/s tensile speed corresponding to the nominal strain rates of 33, 66, 100, 133, 167, 333 and 600/s, respectively.

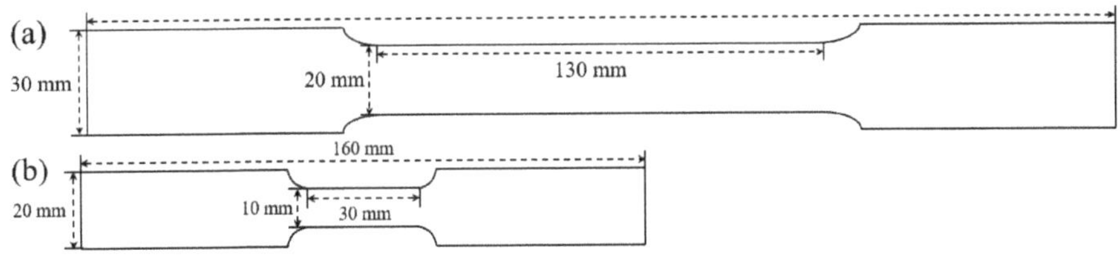

Figure 1. The geometry of (**a**) quasi-static and (**b**) dynamic tensile specimens.

Before the dynamic test, the surface of the deformation area of the samples was sprayed with a white background and then decorated by random black spots to be sure that the speckle patterns were of appropriate gray distribution for recording. That is, the subset size is a 20 × 20 pixel square, the distance between two elements is 12 pixels. According to the image resolution of 40–45 pixels/mm, an approximate 0.5–0.6 mm element size which is equivalent to two-times the subset distance. The tensile deformation was instantaneously recorded by a high-speed camera (Vision Resesarc Inc., Wayne, NJ, USA) [25]. The load data recorded by the weight sensor (ZwickRoell, Ulm, Germany) is used to calculate the engineering stress. The sampling frequency of the weighing sensor matches that of the high-speed camera (60,000 frames per second) to maintain synchronization. The engineering stress–strain curves under each strain rate are obtained by computer software processing.

The microstructures were observed by metallographic microscopy (OM) (Leica, Solms, Germany) at the 1/4-thickness position of the hot rolled sheets along the longitudinal section. The specimens were firstly planning machined to the target plane and then etched after mechanical polishing by using a 4% nital solution. Samples for texture measurements were spark cut with the dimensions of 20 × 10 mm, and a Bruker™ D8 Discover system (Bruker, Karlsruhe, Germany) with a CuKa target were employed under a scanning speed of 8°/min and operating at voltages of 40 kV and 40 mA current.

3. Results and Discussion

3.1. Microstructure Characteristics

Figure 2 shows the microstructures at the 1/4-thickness position along the longitudinal section. All the microstructures exhibit a slab-like bright white phase and black areas. The white areas are of ferrite while the black ones are of pearlite in Q380 and Q420 steels. As for Q460 and Q500, the black areas consist of pearlite and a small quantity of bainite. The ferrite grain of Q380 is more multi-scaled with an average size of 4.59 μm; with the increase of steel grade, the grain becomes finer and finer. The grain sizes of Q420, Q460 and Q500 are about 3.24 μm, 2.82 μm and 2.70 μm, respectively.

According to the Hall–Petch equation [23,24]:

$$\sigma = \sigma_0 + k_{HP} d^{-0.5} \quad (1)$$

where σ is the yield stress, σ_0 is constant and d is the grain size, k_{HP} is the Hall–Petch coefficient related to the grain boundary strengthening effect.

According to the measured yield stress and grain size in this paper and that in the Refs. [18,19], the σ_0 and k_{HP} parameters for different steels can be calculated as listed in Table 2, in which can be seen that the σ_0 and k_{HP} linearly increase with the rise in steel grade.

Table 2. Calculated parameters of Hall–Petch relationship for different steels.

Steel Type	Yield Stress/MPa	Grain Size/μm	σ_0/MPa	k_{HP}/MPa × $\mu m^{0.5}$
Q380	429	4.59	249	386
	407 [18]	5.96 [18]		
Q420	493	3.24	271	401
	455 [18]	4.74 [18]		
Q460	556	2.82	293	442
	565 [19]	2.64 [19]		
Q500	592	2.70	310	464
	537 [18]	4.17 [18]		

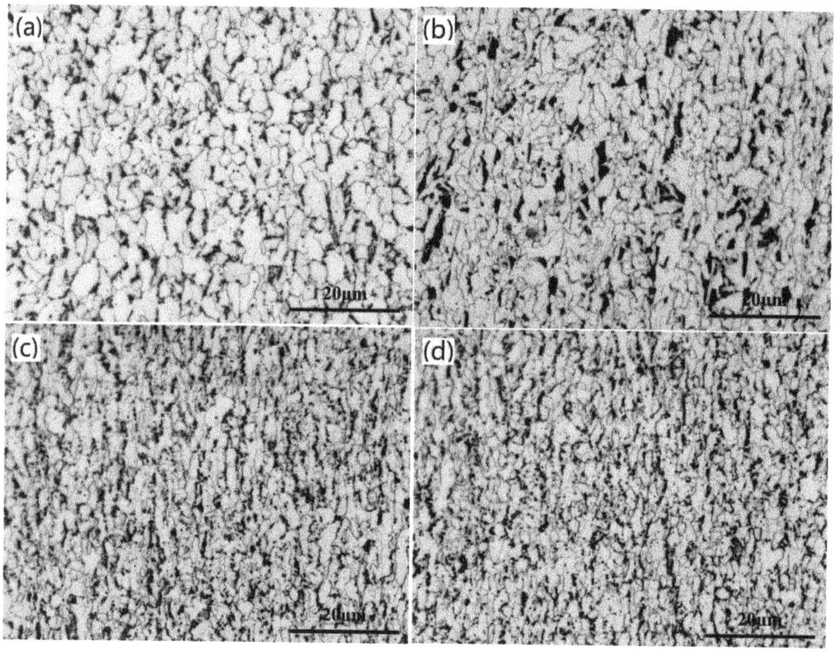

Figure 2. Microstructures of the hot rolled sheets along the rolling direction (**a**) Q380, (**b**) Q420, (**c**) Q460 and (**d**) Q500.

It is easy to understand that the fine grain can more efficiently hinder the dislocation movement and thus owe a better strengthening effect. It has been disclosed that k_{HP} depends on the steel composition [26,27]. Nevertheless, the microstructure distribution, phase constitution and grain boundary characteristic (fraction and distribution of defects and grain boundary misorientation) also contribute to the k_{HP}. It is safe to say that k_{HP} corresponds to the key microstructural parameter on mechanical property of steels.

Figure 3 shows the texture components in the φ_2 = 45° sections of Euler space, where the variation regularity can be clearly identified. The texture constituent of the four steels is similar, and mainly consists of the {011}<111>, {001}<100> and the {111} fiber texture with a maximum intensity of about 2.00. With the increase of steel grade, the intensity of {001}<100> cube texture increases gradually, as well as that of the {011} texture. All steels possess the {111} γ fiber, where {111}<112> texture is obvious for steels Q380, Q420 and Q460, while {111}<110> texture is stronger for Q500. In comparison with the lower steel grades, Q460 and Q500 have stronger {011}<112> and {011}<111>, respectively [28].

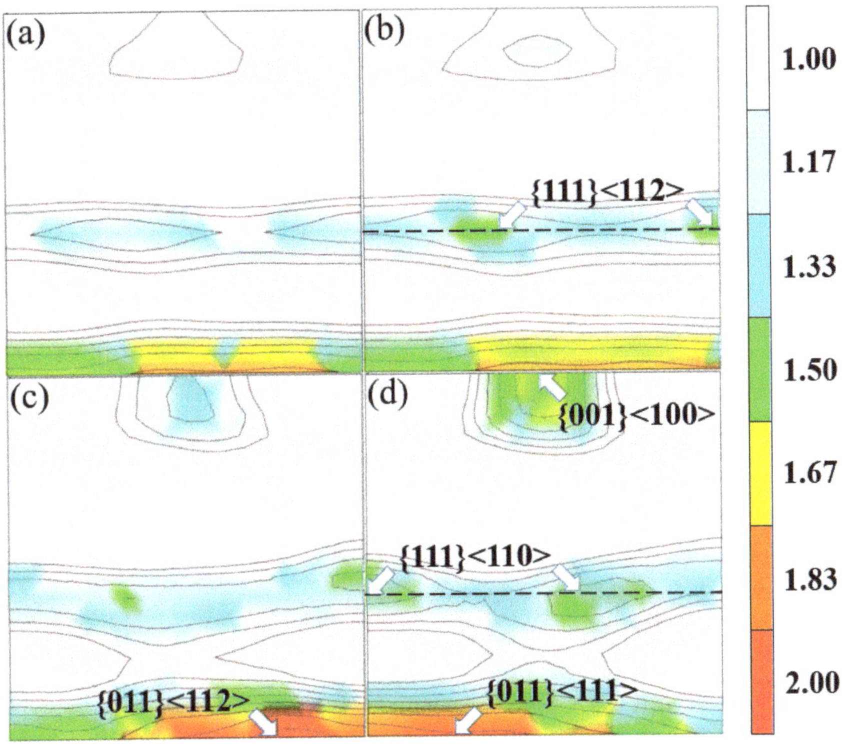

Figure 3. The texture distribution of the hot rolled sheets (**a**) Q380, (**b**) Q420, (**c**) Q460 and (**d**) Q500.

3.2. Tensile Behavior along Rolling Direction

3.2.1. Characteristic of Stress–Strain Curves

Figure 4 represents the stress–strain curves of the steels along a rolling direction under different strain rates from 10^{-3} to 600 s^{-1}. The standard test gives a typical engineering stress–strain curve with an obvious yield plateau. As the strain rate ($\dot{\varepsilon}$) increases, the plastic stage of stress–strain curve shows an overall upward trend of uplift, and appears to fluctuate in a periodic decay mode when the loading strain rate reaches higher than 100 s^{-1}. With the increase of steel grade from Q380 (Figure 4a), to Q420 (Figure 4b), to Q460 (Figure 4c), and then to Q500 (Figure 4d), the variation of stress–strain curves exhibits a similar tendency, except the decrease in total elongation.

Figure 5 shows the characteristic of fluctuation, of which the stress–strain curve can be approximately fitted as a sine attenuation function ($\sigma = A_0 e^{-\xi\varepsilon}\sin(\omega\varepsilon + \phi) + C_0$), where C_0 is the constant, and the distance between the first two crests and the distance between the first crest and the trough is taken as the cycle (T) and the amplitude (A_0), as shown in Figure 5a. In the high speed dynamic tensile test, the stress signal oscillation will inevitably occur owing to the increase of loading rate, and Alabi believed that it was the stress wave generated by the imbalance between internal friction and external force at high strains that caused the loading signal to be noisy [5–7,14,26–29]. Anyway, we assign the first appeared maximum stress σ_U as upper yield stress to consider the strain-rate response of all specimens.

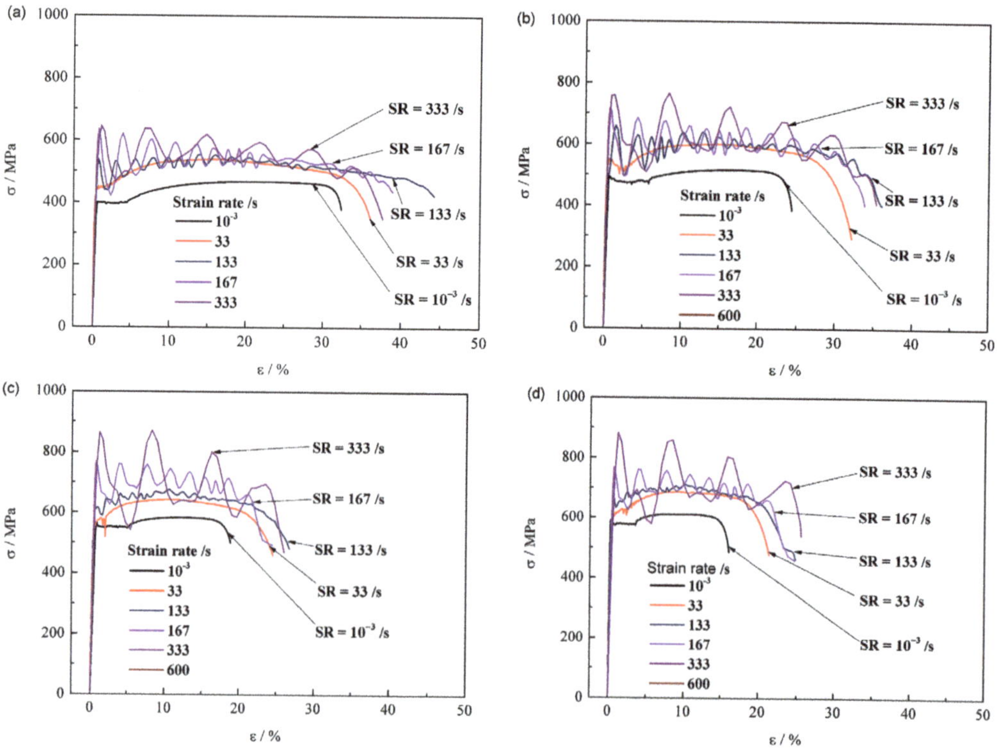

Figure 4. Stress–strain curves of longitudinal (**a**) Q380, (**b**) Q420, (**c**) Q460 and (**d**) Q500 at different strain rates.

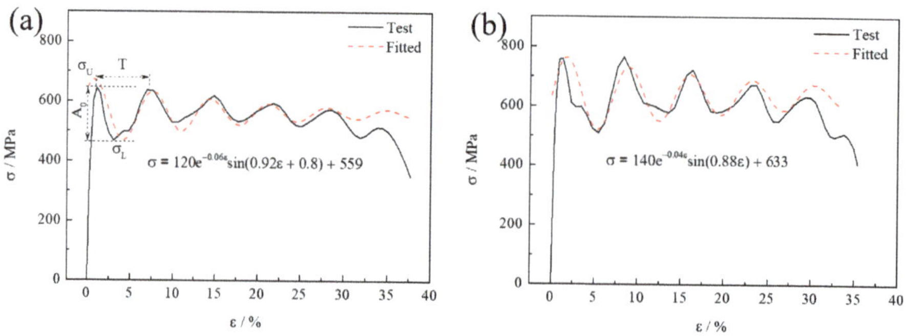

Figure 5. The fluctuation characteristic of stress-strain curves (**a**) Q380 and (**b**) Q420 steels at strain rate of 333 s^{-1}, where the first appeared maximum stress is assigned as σ_U.

3.2.2. Strain Rate Response

Figure 6 represents the dynamic response of the defined σ_U and the total elongation of different steels. It can be seen in Figure 6a that the σ_U value gradually increases to saturation with the increase of strain rate, while the σ_U curve lifts up with the increase of steel grade (Q380: from 400 to 711 MPa, Q420: from 492 to 825 MPa, Q460: from 563 to 896 MPa and Q500: from 591 to 913 MPa). The variation tendency of ε_T exhibits obvious discrepancy, as shown Figure 6b, the overall ε_T curve shifts down with the increase of steel grade. With the increase of $\dot{\varepsilon}$, the ε_T value first rapidly increases and then tends to a maximum value when $\dot{\varepsilon}$ reaches about 100 s^{-1} and then remains stabilized afterward.

The stabilization ε_T is about 37.8% for Q380, 34.7% for Q420, 25.8% for Q460 and 24.6% for Q500, respectively.

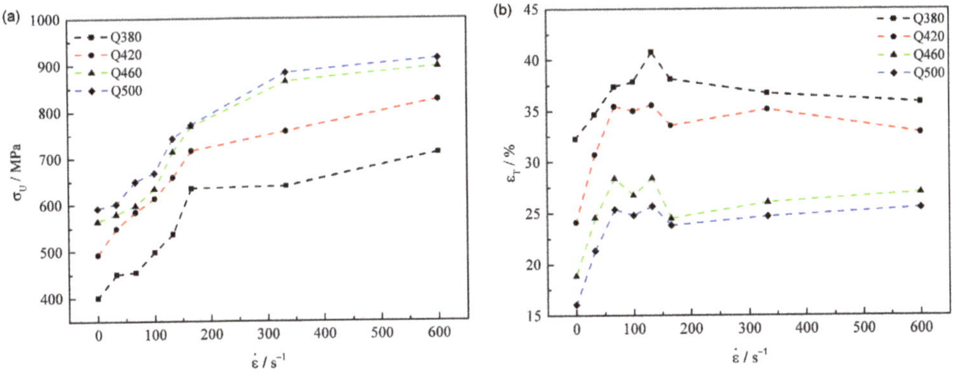

Figure 6. Dynamic response of (**a**) the upper yield stress (σ_U) and (**b**) the total elongation dependence of strain rate for different steels.

Then, the C-S constitutive model is introduced to further describe the above phenomenon. The dynamic increase factor (DIF_y) was defined as the ratio of the yield stress at each intermediate strain rate (σ_i) to the yield stress at the quasi-static strain rate (σ_q):

$$DIF_y = 1 + \left(\frac{\dot{\varepsilon}}{D}\right)^{\frac{1}{p}} \tag{2}$$

where D and p are the coefficient in the relation of stress with strain rate.

Figure 7 shows the longitudinal dynamic increase factor dependence of strain rate. Using the C-S model fitting (Figure 7a), the four steels can be well fitted and are described as $DIF = 1 + (\dot{\varepsilon}/933)^{1/1.81}$ for Q380, $DIF = 1 + (\dot{\varepsilon}/1042)^{1/1.73}$ for Q420, $DIF = 1 + (\dot{\varepsilon}/1139)^{1/1.56}$ for Q460 and $DIF = 1 + (\dot{\varepsilon}/1249)^{1/1.49}$ for Q500 with satisfied accuracy, respectively. It is noted that in the fitting process, the p value is continuously adjusted to seek for the best goodness-of-fit, and the highest R^2 is determined to correspond to the right regression coefficients, as shown in the Figure 7b.

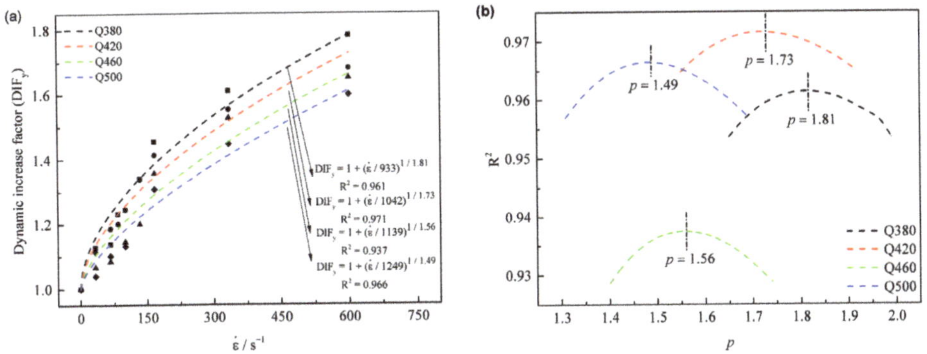

Figure 7. The C-S model fitting curves of Q380, Q420, Q460 and Q500 steels, (**a**) fitting results, (**b**) determination method for the p solution.

3.2.3. Parameters Dependence

The graphs for the linear regression analysis based on the above data depending on steel grade are shown Figure 8a; the essential parameter D is proportional while p is inversely proportional to the steel grade, showing an amazingly strong contrast.

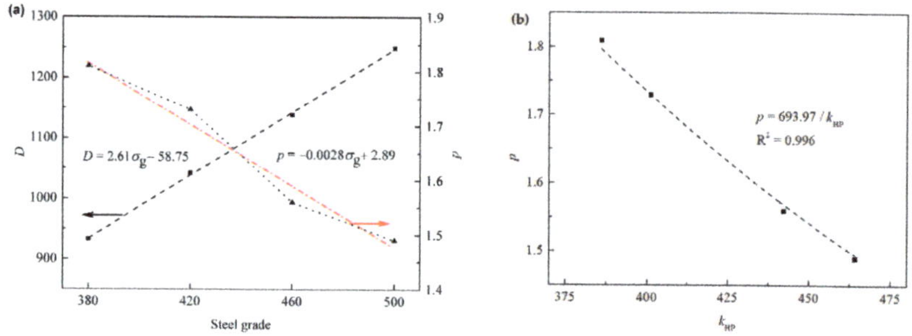

Figure 8. The variation of essential parameters in C-S model, (**a**) D and p dependence of steel grade σ_g where the two arrows are pointed to two Y-axes of D and p, respectively, and (**b**) the p dependence of the Hall-Petch coefficient k_{HP}.

By comparing the parameter p in C-S model, we can see that there is a certain relationship $p = 693.97/k_{HP}$ between them, as shown in Figure 8b.

Since the Hall–Petch equation is very essential for materials, and the k_{HP} is easily available in the literatures, the C-S model can be replaced as:

$$\mathrm{DIF}_y = 1 + \left(\frac{\dot{\varepsilon}}{D}\right)^{\frac{K_{HP}}{673.79}} \tag{3}$$

and then the relationship between the dynamic growth factor of material and the influence coefficient of grain boundary on deformation is established, which shows that the dynamic growth factor may be related to the adhesion between grains and the coordination of grain deformation. The substitutive C-S model can be employed to estimate the dynamic response of a material according to the Hall–Patch coefficient.

3.3. *Anisotropy on Strain Rate Response*

Considering the anisotropy of the material, a tensile test was conducted using specimens prepared along the steel transverse direction to compare the dynamic tensile behavior with that along the longitudinal rolling direction.

Figure 9 shows the stress–strain curves of specimens along a transverse direction at different strain rates from 0.001 s^{-1} to 333 s^{-1}. As for the same steel, the total elongation decreases for the specimen along transverse direction, which is different with that along the rolling direction. Similar to that along the longitudinal direction, with the increase of strain rate, the variation tendency of stress–strain curve is similar.

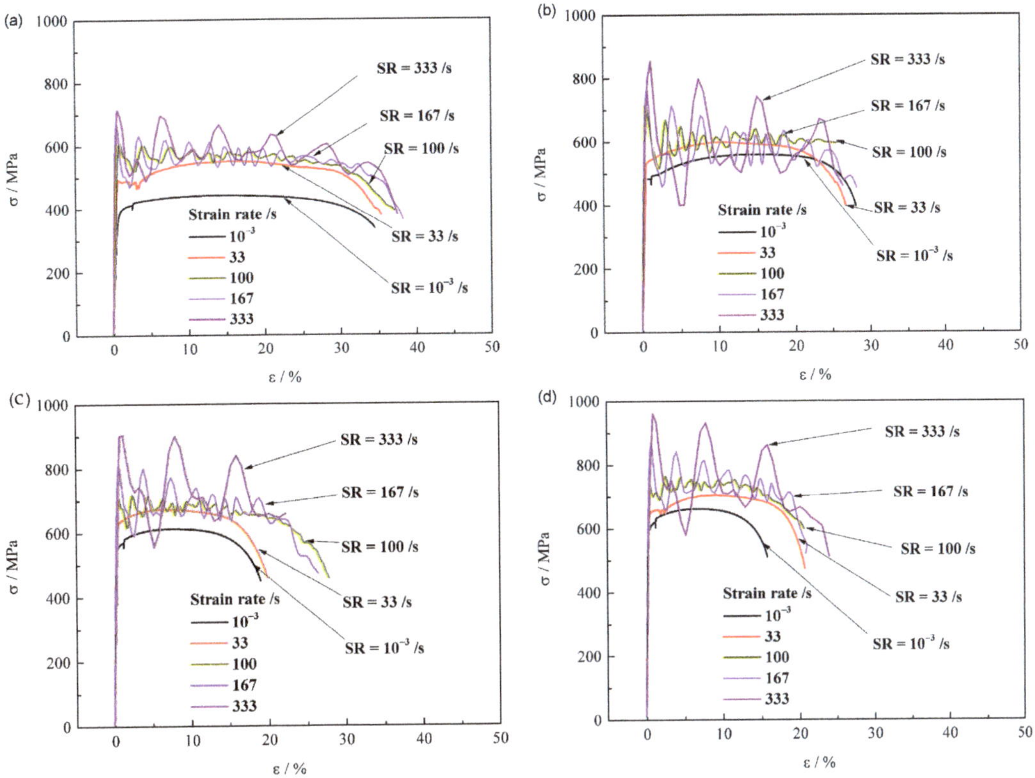

Figure 9. Stress-strain curves along transverse direction for specimens (**a**) Q380, (**b**) Q420, (**c**) Q460 and (**d**) Q500 at different strain rates.

Figure 10 shows the transverse dynamic increase factor dependence of strain rate. Using the C-S model fitting (Figure 10a), the four steels can be well fitted with satisfied accuracy and are described as $\mathrm{DIF}_y = 1 + (\dot{\varepsilon}/527)^{1/2.25}$ for Q380, $\mathrm{DIF}_y = 1 + (\dot{\varepsilon}/581)^{1/1.87}$ for Q420, $\mathrm{DIF}_y = 1 + (\dot{\varepsilon}/644)^{1/1.49}$ for Q460 and $\mathrm{DIF}_y = 1 + (\dot{\varepsilon}/689)^{1/1.37}$ for Q500, respectively.

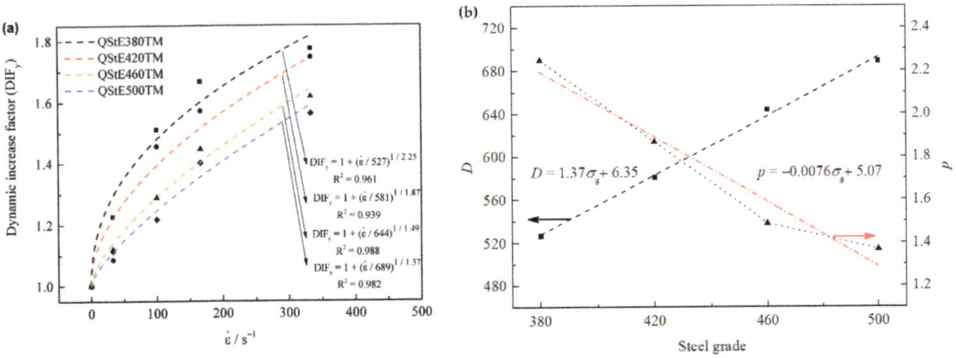

Figure 10. The C-S model fitting curves for steels along transvers direction, (**a**) fitting results, (**b**) D and p dependence of steel grade σ_g.

Based on the linear regression analysis of the above data (Figure 10b), the model parameter D is proportionate to the steel grade, while p is inversely proportionate to the steel grade.

3.4. Factors on Dynamic Deformation

The C-S model has been employed to simulate the dynamic response of different steels, as listed in Table 3 [12–14,17,30]. There exists significant difference in the characteristic parameters D and p, in particular, a vast gap can be found between the D values of different steels. It can be seen in Table 3 that all the steels are of low carbon type, and as for S355 steel, the metallurgical microstructure is of well-developed ferrite and pearlite [17]. Therefore, it is inferred that factors, such as material composition, microstructure and crystallographic texture determine the D and p values of materials.

Table 3. Chemical composition and the C-S model parameters of different steels.

Steels	Chemical Composition								The C-S Modeling	
	C	Si	Mn	P	S	Ni	V	Nb	D	p
Q420 [12]	0.19	0.24	1.35	0.021	0.006				22305	3.9
Q345 [13]	0.16	0.19	1.28	0.025	0.011				3781	4.33
S690 [14]	0.14	0.29	1.19	0.008	<0.002	0.084	0.031	0.016	2.06×10^7	5.18
S960 [14]	0.16	0.21	1.39	0.008	<0.002	0.077	0.021	0.013	4.90×10^7	4.59
S355 [17]	0.22	0.55	1.60	0.025	0.025				4945	2.69
Q235 [30]	0.22	0.35	1.4	0.045	0.05				306	2.75

Note: the parameters for S690 and S960 are deduced from the original data in Ref. [14].

The matrix of all the tested steels here is mainly of ferrite, and the comparison of tensile test along two different directions discloses that the main parameters of the C-S model exhibit a specific relationship with the microstructure, i.e., the parameter D is related to anisotropy while p has close links with the grain boundary characteristics.

In the process of plastic deformation, the stresses on the crystals of different orientations are not consistent owing to the anisotropy of crystal. The crystal grains do not deform at the same time, and their slip system orientation and slip direction are also different. As we all know, the {011} crystal plane of ferrite steel is densely packed, and it is the main slip plane of the bcc crystals, while the <111> direction is its main sliding direction. Therefore, the stronger {011} texture is beneficial to the dislocation slip, and thus is helpful to improve the ductility of the material. In contrast, the Burgers vector *a*<100> is a kind of immobile dislocation, and the texture of {001} is harmful to deformation at most cases. It has been pointed out that the {111} fiber texture is good for material forming but has little effect on improving the elongation of the material [19]. Therefore, there exists a competitive mechanism on plasticity between the textures {011} and {001}, and obviously the effect of {001} texture prevails for the steels in this study, and the increase in the intensity of {001}<100> cube texture leads to the decrease of total elongation with the increase of steel grade (Figure 6b). Additionally, it can be seen that the texture constituent provides a framework for steel to withstand external deformation.

The plastic deformation requires a certain amount of time for the movement of crystal dislocations, for the rotation of the slip plane from the unfavorable to the favorable orientation and the rotation between individual crystals. As large strain rate applies, only elastic deformation of the movement of the atoms away from their equilibrium position occurs owing to the insufficient duration for deformation, resulting in a persistent increase of yield stress with the increase of strain rate [31].

The change of elongation under high strain rate conditions might be related to the combined effect of thermal activation of dislocation movement and suppression of necking. Since plastic strain is mainly carried out by the slip of dislocations, on the one hand, high strain rate constrains the process of thermal activation of dislocations, dislocation multiplication should be dominant to increase the uniform elongation; on the other hand, high strain rate may harden the local constriction zone and expand the subsequent deformation to adjacent area; therefore, the ductility of material increases with the increase of strain rate.

4. Conclusions

A series of Q380, Q420, Q460 and Q500 steels have been prepared to investigate their dynamic deformation behavior. The main results are concluded as follows.

(1) The microstructures for Q380 and Q420 steels consist of ferrite and pearlite, while that for Q460 and Q500 consist of ferrite, pearlite and a small quantity of bainite. All steels mainly consist of the $\{011\}<111>$, $\{001\}<100>$ and the $\{111\}$ fiber texture with a maximum intensity of about 2.00. The ferrite texture constituent provides a framework for steel to withstand external deformation, and the increase in the intensity of $\{001\}<100>$ cube texture leads to the decrease of total elongation with the increase of steel grade;

(2) Specimens for all the steels along both the rolling and the transverse directions show similar tensile behavior. As the strain rate increases, the stress–strain curve fluctuates in a periodic decay mode. The upper-yield stress increases with the increase of strain rate from 10^{-3} to 600 s^{-1}, meanwhile, the total elongation first rapidly increases and then tends to be stabilized at certain values. The C-S model can be employed to well fit the strain rate response of the four steels both along transverse and rolling directions;

(3) Under the specific experiment conditions, the main parameters of the C-S model exhibit a certain relationship with the microstructure, the parameter D is proportional to the steel grade and related to material anisotropy, while p is inversely proportional to the steel grade and has close links with the grain boundary characteristics. The C-S model can be considered to estimate the dynamic response of a material according to the Hall–Patch coefficient.

Author Contributions: Conceptualization, Q.T. and S.Y.; methodology, S.Y. and H.C.; validation, Y.X.; formal analysis, Q.T., C.H. and X.P.; investigation, Y.X., H.C. and Y.W.; data curation, Y.X. and H.C.; writing—original draft preparation, S.Y., Y.X. and Q.T.; writing—review and editing, Q.T., S.Y. and H.C.; supervision, Q.T. All authors have read and agreed to the published version of the manuscript.

Funding: This work was funded by Independent Research and Development Project of State Key Laboratory of Advanced Special Steel, Shanghai Key Laboratory of Advanced Ferrometallurgy, Shanghai University (SKLASS 2020-Z03) and the Science and Technology Commission of Shanghai Municipality (No.19DZ2270200).

Institutional Review Board Statement: Not applicable.

Informed Consent Statement: Not applicable.

Data Availability Statement: The data used to support the findings of this study are included within the article, additional data are available from the corresponding author upon request.

Conflicts of Interest: The authors declare no conflict of interest.

References

1. Bouaziz, O.; Zurob, H.; Huang, M.X. Driving force and logic of development of advanced high strength steels for automotive applications. *Steel Res. Int.* **2013**, *84*, 937–947. [CrossRef]
2. Xue, Y.; Yin, S.; Tian, Q.C.; Pei, X.H. Study on dynamic response model of low alloy automobile structural steel. *J. Net Shape Form. Eng.* **2021**, *13*, 120–126.
3. Sato, K.; Yu, Q.; Hiramoto, J.; Urabe, T.; Yoshitake, A. A method to investigate strain rate effects on necking and fracture behaviors of advanced high-strength steels using digital imaging strain analysis. *Int. J. Impact Eng.* **2015**, *75*, 11–26. [CrossRef]
4. Paul, S.K.; Raj, A.; Biswas, P.; Manikandan, G.; Verma, R.K. Tensile flow behavior of ultra low carbon, low carbon and micro alloyed steel sheets for auto application under low to intermediate strain rate. *Mater. Des.* **2014**, *57*, 211–217. [CrossRef]
5. Jin, H.; Sanborn, B.; Lu, W.Y.; Song, B. Mechanical characterization of 304L-VAR stainless steel in tension with a full coverage of low, intermediate, and high strain rates. *Mech. Mater.* **2021**, *152*, 103654. [CrossRef]
6. Qiao, Y.; Chen, Y.; Cao, F.H.; Wang, H.Y. Dynamic behavior of CrMnFeCoNi high-entropy alloy in impact tension. *Int. J. Imp. Eng.* **2021**, *158*, 104008. [CrossRef]

7. Zhang, T.W.; Ma, S.G.; Zhao, D.; Wu, Y.C. Simultaneous enhancement of strength and ductility in a NiCoCrFe high-entropy alloy upon dynamic tension: Micromechanism and constitutive modeling. *Int. J. Plas* **2020**, *124*, 226–246. [CrossRef]
8. Baumann, G.; Niederkofler, D.; Elldersdorfer, C.; Feist, F. On the development of a release mechanism for a split hopkinson tension an compression bar. *Materials* **2021**, *14*, 7609. [CrossRef] [PubMed]
9. Moshin, M.A.A.; Iannucci, L.; Greenhalgh, E.S. On the dynamic tensile behaviour of thermoplastic composite carbon/polyamide 6.6 using split hopkinson pressure bar. *Materials* **2021**, *14*, 1653. [CrossRef]
10. Xiao, X.R. Dynamic tensile testing of plastic materials. *Polym. Test.* **2008**, *27*, 164–178. [CrossRef]
11. Johnson, G.R.; Cook, W.H. A constitutive model and data for metals subjected to large strains, high strain rates and high temperatures. In Proceedings of the 7th International Symposium on Ballistics, Hague, The Netherlands, 19–21 April 1983.
12. Chen, J.L.; Li, J.W.; Li, Z.X. Experiment research on rate-dependent constitutive model of Q420 steel. *Constr. Build. Mater.* **2017**, *153*, 816–823. [CrossRef]
13. Chen, J.L.; Shu, W.Y.; Li, J.W. Constitutive model of Q345 steel at different intermediate strain rates. *Int. J. Steel Struct.* **2017**, *17*, 127–137. [CrossRef]
14. Alabi, A.A.; Moore, P.L.; Wrobe, L.C.; Campbell, J.C.; He, W. Tensile behaviour of S690QL and S960QL under high strain rate. *J. Constr. Steel Res.* **2018**, *150*, 570–580. [CrossRef]
15. Getter, D.J.; Kantrales, G.C.; Consolazio, G.R.; Eudy, S.; Fallaha, S. Strain rate sensitive steel constitutive models for finite element analysis of vessel-structure impacts. *Mar. Struct.* **2015**, *44*, 171–202. [CrossRef]
16. Cowper, G.R.; Symonds, P.S. *Strain Hardening and Strain Rate Effects in the Impact Loading of Cantilever Beams*; Brown University: Providence, RI, USA, 1957.
17. Forni, D.; Chiaia, B.; Cadoni, E. Strain rate behavior in tension of S355 steel: Base for progressive collapse analysis. *Eng. Struct.* **2016**, *119*, 164–173. [CrossRef]
18. Xue, Y.; Yin, S.; Tian, Q.C. A perspective on the development of automobile steel plate with moderate strength and high plasticity. In Proceedings of the International Seminar on Materials Research and Engineering Technology, Shanghai, China, 18–20 August 2017.
19. Xue, Y.; Yin, S.; Tian, Q.C.; Pei, X.H. Approaching to high elongation of automotive steel by controlled cooling strategy. *Steel Res. Int.* **2020**, *92*, 2000276. [CrossRef]
20. Frew, D.J.; Forrestal, M.J.; Chen, W. A split Hopkinson pressure bar technique to determine compressive stress-strain data for rock materials. *Exp. Mech.* **2001**, *41*, 40–46. [CrossRef]
21. Achenbach, J.D. *Wave Propagation in Elastic Solids*; North-Holland Publishing Company: Amsterdam, The Netherlands, 1973.
22. Peng, S.Y.; Wei, Y.J.; Jin, Z.H.; Yang, W. Supersonic screw dislocations gliding at the shear wave speed. *Phys. Rev. Lett.* **2019**, *122*, 045501. [CrossRef] [PubMed]
23. Petch, N.J. The cleavage strength of polycrystals. *J. Iron Steel Inst.* **1953**, *174*, 25–28.
24. Hall, E.O. The deformation and ageing of mild steel: III discussion of results. *Proc. Phys. Soc. B* **1951**, *64*, 747–753. [CrossRef]
25. Liebold, F.; Heravi, A.A.; Mosig, O.; Curbacn, M.; Mechtcherine, V.; Mass, H.G. Crack propagation velocity determination by high-speed camera image sequence processing. *Materials* **2020**, *13*, 4415. [CrossRef] [PubMed]
26. Astafurov, S.V.; Maier, G.G.; Melnikov, E.V.; Moskvina, V.A.; Panchenko, M.Y.; Astafurova, E.G. The strain-rate dependence of the Hall-Petch effect in two austenitic stainless steels with different stacking fault energies. *Mater. Sci. Eng. A* **2019**, *756*, 365–372. [CrossRef]
27. Zhang, B.; Wang, J.; Wang, Y. Strain rate dependent tensile response of Ti-5Al-2.5Sn alloy. *Materials* **2019**, *12*, 659. [CrossRef]
28. Wang, Q.Q.; Liu, Z.Q. Investigation the effect of strain history on crystallographic texture evolution based on the perspective of macro deformation for high speed machining Ti-6Al-4V. *Mater. Charact.* **2017**, *131*, 331–338. [CrossRef]
29. Li, X.; Wang, S.; Xia, K. Dynamic tensile response of a microwave damaged granitic rock. *Exp. Mech.* **2020**, *10*, 1007. [CrossRef]
30. Chen, J.L.; Shu, W.Y.; Li, J.W. Experimental study on dynamic mechanical property of Q235 steel at different strain rates. *J. Tongji Univ.* **2016**, *44*, 1071–1075. (In Chinese)
31. Kong, X.Z.; Fang, Q.; Hong, J. A new damage-based nonlocal model for danamic tensile failure of concrete material. *Int. J. Imp. Eng.* **2019**, *132*, 103336. [CrossRef]

materials

Article

Understanding Effects of Ultrasonic Vibration on Microstructure Evolution in Hot Forming Process via Cellular Automata Method

Yutong Zhang, Weihua Zhou, Jinyuan Tang and Yuhui He *

State Key Laboratory of High Performance Complex Manufacturing, Central South University, Changsha 410023, China
* Correspondence: z203712162@163.com

Abstract: Compared with traditional forming technology, ultrasonic vibration-assisted plastic forming technology can improve the forming conditions and obtain better surface quality of the workpiece. However, the mechanism and theory of ultrasonic action have not formed a unified understanding. In this paper, ultrasonic-assisted thermal forming technology is taken as the research object. Through experimental research combined with cellular automata methods, based on the dislocation density model, nucleation and growth model, and dynamic recrystallization growth rule, a theoretical model for microstructure simulation of the ultrasonic-assisted thermal forming process was established. By introducing the ultrasonic energy field into the thermal forming process and correcting thermal activation energy and dynamic recovery coefficient, the reasons for flow stress reduction of 9310 steel and the influence of temperature, strain rate, and vibration amplitude on recrystallization were analyzed from the microscopic scale. The results show that the introduction of ultrasonic vibration reduces the dislocation activation energy, promotes dynamic recrystallization behavior, and finally leads to the reduction of flow stress. With an increase in vibration amplitude, the average grain size decreases faster, the recrystallization volume fraction increases faster, the stress decreases larger, and the ultrasonic softening phenomenon becomes more obvious. Decreasing the strain rate will promote the occurrence of dynamic recrystallization, the volume fraction and average grain size of dynamic recrystallization will increase, and the true stress will decrease.

Keywords: ultrasonic-assisted thermal forming; ultrasonic softening effect; stress reduction amplitude; evolution of microstructure

Citation: Zhang, Y.; Zhou, W.; Tang, J.; He, Y. Understanding Effects of Ultrasonic Vibration on Microstructure Evolution in Hot Forming Process via Cellular Automata Method. *Materials* **2022**, *15*, 7359. https://doi.org/10.3390/ma15207359

Academic Editor: Jana Bidulská

Received: 30 September 2022
Accepted: 17 October 2022
Published: 20 October 2022

1. Introduction

Compared with the traditional forming process, ultrasonic-assisted thermal forming can reduce the forming load and flow stress of the material, refine the grain, and obtain a workpiece with good forming quality [1]. Although ultrasonic vibration-assisted plastic-forming technology has been widely used in processing technology, there is no unified understanding of the internal mechanism of ultrasonic vibration [2,3].

After applying ultrasonic vibration, the material softens, and the flow stress decreases, which is known as the ultrasonic softening effect [4,5]. Wen [6] studied the plastic behavior of AZ31 magnesium alloy during the room temperature tensile process under ultrasonic vibration and revealed the volume effect of vibration plastic deformation. The ultrasonic softening effect reduces flow resistance and improves plasticity. When a lower amplitude or vibration energy is applied to the tensile specimen, the softening effect is dominant, resulting in a decrease in the deformation resistance of AZ31 with an increase in formability. Zhou [7] conducted an ultrasonic-assisted compression (UAC) experiment to study the influence of ultrasonic softening on the plasticity of aluminum and titanium by changing vibration amplitudes and frequencies. The experimental results show that ultrasonic vibration can reduce the flow stress of aluminum and titanium in the UAC process. In the range of 20–40 kHz, increasing the vibration amplitude can enhance the ultrasonic softening effect, and increasing the vibration frequency will reduce the ultrasonic softening effect.

The softening effect of ultrasonic is manifested as the decrease in flow stress at the macro level, and it affects the plastic-forming mechanism and microstructure of the material at the micro level [8,9]. Most of the studies on microstructure evolution under the action of ultrasonic vibration use metallographic microscopy or EBSD technology to observe the evolution of microscopic tissues during the application of ultrasonic vibration after ultrasonic vibration physics experiments. Hung [10,11] performed an ultrasonic vibration upsetting experiment and metallographic analysis of an A6061-T6 aluminum alloy. The experimental results showed that the surface grain of the sample is refined when ultrasonic vibration is superimposed on the sample. Zhao [12] studied the influence of ultrasonic vibration applied in three directions on the mechanical behavior and microstructure of steel plates in the process of tensile deformation. The radial vibration results in a significant increase in low-angle grain boundaries and the formation of subgrain boundaries in grains, and the axial and normal vibration results in a decrease in the proportion of low-angle grain boundaries and a decrease in dislocation density.

To study the reason for grain refinement caused by ultrasonic vibration, a large number of scholars have analyzed and explained it through experiments. Currently, a generally accepted explanation is that ultrasonic vibration increases the nucleation rate and promotes dislocation movement. Hu [4] used pure copper as the research object and conducted ultrasonic-assisted compression tests at different amplitudes and found that the stress reduction caused by ultrasonic softening increased with the increase of flow strain or ultrasonic amplitude. EBSD observation shows that when ultrasonic softening occurs, low-angle grain boundaries are randomly distributed in grains, while when no ultrasonic vibration is applied, they show a stacking distribution, which means that acoustic softening improves the movement of small-angle grain boundaries or dislocations, thus reducing flow stress. In addition, at small deformation strains, elongated grains become equiaxed, and the dislocation density is significantly reduced, which may be the result of increased dislocation annihilation due to ultrasound-induced dynamic recovery. However, with the increase in deformation strain, ultrasonic hardening gradually becomes significant, resulting in a greatly reduced effect of acoustic softening on the reduction of dislocation density. Zhou [7] conducted an ultrasonic-assisted compression experiment and an EBSD experiment on aluminum and titanium samples along the compression and transverse directions. The EBSD results show that the upper plate area of the aluminum sample is more sensitive to the influence of ultrasonic vibration than the center of the sample. Transient ultrasonic vibration increases the subgrains in the aluminum sample, and the induced grain boundaries with small angles are often parallel to the vibration direction. For titanium samples, ultrasonic vibration mainly affects the central area of the sample, and transient ultrasonic vibration also increases the generation of subgrains and twinning. This indicates that the refinement of grains is achieved through the proliferation of subgrains and that subgrains can be used as an indicator of dislocation evolution. Liu [13] conducted an ultrasonic vibration upsetting experiment and found that the dislocation movement in the area of severe deformation was intense and that the dislocation was intertwined, resulting in a dislocation wall. Finally, these dislocation walls become grain boundaries with small or large angles, and then the elongated grains are refined into small grains and subgrains. This indicates that ultrasonic vibration promotes dislocation movement. Wang [14,15] studied the evolution characteristics of grain size, orientation angle, and texture under different amplitudes, and found that there was a positive correlation between the number of subgrains, low-angle grain boundaries, and amplitude. EDS, XRD, and EBSD analysis revealed the evolution mechanism of alloy microhardness, and it was found that the microhardness of the Ti-45Nb sample increased first and then decreased with an increase in amplitude.

At present, most studies on microstructure evolution under ultrasonic vibration are based on the analysis of experimental phenomena, and the reasons for grain refinement and rheological stress reduction under ultrasonic vibration have not been unified. Cellular automata (CA) [16] is a grid dynamic model with discrete time, space and state, local spatial interaction, and temporal causality. It can simulate the spatiotemporal evolution of complex systems and can be used to simulate and analyze the process of ultrasonic-assisted thermal forming. Steel 9310 has high hardenability, high hardness, and high fatigue strength. It is widely used for aero engine gear and steam turbine gear. In the rough machining process of gear, high-temperature forging is often adopted, so 9310 steel is selected in this paper as a material for high-temperature deformations. This paper is based on a thermal simulation experiment and an ultrasonic vibration-assisted tensile experiment and studies the microstructure evolution of ultrasonic-assisted thermal forming. The cellular automata simulation model of ultrasonic-assisted thermal forming is established, and the reasons for grain refinement and flow stress reduction are further discussed at the microscopic scale.

2. CA Model

2.1. Grain Growth

In the process of hot compression, when the holding temperature is above the austenitizing temperature, the grains undergo a grain growth process, so the CA model of grain growth is first established. In this paper, the cell size (L_{ca}) is 2 μm, and the simulated space size is 200 × 200; the actual simulated area is 0.4 cm × 0.4 cm. The neighbor type is a six-neighbor. The state variable of the cellular is grain orientation. The number of grain orientations ranged from 1 to 180. In the initial state, all cells in the cell space are given a random integer from 1 to 180 as orientation, which is taken as the initial state of the cellular space. There are three main rules for the transformation of the cellular state:

(1) Thermodynamic energy mechanism. The normal grain growth process is essentially a process of thermal activation. Grain boundary migration can occur only when grain boundary atoms overcome certain energy barriers. When the thermal energy of the cell at the grain boundary is greater than the activation energy of thermal diffusion, its state can change. According to statistical thermodynamic theory, the probability that the thermal energy of the cell at the grain boundary is greater than the activation energy of thermal diffusion can be calculated as follows [17]:

$$P_1 = A \cdot \exp(-\frac{Q_b}{RT}) \tag{1}$$

where A is the material parameter, which can be calculated from $P_1 = 1$ when the temperature reaches the melting point; T is the temperature (K); Q_b is the thermal diffusion activation energy; R is the molar gas constant; and the value is 8.31 J/(mol·K).

(2) Curvature drive mechanism. According to the orientation relationship and grain boundary type of neighbor cells, the current judging cell is regarded as the central cell. If four or more neighbor cells have the same state "A", while the central cell state is different from them, the central cell state will change into "A" state at the next time step.

If there are three neighboring cells in the same state "B", while the central cell is different, then the transition probability $P_2 = 0.4$ will be used to judge. If it is satisfied, the central cell state will change to a "B" state; otherwise, the state is unchanged.

If the currently judged cellular does not satisfy the curvature drive mechanism, the next criterion will be used to judge.

(3) The principle of least energy. According to the principle of energy dissipation, normal grain growth is a process in which the grain boundary expands outward and the grain boundary energy decreases. Since hexagonal cellular can simulate isotropy well, the calculated grain boundary energy also has isotropy, and the Hamiltonian function is used to represent the grain boundary energy [18]:

$$E_i = J\sum_{k}^{N}(1-\delta_{C_iC_k}) \tag{2}$$

where J is the grain boundary energy coefficient, the value is 1; N is the number of neighbors of the cellular, the value is 6; k is the neighbor of the current judged cell; δ is the Kronecher symbol; C_i is the current orientation of the cell; C_k is the orientation value of the neighbor cell of the current judgment cell.

Judging the 6 neighboring cells in random order, the change of grain boundary energy can be calculated as follows:

$$\Delta E_{i\to j} = E_j - E_i = J\sum_{k}^{N}\left(1-\delta_{C_jC_k}\right) - J\sum_{k}^{N}(1-\delta_{C_iC_k}) \tag{3}$$

Whether the state of the center cell changes to the state of the neighbor cell can be determined by judging whether the change in the grain boundary energy of the system is less than 0. If $\Delta E < 0$, the state of the center cell changes to the neighbor cell; If $\Delta E > 0$, the central cell state does not change.

In one time step, the grain boundary cells in the cell space are judged to determine whether the state changes according to the above order.

When all grain boundary cells in the cellular space are judged, update the state variables of each cell under this time step. Then, determine the grain boundary cells according to the orientation value, plot the microstructure image, and calculate the average grain size. Stop when the average grain size meets the set conditions; otherwise, the time step is added to one for the next round of judgment.

2.2. *Dynamic Recrystallization with Ultrasound*

The results obtained by the normal grain growth model are used as the initial cellular space state for the dynamic recrystallization CA simulation. In the simulated dynamic recrystallization process, the state variables of the cellular include grain orientation, dislocation density, number of recrystallizations, state variables reflecting the new and old frames, and state variables regarding whether the cell is located at the boundary.

2.2.1. Dislocation Density Model

To describe the change in dislocation density inside the material, the dislocation density evolution model proposed by KCOKS and MECKING is adopted, as shown in Equation (4) [19]:

$$\frac{d\rho}{d\varepsilon} = k_1\sqrt{\rho} - k_2\rho \tag{4}$$

where ρ is dislocation density; ε is strain; k_1 is the dislocation increment coefficient, k_2 is the dislocation extinction coefficient. The hardening index C can be obtained from the experimental data by Formula (5), and k_1 and k_2 can be obtained by combining Formulas (6) and (7):

$$\frac{d\sigma}{d\varepsilon} = C\left(1-\frac{\sigma}{\sigma_s}\right) \tag{5}$$

$$C = 0.5\mu b k_1/2 \tag{6}$$

$$\sigma_s = 0.5\mu b(k_1/k_2) \tag{7}$$

The flow stress can be calculated using Equation (8):

$$\sigma = \alpha\mu b\sqrt{\overline{\rho}} \tag{8}$$

where σ is the flow stress; α is the material constant, the value is 0.5; μ is the shear modulus; $\overline{\rho}$ is the average dislocation density. The initial dislocation density ρ_0 in the process of thermal simulation can be obtained by substituting the initial yield stress σ_0 into Equation (8).

The time step (Δt) can be calculated as follows [20]:

$$\Delta t = \frac{2{k_1}^2 Lca}{\mu b^2 M {k_2}^2} \tag{9}$$

At each time step, the strain and dislocation density is calculated according to the following equation:

$$\varepsilon^{t+\Delta t} = \varepsilon^t + \dot{\varepsilon}\Delta t \tag{10}$$

$$\rho^{t+\Delta t} = \rho^t + \Delta\rho \tag{11}$$

where $\dot{\varepsilon}$ is the strain rate.

With the increase in strain, the dislocation density increases, and when it reaches the critical dislocation density, dynamic recrystallization grain nucleation occurs at the grain boundary. The critical dislocation density (ρ_{cr}) can be calculated as follows [21]:

$$\rho_{cr} = \left(\frac{20\gamma_i\dot{\varepsilon}}{3blM\tau^2}\right)^{1/3} \tag{12}$$

where γ_i is the grain boundary energy, which can be obtained by Equation (13); M is the grain boundary migration energy, which can be obtained by Equation (15); τ is the dislocation line energy, which can be calculated by Equation (16); and l is the line dislocation energy, which can be obtained by Equation (17).

$$\gamma_i = \begin{cases} \gamma_m, \theta_i \geq 15^\circ \\ \gamma_m \frac{\theta_i}{\theta_m}\left(1 - ln\left(\frac{\theta_i}{\theta_m}\right)\right), \theta_i \leq 15^\circ \end{cases} \tag{13}$$

$$\gamma_m = \frac{\mu b \theta_m}{4\pi(1-v)} \tag{14}$$

where γ_m is the grain boundary energy of large angle grains; θ_m is the orientation angle of large Angle grains; θ_i is the orientation difference; v is Poisson's ratio.

$$M = \frac{D_{ob}b}{k_2 T} e^{-\frac{Q_b}{RT}} \tag{15}$$

$$\tau = 0.5\mu b^2 \tag{16}$$

$$l = \frac{k_1 \mu b}{\sigma} \tag{17}$$

where R is the gas constant; Q_b is the diffusion activation energy of grain boundaries; b is the Berkovian vector; μ is the shear modulus; D_{ob} is the self-diffusion coefficient.

To study the reasons for grain refinement caused by ultrasonic vibration, a large number of scholars have made explanations through experimental phenomena and metallographic analysis. The widely accepted explanation is that ultrasonic vibration improves the nucleation rate and promotes dislocation movement. Ultrasonic vibration acts inside the material, and the defect is easy to absorb ultrasonic energy, so the activation energy required for dislocation movement is reduced after ultrasonic vibration is applied. According to this principle, the influence of ultrasonic was introduced into the thermal activation model, and the energy reduced by ultrasonic was subtracted from the thermal activation energy to obtain the thermal activation model under ultrasonic vibration. Since ultrasonic energy cannot be completely absorbed by the material defect, correction factor C is introduced. The activation energy under ultrasonic vibration can be obtained from the following equation:

$$Q_b' = Q_b - C_2 E_u \tag{18}$$

where Q'_b is the activation energy after applying ultrasonic vibration; C_2 is the correction coefficient, which can be obtained by fitting the results of the ultrasonic-assisted tensile experiment, and is 0.4 in this paper; E_u is ultrasonic energy, which can be calculated by Formula (19):

$$E_u = A^2\omega^2 \rho_1 \tag{19}$$

where A is the ultrasonic vibration amplitude; ω is the vibration circle frequency, and $\omega = 2\pi f$, where f represents the ultrasonic vibration frequency, ρ_1 is the density.

After adjusting the grain boundary migration energy, the expression of the grain boundary migration energy under ultrasonic vibration is obtained as:

$$M_u = \frac{D_{ob}b}{k_2 T} e^{-\frac{Q'_b}{RT}} \tag{20}$$

According to the relevant experimental results of ultrasonic-assisted stretching, the true stress–strain curve after the application of ultrasound is not completely parallel to that without the application of ultrasound, which indicates that the internal structure or microstructure of the metal material changes after the application of ultrasonic vibration. Therefore, the parameters in the dislocation density model were adjusted to obtain the dislocation density model under applied ultrasonic vibration.

Modify the parameter k_2 of dynamic recovery part in the dislocation density model in Equation (4), which is the parameter k_{2u} of the dynamic recovery part under ultrasonic vibration, the calculation formula is:

$$k_{2u} = k_2 \zeta A^{\eta} \tag{21}$$

The ζ and η can be obtained by experimental fitting.

When the strain increases to a certain value, the dislocation density of the material becomes stable, and the dislocation density becomes saturated. The saturation dislocation density (ρ_s) can be obtained from Equation (22).

$$\rho_s = \left(\frac{k_1}{k_2}\right)^2 \tag{22}$$

2.2.2. Dynamic Recrystallization Nucleation and Growth Model

The nucleation model associated with dislocation density was used for the simulation. The nucleation model connects the nucleation rate with dislocation density, as shown in Equations (23) and (24). Combining with the dislocation density model, it can be seen that dislocation density is related to strain, so the nucleation rate changes with strain when the temperature and strain rate are constant.

$$P_N = \frac{\dot{N}_l \Delta t}{N_{CA}} = \dot{N}_l \cdot L_{CA} \cdot \Delta t \tag{23}$$

$$\dot{N}_l = \frac{\sqrt{\rho} - \sqrt{\rho_{cr}}}{\sqrt{\rho_s} - \sqrt{\rho_{cr}}} \tag{24}$$

where P_N is the nucleation probability; $\dot{N}_l$ is the grain boundary nucleation rate, which represents the number of new grains formed per unit length of grain boundaries per unit time when the dislocation density exceeds the critical value.

Dynamic recrystallization can occur many times, and the recrystallization times of newly generated recrystallized grains are increased by one. Recrystallized grains with large recrystallization times will grow into non-recrystallization grains and low-order recrystallized grains.

At each time step, non-recrystallization cells or low-order recrystallized grains located at the grain boundary are taken as the target cells, and the growth distance (L) of the recrystallized grains to the target cells is calculated as follows:

$$L = Vt \tag{25}$$

where V is the growth rate, and the calculation formula is:

$$V = M_u P \tag{26}$$

where P is the driving force and can be calculated as follows [22]:

$$P = \frac{\mu b^2(\rho_m - \rho_{rex})}{2} - \frac{2\gamma_i}{r_i} \tag{27}$$

where ρ_m is the dislocation density of adjacent cells; ρ_{rex} is the dislocation density of recrystallized grains respectively; r_i is the radius of the recrystallization grain.

If the maximum growth distance (L_{max}) in a neighbor cell is larger than a cell size L_{CA}, the target cell will be transformed into the recrystallized cell with a probability of $P_3 = a/6$, where a is the number of cells with the same orientation number among the six neighboring cells of the cell. When the transition is successful, the state of the target cell will be consistent with that of the recrystallized cell, and its dislocation density, recrystallization number, and orientation number will be the same as that of the recrystallized cell.

2.3. Program Flow Chart

The simulation process of dynamic recrystallization is shown in Figure 1.

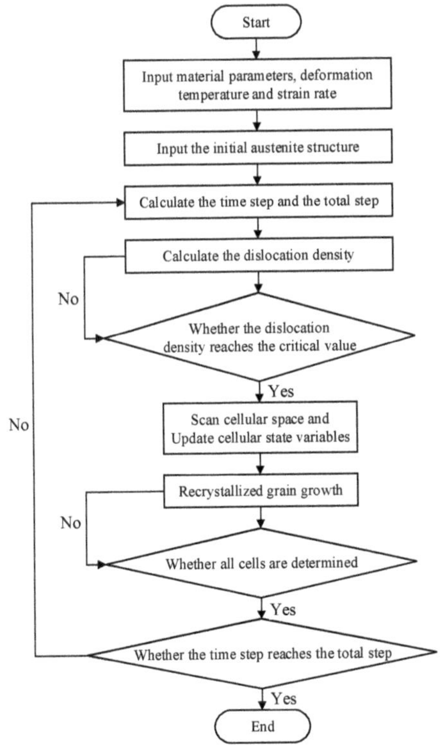

Figure 1. Program flow chart of dynamic recrystallization.

3. Experiment

3.1. *Hot Compression Experiment*

A Gleeble 3180 thermal simulation testing machine was used to conduct a thermal simulation test to obtain the true stress–strain curve. Gleeble 3180 is produced by Dynamic System Inc. in the US. A schematic diagram of hot compression is shown in Figure 2a. Experimental specimens before and after hot compression are shown in Figure 2b.

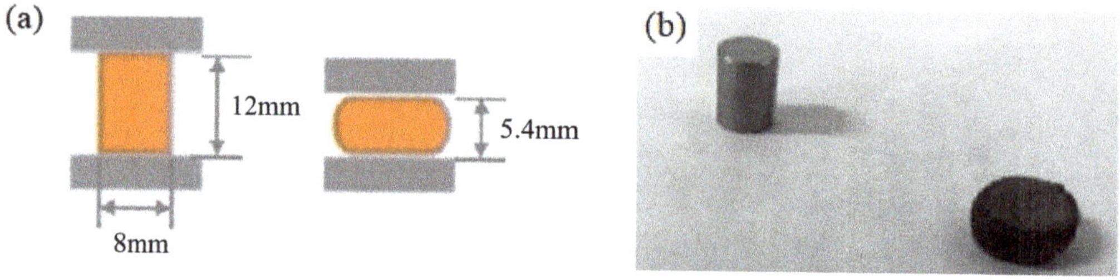

Figure 2. Experimental Equipment: (**a**) Schematic diagram of hot compression; (**b**) Samples before and after compression.

The thermal simulation sample is a cylinder with a diameter of 8 mm and a height of 12 mm. The test material was 9310 steel, and its chemical composition is shown in Table 1.

Table 1. Chemical composition of steel in 9310 (mass fraction, %).

C	Si	Mn	S	P	Cr	Ni	Mo	Cu
0.08	0.20	0.58	<0.015	0.0026	1.31	3.33	0.10	0.015

The first step of the thermal simulation experiment is to heat the 9310 steel cylindrical sample to the deformation temperature at a rate of 5 °C/s. After keeping the deformation temperature for 1 min, let the sample be heated evenly in the heating furnace. Keep the deformation temperature unchanged and compress at a constant strain rate. After finally reaching the set compression amount, the sample is immediately put into water for water quenching to retain the high-temperature tissue of the sample. The deformation temperatures are 900 °C, 1000 °C, and 1100 °C, and the strain rate are $0.01\ s^{-1}$, $0.1\ s^{-1}$, $1\ s^{-1}$.

3.2. *Ultrasonic-Assisted Tensile Experiments*

The ultrasonic-assisted unidirectional tensile experiment was performed on the T-table testing machine MTS 322 which is produced by MTS Systems Company Limited in the US, and the experimental device is shown in Figure 3. During the experiment, the T-type testing machine workbench clamps the lower collet and vise clamps the upper collet. The vise moves up at a certain stretching speed, and the sample is gradually stretched. At the same time, certain vibration frequencies and the amplitude of the workpiece are transferred by the transducer and amplitude rod to realize ultrasonic vibration stretching. Finally, the force and displacement curves under different stretching velocities, amplitudes, and ultrasonic durations were obtained using force sensors and displacement sensors on the test machine.

Figure 3. Ultrasonic-assisted tensile test device.

The tensile samples were designed based on the configuration of the ultrasonic-assisted tensile device and the design standard GB/T 2281.1—2010. The tensile specimen is a rod-length specimen with a diameter of 5 mm and a threaded connection at both ends, as shown in Figure 4. The ultrasonic vibration frequency of the ultrasonic-assisted unidirectional tensile experiment is 20 kHz, and the experimental scheme is shown in the Table 2:

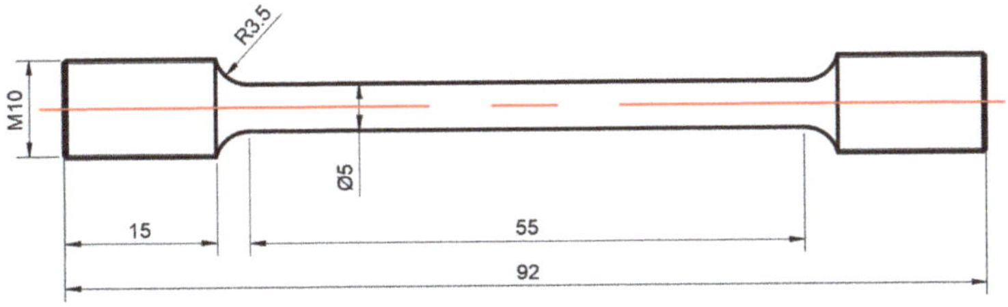

Figure 4. Sample.

Table 2. Ultrasonic-assisted tensile test scheme.

Ultrasound Amplitude/μm	**6.62**	**7.78**	**8.84**	**9.96**
Strain rate/s^{-1}	0.0003	0.003	0.03	0.3

4. Results and Discussion

4.1. Hot Compression Experimental Results

The flow stress curves at different deformation temperatures and strain rates are obtained by hot compression, as shown in Figure 5.

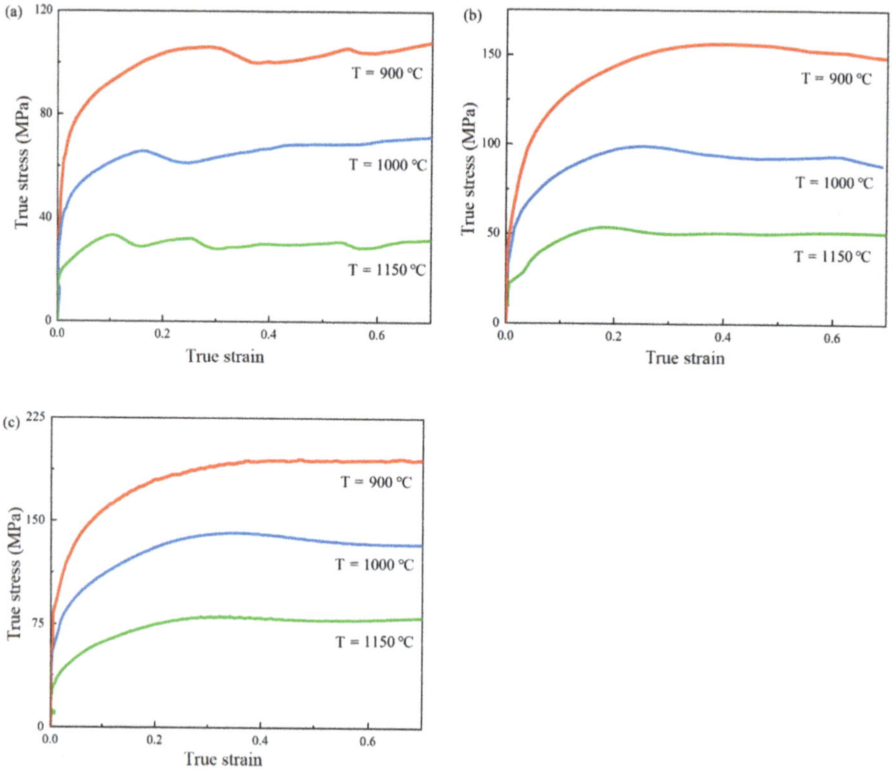

Figure 5. Flow stress curve of 9310 steel. (**a**) $\dot{\varepsilon} = 0.01\ s^{-1}$; (**b**) $\dot{\varepsilon} = 0.1\ s^{-1}$; (**c**) $\dot{\varepsilon} = 1\ s^{-1}$.

It can be seen that the deformation temperature and strain rate have an obvious influence on the flow stress of 9310 steel. With an increase in deformation temperature and a decrease in strain rate, the flow stress decreases. The true stress–strain curve has the characteristics of peak stress, strain softening, and steady flow.

At large strain rates, the flow stress curve first increases and then tends to be stable without single or double peaks. When the strain is small, the flow stress curve increases because of the effect of work hardening. During this period, dislocation multiplication, dislocation density, and flow stress increase. Then, the flow stress increases and slows down because of the softening effect. When the deformation energy storage caused by dislocation multiplication increases to a certain extent, dislocation cells and subgrain boundaries are formed, and softening is mainly based on dynamic recovery. Finally, the steady state is caused by the counterbalance between work hardening and softening and the balance between the multiplication and recovery of dislocation.

When the strain rate decreases, the flow stress curve first increases and then decreases to a steady state with the characteristic of a "single peak." When the strain rate continued to decrease, the true stress–strain curve appeared to be a "multi-peak" phenomenon. The "single peak" characteristic of the flow stress curve is that material softening dominates after the flow stress reaches its maximum value, and the material is significantly softened. At this time, softening has two behaviors: dynamic recovery and recrystallization. Under high temperature and low strain rates, softening is dominant, dynamic recrystallization occurs, the peak stress is significantly reduced, and the softening phenomenon is significant. Since the deformation temperature exceeds the recrystallization temperature, dislocation multiplication at the early stage of deformation promotes the occurrence of recrystallization. When the dislocation proliferates to a certain extent, dynamic recrystallization occurs, and recrystallized grains are generated. The dislocations in the material were reduced, and the flow stress curve was reduced. Finally, the rate of dislocation multiplication and annihilation reaches equilibrium, and the flow stress enters the steady-state stage. The occurrence of "multi-peak" characteristics means that there is continuous dynamic recrystallization in the process.

4.2. Ultrasonic-Assisted Tensile Experiment Results

The data obtained from the results of the ultrasonic-assisted tensile experiment are shown in Figure 6.

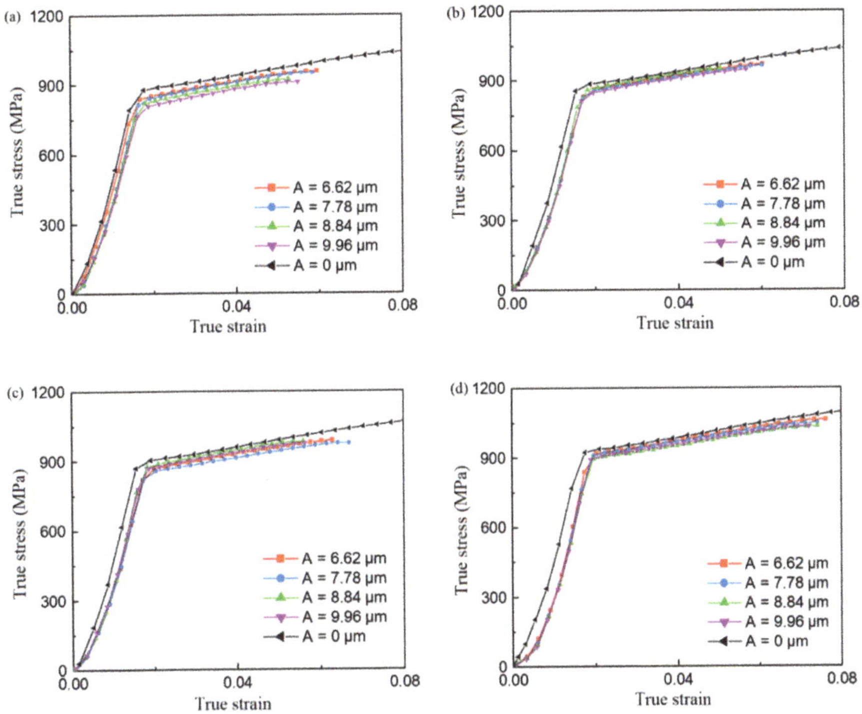

Figure 6. True stress–strain curve (**a**) $\dot{\varepsilon} = 0.0003\ s^{-1}$ (**b**) $\dot{\varepsilon} = 0.003\ s^{-1}$ (**c**) $\dot{\varepsilon} = 0.03\ s^{-1}$ (**d**) $\dot{\varepsilon} = 0.3\ s^{-1}$.

The yield strength, tensile strength, and fracture strength were extracted from the experimental results, and the amplitude of the stress decrease caused by ultrasound was calculated using Equation (28). λ is the average value of the ratio of stress decrease caused by ultrasound at the stage of uniform plastic deformation.

$$\lambda = \frac{\Delta\sigma}{\sigma_C} = \frac{\sigma_C - \sigma_U}{\sigma_C} \tag{28}$$

where σ_C is the flow stress under ordinary tension, and σ_U is the flow stress under ultrasonic-assisted tension.

Figure 7 shows the yield strength, fracture strength, tensile strength, and stress decrease amplitude under different ultrasonic amplitudes and without ultrasonic when the strain rate is 0.0003. It can be seen that ultrasonic treatment effectively reduces yield strength, fracture strength, and tensile strength. With an increase in amplitude, the stress decreases, and the amplitude of stress decreases with an increase in ultrasonic amplitude, and the relationship is linear.

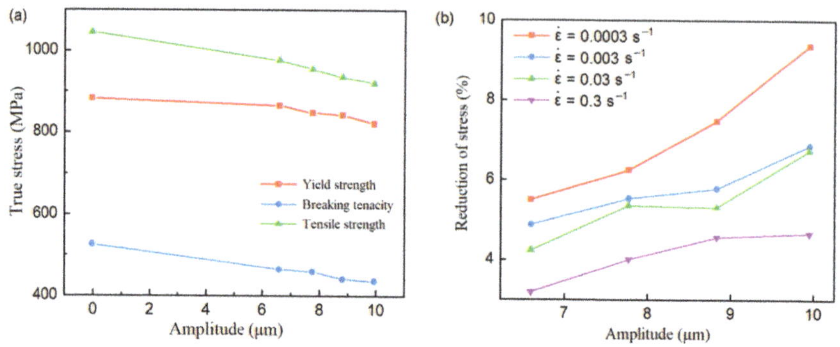

Figure 7. The variation curve of the stress and the stress reduction amplitude with the amplitude. (**a**) Stresses at different amplitudes. (**b**) Stress reduction amplitude at different amplitudes.

Figure 8 shows the variation curve of true stress and stress reduction amplitude with strain rate. It can be seen that all stress values increase with the increase in strain rate. At room temperature, the softening effect of ultrasound decreases with an increase in the strain rate. Thermal activation at room temperature is not sufficient to cause dislocation migration, and work hardening at high strain rates is more significant than that at low strain rates. Therefore, the flow stress at a high strain rate is higher than that at a low strain rate. At high strain rates, ultrasonic vibration can promote more dislocation migration, and higher work hardening can increase flow stress and reduce the ultrasonic softening effect. With the increase in strain rate, the softening effect coefficient of ultrasonic showed a nonlinear decreasing trend.

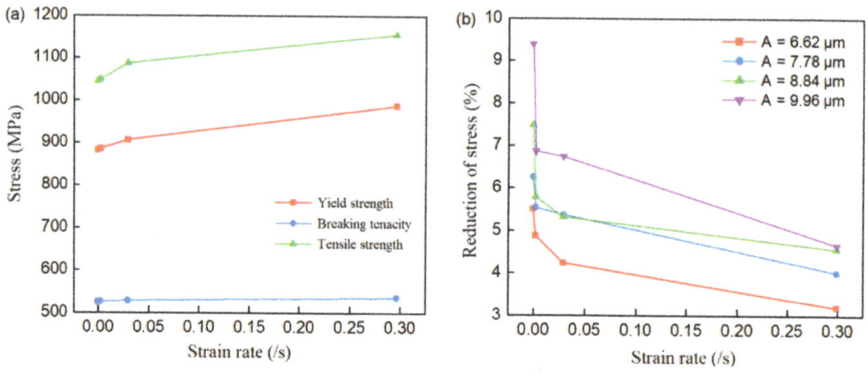

Figure 8. Variation curve of stress values and stress reduction amplitude with strain rate. (**a**) Stresses at different strain rates. (**b**) Stress reduction amplitude at different strain rates.

4.3. Stress Reduction Amplitude

Combined with the theoretical model of ultrasonic softening and the CA model of the metal hot-forming process, the microstructure of ultrasonic-assisted hot forming was simulated. The true stress–strain curves obtained by the CA simulation of ultrasound-assisted thermoforming under different strain rates are shown in Figure 9. The comparison of stress reduction amplitude between the CA simulation and ultrasonic tensile experiment is shown in Figure 10, and the relative errors are shown in Table 3. It can be seen from the figure that the stress reduction amplitude of the ultrasonic-assisted thermal forming simulation and ultrasonic tensile experiment is close. There is a strong linear relationship between them, and the linear correlation coefficient (R^2) is 0.99. The maximum error obtained is 11.60%, and the overall average error is 5.66%. This indicates that stress reduction occurs in the process of compression and tension under the action of ultrasound, and it is found that the amplitude of stress reduction is similar under the same ultrasonic amplitude. It can be seen that when the strain rate is high, the differences between the simulation results and the experimental results are larger than the small strain rate. This is because when the strain rate is high, the strain variable in one time step of the CA simulation will increase, which will lead to the decrease of the total simulation steps, and the decrease of the simulation accuracy.

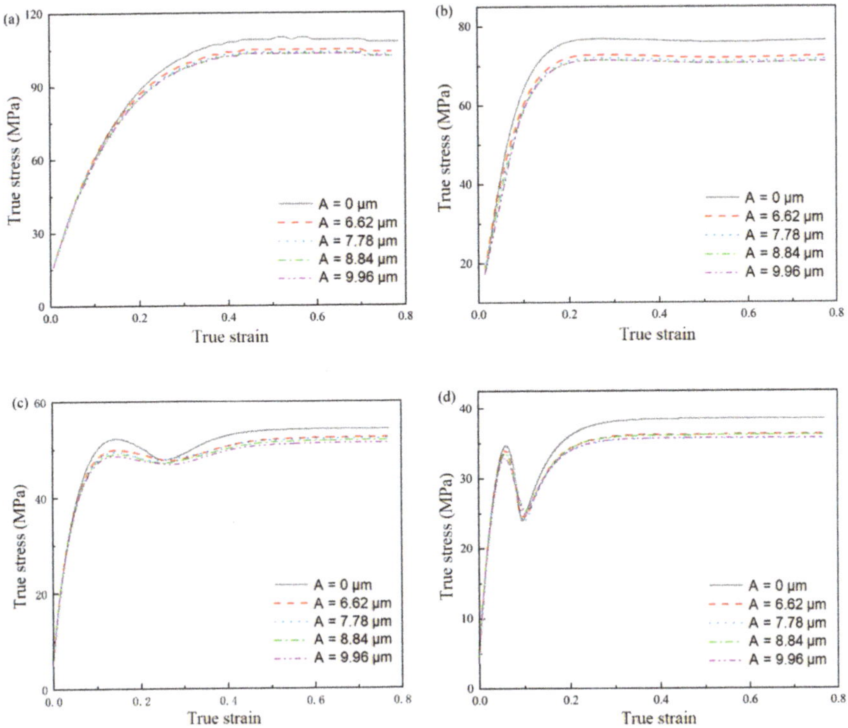

Figure 9. Simulated stress–strain curves at 1000 °C (**a**) $\dot{\varepsilon} = 0.3\ s^{-1}$ (**b**) $\dot{\varepsilon} = 0.03\ s^{-1}$ (**c**) $\dot{\varepsilon} = 0.003\ s^{-1}$ (**d**) $\dot{\varepsilon} = 0.0003\ s^{-1}$.

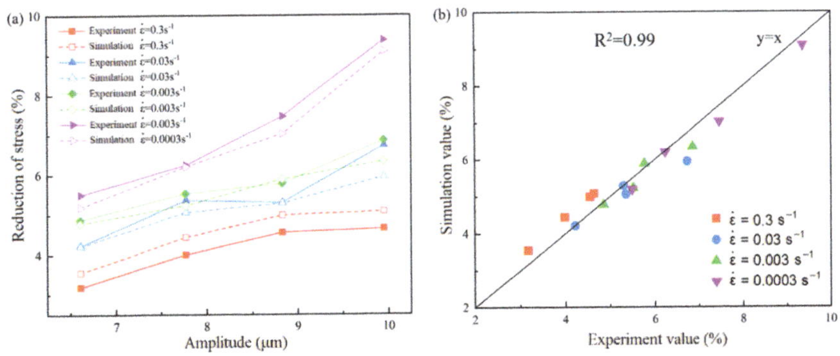

Figure 10. Comparison between simulated and experimental values of stress reduction amplitude. (**a**) Experimental and simulated values for stress reduction amplitude. (**b**) Linear correlation between experimental values and simulated values.

Table 3. Error analysis of stress reduction amplitude.

Strain Rate (s^{-1}).	Ultrasound Amplitude (μm).	Experimental Value (%)	Simulated Value (%)	Relative Error (%)
0.0003	6.62	5.51	5.20	5.57
0.0003	7.78	6.25	6.21	0.73
0.0003	8.84	7.48	7.04	5.97
0.0003	9.96	9.38	9.10	2.99
0.003	6.62	4.88	4.79	1.88
0.003	7.78	5.54	5.24	5.33
0.003	8.84	5.79	5.90	1.79
0.003	9.96	6.87	6.35	7.53
0.03	6.62	4.24	4.22	0.48
0.03	7.78	5.37	5.07	5.53
0.03	8.84	5.32	5.30	0.34
0.03	9.96	6.74	5.96	11.60
0.3	6.62	3.19	3.55	11.07
0.3	7.78	4.01	4.45	11.04
0.3	8.84	4.57	5.00	9.50
0.3	9.96	4.66	5.09	9.22

4.4. CA Simulation Results

4.4.1. Grain Growth Results

In order to study the evolution of microstructure in the process of grain growth, the variation of grain number with time step at different temperatures is shown in Figure 11a. The relationship between average grain size and simulation step size at different temperatures is shown in Figure 11b.

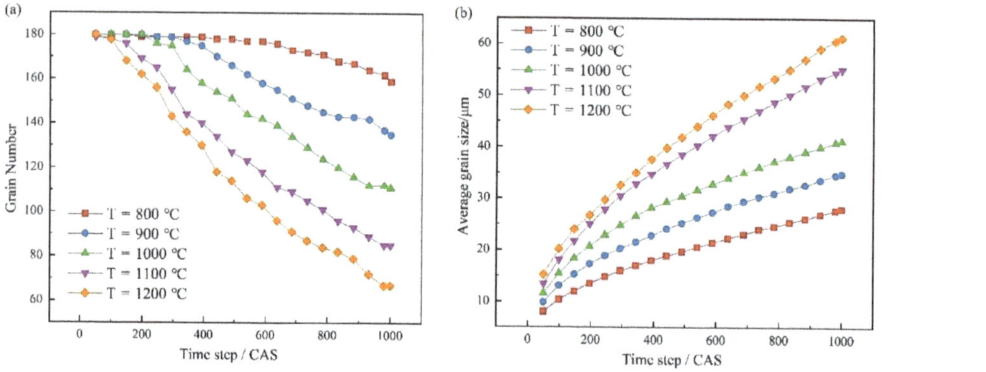

Figure 11. Grain growth results at different temperatures. (**a**) Variation in grain number. (**b**) Variation of average grain size.

It can be seen from Figure 11a that, at the same temperature, the number of grains decreases when the time step increases. At the same time, the number of grains decreases when the temperature increases. This is because grain boundary energy is the driving force of grain growth and can promote grain proliferation. When the grains grow, the grain boundary length per unit area decreases due to a decrease in grain number, which makes the total free energy in the system decrease, and the system is easier to achieve stability.

It can be seen from Figure 11b that under the same deformation temperature, the average grain size increases with an increase in time step, but the growth rate becomes slower. At the same time, the average grain size is larger with a higher deformation temperature. This is because heat can provide energy for grain boundary migration, and grains with higher deformation temperatures grow faster. However, as the grain grows, the driving force of grain boundary migration becomes smaller and smaller, and the growth rate gradually flattens.

4.4.2. Effect of Initial Grain Size

The initial grain size of dynamic recrystallization was set as 30 μm, 50 μm, 70 μm, 90 μm and 110 μm, deformation temperature is 1000 °C, the strain rate is 0.1 s^{-1}, and the curve of average grain size and recrystallization volume fraction with the original grain size is shown in Figure 12. The microstructure image obtained by simulation is shown in Figure 13. The black part represents the grain boundary, the white part represents the initial structure, and the color part represents the grain with dynamic recrystallization.

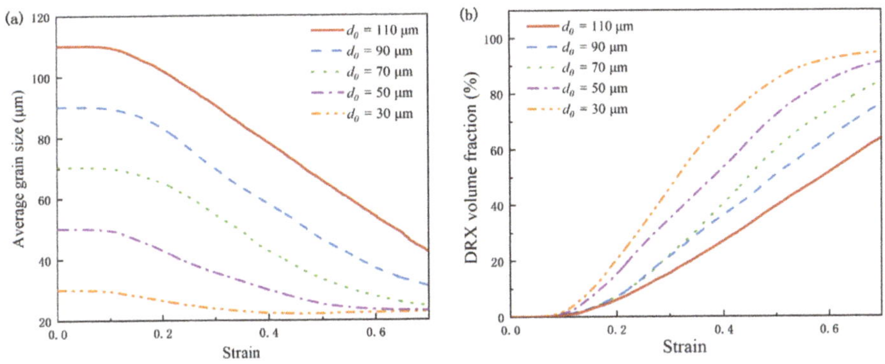

Figure 12. Effect of initial grain size. (**a**) Effect on average grain size. (**b**) Effect on recrystallization volume fraction.

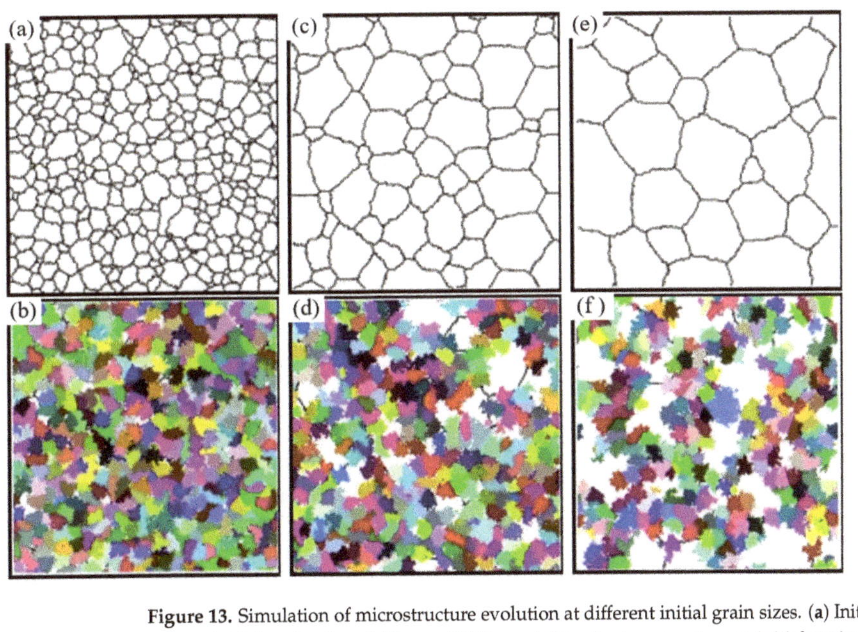

Figure 13. Simulation of microstructure evolution at different initial grain sizes. (**a**) Initial microstructure at initial grain size of 30 μm. (**b**) Recrystallized microstructure at initial grain size of 30 μm. (**c**) Initial microstructure at initial grain size of 70 μm. (**d**) Recrystallized microstructure at initial grain size of 70 μm. (**e**) Initial microstructure at initial grain size of 110 μm. (**f**) Recrystallized microstructure at initial grain size of 110 μm.

It can be seen that with the decrease in the initial grain size, faster dynamic recrystallization occurs, and the faster it reaches stability, the volume fraction of dynamic recrystallization increases, the average grain size decreases, and the more grains have dynamic recrystallization. With the increase of strain, the average grain size first stays constant, then decreases to stable, and the volume fraction of dynamic recrystallization remained constant at first, then increased to stable. This indicates that no dynamic recrystallization occurs when deformation is small. With an increase in deformation, the dislocation density increases. When the dislocation density increases to a certain degree, the dynamic recrystallized grains are first nucleated at the grain boundaries, with a higher dislocation density. Thus, dislocation density and stress are reduced. When the dynamic recrystallized grain grows to a certain extent and the dislocation density reaches saturation, it no longer grows. The manifestation is stress stabilization, which is the softening effect of dynamic recrystallization behavior.

4.4.3. Effects of Temperature and Strain Rate

Figure 14 shows the simulated strain curve of the dynamic recrystallization volume fraction at different temperatures when the strain rate is 0.1 s^{-1}. With the increase in strain, the volume fraction of dynamic recrystallization first remained constant and then increased to stable. It can be seen that dynamic recrystallization easily occurs at high temperatures. Under the same strain, the volume fraction of dynamic recrystallization increases with an increase in temperature. The higher the temperature, the earlier the dynamic recrystallization time, and the shorter the stable time. This is because the growth rate of dislocations is temperature dependent. The increase in temperature leads to the rapid growth of dislocation, the short incubation time of new cells, and the rapid formation of recrystallized grains, so the volume fraction of dynamic recrystallization increases.

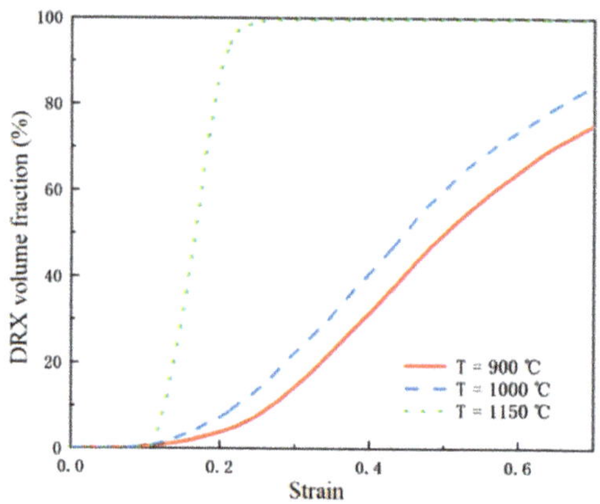

Figure 14. Simulated curve of the dynamic recrystallization volume fraction at different temperatures.

At a temperature of 1000 °C and a vibration amplitude of 7.78 μm, the variation curves of the average grain size and recrystallization volume fraction obtained by simulation with the true strain are shown in Figure 15, and the microstructure images obtained by simulation are shown in Figure 16.

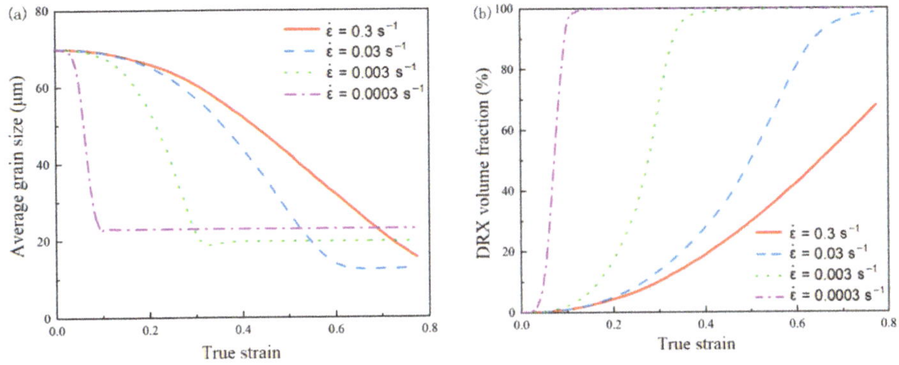

Figure 15. The simulation results at a temperature of 1000 °C and a vibration amplitude of 7.78 μm. (**a**) The curve of the average grain size with the true strain. (**b**) The curve of the recrystallization volume fraction with the true strain.

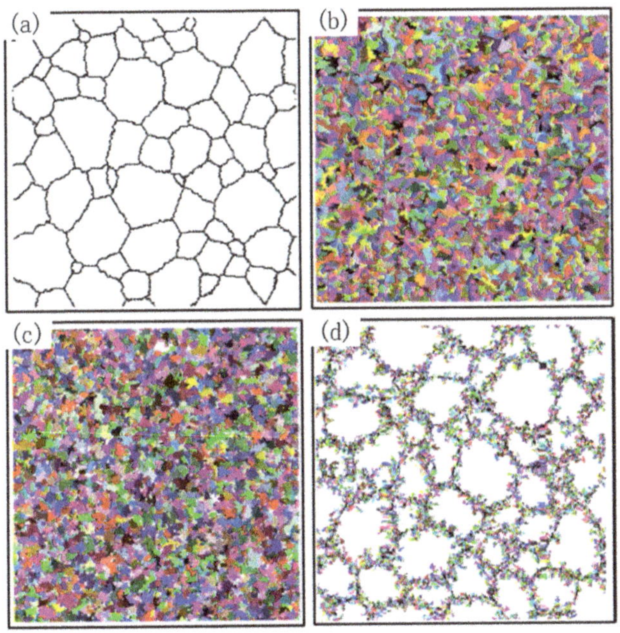

Figure 16. Initial microstructure and simulated microstructure results at different strain rates. (**a**) Initial microstructure. (**b**) $\dot{\varepsilon} = 0.003\ \mathrm{s}^{-1}$ (**c**) $\dot{\varepsilon} = 0.03\ \mathrm{s}^{-1}$ (**d**) $\dot{\varepsilon} = 0.3\ \mathrm{s}^{-1}$.

It can be seen that when the strain rate decreases, the average grain size decreases to stable faster, the final recrystallized grain size is larger, the recrystallized volume fraction increases faster, and the number of recrystallized grains in the final structure is larger.

At a strain rate of 0.3 s^{-1}, the recrystallized grains show a chain-like development. This is because the nucleation rate is dependent on the strain rate. When the strain rate is high, the nucleation probability also increases, and more recrystallization is generated, which makes it easy to contact each other and stop growth. When the strain rate is large, the grain does not have enough time to grow, and the dynamic recrystallization is not sufficient, so the final average grain size is large, and the recrystallization volume fraction is small.

At a temperature of 1000 °C and a vibration amplitude of 7.78 μm, the simulated true stress–strain curves under different strain rates are shown in Figure 17. It can be seen that true stress decreases with a decrease in the strain rate. In addition, when the strain rates are 0.3 s^{-1} and 0.03 s^{-1}, the true stress–strain curves show the characteristics of dynamic recovery, which indicates that material softening mainly depends on dynamic recovery. When the strain rate is small, the true stress–strain curve presents a "single peak" curve characteristic. When the strain rate decreases, the "peak" appears earlier, which indicates that dynamic recrystallization behavior is more likely to occur when the strain rate is small, which is consistent with the test results.

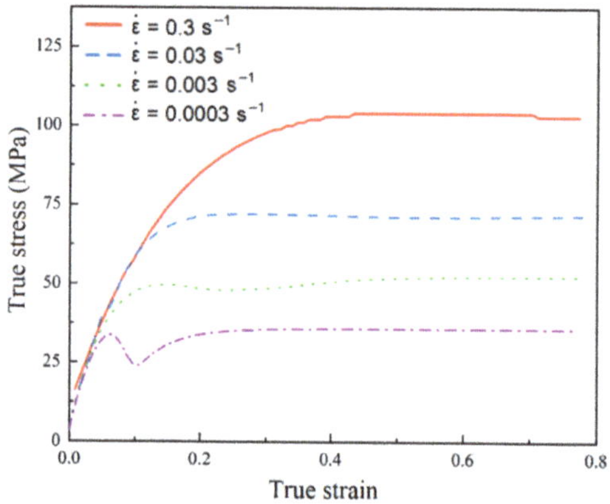

Figure 17. Simulated stress–strain curve at a temperature of 1000 °C and a vibration amplitude of 7.78 μm.

4.4.4. Effect of Ultrasonic Vibration Amplitude

At a strain rate of 0.003 s^{-1} and a temperature of 1000 °C, the simulated true stress–strain curve with vibration amplitude is shown in Figure 18. It can be seen that under different vibration amplitudes, the true stress decreases with the increase in vibration amplitude, the material is softened to different degrees, and the ultrasonic softening degree increases with the increase in vibration amplitude.

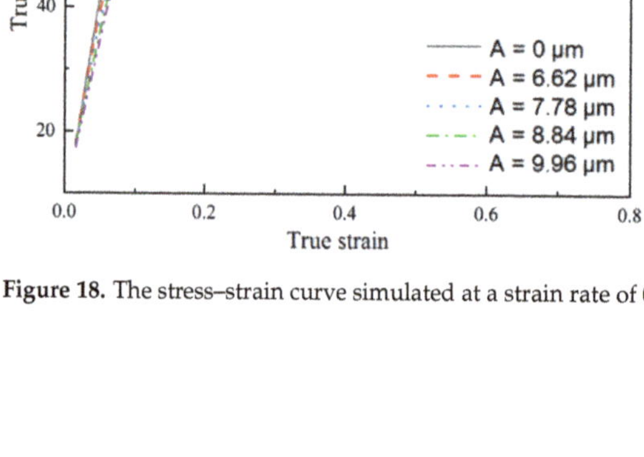

Figure 18. The stress–strain curve simulated at a strain rate of 0.03 s^{-1} and a temperature of 1000 °C.

At a strain rate of 0.003 s^{-1} and a temperature of 1000 °C, the variation curves of average grain size and recrystallization volume fraction with vibration amplitude obtained by simulation are shown in Figure 19, and the microstructure image is shown in Figure 20.

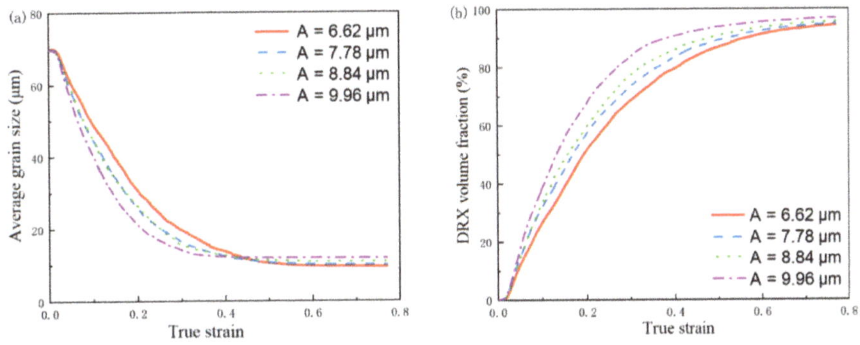

Figure 19. The simulation results at a strain rate of 0.03 s^{-1} and a temperature of 1000 °C. (**a**) The curve of the average grain size with the true strain. (**b**) The curve of the recrystallization volume fraction with the true strain.

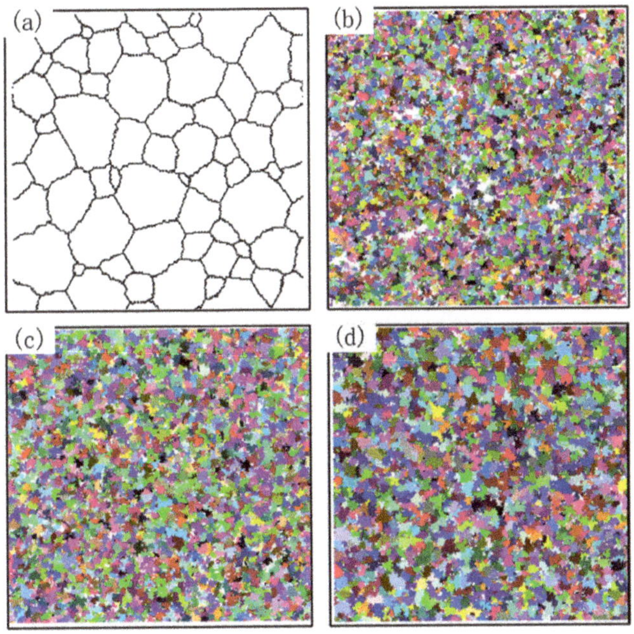

Figure 20. Initial microstructure and simulated microstructure results at different amplitudes. (**a**) Initial microstructure. (**b**) Simulated microstructure result when A = 0. (**c**) Simulated microstructure result when A = 6.62 µm. (**d**) Simulated microstructure result when A = 9.96 µm.

Figure 19a shows that at the early stage of dynamic recrystallization, when the ultrasonic vibration amplitude increases, the average grain size decreases faster, and the average grain size curve reaches stability faster. In the late stage of dynamic recrystallization, the average grain size increases with an increase in vibration amplitude. It can be seen from Figure 19b that with the increase in ultrasonic vibration amplitude, the earlier dynamic recrystallization occurs, the more grains under dynamic recrystallization. Figure 20 shows that the number of recrystallized grains and the degree of recrystallization increase after ultrasonic vibration is applied. When the vibration amplitude is 9.96 μm, the dynamic recrystallization behavior occurs most thoroughly, and the recrystallization volume fraction reaches 99.6%.

This indicates that the introduction of ultrasonic energy reduces the activation energy of the dislocation motion. The increase of vibration amplitude and the increase of ultrasonic energy promote the occurrence of recrystallization, which leads to the faster decrease of average grain size and the faster increase of recrystallization volume fraction so that the grain is refined. Therefore, when the vibration amplitude is large, dynamic recrystallization occurs in advance, the recrystallization time is longer, the grain can grow fully, and the average grain size also increases.

5. Conclusions

In this paper, ultrasonic-assisted thermal forming technology is taken as the research object, based on hot compression experiments, ultrasonic tensile experiments, and cellular automaton simulations. The simulation model of ultrasonic auxiliary hot pier thickness is established based on ultrasonic-assisted thermal forming technology, and the reasons for the reduction of flow stress are explained from the microscopic scale. The main conclusions are as follows:

(1) Based on the dislocation density model, grain growth model, and dynamic recrystallization rule, ultrasonic energy was introduced to modify the thermal activation energy and dynamic recovery coefficient, and the CA model of microstructure simulation of the ultrasonic-assisted thermal forming process was established.
(2) Stress reduction occurs in the process of compression and tension under the action of ultrasound, and the stress reduction amplitude is similar under the same ultrasonic amplitude.
(3) The stress curves at different temperatures and strain rates were obtained by hot compression experiments, and it was found that the true stress–strain curves had the characteristics of peak stress, strain softening, and steady flow. At high temperatures and low strain rates, the softening phenomenon is more significant.
(4) After applying ultrasonic vibration, the introduction of ultrasonic energy reduces the activation energy of dislocation movement, promotes the occurrence of dynamic recrystallization, and finally leads to the reduction of flow stress and material softening. The softening effect is enhanced with an increase in ultrasonic vibration amplitude.
(5) Decreasing the strain rate and increasing the temperature or vibration amplitude will lead to a faster increase in the recrystallization volume fraction, and an increase in the dynamic recrystallization volume fraction and the average grain size.

Author Contributions: Formal analysis, Y.Z.; Funding acquisition, J.T.; Investigation, Y.Z.; Methodology, Y.H.; Project administration, Y.H.; Resources, W.Z.; Software, Y.Z.; Supervision, J.T.; Validation, W.Z.; Visualization, W.Z.; Writing—original draft, Y.Z.; Writing—review & editing, Y.H. All authors have read and agreed to the published version of the manuscript.

Funding: This work is supported by the National Key Defense Basic Scientific Research program of China (Grant No. 2021-JCJQ-ZD-072).

Institutional Review Board Statement: Not applicable.

Informed Consent Statement: Not applicable.

Data Availability Statement: The data presented in this study are available on request from the corresponding author.

Conflicts of Interest: The authors declare no conflict of interest.

Nomenclature

Symbol	Description	Symbol	Description
L_{ca}	Cell size	θ_m	Orientation angle of large Angle grains
A	Material parameter	θ_i	Orientation difference
T	Temperature	b	Berkovian vector
Qb	Thermal diffusion activation energy	Q'_b	Activation energy after applying ultrasonic vibration
R	Molar gas constant		
E_i	Grain boundary energy of austenite	C_2	Correction coefficient of ultrasonic energy
J	Grain boundary energy coefficient	E_u	Ultrasonic energy
N	Number of neighbors of the cellular	A	Ultrasonic vibration amplitude
K	Neighbor of the current judged cell	ω	Vibration circle frequency
δ	Kronecher symbol	f	Ultrasonic vibration frequency
C_i	Current orientation of the cell	ρ_1	Density
C_k	Orientation value of the neighbor cell of the current judgment cell	M_u	Grain boundary migration energy under ultrasonic vibration
ρ	Dislocation density	k_{2u}	Dislocation extinction coefficient under ultrasonic vibration
k_1	Dislocation increment coefficient		
k_2	Dislocation extinction coefficient	ζ	Correlation coefficient of dislocation extinction coefficient under ultrasonic vibration
C	Hardening index		
σ	Flow stress	η	Exponent of the ultrasonic vibration amplitude
α	Material constant of flow stress	ρ_s	Saturation dislocation density
μ	Shear modulus	P_N	Nucleation probability
$\bar{\rho}$	Average dislocation density	$\dot{N}_l$	Grain boundary nucleation rate
ρ_0	Initial dislocation density	L	Growth distance
σ_0	Initial yield stress	V	Growth rate
Δt	Time step	P	Driving force
$\dot{\varepsilon}$	Strain rate	ρ_m	Dislocation density of adjacent cells
γ_i	Grain boundary energy of recrystallization	ρ_{rex}	Dislocation density of recrystallized grains respectively
M	Grain boundary migration energy		
τ	Dislocation line energy	r_i	Radius of the recrystallization grain
D_{ob}	Self-diffusion coefficient	L_{max}	Maximum growth distance
γ_m	Grain boundary energy of large angle grains		

References

1. Xie, J.; Zhou, T.; Liu, Y.; Kuriyagawa, T.; Wang, X. Mechanism study on microgroove forming by ultrasonic vibration assisted hot pressing. *Precis. Eng.* **2016**, *46*, 270–277. [CrossRef]
2. Yang, B.; Ming, Y. Investigation on mechanism of metal foil surface finishing with vibration-assisted micro-forging. *J. Mater. Process. Technol.* **2013**, *213*, 330–336.
3. Zhou, W.; Tang, J.; Shao, W.; Wen, J. Towards understanding the ploughing friction mechanism in ultrasonic assisted grinding with single grain. *Int. J. Mech. Sci.* **2022**, *222*, 107248. [CrossRef]
4. Hu, J.; Shimizu, T.; Yoshino, T.; Shiratori, T.; Yang, M. Evolution of acoustic softening effect on ultrasonic-assisted micro/meso-compression behavior and microstructure. *Ultrasonics* **2020**, *107*, 106107. [CrossRef] [PubMed]
5. Mao, Q.; Coutris, N.; Rack, H.; Fadela, G.; Gibertc, J. Investigating ultrasound-induced acoustic softening in aluminum and its alloys. *Ultrasonics* **2020**, *102*, 106005. [CrossRef] [PubMed]
6. Wen, T.; Wei, L.; Chen, X.; Pei, C.L. Effects of ultrasonic vibration on plastic deformation of AZ31 during the tensile process. *Int. J. Miner. Metall. Mater.* **2011**, *18*, 70–76. [CrossRef]
7. Zhou, H.; Cui, H.; Qin, Q.H. Influence of ultrasonic vibration on the plasticity of metals during compression process. *J. Mater. Process. Technol.* **2018**, *251*, 146–159. [CrossRef]
8. Zhou, H.; Cui, H.; Qin, Q.H.; Wang, H.; Shen, Y. A comparative study of mechanical and microstructural characteristics of aluminium and titanium undergoing ultrasonic assisted compression testing. *Mater. Sci. Eng. A* **2017**, *682*, 376–388. [CrossRef]
9. Liu, T.; Lin, J.; Guan, Y.; Xie, Z.; Zhai, J. Effects of ultrasonic vibration on the compression of pure titanium. *Ultrasonics* **2018**, *89*, 26–33. [CrossRef] [PubMed]

10. Hung, J.-C.; Lin, C.-C. Investigations on the material property changes of ultrasonic-vibration assisted aluminum alloy upsetting. *Mater. Des.* **2013**, *45*, 412–420. [CrossRef]
11. Hung, J.-C.; Tsai, Y.-C. Investigation of the effects of ultrasonic vibration-assisted micro-upsetting on brass. *Mater. Sci. Eng. A* **2013**, *580*, 125–132. [CrossRef]
12. Zhao, X.; Zhao, S.; Han, Q. Effects of vibration direction on the mechanical behavior and microstructure of a metal sheet undergoing vibration-assisted uniaxial tension. *Mater. Sci. Eng. A* **2019**, *743*, 472–481.
13. Liu, Y.; Suslov, S.; Han, Q.; Xu, C.; Hua, L. Microstructure of the pure copper produced by upsetting with ultrasonic vibration. *Mater. Lett.* **2012**, *67*, 52–55. [CrossRef]
14. Wang, X.; Qi, Z.; Chen, W. Investigation of mechanical and microstructural characteristics of Ti–45Nb undergoing transversal ultrasonic vibration-assisted upsetting. *Mater. Sci. Eng. A* **2021**, *813*, 141169. [CrossRef]
15. Wang, X.; Qi, Z.; Chen, W. Investigation of Ti–45Nb alloy's mechanical and microscopic behaviors under transverse ultrasonic vibration-assisted compression. *Mater. Sci. Eng. A* **2022**, *832*, 142401. [CrossRef]
16. Von Neumann, J. *Papers of John Von Neumann on Computing and Computer Theory*; MIT Press: Cambridge, MA, USA, 1986.
17. Humphreys, F.J.; Hatherly, M. *Recryatailization and Related Annealing Phenomena*; Pergamon Press: Oxford, UK, 1995.
18. Duan, C.; Zhang, F.; Qin, S.; Sun, W.; Wang, M. Modeling of dynamic recrystallization in white layer in dry hard cutting by finite element—Cellular automaton method. *J. Mech. Sci. Technol.* **2018**, *32*, 4299–4312. [CrossRef]
19. Mecking, H.; Kocks, U.F. Kinetics of flow and strain-harding. *Acta Metall.* **1981**, *29*, 1865–1875. [CrossRef]
20. Zhang, T.; Lu, S.; Wu, Y.; Gong, H. Optimization of deformation parameters of dynamic recrystallization for 7055 aluminum alloy by cellular automaton. *Trans. Nonferrous Met. Soc. China* **2017**, *27*, 1327–1337. [CrossRef]
21. Yazdipour, N.; Davies, C.H.J.; Hodgson, P.D. Microstructural modeling of dynamic recrystallization using irregular cellular automata. *Comput. Mater. Sci.* **2008**, *44*, 566–576. [CrossRef]
22. Wang, S.; Xu, W.; Wu, H.; Yuan, R.; Jin, X.; Shan, D. Simulation of dynamic recrystallization of a magnesium alloy with a cellular automaton method coupled with adaptive activation energy and matrix deformation topology. *Manuf. Rev.* **2021**, *8*, 11. [CrossRef]

materials

Article

Influence of Mechanical Deformations on the Characteristic Impedance of Sewed Textile Signal Lines

Paweł Kubiak [1,*] and Jacek Leśnikowski [2]

1 Institute of Security Technologies "MORATEX", Marii Sklodowskiej-Curie 3 Street, 90-505 Lodz, Poland
2 Faculty of Material Technologies and Textile Design, Institute of Architecture of Textiles, Lodz University of Technology, Żeromskiego 116, 90-543 Lodz, Poland; jacek.lesnikowski@p.lodz.pl
* Correspondence: pkubiak@moratex.eu

Abstract: The following article describes a new type of textile signal line that can be used in smart clothing. The article presents the structure of this line and the materials used for its construction. The article also presents the results of research on the influence of the line tensile force on the value of its characteristic impedance. The above tests were carried out on lines where the electrically conductive paths do not have the form of straight lines, as is often the case in smart clothing. The article also presents a preliminary statistical analysis, the aim of which was to find those characteristics of the substrate of the line that affect changes in the characteristic impedance during stretching.

Keywords: textile signal lines; textronic systems; data transmission over textiles; smart textiles; characteristic impedance; signal quality tests; smart garments; e-textiles

Citation: Kubiak, P.; Leśnikowski, J. Influence of Mechanical Deformations on the Characteristic Impedance of Sewed Textile Signal Lines. *Materials* **2022**, *15*, 1149. https://doi.org/10.3390/ma15031149

Academic Editors: Lihong Su, Peitang Wei, Hui Wang, Xing Zhao and John T. Kiwi

Received: 10 November 2021
Accepted: 29 January 2022
Published: 2 February 2022

1. Introduction

The contemporary textile industry creates many new, unprecedented fields of functionality and opportunities for defined users, by combining traditional textile elements with other products from the field of sciences which are at first glance unrelated to textiles. By combining elements of textiles with electronics and information technology, it is possible to design and manufacture materials with innovative properties and functionalities. The combination of knowledge from these three areas was defined at the Lodz University of Technology in 2003 as a new discipline of knowledge called textronics [1].

Innovative materials, such as intelligent textiles or functional textronic systems are increasingly used in military, specialized or medical technologies. These products are also used in everyday, casual items, such as sportswear, or textiles with worn electronics.

Nowadays, there is a growing demand for innovative textiles with possibilities and applications far different from traditional ones. One of the main reasons for this is the increasing consumer's awareness, who want modern and fashionable textile products with unusual functions [2]. In consequence, research is focused on innovative materials and solutions increasing the functionality of textile products. Another field of research interest now is the integrated textronic systems combining traditional textiles with electronic elements. These elements allow the processing and transmission of data by electrical signals, their acquisition and management. Also referred to as smart textiles or e-textiles, these systems have a wide range of applications. They can be used, for example, to produce clothing with human physiological parameters monitoring systems [3–6]. Such clothing can be used in monitoring people working in hazardous conditions [7–9], the elderly [10,11] or the chronically ill [12]. Also, textronic clothing used for monitoring the vital parameters of newborns [13,14] is being developed. Textronic applications can also be used in functional clothing for sports [15,16] or clothing for casual users.

Electrically conductive textile materials are used in textronic systems most often to create lines connecting individual electronic systems and elements. These lines can supply low-value electrical energy to these systems or serve as lines to exchange information between them. Lines of this type should have the lowest possible resistance.

The majority of textronic solutions require data exchange between electronic devices of the system. For this purpose, until recently, mainly conventional connections such as wires have been used, but these are elements affecting the ergonomics of the product (stiffness, troublesome maintenance of a textile product with increased weight of the entire product). This resulted in intensive work on an alternative, textile solution. These works concern both the provision of electrical conductivity to textiles and the construction of textile signal lines. The electrical conductivity of textile materials can be achieved in many ways, described, among others, in [17,18].

Such attempts were also made for applying electroconductive layers on the surface of the product (printing, sputtering, embroidering, etc.). The possibility of transmitting electrical signals using electrically conductive textiles is also investigated and developed. The transmission of electrical signals, and in particular high-frequency signals or digital signals, can be a challenge for this type of line. In the presented article, the subject of signal transmission by textile signal lines (TSL) was discussed.

Wired transmission, despite the necessity to use cables between electronic systems, has significant advantages. One of them is simplicity resulting from the lack of necessity to use radio transmitting and receiving systems which usually require an additional power supply and the simplification of electronic circuits and components used. Wired data is also more resistant to naturally occurring signal interference and is also more resistant to eavesdropping. This is important for example in the case of textronic clothing for human physiological parameters monitoring. The advantages of wired data transmission in textronic mean that the construction of such a line made entirely of textile materials is one of the other fields of development. These types of lines should be capable of transmitting signals with a wide frequency spectrum. In addition, they should be resistant to temperature, humidity, mechanical stresses occurring during their use in e-textiles. So far, few published studies of the resistance of textile signal lines to the above factors occurring during the use of e-textiles have been carried out. Leśnikowski [19] investigated the influence of temperature and humidity on textile signal lines in the form of electrically conductive textile strips sewn onto non-conductive fabric. Leśnikowski and Kubiak [20] studied the changes in the characteristic impedance of selected types of textile signal lines during mechanical loads. Leśnikowski [21] also investigated the influence of bending and abrasion of lines on their transmission properties. In all of the above studies, textile signal lines with straight electrically conductive paths were used. In practical applications, textile signal lines must frequently change direction to connect electronic modules placed in different places of the smart garments. Despite this, no results of such studies have been published so far. The article presents studies of the influence of tensile forces acting on the line on its characteristic impedance. As textile signal lines used in smart-garment must change their direction frequently, lines that do not have the form of a straight line were used for the tests. Works related to the subject of data transmission over the surface of textiles also show the importance of adjusting individual elements of the signal path in textronic systems. The appropriate tools, like the Time Domain Reflectometry method [22], for interpreting the quality of the transmitted signal allows the user to identificate potential sources of interference or signal degradation. The article also presents a preliminary statistical analysis, the aim of which was to find those characteristics of the substrate of the line that affect changes in the characteristic impedance during stretching.

2. Materials and Methods

For the tests presented in this article, twenty textile signal lines with curved electroconductive paths were made. Each of the lines consists of a non-conductive substrate in the form of a fabric and sewn strips cut from an electroconductive woven fabric. More details on the construction of the lines can be found in [23]. The dimensions of the signal line made by the sewing method were calculated to gain the assumed characteristic impedance equal to 50 Ω. The dimensions were shown in Figure 1, and a physical example of the constructed line is in Figure 2. The shape of the line resulted from the desire to make a line in which the electrically conductive paths would not have the form of a straight line, and at the same time beginning and end of the line would be on one axis. This feature is necessary for the correct tensioning of the line using the method described later in the article.

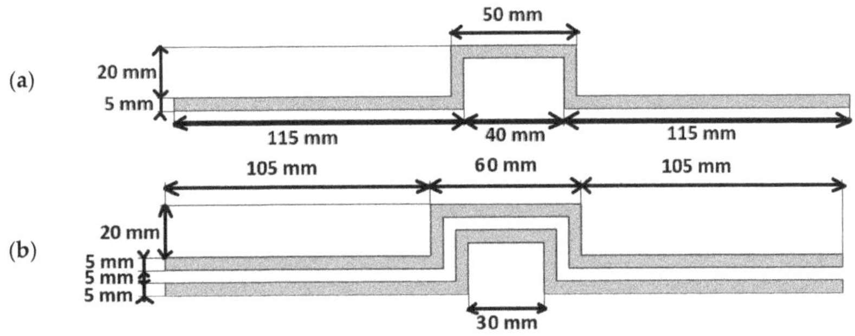

Figure 1. Geometric dimensions of the textile signal line. View from the side of the signal path (**a**), View from the side of the ground paths (**b**).

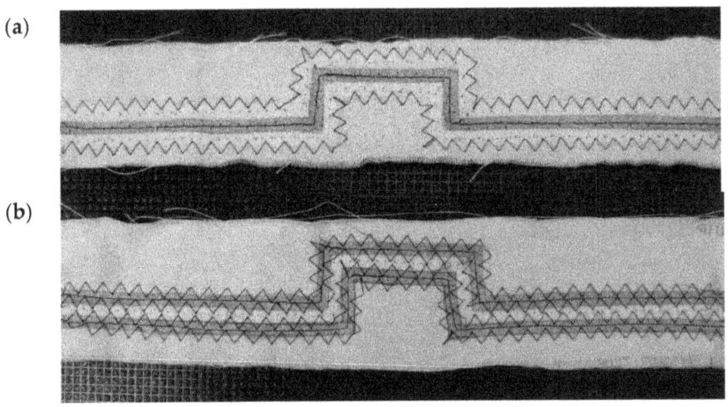

Figure 2. Signal lines produced by sewing technique. View from the side of the signal path (**a**), view from the side of the ground paths (**b**).

The paths of the textile signal lines were made of the electro-conductive Soliani Ponge fabric. These paths were sewn to the non-conductive fabrics, creating the substrate of the lines. The Ponge fabric was chosen due to its excellent conductivity as the main factor required for signal transmission. The basic parameters of the Ponge fabric are presented in Table 1, while the basic parameters of the fabrics used as line substrate are presented in Table 2.

Table 1. The basic parameters of Ponge fabric are used as electro-conductive paths.

Material	Trade Name/Producer	Thickness (mm)	Surface Resistivity (Ohm/sq)	Nickel Amount (g/m^2)	Total Weight (g/m^2)	Weave	Warp Density (Yarns/dm)	Weft Density (Yarns/dm)
Nickel metallised polyester	Ponge/Soliani	0.15	Max. average 0.4	16	60 ± 15	Plain	260	180

Table 2. The basic parameters of fabrics are used as a non-electroconductive substrate.

Line No.	Material	Thickness (mm)	Surface Mass (g/m^2)	Weave	Warp Density (Yarns/dm)	Weft Density (Yarns/dm)
L1	100% Cotton	0.38	89	Plain	25	21
L2	65% Polyester, 35% Cotton,	0.27	170	Plain	42	29
L3	65% Polyester, 35% Cotton,	0.56	290	Twill	40	24
L4	40% Polyester, 35% Cotton, 25% Flax	0.80	230	Plain	22	18
L5	69% Cotton, 31% Polyamide	0.53	194	Satin	73	31
L6	63% Polyester, 33% Cotton, 1% Elastomer	0.42	238	Satin	112	25
L7	60% Polyester, 40% Cotton,	0.50	275	Satin	50	28
L8	100% Polyamide	0.52	211	Satin	42	24
L9	100% viscoze	0.42	147	Plain	45	44
L10	100% Polyester	0.36	158	Plain	56	28
L11	100% Polyamide	0.32	161	Plain	25	22
L12	72% Cotton, 23% Polyester, 5% Elastomer	0.53	163	Plain	48	28
L13	55% Flax, 45% Viscose	0.50	172	Plain	24	14
L14	100% Viscose	0.52	201	Plain	22	18
L15	62% Polyester, 32% Viscoze, 6% Elastomer	0.61	275	Plain	33	31
L16	100% Wool	0.42	287	Plain	24	20
L17	63% Cotton, 37% Polyamide	0.60	168	Twill	59	45
L18	100% Polyester,	0.42	183	Twill	36	34
L19	50% Polyester, 50% Wool	0.64	188	Twill	21	15
L20	50% Polyester, 50% Argon (Viscoze-based)	0.65	243	Twill	27	25

In total, twenty different TSLs were made using the same Soliani Ponge electroconductive material and different material substrates, mainly cotton and polyester to test if any of their substrate parameters may influence the characteristic impedance change. All TSLs are shown in Table 2.

The substrate mass and thickness of the mentioned above, twenty TSL are shown in Figure 3. The lines L1, L2, L3 and L4 were chosen for detailed impedance characteristic waveforms in a further chapter. These lines have a maximum or minimum thickness or surface mass.

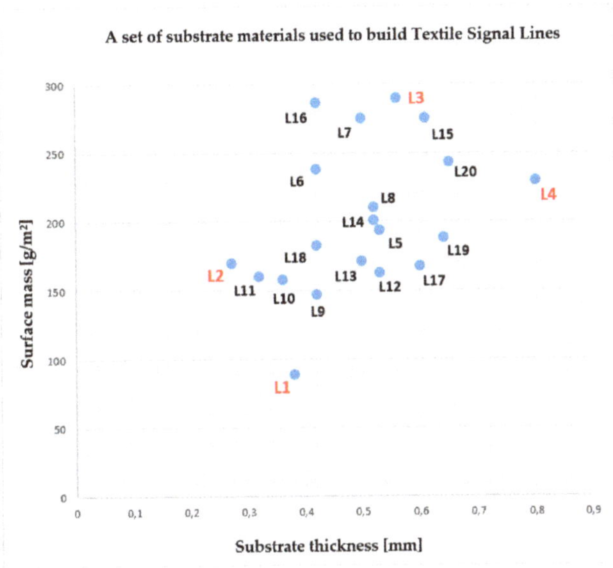

Figure 3. Chart with twenty sewed textile signal lines with their substrate properties.

All twenty TSLs were divided into groups based on the specified substrate parameters: the thickness, the surface mass, and the substrate material weave. The groups of TSLs according to specified substrate properties were shown in Tables 3–5. The plan of dividing the TSLs into groups was to include in each of the three subgroups at least five different lines. Also, the substrates had three specified weaves: plain, twill, and satin, and then all TSL were divided into three groups of the specified weave, the same method used earlier in [20].

Table 3. TSL's population, divided into groups based on their substrate thickness.

Substrate Thickness [mm]	Group Name (Thickness)	Number of Lines in the Group
0.27–0.45	Thin	5
0.45–0.59	Normal	8
0.60–0.80	Thick	7

Table 4. TSL's population, divided into groups based on their substrate surface mass.

Surface Mass [g/m^2]	Group Name (Surface Mass)	Number of Lines in the Group
89–165	Lightweight	8
166–229	Medium	7
230–290	Heavy	5

Table 5. TSLs population, divided into groups based on their substrate weave.

Weave	Number of Lines in the Group
Plain	11
Twill	5
Satin	4

The transmission properties of textile signal lines are characterized by different parameters. These parameters determine the line ability to transmit fast-changing and high-frequency signals. One of the crucial parameters is the line characteristic impedance. It is the main value proving that the signal line is matched to other components of the signal path.

The value of the characteristic impedance Z for an ideal (lossless) line is calculated according to the following formula:

$$Z = \sqrt{\frac{L}{C}} \tag{1}$$

where L is the line inductance and C is the line capacitance.

In the real signal line there are natural changes in series resistance of the conductive elements used, or dielectric losses [24]. Thus, the impedance of the real lines including naturally present losses is calculated by:

$$Z = \sqrt{\frac{R_S + j\omega L}{G + j\omega C}} \tag{2}$$

where Rs is the unit series resistance of the conducting part of the line, G means unit conductance of the dielectric and ω is the pulsation (rad/s).

When designing and developing transmission systems, their elements should be matched to the system and the impedance of each element should be as close to each other as possible. The value of the characteristic impedance of the line depends on its application, e.g., for lines connecting a radio transmitter with a textile antenna, this value is usually 50 Ω. The suitability of signal lines with a characteristic impedance of 50 Ω for signal transmission was discussed by Johnson [23]. These tests showed that the value 50 Ω of the characteristic impedance is a compromise for the impedance value between 30 Ω, at which the maximum power of the transmitted signal occurs, and the value of 77 Ω, where the signal losses are the lowest.

The value of the TSL characteristic impedance depends mainly on the geometrical dimensions of such elements, the dielectric properties of the substrate and the resistance of the material from which the electrically conductive paths were made. Earlier, only straight TSLs were made and tested [20]. In a straight signal line, the electro-conductive paths are in the form of straight lines. In the studies presented in [20], the change of TSL transmission properties was tested before and after subjecting them to tensile loads. Tensile loads were obtained by loading one end of the line with a mass of a certain weight. In smart clothing, it is very often necessary to use TSL, which are not straight lines. This is due to the need to change the direction of the lines to connect the electronic modules that can be placed in different places of the smart garment. As mentioned earlier such studies have not been carried out so far. The research on these types of lines is presented in the article below.

In the research conducted, the so-called characteristic impedance profile of the tested lines has been measured. This profile shows the value of the characteristic impedance of each point of the line as a function of its distance from the beginning of the line. To determine these profiles, the reflectometric method described, among others in [22,25] was used. The measuring stand for testing the TSL's characteristic impedance was built with the Tektronix DSA8200 Digital Serial Analyzer, connected with the 80A02 EOS Electric Charge Protection module, with an additional external 80E08 TDR/Sampling Module.

The block diagram of the stand is shown in Figure 4. The tested line is connected to the TDR module using the terminals described in detail in [26]. These clamps enable the connection of the flat electro-conductive paths of the tested line with concentric measuring connectors of the measuring equipment. The Digital Serial Analyzer, using the 80E08 module, generates a voltage step that is applied to the input of the tested line. The generated wave propagates along the tested line reflecting from places with different impedance. The reflected signal returns to the analyzer. Based on its changes over time, the line impedance profile is determined.

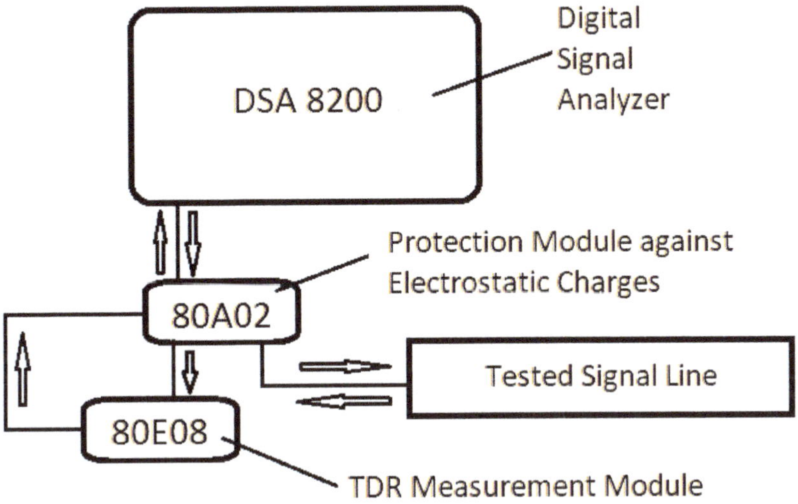

Figure 4. Block diagram of the measuring method for the TSL's characteristic impedance.

The characteristic impedance profiles were collected via the Serial X-Press software extension of the DSA8200 Signal Analyzer. All measurements were made in a standardized, normal climate: 20 °C and air humidity 65% according to the ISO standard [27].The impedance profile of each line connected to the measurement system (Figure 4) was measured in real-time. The measured TSL was mounted vertically, with the end unloaded (m_0 = 0 g) and tested. Then each of the lines was successively loaded with the masses m_1 = 0.414 kg and m_2 = 1.461 kg and tested. The example of a line mounted vertically with load m_2 = 1.461 kg was shown in Figure 5.

These two mass values were experimentally chosen as a simulation of mechanical forces that may occur in real-time deformations in the textiles when wearing. Choosing less mass than proposed results in deformations which are hardly visible, when adding too much mass could lead to permanently deform or even damage constructed TSLs. The reason for vertically stretching the lines was ease of obtaining a reproducible tensile force. For this purpose, test weights with known mass were used.

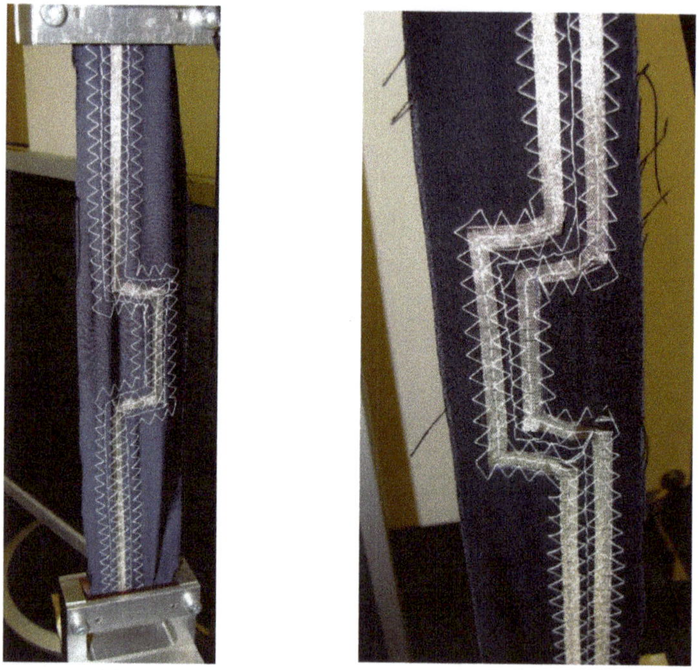

(**a**) (**b**)

Figure 5. Textile signal line mounted vertically with load m_2 = 1.461 kg placed to the end of the line. View from the side of the signal path (**a**), View from the side of the ground paths (**b**).

3. Results and Discussion

The measured characteristic impedance profiles of the transmission lines (Figures 6–9) are characterized by a certain non-uniformity. This phenomenon has a fundamental and direct impact on the quality of the electric signal transmitted through the line. In the case of an ideal line, the value of the characteristic impedance should be constant at every point on the line, but in reality, there will always be some fluctuation in waveforms. The value of these factors also depends on the quality of the implementation of individual line elements.

The observed characteristic impedance waveform for line L1 (with the lowest surface mass) shows some fluctuations in the range of 60 to 80 Ω and some minor differences when loading with mass m_1 or m_2 in comparison to the unloaded line. At the points of curvature of the lines, no significant changes in characteristic impedance under stretching were observed. This means that line bends do not significantly affect the transmission properties of this type of line.

Characteristic impedance waveform for line L2 (with the lowest substrate thickness) shows more significant fluctuations from 60 to 80 Ω and above, at the end of the unloaded line. Also, some major differences, especially when loading with m_2 mass in comparison to the unloaded line, were observed. Significant changes in the characteristic impedance were observed in the places where the electrically conductive paths were bent. This can mean a significant change in the impedance of the line if it is placed where it will be exposed to stretching.

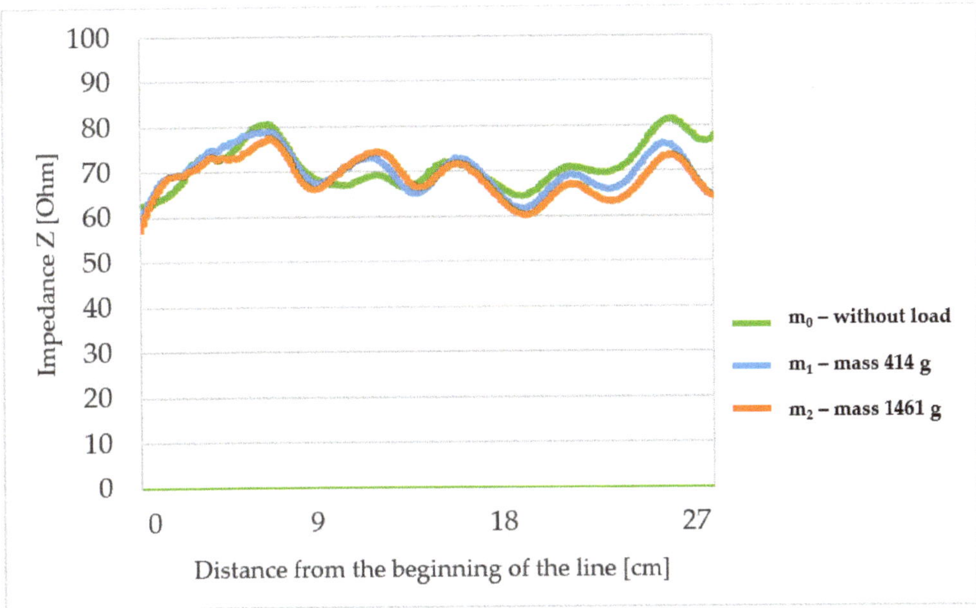

Figure 6. The profiles of the characteristic impedance of the textile signal line L1.

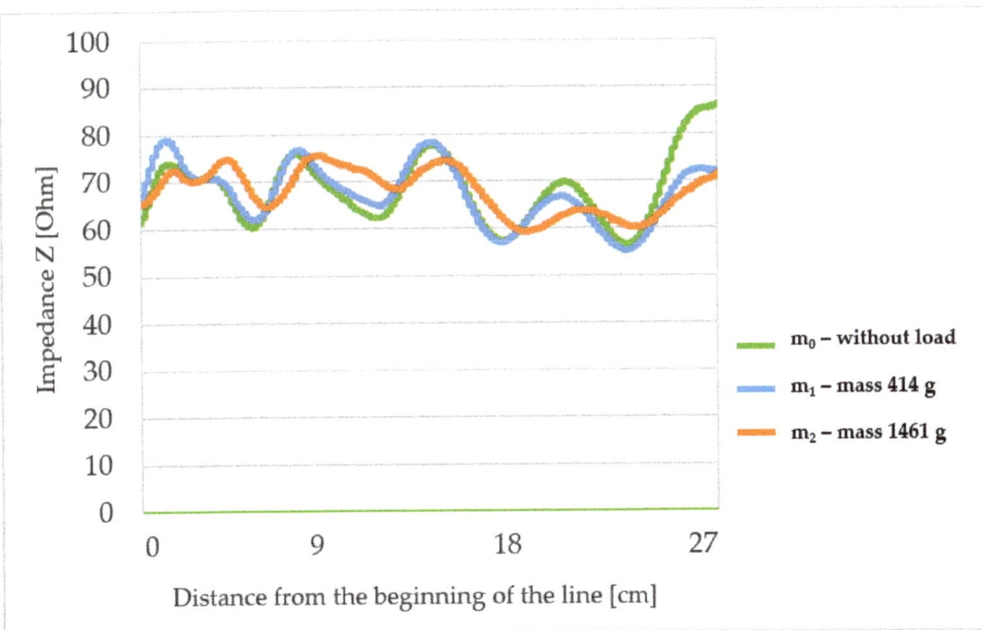

Figure 7. The profiles of the characteristic impedance of the textile signal line L2.

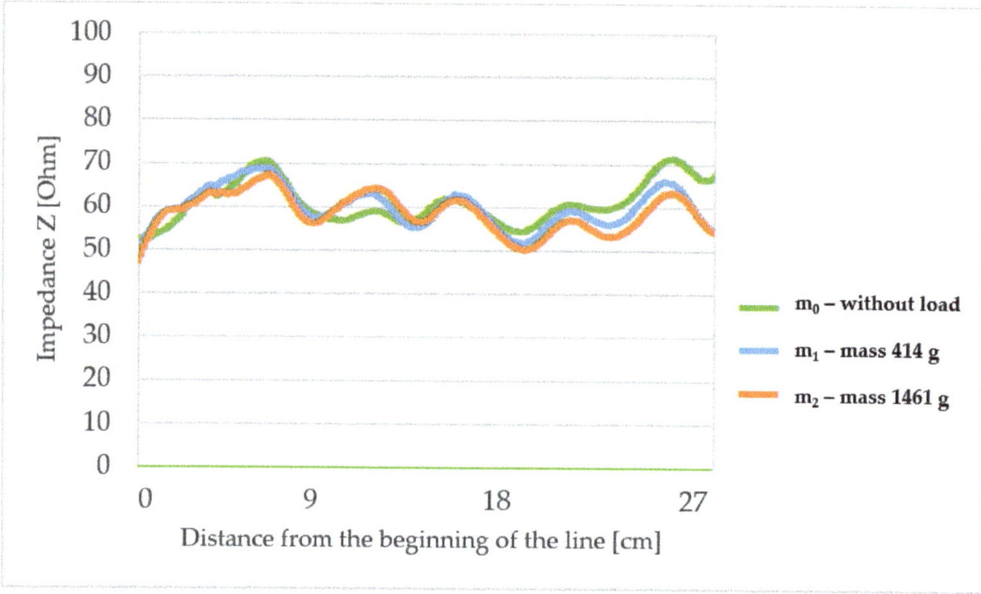

Figure 8. The profiles of the characteristic impedance of the textile signal line L3.

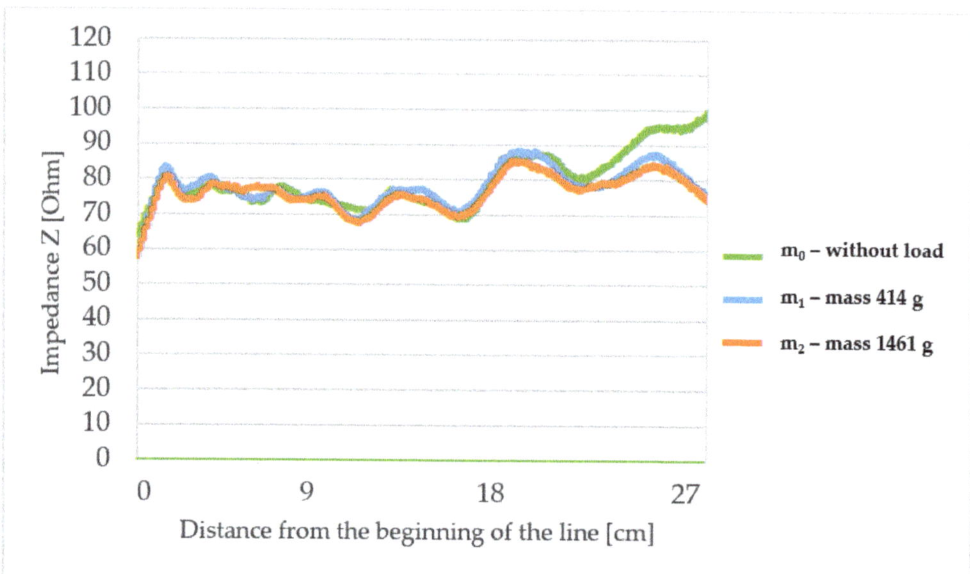

Figure 9. The profiles of the characteristic impedance of the textile signal line L4.

The L3 line (with the highest surface mass) shows the characteristic impedance waveform very similar to L1, although the overall impedance value is lower by about 10 Ω in comparison to other lines. This line also has some fluctuations from 50 to 70 Ω and some minor differences when loading with mass m_1 or m_2 compared to the unloaded line. This line is very similar in its characteristic impedance waveform compared to line L1 and due to the lower values obtained it is the most suitable for applications in textronic systems with the main impedance equal to 50 Ω.

The observed characteristic impedance waveform for line L4 (with the highest substrate thickness) shows only minimal fluctuations from 60 to 85 Ω and some minor differences when loading with mass m_1 or m_2 compared to the unloaded line. This means that the line can correctly transmit signals when placed in areas of the garment where stress occurs.

The value of the characteristic impedance of the tested line depends on the dimensions of its electrically conductive paths and the spacing between them. In particular, the distance between the ground paths significantly influences this value. Stretching the lines longitudinally changes the dimensions of the electrically conductive paths and spacing between them. The degree of these changes may depend on many parameters characterizing the structure of the substrate and the electro-conductive paths of the lines. One of the aims of the research presented in the article below was to check which of the parameters characterizing the line substrate have a significant impact on the change of the characteristic impedance of the line under the influence of its stretching. To assess this, the statistical analysis presented later in the article was performed.

4. Statistical Analysis

For statistical analysis, the average value from each impedance profile was determined according to the formula:

$$Z_{av} = \frac{\sum_{i=1}^{n} Z_i}{n} \quad (3)$$

where Z_{av} is the average line impedance, Zi is the line impedance at the i-th point of the average impedance profile, n is the number of points (measurement values) of which the line impedance profile consists.

The characteristic impedance of each of the twenty tested lines was measured five times in each load condition (unloaded, loaded with mass m_1, loaded with mass m_2). Next, these five impedance waveforms for each load were averaged, resulting in one waveform for specified line without load, for load m_1 and m_2. For the statistical analysis, there was a need to compare fixed values instead of waveforms. Therefore, each averaged waveform was shortened, to one average value, using Equation (3).

A non-parametric test was used instead of the multifactor analysis of variance (ANOVA). This was due to failure to meet the assumptions of analysis of variance, namely a lack of normality of distributions in groups determined by variables, and a lack of homogeneity of variance.

The analyzed parameters of all the signal lines population (twenty lines in total) were divided into more than two groups, thus the statistical test must be appropriate to compare its parameters within several independent groups. For that, the nonparametric Kruskal–Wallis statistical test was used. The significance level at all tests was assumed to be $\alpha = 0.05$. A summary of the collected results showing the characteristic impedance values of all twenty made curved TSL under the influence of various loads is given in Figure 10.

In Figure 10 mean and maximum values of characteristic impedance significantly decrease while loading masses m_1 and m_2. The influence on such results is directly related to the way the TSL is stretched under m_1 and m_2 loads, which significantly alters the mutual arrangement of the electrically conductive strips which are part of the TSL.

Next, the percentage change of characteristic impedance within different groups of TSLs was tested to find if any of the substrate parameters: surface mass, thickness and weaves (Tables 3–5) have a statistical impact. The results are shown in Figures 11–13. The mean value, mean +/− Standard Deviation and Max/Min values are also shown.

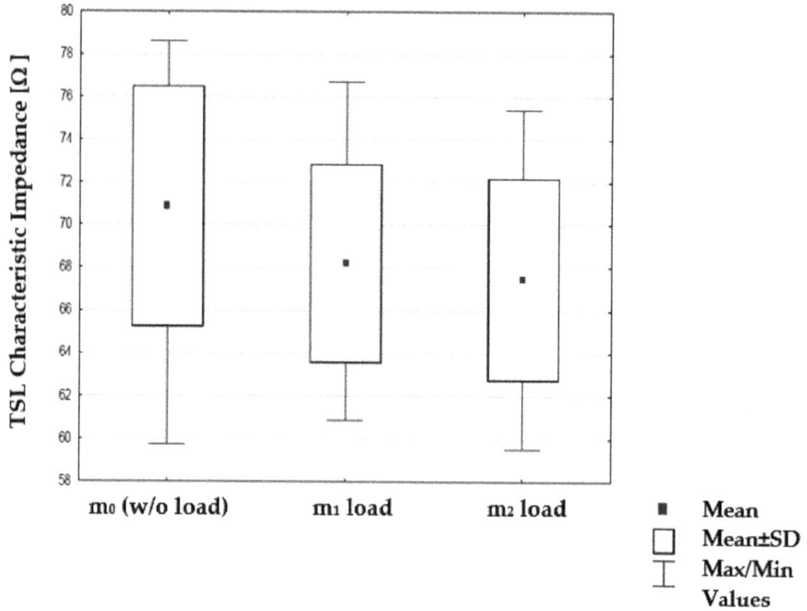

Figure 10. The characteristic impedance of lines under different loads for all sewed lines.

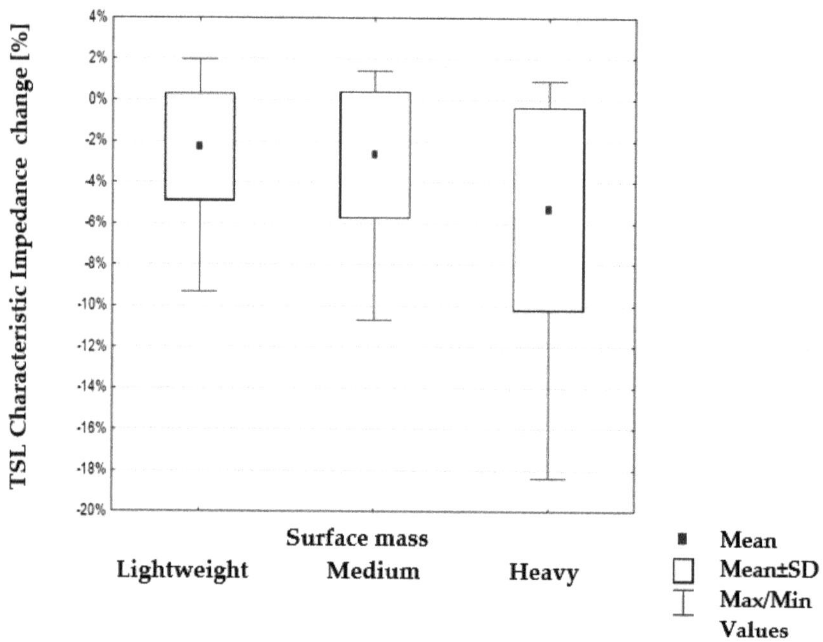

Figure 11. Changes in the line characteristic impedance under load for lines with a different surface mass of the substrate.

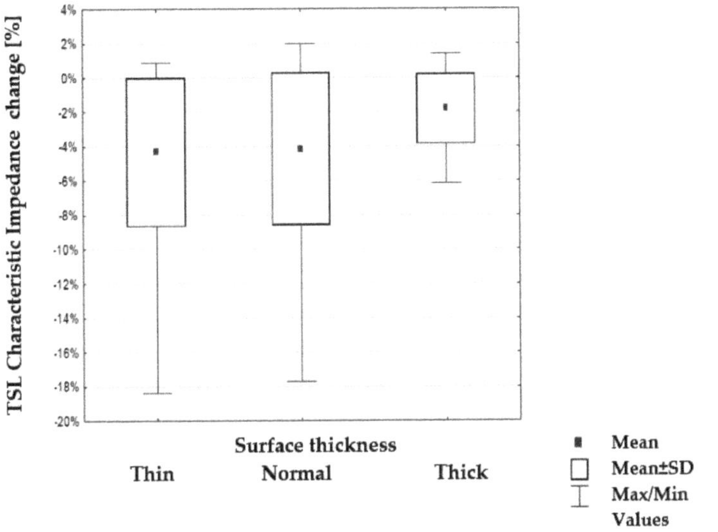

Figure 12. Changes in the line characteristic impedance under load for lines with a different thickness of the substrate.

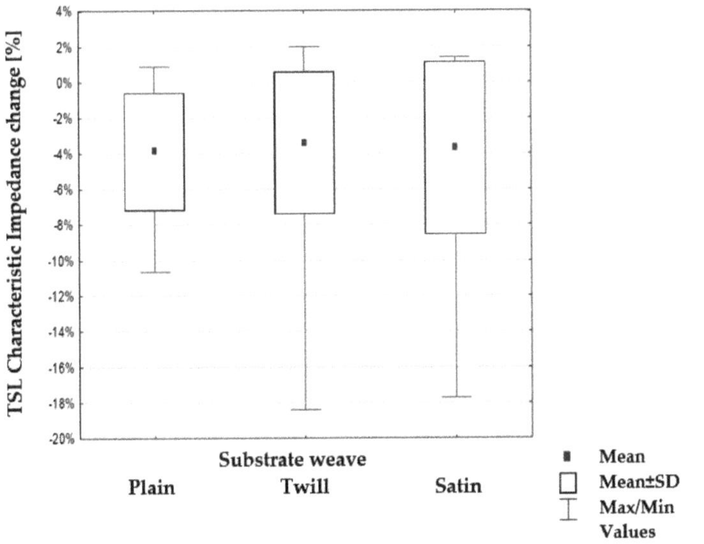

Figure 13. Changes in the line characteristic impedance under load for lines with a different weave of the substrate.

The statistical analysis of the obtained results consisted in assuming the H0 hypothesis assuming the equality of the mean impedance value in the tested groups. Then, the Kruskal–Wallis test was used to verify this hypothesis.

The Kruskal–Wallis statistical test, with its null hypothesis H0, shows the impact on the value of lines characteristic impedance by the mechanical load placed to the line with the weights of m_1 and m_2. Also, the impact from substrate properties—surface mass, thickness and weave to affect the characteristic impedance changes—were tested. The assumed null hypothesis H0 assumes no statistically significant differences between the tested groups of cases. As a result of this test, the probability p-value is obtained that the assumed hypothesis H0 is true. If the probability value p is greater than the assumed significance level ($\alpha = 0.05$), then the H0 hypothesis should be accepted. Otherwise, it would be rejected. For example, the values of the TSL characteristic impedance with mass m_1 and m_2 loads are shown in Figure 10. The figure shows that the mean value of the impedance is lower when the force stretching the lines is higher. This was confirmed by the result of the Kruskal–Wallis test based on which the assumed null hypothesis was rejected (Table 6). It is equivalent to a statistically significant influence of the tested factor on the characteristic impedance of the line.

Table 6. Kruskal-Wallis non-parametrical test summary for curved TSL.

	Characteristic Impedance Z [Ohm] Change between Mass Loaded	
Factor or substrate property tested	Between loads with mass m_0 and m_1	Between loads with mass m_0 and m_2
Mass loading	$p = 0.003$	$p = 0.001$
Surface mass of the substrate	$p = 0.0022$	$p = 0.001$
Substrate thickness	$p = 0.013$	$p = 0.001$
Substrate weave	$\mathbf{p = 0.1176}$	$\mathbf{p = 0.1721}$

The highlighted factors ($p > 0.05$) means that hypotheses H0 that states the mechanical load or substrate properties affects the characteristic impedance changes are accepted. The results of the statistical analysis presented in Table 6 show that the load of the line with the mass m_1 and in particular mass m_2 has a statistically significant influence on the changes of the characteristic impedance. Therefore, a further statistical analysis was carried out to show which property of the line substrate may have a statistically significant impact on impedance changes under the influence of these loads. The results of this analysis are presented in Table 6 which shows the characteristics that have such an effect are the surface mass and the thickness of the substrate material. The weave of the substrate has no effect. Considering collected results, there is a possibility to make a TSL within the limits of predetermined values, and the main difficulty to gain this is the line accuracy and precision in the production method. Constructing other TSLs with different shapes and predetermined values is also possible. As seen in Figures 6–9, the characteristic impedance was around 50–80 Ω. The suitability of lines with such characteristic impedance spread depends on the specific application and data transmission standard.

5. Conclusions

Textronic systems based on electrical elements nowadays have an increasing range of applications. Data transmission via signal lines, constructed with the use of textile elements, will allow improving the products ergonomic by limiting the usage of the conventional elements (cables, connectors, etc.).

Statistically, constructed curved textile signal lines show significant changes in characteristic impedance between unloading state and load with mass m_1 and in particular mass m_2. Curved TSL were made strictly to sewn to textiles and can be placed even in custom shape textiles carried by users. The changes of their shapes don't affect their characteristic impedance, and observed changes were seen when loading with significant mass only. Also, no significant impedance changes were observed at the curvature points of the line in comparison to the straight, non-curved areas of lines. In most cases, no curvature impact for the characteristic impedance change was observed for lines loaded up to mass m_2. Statistically for the sewed TSLs, only the substrate weave does not influence the characteristic impedance change, regardless of the tensile forces. The surface weight and thickness of the fabric from which the line substrate is made has a significant effect on the changes in the characteristic impedance under the influence of stretching. This impact is less for lines with a thick substrate or substrate having a high surface mass. This agrees with our subjective feeling.

The tensile sensitivity of TSL means that these lines in smart clothing should be led and applicable through places that are not very exposed to mechanical deformation. If the line has to run through such places, it should be placed on a thick substrate made of fabric with a large surface mass.

Author Contributions: Conceptualization, P.K. and J.L.; methodology, P.K. and J.L.; formal analysis, P.K. and J.L.; investigation, P.K. and J.L; resources, P.K. and J.L.; writing—original draft preparation, P.K. and J.L.; visualization, P.K. and J.L.; writing—review, P.K., and J.L.; supervision, P.K. and J.L. All authors have read and agreed to the published version of the manuscript.

Funding: This research received no external funding.

Conflicts of Interest: The authors declare no conflict of interest.

References

1. Gniotek, K.; Krucińska, I. The basic problems of textronics. *Fibres Text. East. Eur.* **2004**, *12*, 13–16.
2. Walczak, S. Inteligentne tekstylia—międzynarodowe innowacje w tekstronice. *Acta Innov.* **2012**, *3*, 103–122.
3. Angelucci, A.; Cavicchioli, M.; Cintorrino, I.A.; Lauricella, G.; Rossi, C.; Strati, S.; Aliverti, A. Smart Textiles and Sensorized Garments for Physiological Monitoring: A Review of Available Solutions and Techniques. *Sensors* **2021**, *21*, 814. [CrossRef] [PubMed]
4. Axisa, F.; Dittmar, A.; Delhomme, G. Smart clothes for the monitoring in real time and conditions of physiological, emotional and sensorial reactions of human. In Proceedings of the 25th Annual International Conference of the IEEE Engineering in Medicine and Biology Society (IEEE Cat. No.03CH37439), Cancun, Mexico, 17–21 September 2003; IEEE: Piscataway, NJ, USA, 2004. [CrossRef]
5. Onose, G.; Chendreanu, C.; Neacşu, A.; Grigorean, V.; Strâmbu, V.; Toader, C.; Spânu, A.; Ioana, A.; Anghelescu, A.; Onose, L.; et al. Smart textiles for noninvasive monitoring of physiological signals. *Ind. Text.* **2009**, *60*, 124–133.
6. Kiaghadi, A.; Homayounfar, S.Z.; Gummeson, J.; Andrew, T.L.; Ganesan, D. Phyjama: Physiological Sensing via Fiber-Enhanced Pyjamas. *GetMobile Mob. Comput. Commun.* **2020**, *24*, 33–37. [CrossRef]
7. Gniotek, K.; Gołębiowski, J.; Leśnikowski, J. Temperature Measurements in a Textronic Fireman Suit and Visualisation of the Results. *Fibres Text. East. Eur.* **2009**, *17*, 97–101.
8. Soukup, R.; Blecha, T.; Hamacek, A.; Reboun, J. Smart textile-based protective system for firefighters. In Proceedings of the 5th Electronics System-Integration Technology Conference (ESTC), Helsinki, Finland, 16–18 September 2014; IEEE: Piscataway, NJ, USA, 2014; pp. 1–5.
9. Curone, D.; Secco, E.L.; Tognetti, A.; Loriga, G.; Dudnik, G.; Risatti, M.; Whyte, R.; Bonfiglio, A.; Magenes, G. Smart Garments for Emergency Operators: The ProeTEX Project. *IEEE Trans. Inf. Technol. Biomed.* **2010**, *14*, 694–701. [CrossRef] [PubMed]
10. McCann, J. 9—Smart protective textiles for older people. In *Smart Textiles for Protection, Woodhead Publishing Series in Textiles*; Chapman, R.A., Ed.; Woodhead Publishing: Sawston, UK, 2013; pp. 244–275. ISBN 978-0-85709-056-0.
11. Yang, K.; Isaia, B.; Brown, L.; Beeby, S. E-Textiles for Healthy Ageing. *Sensors* **2019**, *19*, 4463. [CrossRef] [PubMed]
12. Özyangan, V.; Abdulova, V. Utilization of Smart Textiles in Healthcare. *Int. J. Electron. Mech. Mechatron. Eng.* **2015**, *5*, 1025–1033. [CrossRef]
13. Hasan, M.N.U.; Negulescu, I. Wearable Technology for Baby Monitoring: A Review. *J. Text. Eng. Fash. Technol.* **2020**, *6*, 112–120. [CrossRef]
14. Munidasa, S.; Baghaei, P.; Shim, E.; Lin, O.; Ghafar-Zadeh, E. Smart Bedsheet for Baby Monitoring Application: Measurement and Characterization Results. In Proceedings of the 42nd Annual International Conference of the IEEE Engineering in Medicine & Biology Society (EMBC), Montreal, QC, Canada, 20–24 July 2020; IEEE: Piscataway, NJ, USA, 2020; Volume 2020, pp. 4402–4405.
15. Baskan, H.; Acikgoz, H.; Atakan, R.; Eryuruk, H.; Akalın, N.; Kose, H.; Li, Y.; Bahadir, S.K.; Kalaoglu, F. Running functional sport vest and short for e-textile applications. *IOP Conf. Ser. Mater. Sci. Eng.* **2017**, *254*, 072004. [CrossRef]
16. Scataglini, S.; Moorhead, A.; Feletti, F. A Systematic Review of Smart Clothing in Sports: Possible Applications to Extreme Sports. *Muscle Ligaments Tendons J.* **2020**, *10*, 333–342. [CrossRef]
17. Tadesse, M.G.; Mengistie, D.A.; Chen, Y.; Wang, L.; Loghin, M.C.; Nierstrasz, V. Electrically conductive highly elastic polyamide/lycra fabric treated with PEDOT:PSS and polyurethane. *J. Mater. Sci.* **2019**, *54*, 9591–9602. [CrossRef]
18. Chatterjee, K.; Tabor, J.; Ghosh, T. Electrically Conductive Coatings for Fiber-Based E-Textiles. *Fibers* **2019**, *7*, 51. [CrossRef]
19. Leśnikowski, J. Effect of temperature and humidity on the transmission properties of textile signal lines. *J. Text. Inst.* **2019**, *111*, 604–610. [CrossRef]
20. Leśnikowski, J.; Kubiak, P. Changes in the Characteristic Impedance of Textile Signal Lines While Mechanically Loaded. *Autex Res. J.* **2019**, *19*, 375–380. [CrossRef]

21. Leśnikowski, J. Effect of Cyclical Bending and Rubbing on the Characteristic Impedance of Textile Signal Lines. *Materials* **2021**, *14*, 6013. [CrossRef] [PubMed]
22. Hsue, C.-W.; Pan, T.-W. Reconstruction of nonuniform transmission lines from time-domain reflectometry. *IEEE Trans. Microw. Theory Tech.* **1997**, *45*, 32–38. [CrossRef]
23. Johnson, H. Why 50 Ohms? *Johnson Howard EDN Magazine*, 14 September 2000; 30.
24. Leśnikowski, J. New Kind of Textile Transmission Line with an Impedance of 50 Ohms. *Fibres Text. East. Eur.* **2015**, *2*, 51–54.
25. Cataldo, A.; De Benedetto, E.; Cannazza, G. *Broadband Reflectometry for Enhanced Diagnostics and Monitoring Applications*, 1st ed.; Springer: Berlin, Germany, 2011; Volume 1, ISBN 978-3-642-20232-2.
26. Leśnikowski, J. Terminal for Connecting Textile Signal Lines to the Devices for Measuring Electrical Parameters. Patent PL 222920 B1, 2016.
27. Standard ISO 139:2005. *Textiles—Standard Atmospheres for Conditioning and Testing 2005*; International Organization for Standardization: Geneva, Switzerland, 2005.

materials

MDPI

Article

Effect of Multiple Annular Plates on Vibration Characteristics of Laminated Submarine-like Structures

Zhengxiong Chen [1], Rui Zhong [1], Shuangwei Hu [1], Bin Qin [2,3,4,*] and Xing Zhao [1]

1 State Key Laboratory of High Performance Complex Manufacturing, Central South University, Changsha 410083, China
2 Key Laboratory of Traffic Safety on Track, Ministry of Education, School of Traffic & Transportation Engineering, Central South University, Changsha 410075, China
3 Joint International Research Laboratory of Key Technology for Rail Traffic Safety, Central South University, Changsha 410075, China
4 National & Local Joint Engineering Research Center of Safety Technology for Rail Vehicle, Central South University, Changsha 410075, China
* Correspondence: qinbin@csu.edu.cn

Abstract: A numerical model for the prediction of vibration behaviors of a laminated submarine structure consisting of spherical, cylindrical, and cone shells with multiple built-in annular plates is reported in this article. With the aid of the first-order shear deformation theory (FSDT) concerning plates and shells, the energy expressions of each substructure are derived. The displacement functions in the energy functionals are expanded by the employment of Legendre orthogonal polynomials and circumferential Fourier series. Then, the Rayleigh–Ritz procedure is performed to obtain the eigenfrequency and the corresponding eigenmode of the submarine model. The correctness of the structural model is examined by comparing the results with existing papers and the finite element method, and the maximum deviation is not more than 2.07%. Additionally, the influence of the plate's thickness, position, inner diameter, as well as the laying angle on the intrinsic vibration characteristics of laminated submarine-like structure is determined. The results reveal that rational geometry design and assemblage benefit the vibration performance of the combination. Increasing the thickness of all the annular plates, decreasing the inner radius, and regulating the laminated scheme, make remarkable influence on structural free vibration, with the maximum relative changing rate of frequency exceeding 97%, 16%, and 23%, respectively.

Keywords: Rayleigh–Ritz technology; laminated submarine-like model; annular plate; vibration analysis

Citation: Chen, Z.; Zhong, R.; Hu, S.; Qin, B.; Zhao, X. Effect of Multiple Annular Plates on Vibration Characteristics of Laminated Submarine-like Structures. *Materials* **2022**, *15*, 6357. https://doi.org/10.3390/ma15186357

Academic Editor: Nicholas Fantuzzi

Received: 14 August 2022
Accepted: 8 September 2022
Published: 13 September 2022

1. Introduction

Submarines, with typical combined conical–cylindrical–spherical structures, play an important role in military reconnaissance, deep-sea exploration, marine scientific research, etc. In terms of specific engineering requirements, such as improving a submarine's rigidity, vibration suppression, outfitting, and multiple annular plates are always introduced to the coupled structure, which may cause significant changes in the vibration characteristics of the coupled structure. For this reason, a reasonable and clear understanding of vibration characteristics for coupled laminated submarine-like structures considering multiple built-in annular plates is of great significance in terms of their multi-functional design.

On basis of an investigation into the dynamic characteristics of plate and shell structures [1–7], with the continuous and in-depth research into the vibration mechanism of revolution structures, a series of numerical methods have been developed, including the well-known Ritz technology [8–10], the domain decomposition method [11,12], the differential quadrature method [13,14], and other methods [15–17]. Parashar et al. [18] used polynomial functions with orthogonality to express the displacement components of a piezoelectric ceramic cylindrical shell and used the Rayleigh–Ritz method to obtain its modal performances. On the basis of the unified Rayleigh–Ritz scheme, Du et al. [8] reported the inherent mechanical properties of a rotating cylindrical shell with a hard coating. Li et al. [9] derived the geometric equations of a conical shell with the help of Reissner's shell theory, constructed the displacement functions with improved Fourier series, and finally solved the modal features of the conical shell under elastic constraints. To clarify the vibration mechanism of the coupled shell, An et al. [11] used the domain decomposition method to establish a structural analysis model. Choe et al. [12] solved the elastodynamic problems of functionally graded elliptic shells with elastic constraints by using the semi-analytical domain decomposition method. The generalized differential quadrature method was employed by Li et al. [13] to construct a numerical vibration model for a conical shell made of metal foam material under the elastic boundary. Within the framework of the classical plate theory, Saini et al. [14] derived the motion equation of a non-uniform, functionally graded circular plate by using the energy principle and obtained the frequency equation of the circular plate under the classical boundary. Zhang et al. [15] used the dynamic stiffness approach to calculate the vibration properties of cylindrical-conical composite structures with reinforced ribs under multiple classical boundary conditions. Araki et al. [16] studied the free vibration of a cylindrical shell via the Galerkin method, where the displacement functions of the so-called spectral nodes were expanded by Fourier series. Kim et al. [17] analyzed the free vibration of a coupled shell composed of a cylindrical shell, parabolic shell, and hyperbolic shell by using the Haar wavelet discrete method.

According to the above literature survey, investigations into the vibration behaviors of assembled structures are mostly concerned with numerical algorithms and assembled shell structures. Based on this foundation, the mechanical behaviors of shell–plate coupling structures have been observed. Cao et al. [19] presented an accurate solution to the cylindrical shell–circular plate combination. By means of the modified Fourier series method and considering coupling springs, parametric analyses including the stiffness and conditions of coupling positions and the boundaries were carried out. An improved Fourier series was also applied by Chen et al. [20] to develop a unified theoretical model of an open cylindrical shell–rectangular plate assembled structure with laminated materials in the framework of the first-order shear deformation theory (FSDT). Then, the solutions were obtained by means of the Rayleigh–Ritz method, and comprehensive discussions on the impacts of the main factors on the structural vibration performance were presented, including the geometric properties, coupling positions, and the stiffness of coupling springs. Jin et al. [21] analyzed the vibro-acoustic behaviors of a submarine hull which was symbolled by a conical–cylindrical–hemispherical shell combination and surrounded by heavy fluid. Kim et al. [22] used the Haar wavelet discretization method to investigate the free vibration of conical–cylindrical coupling structures with laminated materials. To reveal the mechanism of structural vibro-acoustic performance, Qu et al. [23] built an immersed submarine model composed of a rigid propeller, a main shaft, two bearings, and an orthogonally stiffened pressure hull. By introducing Fourier series and Chebyshev orthogonal polynomials, the structural displacement and pressure were expressed with those series. The contributions of the stiffness of the ring and the bearing to acoustic behaviors were studied. With experiments and the finite elements method (FEM) simulations, Wang et al. [24] conducted a study about the vibro-acoustic characteristics of a submarine-like system. Xie et al. [25] discussed the main parametric influence on the free and forced vibration of annular plate–cylindrical shell elastically coupled structures with Flügge shell theory [26] and thin-plate theory. Chen et al. [27] studied the vibrational characteristics of stepped cylindrical shell–annular plate coupling structures considering temperature with Chebyshev polynomials and Fourier series. Zhang et al. [28,29] analyzed the vibration characteristics of shell–plate structures. Sobhani et al. [30,31] investigated the vibration behavior of the combination consisting of multiple shells made of composite materials using the Generalized Differential Quadrature method and FSDT. Bagheri et al. [32] analyzed the free vibration of a joined shell structure composed of cylindrical and spherical shells with functionally graded material. Shi et al. [33] studied the vibration of conical-cylindrical shell assembled structure made of functionally graded materials with respective to environment.

The aforementioned papers are almost concerned with shell–shell or shell–plate coupling structures; studies regarding submarine-like structures are scarce. In addition, submarine-like structures made of laminated materials are widely used in reality due to the advantages of laminated materials compared to conventional materials, such as their strength-to-weight ratio and heat and corrosion resistance. Nevertheless, studies regarding the vibration characteristics of submarine-like structures with laminated materials have not been carried out yet. Generally, a submarine-like structure can be treated as a cylindrical shell–spherical shell–conical shell combination, which is divided into several sub-compartments by annular plates according to actual requirements, for example, America's Ohio-class nuclear submarine and Russia's Borei-class/Dolgorukiy-class nuclear submarine, etc. Consequently, the unified Rayleigh–Ritz technology in conjunction with the FSDT assumption is applied in the current work to investigate the vibration mechanism of such a combined structure with laminated materials. Furthermore, the effect of the annular plate on the vibration characteristics of the submarine structure is analyzed in detail.

2. Analysis Model

2.1. Description of the Model

Figure 1 describes the structural element and coordinate system of a general shell structure. Here, it is assumed that the coordinate system O-αβz is on the middle plane of the shell. Rα and Rβ separately represent the radius of curvature of the shell along α and β directions, when Lα and Lβ are the corresponding lengths of the shell element. h denotes the structural thickness. In addition, artificial spring technology is introduced to simulate boundary conditions. Figure 2 presents a laminated submarine-like model composed of spherical, cylindrical, and conical shells, where J annular plates are arranged. The cone angle of the conical shell is denoted by the symbol α0, and the length of the conical shell along the meridian direction is represented by the symbol Lc. The large end of the conical shell is assembled with a cylindrical shell with length L and radius R1. The right end of the cylindrical shell is connected to the closed hemispherical shell, which means that the radius of the hemispherical shell is consistent with that of the cylindrical shell. The inner radius and thickness of the j-th annular plate are, respectively, expressed by Rj 2 and hj r. The coupling relationship between adjacent substructures is simulated by coupling spring technology.

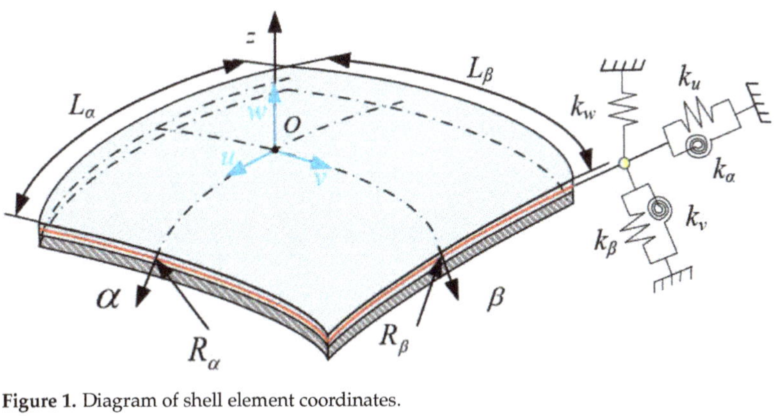

Figure 1. Diagram of shell element coordinates.

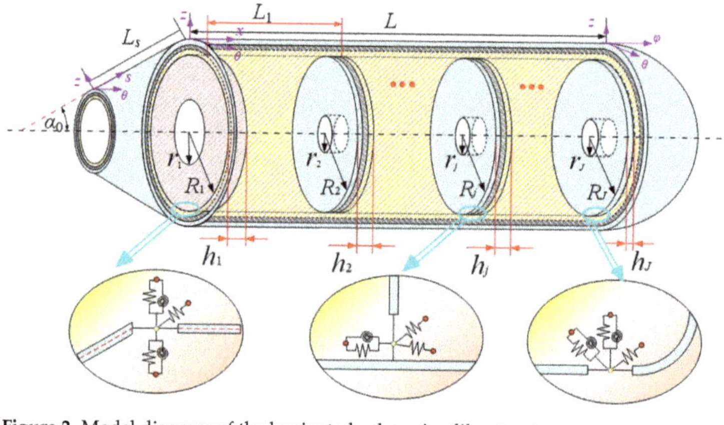

Figure 2. Model diagram of the laminated submarine-like structure.

2.2. Kinematic Relations and Stress Resultants

Given that the physical model is provided with geometric features of a medium–thick plate or shell, under the FSDT's [30] frame, the displacement variables (U, V, and W) of arbitrary points of the kth layer of structures can be written as

$$\begin{aligned} U(\alpha,\beta,z,t) &= u_0(\alpha,\beta,t) + z\xi_\alpha(\alpha,\beta,t) \\ V(\alpha,\beta,z,t) &= v_0(\alpha,\beta,t) + z\xi_\beta(\alpha,\beta,t) \\ W(\alpha,\beta,z,t) &= w_0(\alpha,\beta,t) \end{aligned} \tag{1}$$

in which the symbols (u_0, v_0, and w_0) are the displacement components of the kth midplane. ξ_α and ξ_β are the angular displacement with regard to the α-z plane and β-z plane. The strain components (ε_α, ε_β, $\gamma_{\alpha\beta}$, $\gamma_{\alpha z}$, and $\gamma_{\beta z}$) considered in the kth midplane can be written as follows:

$$\begin{cases} l\varepsilon_\alpha = \varepsilon_\alpha^0 + z\chi_\alpha, \ \gamma_{\alpha z} = \gamma_{\alpha z}^0 \\ \varepsilon_\beta = \varepsilon_\beta^0 + z\chi_\beta, \ \gamma_{\beta z} = \gamma_{\beta z}^0 \\ \gamma_{\alpha\beta} = \gamma_{\alpha\beta}^0 + z\chi_{\alpha\beta} \end{cases} \tag{2}$$

where the strains (ε_α^0, ε_β^0, $\gamma_{\alpha\beta}^0$, $\gamma_{\alpha z}^0$, $\gamma_{\beta z}^0$) on the kth midplane, the curvatures (χ_α and χ_β), and the related twist changes ($\chi_{\alpha\beta}$) are stated as

$$\begin{aligned} &\varepsilon_\alpha^0 = \tfrac{1}{A}\tfrac{\partial u}{\partial \alpha} + \tfrac{v}{AB}\tfrac{\partial A}{\partial \beta} + \tfrac{w}{R_\alpha}, \ \varepsilon_\beta^0 = \tfrac{\partial v}{\partial \beta}\tfrac{1}{B} + \tfrac{\partial B}{\partial \alpha}\tfrac{u}{AB} + \tfrac{w}{R_\beta}, \\ &\gamma_{\alpha\beta}^0 = \tfrac{B}{A}\tfrac{\partial}{\partial \alpha}\left(\tfrac{v}{B}\right) + \tfrac{A}{B}\tfrac{\partial}{\partial \beta}\left(\tfrac{u}{A}\right), \ \chi_\alpha = \tfrac{1}{A}\tfrac{\partial \xi_\alpha}{\partial \alpha} + \tfrac{\xi_\beta}{AB}\tfrac{\partial A}{\partial \beta}, \\ &\chi_\beta = \tfrac{\partial \xi_\beta}{\partial \beta}\tfrac{1}{B} + \tfrac{\partial B}{\partial \alpha}\tfrac{\xi_\alpha}{AB}, \ \chi_{\alpha\beta} = \tfrac{B}{A}\tfrac{\partial}{\partial \alpha}\left(\tfrac{\xi_\beta}{B}\right) + \tfrac{A}{B}\tfrac{\partial}{\partial \beta}\left(\tfrac{\xi_\alpha}{A}\right), \\ &\gamma_{\alpha z}^0 = -\tfrac{u}{R_\alpha} + \tfrac{1}{A}\tfrac{\partial w}{\partial \alpha} + \xi_\alpha, \ \gamma_{\beta z}^0 = -\tfrac{v}{R_\beta} + \tfrac{1}{B}\tfrac{\partial w}{\partial \beta} + \xi_\beta \end{aligned} \tag{3}$$

herein, A and B represent the Lamé parameters of substructures: (a) cylindrical shell: $\alpha = x$, $\beta = \theta$, $A = 1$, $B = R_1$; (b) conical shell: $\alpha = s$, $\beta = \theta$, $A = 1$, $B = s\cdot\sin\alpha_0$; (c) spherical shells: $\alpha = \varphi$, $\beta = \theta$, $A = R_1$, $B = R_1\sin\varphi$; (d) annular plate: $\alpha = r$, $\beta = \theta$, $A = 1$, $B = s\cdot\sin\alpha_0$, $\alpha_0 = \pi/2$.

According to the generalized Hooke's law, the relationships between the stresses (σ_α, σ_β, $\tau_{\alpha\beta}$, $\tau_{\alpha\beta}$, and $\tau_{\alpha\beta}$) and strains in terms of linearly elastic materials are given by [34]:

$$\begin{bmatrix} \sigma_\alpha \\ \sigma_\beta \\ \tau_{\beta z} \\ \tau_{\alpha z} \\ \tau_{\alpha\beta} \end{bmatrix} = \begin{bmatrix} \overline{Q}_{11}^k & \overline{Q}_{12}^k & \overline{Q}_{16}^k & 0 & 0 \\ \overline{Q}_{12}^k & \overline{Q}_{22}^k & \overline{Q}_{26}^k & 0 & 0 \\ \overline{Q}_{16}^k & \overline{Q}_{26}^k & \overline{Q}_{66}^k & 0 & 0 \\ 0 & 0 & 0 & \overline{Q}_{44}^k & \overline{Q}_{45}^k \\ 0 & 0 & 0 & \overline{Q}_{45}^k & \overline{Q}_{55}^k \end{bmatrix} \begin{bmatrix} \varepsilon_\alpha \\ \varepsilon_\beta \\ \gamma_{\beta z} \\ \gamma_{\alpha z} \\ \gamma_{\alpha\beta} \end{bmatrix} \tag{4}$$

where $\overline{Q}_{ij}^k (i, j = 1, 2, 4, 5, 6)$ represents the elastic stiffness coefficient of the kth layer material, which is closely related to Poisson's ratio and the Young's modulus of the material. Their expressions are as follows:

$$\begin{bmatrix} \overline{Q}_{11}^k & \overline{Q}_{12}^k & \overline{Q}_{16}^k & 0 & 0 \\ \overline{Q}_{12}^k & \overline{Q}_{22}^k & \overline{Q}_{26}^k & 0 & 0 \\ \overline{Q}_{16}^k & \overline{Q}_{26}^k & \overline{Q}_{66}^k & 0 & 0 \\ 0 & 0 & 0 & \overline{Q}_{44}^k & \overline{Q}_{45}^k \\ 0 & 0 & 0 & \overline{Q}_{45}^k & \overline{Q}_{55}^k \end{bmatrix} = T \begin{bmatrix} Q_{11}^k & Q_{12}^k & 0 & 0 & 0 \\ Q_{12}^k & Q_{22}^k & 0 & 0 & 0 \\ 0 & 0 & Q_{44}^k & 0 & 0 \\ 0 & 0 & 0 & Q_{55}^k & 0 \\ 0 & 0 & 0 & 0 & Q_{66}^k \end{bmatrix} T^T \tag{5}$$

$$T = \begin{bmatrix} c^2 & s^2 & 0 & 0 & -2sc \\ s^2 & c^2 & 0 & 0 & 2sc \\ 0 & 0 & c & s & 0 \\ 0 & 0 & -s & c & 0 \\ sc & -sc & 0 & 0 & c^2 - s^2 \end{bmatrix}, s = \sin\theta_k, c = \cos\theta_k \tag{6}$$

where θ_k represents the fiber-laying angle of the kth layer of laminated structures; $Q_{ij}^k (i, j = 1, 2, 4, 5, 6)$ denote the elastic parameters of materials, which can be obtained by [35]:

$$Q_{11}^k = \frac{E_1^k}{1 - \mu_{12}^k \mu_{21}^k},\ Q_{12}^k = \mu_{21}^k Q_{11}^k,\ Q_{22}^k = \frac{E_2^k}{1 - \mu_{12}^k \mu_{21}^k},\ Q_{44}^k = G_{23}^k,\ Q_{55}^k = G_{13}^k,\ Q_{66}^k = G_{12}^k \tag{7}$$

By integrating the stresses along thickness, the force and moment resultant of laminated thick shells can be obtained as:

$$\mathbf{N} = \begin{bmatrix} \mathbf{A} & \mathbf{B} & 0 \\ \mathbf{B} & \mathbf{D} & 0 \\ 0 & 0 & \mathbf{A_s} \end{bmatrix} \varepsilon \tag{8}$$

where **A**, **B**, and **D** are the tensile stiffness matrix, coupling stiffness matrix, and bending stiffness matrix, respectively. $\mathbf{A_s}$ is the shear stiffness matrix. The internal element expressions have been reported by Guo et al. [36].

2.3. Energy Expressions

The strain energy of the substructure is expressed as follows [10]:

$$U_V = \frac{1}{2} \int_S \begin{bmatrix} N_\alpha \varepsilon_\alpha^0 + N_\beta \varepsilon_\beta^0 + N_{\alpha\beta} \varepsilon_{\alpha\beta}^0 + M_\alpha \gamma_\alpha + \\ M_\beta \gamma_\beta + M_{\alpha\beta} \gamma_{\alpha\beta} + Q_\alpha \gamma_{\alpha z} + Q_\beta \gamma_{\beta z} \end{bmatrix} dS = U_S + U_{BC} + U_B \tag{9}$$

The strain energy of the structure consists of the potential energy with regard to structure stretching, bending, and stretching–bending coupling. By substituting (3) and (7) into (8), the unified expressions of the strain energy of each substructure can be obtained, of which the expressions are below:

$$U_S = \frac{1}{2}\iint_S \left\{ \begin{array}{l} A_{11}\left(\frac{1}{A}\frac{\partial u_0}{\partial \alpha} + \frac{v_0}{AB}\frac{\partial A}{\partial \beta} + \frac{w_0}{R_\alpha}\right)^2 + A_{11}\left(\frac{1}{B}\frac{\partial v_0}{\partial \beta} + \frac{u_0}{AB}\frac{\partial B}{\partial \alpha} + \frac{w_0}{R_\beta}\right)^2 + \\ \kappa A_{66}\left(\xi_\alpha - \frac{u_0}{R_\alpha} + \frac{1}{A}\frac{\partial w_0}{\partial \alpha}\right)^2 + A_{66}\left(\frac{B}{A}\frac{\partial}{\partial \alpha}\left(\frac{v_0}{B}\right) + \frac{A}{B}\frac{\partial}{\partial \beta}\left(\frac{u_0}{A}\right)\right)^2 + \\ 2A_{12}\left(\frac{1}{A}\frac{\partial u_0}{\partial \alpha} + \frac{v_0}{AB}\frac{\partial A}{\partial \beta} + \frac{w_0}{R_\alpha}\right)\left(\frac{1}{B}\frac{\partial v_0}{\partial \beta} + \frac{u_0}{AB}\frac{\partial B}{\partial \alpha} + \frac{w_0}{R_\beta}\right) \\ +\kappa A_{66}\left(\xi_\beta - \frac{v_0}{R_\beta} + \frac{1}{B}\frac{\partial w_0}{\partial \beta}\right)^2 \end{array} \right\} dS \quad (10)$$

$$U_{BC} = \iint_S \left\{ \begin{array}{l} B_{11}\left(\frac{1}{A}\frac{\partial u_0}{\partial \alpha} + \frac{v_0}{AB}\frac{\partial A}{\partial \beta} + \frac{w_0}{R_\alpha}\right)\left(\frac{1}{A}\frac{\partial \xi_\alpha}{\partial \alpha} + \frac{\xi_\beta}{AB}\frac{\partial A}{\partial \beta}\right) + \\ B_{11}\left(\frac{1}{B}\frac{\partial v_0}{\partial \beta} + \frac{u_0}{AB}\frac{\partial B}{\partial \alpha} + \frac{w_0}{R_\beta}\right)\left(\frac{1}{B}\frac{\partial \xi_\beta}{\partial \beta} + \frac{\xi_\alpha}{AB}\frac{\partial B}{\partial \alpha}\right) \\ +B_{12}\left(\frac{1}{A}\frac{\partial u_0}{\partial \alpha} + \frac{v_0}{AB}\frac{\partial A}{\partial \beta} + \frac{w_0}{R_\alpha}\right)\left(\frac{1}{B}\frac{\partial \xi_\beta}{\partial \beta} + \frac{\xi_\alpha}{AB}\frac{\partial B}{\partial \alpha}\right) + \\ B_{12}\left(\frac{1}{B}\frac{\partial v_0}{\partial \beta} + \frac{u_0}{AB}\frac{\partial B}{\partial \alpha} + \frac{w_0}{R_\beta}\right)\left(\frac{1}{A}\frac{\partial \xi_\alpha}{\partial \alpha} + \frac{\xi_\beta}{AB}\frac{\partial A}{\partial \beta}\right) + \\ B_{66}\left(\frac{B}{A}\frac{\partial}{\partial \alpha}\left(\frac{v_0}{B}\right) + \frac{A}{B}\frac{\partial}{\partial \beta}\left(\frac{u_0}{A}\right)\right)\left(\frac{B}{A}\frac{\partial}{\partial \alpha}\left(\frac{\xi_\beta}{B}\right) + \frac{A}{B}\frac{\partial}{\partial \beta}\left(\frac{\xi_\alpha}{A}\right)\right) \end{array} \right\} dS \quad (11)$$

$$U_B = \frac{1}{2}\iint_S \left\{ \begin{array}{l} D_{11}\left(\frac{1}{A}\frac{\partial \xi_\alpha}{\partial \alpha} + \frac{\xi_\beta}{AB}\frac{\partial A}{\partial \beta}\right)^2 + D_{11}\left(\frac{1}{B}\frac{\partial \xi_\beta}{\partial \beta} + \frac{\xi_\alpha}{AB}\frac{\partial B}{\partial \alpha}\right)^2 \\ +2D_{12}\left(\frac{1}{A}\frac{\partial \xi_\alpha}{\partial \alpha} + \frac{\xi_\beta}{AB}\frac{\partial A}{\partial \beta}\right)\left(\frac{1}{B}\frac{\partial \xi_\beta}{\partial \beta} + \frac{\xi_\alpha}{AB}\frac{\partial B}{\partial \alpha}\right) \\ +D_{66}\left(\frac{B}{A}\frac{\partial}{\partial \alpha}\left(\frac{\xi_\beta}{B}\right) + \frac{A}{B}\frac{\partial}{\partial \beta}\left(\frac{\xi_\alpha}{A}\right)\right)^2 \end{array} \right\} dS \quad (12)$$

Accordingly, the kinetic energy of the structure can be expressed as:

$$T = \frac{1}{2}\iint_S \left\{ \begin{array}{l} I_0\left[\left(\frac{\partial u_0}{\partial t}\right)^2 + \left(\frac{\partial v_0}{\partial t}\right)^2 + \left(\frac{\partial w_0}{\partial t}\right)^2\right] + \\ 2I_1\left(\frac{\partial u_0}{\partial t}\frac{\partial \xi_\alpha}{\partial t} + \frac{\partial v_0}{\partial t}\frac{\partial \xi_\beta}{\partial t}\right) + I_2\left[\left(\frac{\partial \xi_\alpha}{\partial t}\right)^2 + \left(\frac{\partial \xi_\beta}{\partial t}\right)^2\right] \end{array} \right\} dS \quad (13)$$

where

$$(I_0, I_1, I_2) = \int_{-h/2}^{h/2} \rho\left(1, z, z^2\right) dz \quad (14)$$

As stated earlier, the artificial spring technique is introduced to simulate boundary conditions and coupling relationships. Thus, the elastic potential energy stored in the boundary springs is expressed as follows:

$$U_{sp} = \frac{1}{2}\int_{-\frac{h}{2}}^{\frac{h}{2}}\int_0^{2\pi} \left\{ \begin{array}{l} \left[k_{u0}u_0^2 + k_{v0}v_0^2 + k_{w0}w_0^2 + k_{\varphi 0}\xi_\alpha^2 + k_{\theta 0}\xi_\beta^2\right]_{\alpha=0} \\ \left[k_{u1}u_0^2 + k_{v1}v_0^2 + k_{w1}w_0^2 + k_{\varphi 1}\xi_\alpha^2 + k_{\theta 1}\xi_\beta^2\right]_{\alpha=L_\alpha} \end{array} \right\} Bd\theta dz \quad (15)$$

The coupling potential energy due to coupling springs can be written as:

$$U_{cp}^i = \frac{1}{2}\int_{-h_s/2}^{h_s/2}\int_0^{\phi} \left\{ \begin{array}{l} k_u\left(u_0^i - u_0^{i+1}\right)^2 + k_v\left(v_0^i - v_0^{i+1}\right)^2 + k_w\left(w_0^i - w_0^{i+1}\right)^2 \\ +k_\varphi\left(\xi_x^i - \xi_r^{i+1}\right)^2 + k_\theta\left(\xi_\theta^i - \xi_\theta^{i+1}\right)^2 \end{array} \right\} R_s d\theta_s dz_s \quad (16)$$

$$U_{scp}^i = \frac{h_r^j}{2}\int_0^{\phi} \left\{ \begin{array}{l} k_u\left(u_0^j - u_0^{c,j}\right)^2 + k_v\left(v_0^2 - v_0^{c,j}\right)^2 + k_w\left(w_0^2 - w_0^{c,j}\right)^2 \\ +k_\varphi\left(\xi_x^2 - \xi_r^{c,j}\right)^2 + k_\theta\left(\xi_\theta^2 - \xi_\theta^{c,j}\right)^2 \end{array} \right\} R_s \quad (17)$$

As for the overall substructure, the spring potential energy can be expressed as:

$$U_{BC} = U_{sp} + \sum_{i=1}^{2} U_{cp}^i + \sum_{j=1}^{J}\left(U_{scp}^j + U_{cp}^j\right) \quad (18)$$

2.4. Displacement Admissible Functions and Solution Process

Appropriate displacement admissible functions play an important role in the accuracy of the vibration solutions. Here, the displacement components of the structure in the circumferential direction are expressed as trigonometric series expansions. Additionally, the Legendre polynomials are introduced to ensure and accelerate the convergence of the solution. Thus, the displacement variables are stated as

$$\begin{aligned}
u(\alpha,\beta,t) &= \sum_{m=0}^{M}\sum_{n=0}^{N} T_m(\alpha)[\cos(n\beta)\overline{u}_{mn}(t) + \sin(n\beta)\overline{u}_{mn}(t)] = \mathbf{U}(\alpha,\beta)\mathbf{u}(t) \\
v(\alpha,\beta,t) &= \sum_{m=0}^{M}\sum_{n=0}^{N} T_m(\alpha)[\sin(n\beta)\overline{v}_{mn}(t) + \cos(n\beta)\overline{v}_{mn}(t)] = \mathbf{V}(\alpha,\beta)\mathbf{v}(t) \\
w(\alpha,\beta,t) &= \sum_{m=0}^{M}\sum_{n=0}^{N} T_m(\alpha)[\cos(n\beta)\overline{w}_{mn}(t) + \sin(n\beta)\overline{w}_{mn}(t)] = \mathbf{W}(\alpha,\beta)\mathbf{w}(t) \\
\xi_\alpha(\alpha,\beta,t) &= \sum_{m=0}^{M}\sum_{n=0}^{N} T_m(\alpha)\left[\cos(n\beta)\overline{\xi}_{\alpha,mn}(t) + \sin(n\beta)\overline{\xi}_{\alpha,mn}(t)\right] = \mathbf{\Psi}_\alpha(\alpha,\beta)\boldsymbol{\xi}_\alpha(t) \\
\xi_\theta(\alpha,\beta,t) &= \sum_{m=0}^{M}\sum_{n=0}^{N} T_m(\alpha)\left[\sin(n\beta)\overline{\xi}_{\beta,mn}(t) + \cos(n\beta)\overline{\xi}_{\beta,mn}(t)\right] = \mathbf{\Psi}_\beta(\alpha,\beta)\boldsymbol{\xi}_\beta(t)
\end{aligned} \tag{19}$$

where N represents the maximum value of the calculated wave number n; T_m is the m-th polynomial; M is the highest order of the polynomial, also known as truncated values of axial/radial displacement expansions; $\mathbf{U}$, $\mathbf{V}$, $\mathbf{W}$, $\mathbf{\Psi}_\alpha$, and $\mathbf{\Psi}_\beta$ represent the function vectors; $\mathbf{u}$, $\mathbf{v}$, $\mathbf{w}$, $\boldsymbol{\xi}_\alpha$, and $\boldsymbol{\xi}_\beta$ are the coordinate vectors composed of unknown coefficients $(\overline{u}_{mn}, \overline{v}_{mn}, \overline{w}_{mn}, \overline{\xi}_{\alpha,mn}, \overline{\xi}_{\beta,mn})$.

The employed Legendre orthogonal polynomials are as follows [37]:

$$\begin{gathered}
T_0(\alpha) = 1, T_1(\alpha) = \alpha \\
(p+1)T_{p+1}(\alpha) = (2p+1)\alpha T_p(\alpha) - pT_{p-1}(\alpha),\ p \geq 2,\ \alpha \in [-1,1]
\end{gathered} \tag{20}$$

The energy functional of the laminated submarine-like structure is:

$$L = \left(T_S + T_C + T_L + \sum_{j=1}^{J} T_r^j\right) - \left(U_{S,S} + U_{S,C} + U_{S,L} + \sum_{j=1}^{J} U_r^j\right) - U_{BC} \tag{21}$$

Substituting Equation (18) into Equation (20) and performing a partial derivative operation on unknown coefficients, the vibration equation of the overall model is illustrated as

$$\left(\mathbf{K} - \omega^2\mathbf{M}\right)\vartheta = 0 \tag{22}$$

where $\mathbf{K}$ is the stiffness matrix and $\mathbf{M}$ is the mass matrix. ϑ is the global coordinate vector containing the unknown expansion coefficients of all substructures. By solving the eigenvalues and eigenvectors in (22), the vibration characteristics of the laminated submarine-like model can be obtained.

3. Numerical Calculation and Analysis

The free vibration characteristics of the assembled laminated submarine-like structure are further studied based on the established vibration model. To simplify the presentation, three annular plates (i.e., $J = 3$) are used in the follow-up analysis. The geometric parameters of the assembled structure are given as: $L = 6$m, $L_s = 1.5$m, $\alpha = 30°$, $R_i = 1$ m, $r_i = 0.4$ m, $h = hj\ r = 0.05$ m, $L_1 = 0$, $L_2 = L/2$, and $L_3 = L$, in which L_i represents the position of the ith annular plate in the axial directions of cylindrical shell. If there is no other explanation, all parameters remain as their initial value. In this article, the dynamic characteristics of structure subject to different classical boundary constraints are analyzed. Here, the spring stiffness related to the classical boundary is predefined as: free boundary (F): $k_u = 0$, $k_v = 0$, $k_w = 0$, $k_\alpha = 0$, $k_\beta = 0$; simply supported boundary (S): $k_u = 10^{14}$, $k_v = 10^{14}$, $k_w = 10^{14}$,

$k_\alpha = 0, k_\beta = 10^{14}$; shear–diaphragm (SD): $k_u = 0, k_v = 10^{14}, k_w = 10^{14}, k_\alpha = 0, k_\beta = 0$; clamped boundary (C): $k_u = k_v = k_w = k_\alpha = k_\beta = 10^{14}$. For the purpose of ensuring the convergence and solution accuracy of the developed model, without the loss of generality, here, the truncation values are configured as: $M_2 = 35$ for the cylindrical shell, $M_1 = M_3 = 20$ for spherical and cone shells, and $M_j = 18$ for all annular plates.

3.1. Numerical Verifications

In order to test the correctness and accuracy of the current method, several numerical examples are carried out in the following works. Table 1 exhibits the comparisons between results with the current method and FEM with regard to the laminated cylindrical shell, spherical shell, and conical shell. The relative parameters are: $L/R = 5$, $h/R = 0.05$, $[0°/90°/0°]$ for laminated schemes, and $E_2 = 10.6$ GPa, $E_1 = 138$ GPa, $G_{12} = 6$ GPa, $G_{13} = G_{23} = 3.9$ GPa $\mu_{12} = 0.28$, and $\rho = 1500$ kg/m^3 for the spherical shell; $[0°/90°]$ for laminated schemes, and $E_2 = 10$ GPa, $E_1 = 15\,E_2$, $G_{12} = 0.6\,E_2$, $G_{13} = 0.6\,E_2$, $G_{23} = 0.5\,E_2$, $\mu_{12} = 0.25$, and $\rho = 1500$ kg/m^3 for cylindrical and conical shells; $R_0 = 1$ m, $L = 3$ m, and $h = 0.05$ m for the cylindrical shell, and $R_s = 1$ m, $L_S = 3$ m, $h = 0.05$ m, $\alpha = 30°$. It is noted that the ABAQUS simulation models herein are divided into S4R elements of a 0.04 m × 0.04 m approximate global size. From Table 1, it is visible that results obtained with various methods are in good consistence with each other, demonstrating the correctness and accuracy of this method.

Table 1. Comparison of calculation frequency of spherical, conical, and cylindrical shell with laminated material.

BC	Method	Mode Number							
		1	2	3	4	5	6	7	8
		Spherical shell							
F	Present	419.71	508.58	522.89	568.42	603.55	614.46	635.78	670.55
	FEM	417.43	507.55	528.50	557.09	606.15	613.24	637.32	676.48
		Conical shell							
F	Present	21.20	53.97	59.74	95.55	135.80	145.74	197.64	204.71
	FEM	21.98	55.83	59.69	62.99	137.69	149.02	194.30	203.63
		Cylindrical shell							
C-C	Present	134.31	150.77	176.98	231.98	232.14	237.70	259.37	294.64
	FEM	134.23	150.74	177.04	231.97	232.05	237.47	259.80	294.65
F-F	Present	28.81	31.98	80.96	84.80	154.17	158.21	195.66	206.97
	FEM	28.82	30.90	81.04	84.30	154.42	158.09	195.24	206.44

Table 2 shows the natural frequencies of assembled structures corresponding to multiple circumferential wave numbers (i.e., $n = 1, 2, 3, 4$) under classical conditions. The material parameters of the coupling structure made of steel are as follows: $E_1 = E_2 = 206$ GPa, $\rho = 7800$ kg/m^3, and $\mu_{12} = \mu_{21} = 0.3$. The results are compared with those from the ABAQUS simulation model divided into 24454 S4R elements of a 0.04 m × 0.04 m approximate global size, of which the overall and axial section views are shown as Figure 3. Here, the abbreviated symbols (namely, C, S, and F) only represent the boundary constraints at the small-diameter edge of the cone shell, and the inner diameter of all plates are unconstrainted. As can be seen from Table 2, the maximum deviation between the frequency results computed by the two numerical methods is 2.07%, which reveals that the presented model can be effectively applied to analyze the frequency characteristics of the coupled laminated submarine-like structure within an acceptable error boundary. On the other hand, the mode shapes corresponding to the frequency results considering the rigidly clamped boundary in Table 2 are depicted in Figure 4. Note that, in view of the fact that annular plates are arranged inside the structure, to facilitate the observation of the mode deformation located on the annular plates, the modal shapes are given in the form of a half-section view. Clearly, the modal shapes obtained by the two methods are highly consistent with each other. For further validation, Figure 5 is shown to compare results obtained by the current method and FEM, taking laminate materials into account, of which the material properties are: E_2 = 10 GPa, E_1 = 15 E_2, G_{12} = 0.6 E_2, G_{13} = 0.6 E_2, G_{23} = 0.5 E_2, μ_{12}= 0.25, and ρ = 1500 kg/m^3, and the laminated scheme is [0°/90°]. From Figure 5, it is apparent that the numerical results from the two methods share good agreement with each other.

Table 2. Comparison of calculation frequency of laminated submarine-like structure under various boundary constraints.

n	m	C			S			F		
		Present	FEM	Deviation (%)	Present	FEM	Deviation (%)	Present	FEM	Deviation (%)
1	1	6.92	6.78	2.07	6.68	6.54	2.05	143.66	143.56	0.07
	2	94.52	94.46	0.06	94.17	94.11	0.06	213.96	214.28	0.15
	3	213.19	213.46	0.12	213.10	213.36	0.12	215.22	215.48	0.12
	4	214.37	214.67	0.14	214.37	214.67	0.14	304.28	304.32	0.01
2	1	158.15	157.92	0.15	158.15	157.92	0.15	157.93	157.68	0.16
	2	165.67	164.44	0.75	165.67	164.44	0.75	165.42	164.18	0.76
	3	323.77	322.63	0.35	320.16	320.06	0.03	256.37	256.40	0.01
	4	345.01	345.01	0.00	343.93	343.84	0.03	337.88	336.27	0.48
3	1	141.73	141.27	0.33	141.73	141.27	0.32	141.72	141.26	0.32
	2	159.32	158.62	0.44	159.32	158.61	0.45	159.31	158.60	0.45
	3	294.35	293.89	0.16	294.34	293.88	0.16	294.33	293.87	0.16
	4	300.91	298.65	0.76	300.90	298.65	0.75	300.89	298.65	0.75
4	1	201.39	201.38	0.00	201.39	201.38	0.00	201.39	201.38	0.00
	2	209.32	209.28	0.02	209.32	209.28	0.02	209.32	209.28	0.02
	3	292.68	292.03	0.22	292.68	292.03	0.22	292.68	292.03	0.22
	4	304.45	303.11	0.44	304.45	303.11	0.44	304.45	303.11	0.44

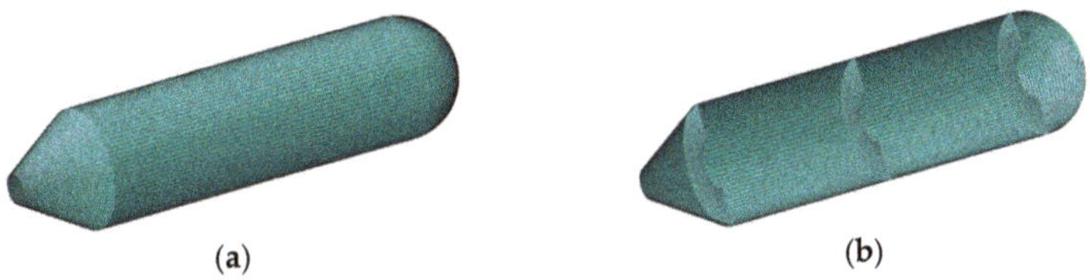

Figure 3. FEM model of submarine-like structure. (**a**) Overall view of FEM model of submarine-like structure; (**b**) Axial section view of FEM model of submarine-like structure.

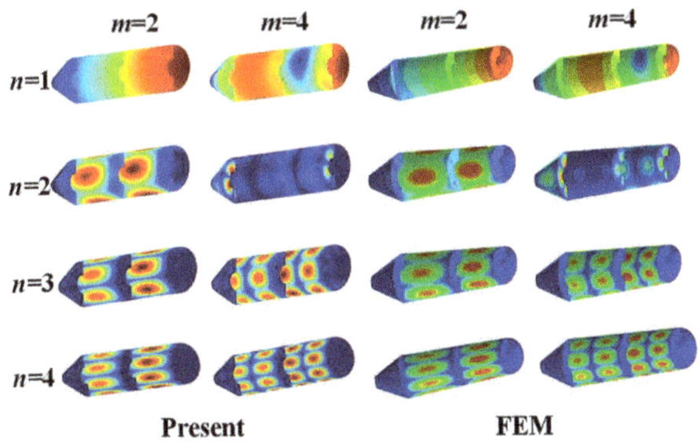

Figure 4. Comparison of modal shapes obtained using the two methods.

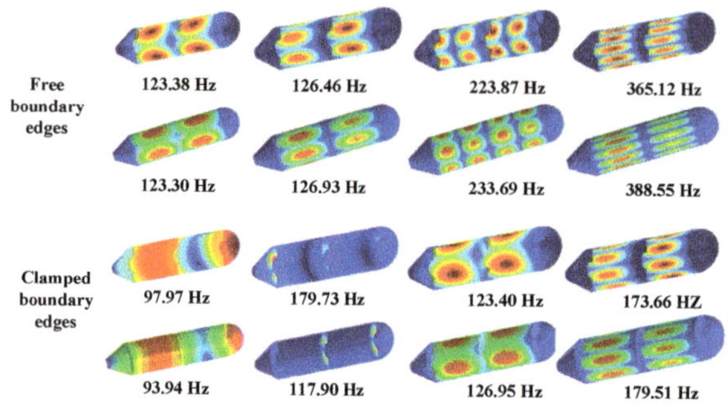

Figure 5. Comparison of modal shapes and corresponding frequencies obtained by the two methods.

3.2. Parametric Study

In what follows, the validated analysis model is employed in this section to analyze the effect of geometric parameters associated with the annular plates on the internal characteristics of the overall laminated submarine-like structure.

First, let us focus on the influence of the position and thickness of the annular plate in the middle of the coupling structure on the inherent characteristics of the coupling structure. Note that only the position of the middle annular plate is changed, whereas the positions of the plates at the first and last ends of the cylindrical shell remain unchanged. Figure 6 shows the variation in the first frequency values, which vary with plate's geometric parameters, and four constraints are included, with circumferential wave numbers of $n = 1$. It is worth noting that L_1 varying from 1 to 5 with a calculation step of 0.2 is the distance between the middle annular plate and the first fixed annular plate located at the first end of shell. h_2 is the thickness changing with 0.005 increments of the intermediate round plate, which is limited at the interval of [0.05, 0.15]. The material properties are as shown in Figure 5, and the laminated scheme is [0°/90°/0°/90°]. In order to facilitate the analysis, the first mode shapes with L_1 = 5 m and h_2 = 0.15 m are exhibited in Figure 7.

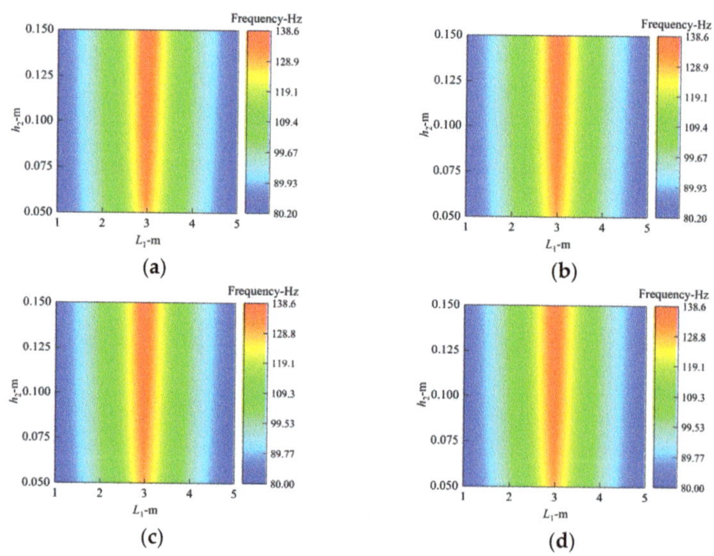

Figure 6. The first frequency value with respect to thickness and position of middle annular plate. (**a**) Clamped boundary. (**b**) Simply supported boundary. (**c**) Shear–diaphragm boundary. (**d**) Free boundary.

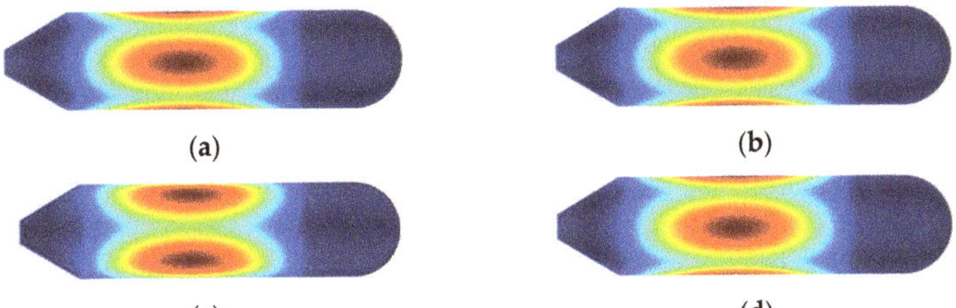

(**a**) (**b**) (**c**) (**d**)

Figure 7. The first mode shapes with L_1 = 5 m and h_2 = 0.15 m. (**a**) Clamped boundary. (**b**) Simply supported boundary. (**c**) Shear–diaphragm boundary. (**d**) Free boundary.

From Figure 6, no matter what the boundary conditions, the natural frequency of the structure will increase with the thickness of the annular plate. This may be explained by the fact that the growth in the thickness of the annular plate contributes to the improvement of the overall rigidity of the structure in the form of reinforcing ribs, which can increase the associated frequencies. Similarly, changes in the position of the annular plate will also have an important impact on the overall rigidity of the structure. For all the boundary conditions herein, when the second annular plate is located in the middle (L_1 = 3 m), the frequency of the structure is the largest, with an increment of over 72% compared to those of L_1 = 1 m or 5 m. When the annular plate moves to both ends, the frequency result decreases. In addition, one phenomenon can be easily observed that the value and trend of the first frequencies are similar for those in Figure 6a–d; the reason may be that the assembled structure owes strong structural stiffness to the coupling relationships between the substructures for one thing; for another, the mode shapes which are presented in Figure 6 and correspond to the first structural frequency are located in the cylindrical shell, of which both ends are combined to other substructures, resulting in the first structural frequency being insensitive to boundaries imposed on the conical shell.

Next, we study the comprehensive influence of the thickness and inner diameter of the three annular plates on the vibration characteristics of the structure in Figure 8. The circumferential wave number is set as n = 2, the material and frequency orders discussed here are consistent with Figure 6, the laminated scheme is [0°/90°/0°], and the boundaries are clamped and free boundaries. The thickness of the annular plates varies from 0.16 m to 0.01m with the change step being 0.01 m. The variation range of the inner diameter of the ring plate is [0.3, 0.6], and the step distance is set as 0.02 m. It can be clearly seen from Figure 8 that, for the same modal order, a similar frequency change trend is observed under the free and clamped conditions. In other words, as the annular plates become thicker and the inner diameter decreases, the natural frequency always continues to increase, and the maximum change rates exceed 97% and 16%. Moreover, the first frequency is similar to the same thickness of the annular plate and inner diameter, regardless of the impact of boundaries; the explanation for this is revealed in the discussion about Figure 6.

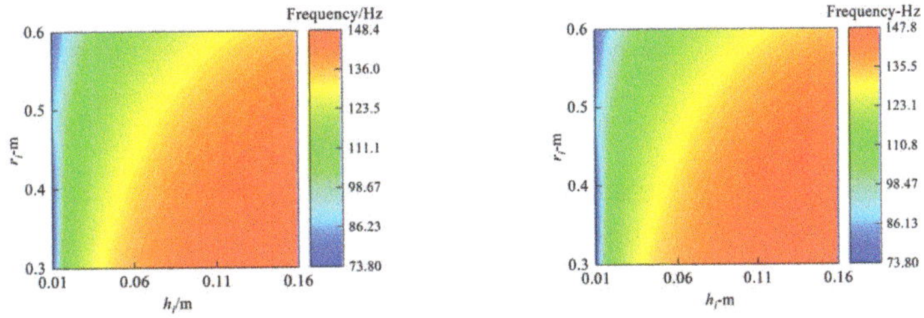

(**a**) (**b**)

Figure 8. Comprehensive influence of thickness and inner diameter of annular plates. (**a**) Clamped boundary. (**b**) Free boundary.

Finally, the influence of the laying angle on the first eight natural frequency is discussed in detail. As shown in Figure 9, the laminated scheme of the overall structure is $[0°/\theta°/0°]$, of which θ varies from -90 to 90 with an equal step of 10. The circumferential wave number is set as $n = 2$, and the other geometry and material parameters are as shown in Figure 8. From Figure 9, it is apparent that the trend of structural frequency is symmetrical at about $\theta = 0$; in the case of θ being below zero, the frequency increases over 23% with decreasing θ to about -45 and then decreases as θ varies from -45 to -90. Similarly, this trend appears as θ is over zero.

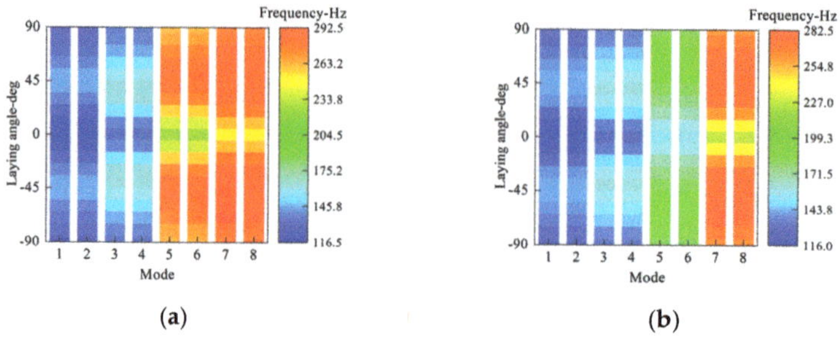

Figure 9. The first eight frequency values with respect to thickness and laminated scheme of the middle annular plate. (**a**) Clamped boundary. (**b**) Free boundary.

4. Conclusions

This article investigated the vibration performances of a laminated submarine-like structure assembled with multiple built-in annular plates. In view of the well-known FSDT of a shell, the unified energy expressions of structural components, including laminated cylindrical, conical, and spherical shells, as well as annular plates, were deduced. By means of the superposition principle, the energy expression of the assembled structure considering various circular plates was obtained. Finally, the vibration equation was obtained by implementing the Rayleigh–Ritz method. Verification analysis was carried out in numerical examples to present the correctness of the method. Furthermore, the influence mechanism of the annular plate on the dynamic characteristics of the assembled laminated submarine-like structure was investigated by placing the research variables on the geometric parameters of the annular plate. Thus, the following conclusions can be drawn:

(1) The established model has good predictive ability regarding the vibration characteristics of laminated submarine-like structures combined with multiple annular plates; the maximum deviation of submarine-like structures is only 2.07%.
(2) The influence of the plate's position on the inherent mechanical properties of the structure is closely related to the modal order, and the frequency of the structure is the largest when the annular plate is set in the middle of the cylindrical shell.
(3) The distribution of the laying angle is symmetric to about zero; structural frequencies increase first and then decrease with the increment of the laying angle to some extent.
(4) Boundary constraints imposed on the conical shell have little influence on structural vibration as mode shapes relate to the cylindrical shell.
(5) The reasonable design of the geometric parameters of the annular plate can effectively improve the rigidity of the structure. Increasing the thickness of all the annular plates, decreasing the inner radius, and regulating the laminated scheme have a remarkable influence on structural free vibration, and the maximum relative changing rates of frequency exceed 97%, 16%, and 23%, respectively.

Author Contributions: Z.C. obtained some of the numerical results and edited the manuscript. R.Z. presented the writing—original draft. S.H. analyzed the sequencing data and developed analysis tools. B.Q. carried out data curation. X.Z. reviewed this paper. All authors have read and agreed to the published version of the manuscript.

Funding: This research was funded by the National Natural Science Foundation of China (Grant No. 51705537). The authors gratefully acknowledge the financial support from the Postgraduate Scientific Research Innovation Project of Hunan Province. This research is also supported by the Fundamental Research Funds for the Central Universities of Central South University (Grant No. 2022ZZTS0174).

Institutional Review Board Statement: Not applicable.

Informed Consent Statement: Not applicable.

Data Availability Statement: Not applicable.

Conflicts of Interest: The authors declare no conflict of interest. The funders had no role in the design of the study; in the collection, analyses, or interpretation of data; in the writing of the manuscript, or in the decision to publish the results.

References

1. Zhong, R.; Wang, Q.; Hu, S.; Qin, B.; Shuai, C. Spectral element modeling and experimental investigations on vibration behaviors of imperfect plate considering irregular hole and curved crack. *J. Sound Vib.* **2022**, *529*, 116924. [CrossRef]
2. Xue, Y.; Jin, G.; Ma, X.; Chen, H.; Ye, T.; Chen, M.; Zhang, Y. Free vibration analysis of porous plates with porosity distributions in the thickness and in-plane directions using isogeometric approach. *Int. J. Mech. Sci.* **2019**, *152*, 346–362. [CrossRef]
3. Yin, Z.; Zhang, J.; Yang, F.; Ye, W.; Liu, J.; Lin, G. An efficient scaled boundary finite element approach in bending and bucking analysis of functionally graded piezoelectric plates. *Eng. Anal. Bound. Elem.* **2021**, *132*, 168–181. [CrossRef]
4. Zhao, J.; Choe, K.; Xie, F.; Wang, A.; Shuai, C.; Wang, Q. Three-dimensional exact solution for vibration analysis of thick functionally graded porous (FGP) rectangular plates with arbitrary boundary conditions. *Compos. Part B Eng.* **2018**, *155*, 369–381. [CrossRef]
5. Zhao, J.; Wang, Q.; Deng, X.; Choe, K.; Zhong, R.; Shuai, C. Free vibrations of functionally graded porous rectangular plate with uniform elastic boundary conditions. *Compos. Part B Eng.* **2019**, *168*, 106–120. [CrossRef]
6. Chen, Z.; Qin, B.; Zhong, R.; Wang, Q. Free in-plane vibration analysis of elastically restrained functionally graded porous plates with porosity distributions in the thickness and in-plane directions. *Eur. Phys. J. Plus* **2022**, *137*, 158. [CrossRef]
7. Liu, T.; Wang, Q.; Qin, B.; Wang, A. Modeling and experimental study on free vibration of plates with curved edges. *J. Sound Vib.* **2021**, *513*, 116417. [CrossRef]
8. Du, D.X.; Sun, W.; Yan, X.F.; Xu, K.P. Free vibration analysis of rotating thin-walled cylindrical shells with hard coating based on Rayleigh-Ritz method. *Proc. Inst. Mech. Eng. Part G J. Aerosp. Eng.* **2021**, *235*, 1170–1186. [CrossRef]
9. Li, X.; Wang, T.; Zhao, Y.; Wang, H. Free vibration of conical shells under arbitrary elastic boundary and simulation verification. *J. Phys. Conf. Ser.* **2021**, *1885*, 032059. [CrossRef]
10. Chen, Z.; Wang, A.; Qin, B.; Wang, Q.; Zhong, R. Investigation on free vibration and transient response of functionally graded graphene platelets reinforced cylindrical shell resting on elastic foundation. *Eur. Phys. J. Plus* **2020**, *135*, 582. [CrossRef]
11. An, K.; Kim, C.; Kim, K.; Kwak, S.; Ri, K.; Kim, H. Dynamic analysis of coupled composite laminated shells with elastic boundary condition using a domain decomposition method. *Eng. Res. Express* **2021**, *3*, 26. [CrossRef]
12. Choe, K.; Ri, K.; Zhang, Z.Y.; Shuai, C.J.; Wang, Q.S. A domain decomposition method for elastodynamic problems of functionally graded elliptic shells and panels with elastic constraints. *Thin-Walled Struct.* **2019**, *142*, 262–276. [CrossRef]
13. Li, H.; Hao, Y.; Zhang, W.; Liu, L.; Yang, S.; Wang, D. Vibration analysis of porous metal foam truncated conical shells with general boundary conditions using GDQ. *Compos. Struct.* **2021**, *269*, 114036. [CrossRef]
14. Saini, R.; Lal, R. Effect of Thermal Environment and Peripheral Loading on Axisymmetric Vibrations of Non-uniform FG Circular Plates via Generalized Differential Quadrature Method. *J. Vib. Eng. Technol.* **2021**, *9*, 873–886. [CrossRef]
15. Zhang, C.; Jin, G.; Wang, Z.; Sun, Y. Dynamic stiffness formulation and vibration analysis of coupled conical-ribbed cylindrical-conical shell structure with general boundary condition. *Ocean Eng.* **2021**, *234*, 109294. [CrossRef]
16. Araki, Y.; Samejima, T. Fourier series expansion type of spectral nodal Galerkin method for vibration analysis of cylindrical shells: Formulation and trial calculation. *Acoust. Sci. Technol.* **2016**, *37*, 254–257. [CrossRef]
17. Kim, K.; Choe, D.; Kwak, S.; Ri, Y.; Kim, C. Haar wavelet method for frequency analysis of the combined functionally graded shells with elastic boundary condition. *Thin-Walled Struct.* **2021**, *169*, 108340. [CrossRef]
18. Parashar, S.K.; Kumar, A. Three-dimensional analytical modeling of vibration behavior of piezoceramic cylindrical shells. *Arch. Appl. Mech.* **2015**, *85*, 641–656. [CrossRef]
19. Cao, Y.; Zhang, R.; Zhang, W.; Wang, J. Vibration Characteristics Analysis of Cylindrical Shell-Plate Coupled Structure Using an Improved Fourier Series Method. *Shock. Vib.* **2018**, *2018*, 9214189. [CrossRef]
20. Chen, Z.; Qin, B.; Wang, Q.; Zhong, R.; Wang, A. Vibration analysis of laminated open cylindrical shell coupled with rectangular plates. *Compos. Struct.* **2022**, *292*, 115607. [CrossRef]
21. Jin, G.; Ma, X.; Wang, W.; Liu, Z. An energy-based formulation for vibro-acoustic analysis of submerged submarine hull structures. *Ocean Eng.* **2018**, *164*, 402–413. [CrossRef]
22. Kim, K.; Kwak, S.; Choe, K.; Han, W.; Ri, Y.; Ri, K. Application of Haar wavelet method for free vibration of laminated composite conical–cylindrical coupled shells with elastic boundary condition. *Phys. Scr.* **2021**, *96*, 035223. [CrossRef]

23. Qu, Y.; Su, J.; Hua, H.; Meng, G. Structural vibration and acoustic radiation of coupled propeller-shafting and submarine hull system due to propeller forces. *J. Sound Vib.* **2017**, *401*, 76–93. [CrossRef]
24. Wang, J.; Yang, H.; Hua, H. Investigations on the vibrational and acoustic characteristics of a submarine-like system by experiments and simulations. *Proc. Inst. Mech. Eng. Part M J. Eng. Marit. Environ.* **2017**, *233*, 3–13. [CrossRef]
25. Xie, K.; Chen, M.; Zhang, L.; Xie, D. Wave based method for vibration analysis of elastically coupled annular plate and cylindrical shell structures. *Appl. Acoust.* **2017**, *123*, 107–122. [CrossRef]
26. Wang, Y.Q.; Liang, C.; Zu, J.W. Wave propagation in functionally graded cylindrical nanoshells based on nonlocal Flügge shell theory. *Eur. Phys. J. Plus* **2019**, *134*, 233. [CrossRef]
27. Chen, Z.; Wang, A.; Qin, B.; Wang, Q.; Zhong, R. Investigation on vibration of the functionally graded material–stepped cylindrical shell coupled with annular plate in thermal environment. *J. Low Freq. Noise Vib. Act. Control* **2022**, *41*, 85–111. [CrossRef]
28. Zhang, Y.; Shi, D.; He, D. A Spectro-geometric solution of dynamic characteristics of submarine-like structures. *Compos. Struct.* **2022**, *299*, 115976. [CrossRef]
29. Zhang, Y.; Shi, D.; He, D. Vibration Characteristics of a Laminated Composite Double-Cylindrical Shell System Coupled with a Variable Number of Annular Plates. *Materials* **2022**, *15*, 4246. [CrossRef]
30. Sobhani, E.; Masoodi, A.R.; Civalek, O.; Ahmadi-Pari, A.R. Agglomerated impact of CNT vs. GNP nanofillers on hybridization of polymer matrix for vibration of coupled hemispherical-conical-conical shells. *Aerosp. Sci. Technol.* **2022**, *120*, 107257. [CrossRef]
31. Sobhani, E.; Masoodi, A.R. Natural frequency responses of hybrid polymer/carbon fiber/FG-GNP nanocomposites paraboloidal and hyperboloidal shells based on multiscale approaches. *Aerosp. Sci. Technol.* **2021**, *119*, 107111. [CrossRef]
32. Bagheri, H.; Kiani, Y.; Bagheri, N.; Eslami, M.R. Free vibration of joined cylindrical–hemispherical FGM shells. *Arch. Appl. Mech.* **2020**, *90*, 2185–2199. [CrossRef]
33. Shi, X.; Zuo, P.; Zhong, R.; Guo, C.; Wang, Q. Thermal vibration analysis of functionally graded conical-cylindrical coupled shell based on spectro-geometric method. *Thin-Walled Structures* **2022**, *175*, 109138. [CrossRef]
34. Shao, D.; Wang, Q.; Shuai, C.; Gu, J. Investigation on dynamic performances of a set of composite laminated plate system under the influences of boundary and coupling conditions. *Mech. Syst. Signal Process.* **2019**, *132*, 721–747. [CrossRef]
35. Xie, X.; Zheng, H.; Jin, G. Integrated orthogonal polynomials based spectral collocation method for vibration analysis of coupled laminated shell structures. *Int. J. Mech. Sci.* **2015**, *98*, 132–143. [CrossRef]
36. Guo, C.; Liu, T.; Wang, Q.; Qin, B.; Wang, A. A unified strong spectral Tchebychev solution for predicting the free vibration characteristics of cylindrical shells with stepped-thickness and internal–external stiffeners. *Thin-Walled Struct.* **2021**, *168*, 108307. [CrossRef]
37. Baghaee, M.; Farrokhabadi, A.; Jafari-Talookolaei, R.-A. A solution method based on Lagrange multipliers and Legendre polynomial series for free vibration analysis of laminated plates sandwiched by two MFC layers. *J. Sound Vib.* **2019**, *447*, 42–60. [CrossRef]

materials

Article

Meshless Chebyshev RPIM Solution for Free Vibration of Rotating Cross-Ply Laminated Combined Cylindrical-Conical Shells in Thermal Environment

Zhen Li [1,2], Shuangwei Hu [1,2], Rui Zhong [1,2], Bin Qin [3,4,5] and Xing Zhao [1,2,*]

1 College of Mechanical and Electrical Engineering, Central South University, Changsha 410083, China
2 State Key Laboratory of High Performance Complex Manufacturing, Central South University, Changsha 410083, China
3 Key Laboratory of Traffic Safety on Track, Ministry of Education, School of Traffic & Transportation Engineering, Central South University, Changsha 410075, China
4 Joint International Research Laboratory of Key Technology for Rail Traffic Safety, Central South University, Changsha 410075, China
5 National & Local Joint Engineering Research Center of Safety Technology for Rail Vehicle, Central South University, Changsha 410075, China
* Correspondence: xingzhao@csu.edu.cn

Abstract: This paper provides a numerical solution to the vibration of a rotating cross-ply laminated combined conical-cylindrical shell in the thermal environment. Its numerical discrete solution method uses the meshless method. The combined shell assumed the temperature independence of material property is divided to the fundamental conical and cylindrical shell substructures, and the theoretical formulation for each substructure is derived based on the first order shear deformation theory (FSDT) and Hamilton's principle. The effects of the initial hoop tension and temperature change are considered through the kinetic energy reflecting the effects of centrifugal and Coriolis forces and additional strain energy by the nonlinear part of the Green–Lagrange strains. The substructures are then assembled according to the continuity conditions. The boundary and continuity conditions are simulated by introducing artificial virtual spring technology. The displacement component in the theoretical formulation is approximated using a meshless Chebyshev-RPIM shape function. The reliability of the method is verified by comparing with mature and reliable results. The free vibration characteristics of the rotating combined conical-cylindrical shell structure under various sizes, speeds and temperatures are given by numerical examples.

Keywords: meshfree method; laminated composite shell; rotating shell; free vibration analysis; thermal effect

Citation: Li, Z.; Hu, S.; Zhong, R.; Qin, B.; Zhao, X. Meshless Chebyshev RPIM Solution for Free Vibration of Rotating Cross-Ply Laminated Combined Cylindrical-Conical Shells in Thermal Environment. *Materials* **2022**, *15*, 6177. https://doi.org/10.3390/ma15176177

Academic Editor: Michele Bacciocchi

Received: 21 July 2022
Accepted: 2 September 2022
Published: 5 September 2022

1. Introduction

In the aerospace field, laminated shell structures are widely used in the shell structures of gas turbines and high-power aircraft engines [1–5]. In these high-end fields, the vibration of the structure will bring huge economic losses, so it is necessary to study its free vibration behavior before designing such a structure.

With the progress of computational science, many different methods such as the Haar wavelet discretization method [6], geometric analysis (IGA) method [7], spectral-Tchebychev solution technique [8], Ritz method [9,10] and finite element method [11–13] are employed for dynamic characteristics analysis of the composite structures. Ye et al. [14] derived the classical open shell formula on the basis of FSDT, and used Chebyshev polynomial to construct the displacement shape function, and solved the free vibration frequency of the open shell through the Rayleigh Ritz program. Caresta and Kessissoglou [15] reported a wave solution for the free vibrational frequencies of a homogeneous composite conical-cylindrical shell, where the displacement component was approximated by a power series. Tornabene et al. [16] reported a method for dynamic analysis of laminated hyperbolic shells and rotating panels on elastic foundations using GDQ. Li et al. [17] reported a Jacobi Ritz method for solving the free vibrations of laminated hyperbolic rotating shells with general boundary constraints. In addition, in recent decades, several studies on the dynamic mechanical properties of rotating structures in thermal environments have been developed. Shakouri et al. [18] reported the vibrational behavior of a conical shell of a functionally graded material with temperature-dependent material properties during rotation. Afshari [19] extended the generalized differential quadrature method to the solution of free vibration of a rotating conical shell reinforced by graphene nanomaterials. Bhangale et al. [20] reported a finite element method for analyzing the dynamics of functionally graded conical shells operating in high temperature environments. Tian et al. [21] obtained the free vibration and forced vibration solutions of the combined conical cylindrical shell by the dynamic stiffness method. Qin et al. [22] used the Rayleigh-Ritz method which based on the energy variation principle to solve the free vibration problem of a cylindrical shell-ring-plate coupling system. Singha et al. [23] analyzed the free vibration characteristics of rotating pretwisted sandwich conical shells in a thermal environment based on high-order shear deformation theory by using a finite element method. Talebitooti et al. [24] investigated the frequency behaviors of the joined conical-conical panel structures based on FSDT by applying Hamilton's principle. Soureshjani et al. [25] investigated the free vibration behaviors of composite joined conical-conical shell in the thermal environment by using a generalized differential quadrature method. Shi et al. [26] proposed an analytical model for investigating the vibration characteristics of a functionally graded conical-cylindrical coupled shell structure by using a spectro-geometric method. Ghasemi et al. [27] investigated the influences of distribution, mass and volume fractions of fiber, boundary conditions and lay-ups on the sensitivity of vibration behaviors of hybrid laminates cylindrical shell according to Kirchhoff Love's first approximation shell theory. Liu et al. [28] focused on the influences of rotation on the frequencies and critical speed of CNTs/fiber/polymer/metal laminates cylindrical shell based on Love's first approximation shell theory. Semnani et al. [29,30] analyzed the vibration behaviors of microshell under varied working conditions by using a finite element method.

In addition to the above methods, the development of meshless theory provides a brand-new idea for plate-shell vibration analysis. Based on the three-dimensional elastic theory, Kwak et al. [31,32] proposed a meshless strong-form solution for the free vibration of laminate shells. In their method, Chebyshev polynomials are introduced as basis functions in the construction of shape functions. In the meshless approach, the establishment of the system algebraic equations of the problem domain does not use a pre-defined mesh for domain discretization, but instead uses nodes [33]. Zarei et al. [34] constructed the displacement function of a prestressed laminate using meshless radial basis point interpolation and analyzed its vibration characteristics. Mellouli et al. [35] used the same method to build a vibrational analysis model of functionally graded carbon nanotube-reinforced shells. Zhang et al. [36] introduced the vibrational behavior of carbon nanotube-enhanced functionally graded triangular plates using a meshless method. Fallah and Delzendeh [37] studied the free vibration of laminates with meshless finite volume method (MFV) as the model solution method and moving least squares approximation to approximate the displacement component. Kwak et al. [38] combined the Chebyshev polynomial with the radial basis point interpolation method to construct the displacement shape function of the open laminated cylindrical shell with elliptical section, and solved its natural frequency.

The purpose of this paper is to study the vibration properties of a rotating cross-laminated conical-cylindrical shell in a thermal environment using meshless theory, considering that the combined structure is divided into cylindrical shell and conical shell structure, and the cylindrical shell is a special conical shell. Therefore, the equations of motion suitable for rotating conical shells are first established within the FSDT framework. Then, the two substructures are assembled by the continuity equation to obtain the equation of the overall structure. The effects of centrifugal force, Coriolis force, and temperature are considered in the equations of motion, and the displacement components involved are approximated using a meshless TRPIM shape function. The accuracy and reliability of the proposed method are verified through the convergence study and comparing with the results of the literature and ABAQUS. Finally, the effects of parameters such as geometry, temperature difference and rotational speed on the free vibration of the cross-ply laminated composite conical-cylindrical shell structure are studied. To sum up, the investigations of this paper can analyze the variation tendency of vibration characteristics of rotating cross-laminated conical-cylindrical shell in the thermal environment and provide the theoretical basis for the designation and manufacture of rotating cross-laminated conical-cylindrical shell structures which are used in aircraft, missiles, submarines, etc.

2. Theoretical Formulations

2.1. Description of the Model

Figure 1 shows a model of laminated combined conical-cylindrical shell rotating with rotating angular velocity Ω under the influence of temperature difference ΔT. The symbols L_1 and L_2 denote the lengths of the two meridians. The thickness of the combined shell is uniformly set to h. φ represent the semi-vertex angle of conical shell. The symbols R_1 and R_2 represent the radii at both ends of the conical shell, respectively. The cylindrical shell is connected at the big end of the conical shell, so the radius of the cylindrical shell is also R_2. The orthogonal curvilinear coordinate system (x, θ, z) is introduced into the middle surface of each substructure. The orthogonal coordinate system o-$x\theta z$ is established on the middle surface of the substructure, then the radius R of the random position on the conical shell is as follows:

$$R = R_1 + x \sin \varphi \tag{1}$$

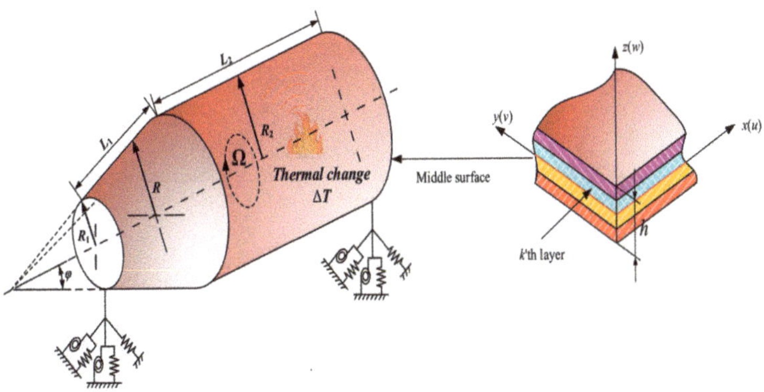

Figure 1. Geometry of rotating cross-ply combined conical-cylindrical shell in thermal environment.

2.2. Governing Equations and Boundary Conditions

According to the assumption of first-order shear deformation [39], the displacement $(\overline{u}, \overline{v}, \overline{w})$ of any position on the elastic structure can be represented by the displacement $(u, v, w, \psi_x, \psi_\theta)$ of the mid-plane.

$$\begin{cases} \overline{u}(x,\theta,z,t) = u(x,\theta,t) + z\psi_x(x,\theta,t) \\ \overline{v}(x,\theta,z,t) = v(x,\theta,t) + z\psi_\theta(x,\theta,t) \\ \overline{w}(x,\theta,z,t) = w(x,\theta,t) \end{cases} \tag{2}$$

Combined with the linear elasticity theory, the relationship between the stress and displacement of the shell is defined:

$$\begin{Bmatrix} \varepsilon_\alpha^0 \\ \varepsilon_\beta^0 \\ \gamma_{\alpha\beta}^0 \end{Bmatrix} = \begin{Bmatrix} \frac{1}{A}\frac{\partial u}{\partial \alpha} + \frac{v}{AB}\frac{\partial A}{\partial \beta} + \frac{w}{R_\alpha} \\ \frac{1}{B}\frac{\partial v}{\partial \beta} + \frac{u}{AB}\frac{\partial B}{\partial \alpha} + \frac{w}{R_\beta} \\ \frac{1}{A}\frac{\partial v}{\partial \alpha} - \frac{u}{AB}\frac{\partial A}{\partial \beta} + \frac{1}{B}\frac{\partial u}{\partial \beta} - \frac{v}{AB}\frac{\partial B}{\partial \alpha} \end{Bmatrix},$$
$$\begin{Bmatrix} \chi_\alpha \\ \chi_\beta \\ \chi_{\alpha\beta} \end{Bmatrix} = \begin{Bmatrix} \frac{1}{A}\frac{\partial \psi_\alpha}{\partial \alpha} + \frac{\psi_\beta}{AB}\frac{\partial A}{\partial \beta} \\ \frac{1}{B}\frac{\partial \psi_\beta}{\partial \beta} + \frac{\psi_\alpha}{AB}\frac{\partial B}{\partial \alpha} \\ \frac{1}{A}\frac{\partial \psi_\beta}{\partial \alpha} - \frac{\psi_\alpha}{AB}\frac{\partial A}{\partial \beta} + \frac{1}{B}\frac{\partial \psi_\alpha}{\partial \beta} - \frac{\psi_\beta}{AB}\frac{\partial B}{\partial \alpha} \end{Bmatrix}, \begin{Bmatrix} \gamma_{\beta z}^0 \\ \gamma_{\alpha z}^0 \end{Bmatrix} = \begin{Bmatrix} \frac{1}{B}\frac{\partial w}{\partial \beta} + \psi_\beta \\ \frac{1}{A}\frac{\partial w}{\partial \alpha} + \psi_\alpha \end{Bmatrix} \tag{3}$$

where $\varepsilon = \left\{\varepsilon_\alpha^0, \varepsilon_\beta^0, \gamma_{\alpha\beta}^0\right\}^T$ represents the normal strain and shear strain of the elastic element, and $\chi = \{\chi_\alpha, \chi_\beta, \chi_{\alpha\beta}\}^T$ represents the bending and torsional curvature changes of the elastic body. $\gamma = \left\{\gamma_{\beta z}^0, \gamma_{\alpha z}^0\right\}^T$ denotes transverse shear strain. A and B denote the Lamé parameters.

$$\begin{aligned} &\text{conicalshell}: \alpha = x, \beta = \theta, A = 1, B = R, R_\alpha = \infty, R_\beta = R/\cos\varphi \\ &\text{cylindricalshell}: \alpha = x, \beta = \theta, A = 1, B = R, R_\alpha = \infty, R_\beta = R \end{aligned} \tag{4}$$

The matrix form of the stress resultants-strain relationship of moderately thick cross-ply conical shell is as

$$\begin{bmatrix} \mathbf{N} \\ \mathbf{M} \\ \mathbf{Q} \end{bmatrix} = \begin{bmatrix} \mathbf{A} & \mathbf{B} & 0 \\ \mathbf{B} & \mathbf{D} & 0 \\ 0 & 0 & \mathbf{A_c} \end{bmatrix} \begin{bmatrix} \varepsilon \\ \chi \\ \gamma \end{bmatrix} \tag{5}$$

where $N = \{N_\alpha, N_\beta, N_{\alpha\beta}\}^T$, the internal element represents the in-plane force. $\boldsymbol{M} = \{M_\alpha, M_\beta, M_{\alpha\beta}\}^T$, the element represents the bending moment, and $\boldsymbol{Q} = \{Q_\beta, Q_\alpha\}^T$ is the shear force vector. $\boldsymbol{A}$ represents the tensile stiffness matrix, $\boldsymbol{B}$ is the bending stiffness matrix, and $\boldsymbol{D}$ is the coupled tensile bending stiffness matrix. $\boldsymbol{A_c}$ denotes the shear stiffness matrix. Their specific form is:

$$\boldsymbol{A} = \sum_{k=1}^{N} \begin{bmatrix} \overline{Q}_{11}^k & \overline{Q}_{12}^k & \overline{Q}_{16}^k \\ \overline{Q}_{12}^k & \overline{Q}_{22}^k & \overline{Q}_{26}^k \\ \overline{Q}_{16}^k & \overline{Q}_{26}^k & \overline{Q}_{66}^k \end{bmatrix} (z_{k+1} - z_k), \boldsymbol{B} = \frac{1}{2}\sum_{k=1}^{N} \begin{bmatrix} \overline{Q}_{11}^k & \overline{Q}_{12}^k & \overline{Q}_{16}^k \\ \overline{Q}_{12}^k & \overline{Q}_{22}^k & \overline{Q}_{26}^k \\ \overline{Q}_{16}^k & \overline{Q}_{26}^k & \overline{Q}_{66}^k \end{bmatrix} \left(z_{k+1}^2 - z_k^2\right)$$
$$\boldsymbol{D} = \frac{1}{3}\sum_{k=1}^{N} \begin{bmatrix} \overline{Q}_{11}^k & \overline{Q}_{12}^k & \overline{Q}_{16}^k \\ \overline{Q}_{12}^k & \overline{Q}_{22}^k & \overline{Q}_{26}^k \\ \overline{Q}_{16}^k & \overline{Q}_{26}^k & \overline{Q}_{66}^k \end{bmatrix} \left(z_{k+1}^3 - z_k^3\right), \boldsymbol{A_c} = k_c \sum_{k=1}^{N} \begin{bmatrix} \overline{Q}_{44}^k & \overline{Q}_{45}^k \\ \overline{Q}_{45}^k & \overline{Q}_{55}^k \end{bmatrix} (z_{k+1} - z_k) \tag{6}$$

$$\begin{aligned} &Q_{11}^k = \frac{E_1^k}{1-\mu_{12}^k\mu_{21}^k}, Q_{12}^k = \mu_{21}^k Q_{11}^k, Q_{22}^k = \frac{E_2^k}{1-\mu_{12}^k\mu_{21}^k}, Q_{44}^k = G_{23}^k \\ &Q_{55}^k = G_{13}^k, Q_{66}^k = G_{12}^k \end{aligned}$$

where N denotes the number of laying layers of the laminate, $k_c = 5/6$ is the shear correction coefficient and the symbol $\overline{Q}_{ij}^k$ denotes the elastic stiffness coefficient [38].

In the thermal environment, the thermal stress of the kth layer in the cross-layer is expressed as follows:

$$\begin{Bmatrix} \sigma_x^{Tk} \\ \sigma_\theta^{Tk} \\ \tau_{x\theta}^{Tk} \end{Bmatrix} = -\begin{bmatrix} Q_{11}^k & Q_{12}^k & 0 \\ Q_{12}^k & Q_{22}^k & 0 \\ 0 & 0 & Q_{66}^k \end{bmatrix} \begin{Bmatrix} \alpha_{11}^k \Delta T \\ \alpha_{22}^k \Delta T \\ \alpha_{12}^k \Delta T \end{Bmatrix} \begin{bmatrix} Q_{cxx} \\ Q_{c\theta\theta} \end{bmatrix} = \kappa_s \begin{bmatrix} A_{44} & A_{45} \\ A_{45} & A_{55} \end{bmatrix} \begin{Bmatrix} \gamma_{\theta r0}^p \\ \gamma_{xr0}^p \end{Bmatrix} \tag{7}$$

where the thermal expansion coefficients α_{ij}^{k} of the kth layer are given by

$$\{A_{ij}, B_{ij}, D_{ij}\} = \sum_{k=1}^{N_k} \int_{z_k}^{z_{k+1}} \overline{Q}_{ij}^{k}\{1, z, z^2\}dz \tag{8}$$

where δ_k and α_{ij} denote the fiber angle of the kth layer and linear thermal expansion coefficients along the principal axes of a layer, respectively.

The thermal strain can be written as nonlinear part of Green–Lagrange strain.

$$\begin{aligned}
\varepsilon_{\alpha}^{NL} &= \left(\frac{1}{A}\frac{\partial \overline{u}}{\partial \alpha} + \frac{1}{AB}\frac{\partial A}{\partial \beta}\overline{v} + \frac{\overline{w}}{R_\alpha}\right)^2 + \left(\frac{1}{A}\frac{\partial \overline{v}}{\partial \alpha} - \frac{1}{AB}\frac{\partial A}{\partial \beta}\overline{u}\right)^2 + \left(\frac{1}{A}\frac{\partial \overline{w}}{\partial \alpha} - \frac{\overline{u}}{R_\alpha}\right)^2 \\
\varepsilon_{\beta}^{NL} &= \left(\frac{1}{B}\frac{\partial \overline{u}}{\partial \beta} - \frac{1}{AB}\frac{\partial B}{\partial \alpha}\overline{v}\right)^2 + \left(\frac{1}{B}\frac{\partial \overline{v}}{\partial \beta} + \frac{1}{AB}\frac{\partial B}{\partial \alpha}\overline{u} + \frac{\overline{w}}{R_\beta}\right)^2 + \left(\frac{1}{B}\frac{\partial \overline{w}}{\partial \beta} - \frac{\overline{v}}{R_\beta}\right)^2 \\
\gamma_{\alpha\beta}^{NL} &= \left(\frac{1}{A}\frac{\partial \overline{u}}{\partial \alpha} + \frac{1}{AB}\frac{\partial A}{\partial \beta}\overline{v} + \frac{\overline{w}}{R_\alpha}\right)\left(\frac{1}{B}\frac{\partial \overline{u}}{\partial \beta} - \frac{1}{AB}\frac{\partial B}{\partial \alpha}\overline{v}\right) + \\
&\left(\frac{1}{A}\frac{\partial \overline{v}}{\partial \alpha} - \frac{1}{AB}\frac{\partial A}{\partial \beta}\overline{u}\right)\left(\frac{1}{B}\frac{\partial \overline{v}}{\partial \beta} + \frac{1}{AB}\frac{\partial B}{\partial \alpha}\overline{u} + \frac{\overline{w}}{R_\beta}\right) + \left(\frac{1}{A}\frac{\partial \overline{w}}{\partial \alpha} - \frac{\overline{u}}{R_\alpha}\right)\left(\frac{1}{B}\frac{\partial \overline{w}}{\partial \beta} - \frac{\overline{v}}{R_\beta}\right)
\end{aligned} \tag{9}$$

Substituting Equations (2) and (14) into Equation (13)

$$\varepsilon_x^{NL} = \left(\frac{\partial u}{\partial x}\right)^2 + \left(\frac{\partial v}{\partial x}\right)^2 + \left(\frac{\partial w}{\partial x}\right)^2 + z^2\left(\frac{\partial \psi_x}{\partial x}\right)^2 + z^2\left(\frac{\partial \psi_\theta}{\partial x}\right)^2 + 2z\frac{\partial u}{\partial x}\frac{\partial \psi_x}{\partial x} + 2z\frac{\partial v}{\partial x}\frac{\partial \psi_\theta}{\partial x} \tag{10}$$

$$\begin{aligned}
\varepsilon_\theta^{NL} = \frac{1}{R^2}\Bigg[&\left(\frac{\partial u}{\partial \theta}\right)^2 + \left(\frac{\partial v}{\partial \theta}\right)^2 + \left(\frac{\partial w}{\partial \theta}\right)^2 + 2\sin\varphi\left(\frac{u\partial v}{\partial \theta} - \frac{v\partial u}{\partial \theta}\right) + 2\cos\varphi\left(\frac{w\partial v}{\partial \theta} - \frac{v\partial w}{\partial \theta}\right) \\
&+u^2\sin^2\varphi + v^2 + w^2\cos^2\varphi + 2uw\sin\varphi\cos\varphi + 2z\left(\frac{\partial u}{\partial \theta}\frac{\partial \psi_x}{\partial \theta} + \frac{\partial v}{\partial \theta}\frac{\partial \psi_\theta}{\partial \theta}\right) \\
&+2z\sin\varphi\left(\frac{u\partial \psi_\theta}{\partial \theta} - \frac{\psi_\theta\partial u}{\partial \theta} + \frac{\psi_x\partial v}{\partial \theta} - \frac{v\partial \psi_x}{\partial \theta}\right) + 2z\cos\varphi\left(\frac{w\partial \psi_\theta}{\partial \theta} - \frac{\psi_\theta\partial w}{\partial \theta}\right) + 2zu\psi_x\sin^2\varphi \\
&+2zv\psi_\theta + 2zw\psi_x\sin\varphi\cos\varphi + z^2\left(\frac{\partial \psi_x}{\partial \theta}\right)^2 + z^2\left(\frac{\partial \psi_\theta}{\partial \theta}\right)^2 + 2z^2\sin\varphi\left(\frac{\psi_x\partial \psi_\theta}{\partial \theta} - \frac{\psi_\theta\partial \psi_x}{\partial \theta}\right) \\
&+z^2\psi_x^2\sin^2\varphi + z^2\psi_\theta^2\Bigg]
\end{aligned}$$

$$\begin{aligned}
\gamma_{x\theta}^{NL} = \frac{1}{R}\Bigg[&\frac{\partial u}{\partial x}\frac{\partial u}{\partial \theta} + \frac{\partial v}{\partial x}\frac{\partial v}{\partial \theta} + \frac{\partial w}{\partial x}\frac{\partial w}{\partial \theta} + z\frac{\partial u}{\partial x}\frac{\partial \psi_x}{\partial \theta} + z\frac{\partial u}{\partial \theta}\frac{\partial \psi_x}{\partial x} + z\frac{\partial v}{\partial x}\frac{\partial \psi_\theta}{\partial \theta} + z\frac{\partial v}{\partial \theta}\frac{\partial \psi_\theta}{\partial x} \\
&+z^2\frac{\partial \psi_x}{\partial x}\frac{\partial \psi_x}{\partial \theta} + z^2\frac{\partial \psi_\theta}{\partial x}\frac{\partial \psi_\theta}{\partial \theta} + \sin\varphi\frac{u\partial v}{\partial x} - \sin\varphi\frac{v\partial u}{\partial x} - \cos\varphi\frac{v\partial w}{\partial x} + \cos\varphi\frac{w\partial v}{\partial x} \\
&+z\sin\varphi\frac{u\partial \psi_\theta}{\partial x} - z\sin\varphi\frac{\psi_\theta\partial u}{\partial x} + z\sin\varphi\frac{\psi_x\partial v}{\partial x} - z\sin\varphi\frac{v\partial \psi_x}{\partial x} + z\cos\varphi\frac{w\partial \psi_\theta}{\partial x} \\
&-z\cos\varphi\frac{\psi_\theta\partial w}{\partial x} + z^2\sin\varphi\frac{\psi_x\partial \psi_\theta}{\partial x} - z^2\sin\varphi\frac{\psi_\theta\partial \psi_x}{\partial x}\Bigg]
\end{aligned}$$

In the thermal field, the strain energy of the structure is expressed as:

$$\begin{aligned}
U = U_e + U_T = &\frac{1}{2}\iint_\Omega \begin{pmatrix} N_x\varepsilon_x^0 + N_\theta\varepsilon_\theta^0 + N_{x\theta}\gamma_{x\theta}^0 + M_x\chi_x \\ +M_\theta\chi_\theta + M_{x\theta}\chi_{x\theta} + Q_\theta\gamma_{\theta z} + Q_x\gamma_{xz} \end{pmatrix} Rdxd\theta \\
&+\frac{1}{2}\iint_\Omega \sum_{k=1}^{N_k} \int_{Z_k}^{Z_{k+1}} \left(\sigma_x^{Tk}\varepsilon_x^{NL} + \sigma_\theta^{Tk}\varepsilon_\theta^{NL} + 2\tau_{x\theta}^{Tk}\gamma_{x\theta}^{NL}\right) Rdzdxd\theta
\end{aligned} \tag{11}$$

Meanwhile, when the shell rotates, an initial hoop tension will be generated due to centrifugal force, which will generate a part of the strain energy.

$$U_h = \frac{1}{2}\int_0^{2\pi}\int_0^{L} N_\theta^0 \varepsilon_\theta^{NL} R dx d\theta \tag{12}$$

where $N_\theta^0 = \rho h \Omega^2 R^2$ is the initial hoop tension, the unit of ρ is kg/m^3.

The kinetic energy is

$$T = \frac{1}{2}\int_0^{h}\int_0^{2\pi}\int_0^{L} \rho \vec{v} \cdot \vec{v} R dx d\theta dz \tag{13}$$

where $\vec{v}$ is absolute velocity vector.

$$\vec{v} = \dot{\vec{r}} + \Omega(-\cos\varphi \vec{i} + \sin\varphi \vec{k}) \times \vec{r} \tag{14}$$

where $\vec{r} = \overline{U}\vec{i} + \overline{V}\vec{j} + \overline{W}\vec{k}$ is the displacement vector.

When the shell is not affected by external force, according to the variational principle, the equilibrium equation and boundary conditions of the heated rotating cross-layer shell are deduced.

$$\delta \int_{t_1}^{t_2} (T - U - U_h) dt = 0 \tag{15}$$

The obtained governing equations are expressed as:

$$\boldsymbol{K}\boldsymbol{u} + \boldsymbol{C}\dot{\boldsymbol{u}} + \boldsymbol{M}\ddot{\boldsymbol{u}} = 0 \tag{16}$$

where the matrices $\boldsymbol{C}$ and $\boldsymbol{M}$ are expressed as:

$$\boldsymbol{M} = \begin{bmatrix} -I_0 & 0 & 0 & -I_1 & 0 \\ 0 & -I_0 & 0 & 0 & -I_1 \\ 0 & 0 & -I_0 & 0 & 0 \\ -I_1 & 0 & 0 & -I_2 & 0 \\ 0 & -I_1 & 0 & 0 & -I_2 \end{bmatrix}$$
$$\boldsymbol{C} = \begin{bmatrix} 0 & 2I_0\Omega\sin\varphi & 0 & 0 & 2I_1\Omega\sin\varphi \\ -2I_0\Omega\sin\varphi & 0 & -2I_0\Omega\cos\varphi & -2I_1\Omega\sin\varphi & 0 \\ 0 & 2I_0\Omega\cos\varphi & 0 & 0 & 2I_1\Omega\cos\varphi \\ 0 & 2I_1\Omega\sin\varphi & 0 & 0 & 2I_2\Omega\sin\varphi \\ -2I_1\Omega\sin\varphi & 0 & -2I_1\Omega\cos\varphi & -2I_2\Omega\sin\varphi & 0 \end{bmatrix} \tag{17}$$

where the inertia terms are

$$[I_0, I_1, I_2] = \int_{-h/2}^{h/2} \rho[1, z, z^2] dz \tag{18}$$

The boundary conditions obtained from Hamilton's principle are expressed as:

$$\boldsymbol{B}_c \boldsymbol{u} = \boldsymbol{0} \tag{19}$$

Combining Equations (21), (24) and (25), we can derive the governing equations and boundary conditions for the cross-layer shell in the thermal physics field.

$$\begin{cases} u\,(x,\theta,t) = U(x)\cos(n\theta+\omega t) \\ v\,(x,\theta,t) = V(x)\sin(n\theta+\omega t) \\ w\,(x,\theta,t) = W(x)\cos(n\theta+\omega t) \\ \psi_x\,(x,\theta,t) = \Psi_x(x)\cos(n\theta+\omega t) \\ \psi_\theta\,(x,\theta,t) = \Psi_\theta(x)\sin(n\theta+\omega t) \end{cases} \tag{20}$$

where ω and n denote the natural frequency and circumferential wave number, respectively.

Substituting Equation (20) into Equations (21) and (24), the one-dimensional governing equations and boundary conditions of rotating cross-ply conical shell in thermal environment are obtained.

$$\left(\boldsymbol{K}_x + \omega\boldsymbol{C}_x - \omega^2\boldsymbol{m}\right)\boldsymbol{U} = 0 \tag{21}$$

$$\boldsymbol{B}_x\boldsymbol{U} = 0 \tag{22}$$

where

$$U = [U(x)\ V(x)\ W(x)\ \Psi_x(x)\ \Psi_\theta(x)]^T \tag{23}$$

$$\boldsymbol{C}_x = \begin{bmatrix} 0 & -2I_0\Omega\sin\varphi & 0 & 0 & -2I_1\Omega\sin\varphi \\ -2I_0\Omega\sin\varphi & 0 & -2I_0\Omega\cos\varphi & -2I_1\Omega\sin\varphi & 0 \\ 0 & -2I_0\Omega\cos\varphi & 0 & 0 & -2I_1\Omega\cos\varphi \\ 0 & -2I_1\Omega\sin\varphi & 0 & 0 & -2I_2\Omega\sin\varphi \\ -2I_1\Omega\sin\varphi & 0 & -2I_1\Omega\cos\varphi & -2I_2\Omega\sin\varphi & 0 \end{bmatrix} \tag{24}$$

2.3. Meshfree TRPIM Shape Function

The radial point interpolation method is a newly developed meshless method, which is an important and widely used method for solving partial differential equations. The unknown displacement function $u(x)$ is approximated by using the RPIM difference of the polynomials and can be defined as in [38].

$$u(\boldsymbol{x}) = \sum_{i=1}^{n_r} R_i(\boldsymbol{x})a_i + \sum_{j=1}^{n_p} p_j(\boldsymbol{x})b_j = \boldsymbol{R}^T(\boldsymbol{x})\boldsymbol{a} + \boldsymbol{P}^T(\boldsymbol{x})\boldsymbol{b} \tag{25}$$

where $R_i(\boldsymbol{x})$ is the radial basis function (RBFS), and n_r is the number of nodes of the point x in the support domain. $p_j(\boldsymbol{x})$ is the polynomial in the space coordinate $\boldsymbol{x}^T = (x, y)$, and n_p represents the number of polynomials. If $n_p = 0$, it is a single radial basis function (RBFS), otherwise it is an RBF with n_p polynomial basis functions added. Generally, for a one-dimensional problem, the basis function of the polynomial is $p_j(x) = [1,x,\ldots,x^{np}]^T$, and in a two-dimensional problem, the polynomial basis is $p_j(x) = [1,x,y,\ldots,x^{np},xy^{np-1},\ldots,yx^{np-1},y^{np}]^T$. However, using a power function polynomial basis is often inaccurate in solving differential equations. Chebyshev polynomials have important applications in approximation theory. Corresponding interpolation polynomials minimize the Longo phenomenon and provide the best consistent approximation of polynomials in continuous functions. Therefore, this study uses Chebyshev polynomials as interpolation basis functions.

$$\boldsymbol{P}(\boldsymbol{x}) = \boldsymbol{T}(\boldsymbol{x}) = \{T_0(x)\quad T_1(x)\quad \cdots\quad T_p(x)\quad \cdots\}^T \tag{26}$$

where

$$T_p(x) = \cos[p\cos^{-1}(x)]\,,\ p = 0,\ 1,\ 2\cdots\mathrm{v} \tag{27}$$

The multi-quadrics (MQ) radial function with shape parameters αc and q are used in this paper.

$$\mathrm{v}R_i(x) = \left[r_i^2 + (\alpha_c d_c)^2\right]^q \tag{28}$$

where r_i denotes the distance between the supported point x_J (J = 1,2,n_r) in the supported domain and calculated node x_I. For the one-dimensional problem in this paper r_i = ∣x_J-x_I∣, d_c is a characteristic length related to the node spacing in the support domain of the compute node. When the nodes are evenly distributed, d_c is the distance between adjacent nodes. Otherwise, dc is the average node spacing within the node distribution domain.

In meshless theory, the size of the local support domain will affect the interpolation accuracy, and a suitable size of the supported domain should be selected [32]. The size of the supported domain of the calculated node can be characterized as follows.

$$\mathrm{v}d_s = \alpha_s d_c \tag{29}$$

where α_s represents the scale factor of the support domain.

In order to determine the coefficient vectors $\boldsymbol{a}$ and $\boldsymbol{b}$ in Equation (30), a support field for calculated node x_I needs to be formed, which includes n_r field nodes. Let Equation (30) satisfy the calculation of n node values around point x_I, which yields n_r linear equations. The matrix of these equations can be expressed as the following form.

$$\boldsymbol{U}_s = \boldsymbol{R}_0\boldsymbol{a} + \boldsymbol{T}_{n_t}\boldsymbol{b} \tag{30}$$

where $\boldsymbol{R}_0$ represents the RBFs matrix and $\boldsymbol{T}_{nt}$ is the Chebyshev polynomial matrix [30]. The coefficient vector a of RBFs is expressed as follow.

$$\boldsymbol{a} = \begin{Bmatrix} a_1 & a_2 & \cdots & a_{n_r} \end{Bmatrix}^T \tag{31}$$

The coefficient vector $\boldsymbol{b}$ of the Chebyshev polynomial basis function is written as follow:

$$\boldsymbol{b} = \begin{Bmatrix} b_1 & b_2 & \cdots & b_{n_t} \end{Bmatrix}^T \tag{32}$$

Since there are n_r + n_t unknowns in Equation (35), a unique solution cannot be obtained, so it is necessary to add n_r equations through the following constraints to make the coefficient matrix of the eqiuation system full rank.

$$\sum_{i=1}^{n_r} T_j(x_i)a_i = \boldsymbol{T}_{n_t}^T\boldsymbol{a} = 0\ ,\ j = 1,2,\cdots,n_t \tag{33}$$

Combining Equations (35) and (38), the matrix representation of the following system of equations can be generated.

$$\overline{\boldsymbol{U}}_s = \begin{Bmatrix} \boldsymbol{U}_s \\ \boldsymbol{0} \end{Bmatrix} = \begin{bmatrix} \boldsymbol{R}_0 & \boldsymbol{T}_{n_t} \\ \boldsymbol{T}_{n_t}^T & \boldsymbol{0} \end{bmatrix} \begin{Bmatrix} \boldsymbol{a} \\ \boldsymbol{b} \end{Bmatrix} = \boldsymbol{G}\boldsymbol{a}_0 \tag{34}$$

where

$$\boldsymbol{a}_0 = \begin{Bmatrix} a_1 & a_2 & \cdots & a_{n_r} & b_1 & b_2 & \cdots & b_{n_t} \end{Bmatrix}^T \tag{35}$$

$$\overline{\boldsymbol{U}}_s = \begin{Bmatrix} u_1 & u_2 & \cdots & u_{n_r} & 0 & \cdots & 0 \end{Bmatrix}^T \tag{36}$$

From Equation (39)

$$\boldsymbol{a}_0 = \begin{Bmatrix} \boldsymbol{a} \\ \boldsymbol{b} \end{Bmatrix} = \boldsymbol{G}^{-1}\overline{\boldsymbol{U}}_s \tag{37}$$

Substituting Equation (42) into Equation (30)

$$\begin{aligned} u(\boldsymbol{x}) &= \boldsymbol{R}^T(\boldsymbol{x})\boldsymbol{a} + \boldsymbol{T}^T(\boldsymbol{x})\boldsymbol{b} = \{\ \boldsymbol{R}^T(\boldsymbol{x})\quad \boldsymbol{T}^T(\boldsymbol{x})\ \}\begin{Bmatrix} a \\ b \end{Bmatrix} \\ &= \{\ \boldsymbol{R}^T(\boldsymbol{x})\quad \boldsymbol{T}^T(\boldsymbol{x})\ \}\boldsymbol{G}^{-1}\overline{\boldsymbol{U}}_s = \boldsymbol{\Phi}^{-T}(\boldsymbol{x})\overline{\boldsymbol{U}}_s \end{aligned} \tag{38}$$

Then, the Chebyshev-RPIM shape function is expressed as follow.

$$\begin{aligned}\overline{\boldsymbol{\Phi}}^T(\boldsymbol{x}) &= \{\ \boldsymbol{R}^T(\boldsymbol{x}) \quad \boldsymbol{T}^T(\boldsymbol{x})\ \}\boldsymbol{G}^{-1} \\ &= \{\ \Phi_1(\boldsymbol{x}) \quad \Phi_2(\boldsymbol{x}) \quad \cdots \quad \Phi_{n_r}(\boldsymbol{x}) \quad \Phi_{n_r+1}(\boldsymbol{x}) \quad \cdots \quad \Phi_{n_r+n_t}(\boldsymbol{x})\ \}\end{aligned} \tag{39}$$

Delete unnecessary terms in the Chebyshev-RPIM shape function above, and obtain the Chebyshev-RPIM shape function corresponding to the final node displacement.

$$\boldsymbol{\Phi}^T(\boldsymbol{x}) = \{\Phi_1(\boldsymbol{x}) \quad \Phi_2(\boldsymbol{x}) \quad \cdots \quad \Phi_{n_r}(\boldsymbol{x})\} \tag{40}$$

Through the above derivation, the displacement components of the nodes can be expressed as follows.

$$u(\boldsymbol{x}) = \boldsymbol{\Phi}^T(\boldsymbol{x})\boldsymbol{U}_s = \sum_{i=1}^{n_r} \Phi_i u_i \tag{41}$$

$$\boldsymbol{U}_s = \{u_1, u_2, \cdots, u_{n_r}\} \tag{42}$$

2.4. Discretization of Governing Equations and Boundary Conditions

The substructure of the combined structure is discretized using N nodes, and the displacement approximation function at node x_I is represented by a Chebyshev-RPIM shape function.

$$\boldsymbol{U}(x_I) = \{u_I \quad v_I \quad w_I \quad \psi_{xI} \quad \psi_{\theta I}\}^T = \boldsymbol{\Phi}^T(x_I)\boldsymbol{U}_s \tag{43}$$

$$\boldsymbol{\Phi}^T(x_I) = [\Phi_1 \boldsymbol{I}_5 \quad \Phi_2 \boldsymbol{I}_5 \quad \cdots \quad \Phi_{Ns}\boldsymbol{I}_5] \tag{44}$$

$$\boldsymbol{U}_s = [u_1 \quad v_1 \quad w_1 \quad \psi_{x1} \quad \psi_{\theta 1} \quad \cdots \quad u_{Ns} \quad v_{Ns} \quad w_{Ns} \quad \psi_{xNs} \quad \psi_{\theta Ns}]^T \tag{45}$$

where N_s represents the number of nodes covered by the support domain, $\boldsymbol{I}_5$ represents a 5 × 5 identity matrix.

Substituting Equation (47) into Equation (26) to obtain the discretized governing equation represented by node information.

$$\left(\boldsymbol{K}_{xI} + \omega \boldsymbol{C}_{xI} - \omega^2 \boldsymbol{m}_I\right)\boldsymbol{U}_s = \boldsymbol{0} \tag{46}$$

where the nodal matrices $\boldsymbol{K}_{xI}$, $\boldsymbol{C}_{xI}$ and $\boldsymbol{m}_I$ are as follows.

$$\boldsymbol{K}_{xI} = \boldsymbol{K}_x\boldsymbol{\Phi}_I^T,\ \boldsymbol{C}_{xI} = \boldsymbol{C}_x\boldsymbol{\Phi}_I^T,\ \boldsymbol{m}_{xI} = \boldsymbol{m}_x\boldsymbol{\Phi}_I^T\,(x = \mathrm{co},\ \mathrm{cy}) \tag{47}$$

Similarly, the discrete equations for whole system are obtained by assembling those of each node according to the node number [40].

Substituting Equation (47) into Equation (27) to discretize boundary condition.

$$\boldsymbol{B}_x\boldsymbol{\Phi}_I^T\boldsymbol{U}_s = 0 \tag{48}$$

2.5. Continuous Condition

The governing equations and boundary equations of the substructure have been deduced and discretized before, but a complete solution system has not been established. The combined structure can be divided into conical shell and cylindrical shell. According to their geometric characteristics, considering their displacement continuity and physical coordination, the right boundary of the conical shell and the left boundary of the cylindrical shell can be modified as follows.

$$\begin{cases} N_x + k_b(u_{co} - u_{cy}\cos\varphi - w_{cy}\sin\varphi) = 0 \\ N_{x\theta} + k_b(v_{co} - v_{cy}) = 0 \\ Q_x + k_b(w_{co} + u_{cy}\sin\varphi - w_{cy}\cos\varphi) = 0 \quad : \text{Right boundary of conical shell} \\ M_x + k_b(\psi_{xco} - \psi_{xcy}) = 0 \\ M_{x\theta} + k_b(\psi_{\theta co} - \psi_{\theta cy}) = 0 \end{cases}$$
$$\begin{cases} N_x - k_b(u_{cy} - u_{co}\cos\varphi + w_{co}\sin\varphi) = 0 \\ N_{x\theta} - k_b(v_{cy} - v_{co}) = 0 \\ Q_x - k_b(w_{cy} - u_{co}\sin\varphi - w_{co}\cos\varphi) = 0 \quad : \text{Left boundary of cylindrical shell} \\ M_x - k_b(\psi_{xcy} - \psi_{xco}) = 0 \\ M_{x\theta} - k_b(\psi_{\theta cy} - \psi_{\theta co}) = 0 \end{cases} \tag{49}$$

where k_b denotes the connection stiffness between substructures, and symbols co and cy denote conical and cylindrical shells, respectively. The matrix form of the continuous condition can be written as follows.

$$\begin{aligned} &\boldsymbol{B}_{xco}\boldsymbol{\Phi}_{co}^T U_{sco} + \boldsymbol{K}_{12}\boldsymbol{\Phi}_{cy}^T \boldsymbol{U}_{scy} = \boldsymbol{0} : \text{Right boundary of conical shell} \\ &\boldsymbol{B}_{xcy}\boldsymbol{\Phi}_{cy}^T \boldsymbol{U}_{scy} + \boldsymbol{K}_{12}\boldsymbol{\Phi}_{co}^T \boldsymbol{U}_{sco} = \boldsymbol{0} : \text{Left boundary of cylindrical shell} \end{aligned} \tag{50}$$

where $\boldsymbol{U}_{sco}$ and $\boldsymbol{U}_{scy}$ are the displacement vectors of the nodes of the cylindrical shell and the conical shell on the coupling interface, respectively. The coupled stiffness matrices $\boldsymbol{K}_{12}$ and $\boldsymbol{K}_{21}$ are as follows.

$$\boldsymbol{K}_{12} = \begin{bmatrix} -k_b\cos\varphi & 0 & -k_b\sin\varphi & 0 & 0 \\ 0 & -k_b & 0 & 0 & 0 \\ k_b\sin\varphi & 0 & -k_b\cos\varphi & 0 & 0 \\ 0 & 0 & 0 & -k_b & 0 \\ 0 & 0 & 0 & 0 & -k_b \end{bmatrix} \tag{51}$$

$$\boldsymbol{K}_{21} = \begin{bmatrix} k_b\cos\varphi & 0 & -k_b\sin\varphi & 0 & 0 \\ 0 & k_b & 0 & 0 & 0 \\ k_b\sin\varphi & 0 & k_b\cos\varphi & 0 & 0 \\ 0 & 0 & 0 & k_b & 0 \\ 0 & 0 & 0 & 0 & k_b \end{bmatrix} \tag{52}$$

Finally, matrix assembly is performed to obtain the vibration control equation of the overall structure.

$$(\boldsymbol{K} + \omega \boldsymbol{C} - \omega^2 \boldsymbol{m})\boldsymbol{U} = \boldsymbol{0} \tag{53}$$

where

$$\boldsymbol{K} = \begin{bmatrix} \boldsymbol{K}_{coI} & \boldsymbol{K}_{12}\boldsymbol{\Phi}_{co}^T \\ \boldsymbol{K}_{21}\boldsymbol{\Phi}_{cy}^T & \boldsymbol{K}_{cyI} \end{bmatrix} \tag{54}$$

$$\boldsymbol{C} = \begin{bmatrix} \boldsymbol{C}_{coI} & \boldsymbol{0} \\ \boldsymbol{0} & \boldsymbol{C}_{cyI} \end{bmatrix} \tag{55}$$

$$\boldsymbol{m} = \begin{bmatrix} \boldsymbol{m}_{coI} & \boldsymbol{0} \\ \boldsymbol{0} & \boldsymbol{m}_{cyI} \end{bmatrix} \tag{56}$$

The natural frequency of the conical-cylindrical composite structure in the thermal environment is obtained by the harmonic response method.

3. Numerical Results and Discussions

This paper provides a meshless free vibration analysis model of a rotating combined conical-cylindrical shell structure in a thermal environment. The proposed method is compiled with MATLAB software. The number of nodes and the size of the support domain will affect the convergence effect of the algorithm. After obtaining the appropriate support domain size and number of nodes through convergence analysis, the numerical results are compared with finite element software or published literature to ensure the reliability and accuracy of the proposed method. Then, focusing on structural characteristic parameters and the effect of external physics on structural frequencies, some parametric study cases are provided. Unless otherwise stated, the natural frequencies of the considered combined shells are expressed in the dimensionless parameters as $\omega^* = \omega R_1\sqrt{\rho/E_2}$ and the material properties of the layers are given as: E_1 = 175 GPa, E_2 = 32 GPa, μ = 0.25, $G_{12} = G_{13}$ = 12 GPa, G_{23} = 5.7 GPa, ρ = 1760 kg/m^3, $\alpha_{11} = 1.2 \times 10^{-6}$, $\alpha_{22} = 2.3 \times 10^{-6}$ and $\alpha_{12} = 0$. The symbols C, F, and S are used to represent the tightened boundary conditions, free boundary conditions and simply supported boundary conditions, respectively. The corresponding boundaries are described as follows: C: $k = k_v = k_w = k_x = k_\theta = 10^{14}$. S: $k_u = k_v = k_w = k_\theta = 10^{14}$. $k_x = 0$, F: $k_u = k_v = k_w = k_x = k_\theta = 0$. Then define boundary rules. For example, CF represents that the boundary of the conical shell segment is a fixed boundary, and the boundary of the cylindrical shell segment is a free boundary.

3.1. Verification and Convergence Study

First, according to the basic theory of the meshless method, the key factor affecting the convergence of numerical results is the number of nodes. Therefore, before the numerical comparison and parametric analysis, the advanced convergence analysis is carried out to ensure that the obtained calculation results are stable. Table 1 shows the convergence results of the frequency parameter Ω^* ($n = m = 1$) of the non-rotating cross-layer cylindrical-conical shell under the classical boundary conditions, and the corresponding geometric dimensions are: R_1 = 0.5 m, L_1 = 1 m, L_2 = 2 m, h = 0.05 m, $\varphi = 30°$, ΔT = 0 K; the lamination scheme is $\delta_k = [0°/90°]$. The research results show that, no matter what kind of boundary, when $N \geq 9$ (N is node number), the numerical results are stable and the convergence speed is faster.

Table 1. Variation of dimensionless frequencies on number of nodes (m = 1).

N	CC		SS		FC		CF	
	n = 1	n = 2	n = 1	n = 2	n = 1	n = 2	n = 1	n = 2
5	0.2328	0.1812	0.2232	0.1754	0.1171	0.0803	0.0494	0.0339
6	0.2284	0.1790	0.2262	0.1761	0.1127	0.0825	0.0464	0.0317
7	0.2306	0.1805	0.2255	0.1754	0.1142	0.0833	0.0457	0.0317
8	0.2299	0.1798	0.2262	0.1754	0.1135	0.0840	0.0464	0.0324
9	0.2299	0.1798	0.2262	0.1754	0.1135	0.0840	0.0457	0.0317
10	0.2299	0.1798	0.2262	0.1754	0.1135	0.0840	0.0457	0.0317
11	0.2299	0.1798	0.2262	0.1754	0.1135	0.0840	0.0457	0.0317
12	0.2299	0.1798	0.2262	0.1754	0.1135	0.0840	0.0457	0.0317
13	0.2299	0.1798	0.2262	0.1754	0.1135	0.0840	0.0457	0.0317
14	0.2299	0.1798	0.2262	0.1754	0.1135	0.0840	0.0457	0.0317
15	0.2299	0.1798	0.2262	0.1754	0.1135	0.0840	0.0457	0.0317
16	0.2299	0.1798	0.2262	0.1754	0.1135	0.0840	0.0457	0.0317
17	0.2299	0.1798	0.2262	0.1754	0.1135	0.0840	0.0457	0.0317
18	0.2299	0.1798	0.2262	0.1754	0.1135	0.0840	0.0457	0.0317
19	0.2299	0.1798	0.2262	0.1754	0.1135	0.0840	0.0457	0.0317
20	0.2299	0.1798	0.2262	0.1754	0.1135	0.0840	0.0457	0.0317
21	0.2299	0.1798	0.2262	0.1754	0.1135	0.0840	0.0457	0.0317

In the previous convergence analysis, it has been determined that the meshless theory is applied to the structural vibration analysis, and the obtained results have good stability.

However, it has not been demonstrated whether the obtained results have a high level of confidence. Therefore, it is necessary to further compare the results obtained in this study with the existing publications or the results obtained by the finite element software ABAQUS. Table 2 compares the vibration frequencies of the non-rotating combined conical-cylindrical shell, considering no temperature difference between the inside and outside of the shell. The dimensions of the structure are:R_1 = 0.4226 m, $\varphi = \pi/6$, $L_2 = R_2 = 1$ m and h = 0.01 m. The material properties are: E = 211 GPa, ρ = 7800 kg/m^3, μ = 0.3. The dimensionless frequency of a non-rotating combined conical-cylindrical shell structure is defined as: $\omega^* = \omega R_2\sqrt{\rho(1-\mu^2)/E}$. The results obtained by the meshless method are compared with the published literature [10] and [15], and the difference between the results obtained by the meshless method and the literature is very small. Table 3 compares the frequency results obtained by different numerical methods for rotating isotropic combined conical-cylindrical shells. The boundary conditions, geometry and Poisson's ratio of the combined structure are the same as those in Table 2, and the rotational speeds considered are 0.01 rad/s, 100 rad/s, and 500 rad/s, respectively. The comparison results show that the method in this paper is in good agreement with the results in the literature. Finally, it is verified that the model established in this paper can be applied to the structural vibration solution in a thermal environment. In Table 4, the vibration frequency of the non-rotating laminated combined conical-cylindrical shell structure in a thermal environment is analyzed using the finite element software ABAQUS and the method in this paper, respectively. The considered structural geometry is: R_1 = 0.5 m, R_2 = 1.5 m, L_2 = 2 m, h = 0.1 m, N_k = [0°/90°/0°]. The temperature change is 50K. The frequencies obtained by these two methods are in good agreement. Figures 2 and 3 represent the mode shapes of laminated combined conical-cylindrical shell, corresponding to the natural frequencies from Table 4. Meanwhile, it is necessary to point out the fact that the following numerical discussion illustrates that this method can be used to analyze structural vibration behavior in thermal environments. All in all, after sufficient comparison, it is proved that the method established in this paper can be applied to the vibration analysis of the rotating composite conical shell and cylindrical shell in a thermal environment.

Table 2. Comparison of dimensionless frequencies for non-rotating isotropic combined conical-cylindrical shell with F-C boundary condition (μ = 0.3).

m	$n = 0$		$n = 1$		$n = 2$	
	FEM	**Present**	**FEM**	**Present**	**FEM**	**Present**
1	0.50375	0.50305	0.29287	0.29279	0.10203	0.09996
2	0.60986	0.60985	0.63581	0.63506	0.50290	0.50217
3	0.93092	0.93082	0.81123	0.81141	0.69148	0.69116
4	0.95632	0.95612	0.93088	0.93137	0.85890	0.85888
5	0.97160	0.97134	0.94850	0.95183	0.91607	0.91544
6	1.01188	1.01142	0.99145	0.99156	0.96048	0.96007
	$n = 3$		$n = 4$		$n = 5$	
	FEM	**Present**	**FEM**	**Present**	**FEM**	**Present**
1	0.09377	0.08750	0.14460	0.14441	0.20390	0.19930
2	0.39220	0.39115	0.33034	0.32996	0.29633	0.29579
3	0.51518	0.51434	0.39562	0.39537	0.37623	0.37013
4	0.75359	0.75289	0.64458	0.64594	0.58167	0.57874
5	0.79698	0.79629	0.69114	0.69248	0.61422	0.61285
6	0.91939	0.91893	0.87194	0.87098	0.81980	0.81642

Table 3. Comparison of dimensionless frequencies for a rotating isotropic combined conical-cylindrical shell.

Ω^*	n	FEM		Present	
		w^*_b	w^*_f	w^*_b	w^*_f
0.01 rad/s	1	0.5264	0.5264	0.5267	0.5267
	2	0.3769	0.3769	0.3774	0.3774
	3	0.2873	0.2873	0.2869	0.2869
	4	0.236	0.236	0.2363	0.2363
	5	0.2231	0.2231	0.2246	0.2246
	6	0.2474	0.2474	0.2469	0.2469
100 rad/s	1	0.5430	0.5097	0.5432	0.5103
	2	0.3906	0.3648	0.3904	0.3645
	3	0.3005	0.2816	0.3010	0.2822
	4	0.2527	0.2383	0.2528	0.2387
	5	0.2455	0.234	0.2469	0.2352
	6	0.2747	0.2647	0.2740	0.2645
500 rad/s	1	0.6085	0.4422	0.6090	0.4421
	2	0.4605	0.3308	0.4609	0.3304
	3	0.4174	0.322	0.4174	0.3222
	4	0.4484	0.3756	0.4480	0.3762
	5	0.5212	0.4629	0.5220	0.4633
	6	0.6157	0.5612	0.6161	0.5608

Table 4. Comparison of natural frequencies for non-rotating isotropic combined conical-cylindrical shell in thermal environment (ΔT = 50 K).

φ	Mode	CC			CS			FC		
		FEM	Present	Diff,%	FEM	Present	Diff,%	FEM	Present	Diff,%
$\pi/6$	1	234.96	237.01	0.872	227.4	228.08	0.299	120.2	119.54	−0.549
	2	250.73	251.66	0.371	227.84	228.85	0.443	133.66	133.66	0
	3	252.19	254.21	0.801	244.34	243.77	−0.233	234.47	234.16	−0.132
	4	265.13	264.69	−0.166	247.19	247.93	0.299	240.92	241.13	0.087
	5	272.86	274.39	0.561	265.13	264.69	−0.166	270.66	270.72	0.022
	6	285.89	286.71	0.287	281.3	282.09	0.281	272.89	274.28	0.509
$\pi/4$	1	270.23	272.39	0.799	239.64	239.73	0.038	145.93	145.81	−0.082
	2	281.53	283.4	0.664	250.99	250.54	−0.179	152.71	152.7	−0.007
	3	293.78	293.32	−0.157	271.81	272.52	0.261	256.96	256.67	−0.113
	4	294.13	296.48	0.799	293.78	293.32	−0.157	268.27	270.41	0.798
	5	319.31	321.55	0.702	302.2	300.76	−0.477	281.31	283.08	0.629
	6	328.18	329.1	0.28	312.49	314.21	0.55	305.66	306.68	0.334

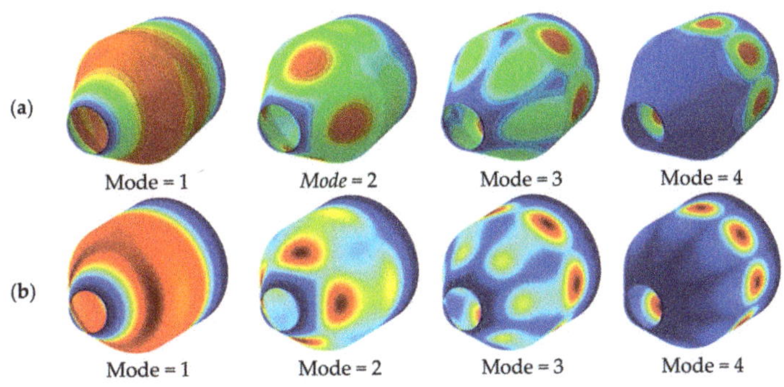

Figure 2. Mode shapes of laminated combined conical-cylindrical shell with CC boundary condition (m = 1, $\varphi = \pi/6$) (**a**) ABAQUS (**b**) Present.

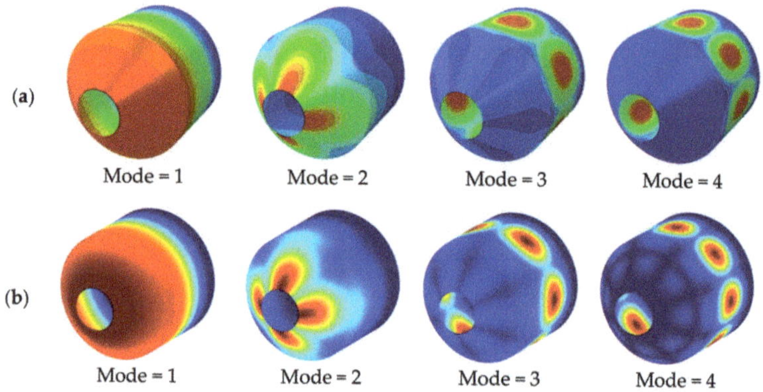

Figure 3. Mode shapes of laminated combined conical-cylindrical shell with FC boundary condition ($m = 1$, $\varphi = \pi/4$) (**a**) ABAQUS (**b**) Present.

3.2. Numerical Examples

In Section 3.2, we conduct sufficient comparative verifications to demonstrate that this method can analyze the vibrational behavior of the rotating composite cone-column structure in a thermal environment. First, the effect of semi-vertex of the combined conical-cylindrical shell structure with $\triangle T = 0$ K on the natural frequency is studied. The structure shape is: $R_1 = 0.5$ m, $L_1 = 2$ m, $L_2 = 1$ m, $h = 0.05$ m, $\delta_k = [0°/90°/0°/90°]$. It can be seen from Figure 4a,c,d that as the half-apex angle of the conical shell increases, the frequency of the combined structure gradually increases slightly first and then decreases significantly under the CC, SS and CS boundary conditions, respectively. At the same time, the difference between the forward wave frequency and the backward wave frequency of the conical-cylindrical composite shell with different rotational speeds is getting smaller and smaller, and the influence of rotational speed is also weakened. As on can see from Figure 4b, under the CF boundary condition, the backward wave frequency of the composite structure corresponding to $\Omega = 150$ rad/s and $\Omega = 200$ rad/s will decrease first and then increase with the increase of the half apex angle, and the rest of the natural frequency change curves all decrease. Likewise, with the same rotational speed, the gap between the forward traveling wave and the backward traveling wave of the structure also decreases.

Secondly, Figure 5 studies the variation of forward wave frequency with temperature for a combined conical-cylindrical shell with a rotational speed of 50 rad/s. In Figure 5a,c,d we selected the forward wave frequency with the circumferential wave number $n = 1$~4 and the axial half-wave number $m = 1$ as the research object to discuss its variation with ΔT. The variation interval of the temperature difference is [0 K, 500 K], and the boundary conditions of the studied structures are CC, CF, SS and CS boundary conditions, respectively. The structural geometry parameters are the same as in Figure 4. It is clear from Figure 5b that for the combined structure under the CC, CF, SS and CS boundary conditions, respectively, when $n = 0$ and $n = 1$, the curve representing the relationship between frequency and temperature difference approaches the horizontal line. At this time, the effect of temperature difference on them is minimal. However, for $n = 2$~4, the natural frequency of the structure decreases with increasing temperature difference. Under the CF boundary condition, for $n = 2$ and $n = 3$ with the increase of the temperature difference, the frequency value first decreases, and then does not change. At this time, at the turning point of the curve, the structure undergoes thermal buckling.

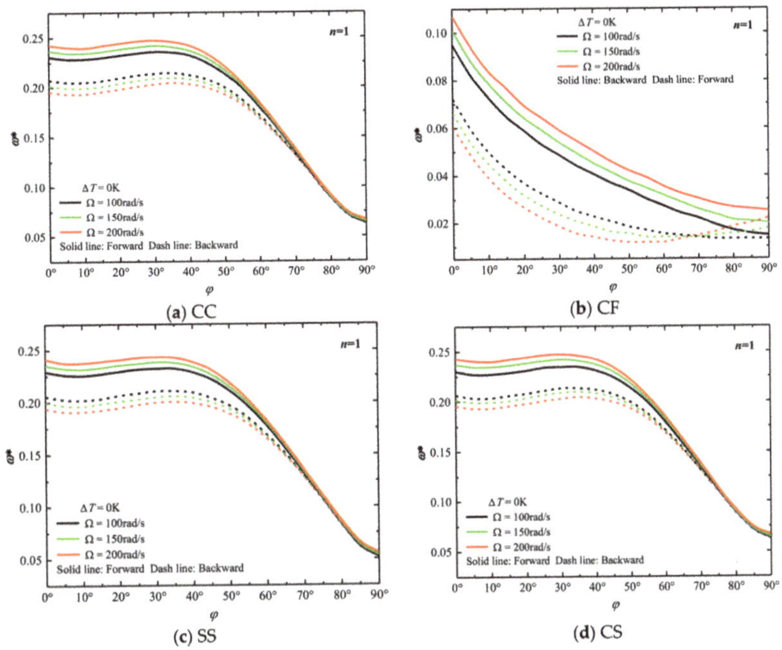

Figure 4. Variation of dimensionless frequencies ω^* of rotating laminated combined conical-cylindrical shell with different semi-vertex angle (m = 1).

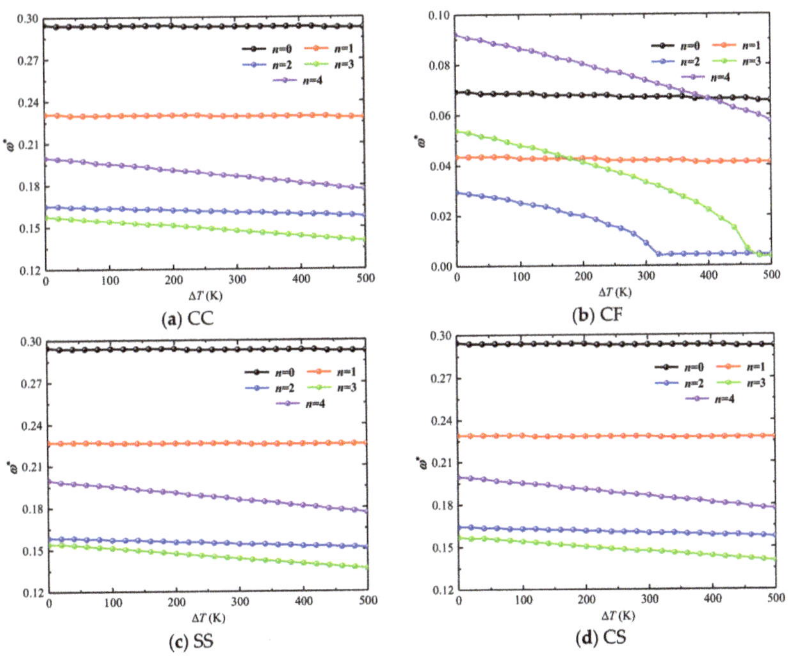

Figure 5. Variation of dimensionless frequencies ω^* of non-rotating laminated combined conical-cylindrical shell subjected to thermal effect (m = 1).

In addition, the effect of the rotational speed on the frequency of the combined structure in the thermal environment is also studied, where three temperature differences are selected as 0 K, 200 K and 1000 K, respectively, the rotational speed variation interval is 0,500], and the boundary condition is CC, CF, SS and CS, respectively. Following the structure of Figure 5 for analysis, the research results are shown in Figure 6. Meanwhile, it is necessary to point out that the $\omega^*_{f(n=1,m=1)}$ and $\omega^*_{f(n=3,m=1)}$ are considered in the following discussion. As can be seen in Figure 6a,c,d, the frequency of all intercepted backward wave frequencies under CC, SS, CS boundary conditions, respectively, increases as the rotational speed increases. However, under CC, CF, SS, CS boundary conditions, respectively, the forward wave frequencies for $\omega^*_{f(n=1,m=1)}$ decrease and for $\omega^*_{f(n=3,m=1)}$ increase with increased rotating speed. As shown in Figure 6b, under CF boundary conditions, the variation tendencies of forward and backward wave frequency are the same as the above other boundary conditions. However, the forward and backward wave frequencies produce model jumping with temperature differences increased, and the above phenomenon can be attributed to thermal buckling.

Finally, Tables 5 and 6 show the dimensionless frequencies of rotating cross-ply combined conical-cylindrical shell with various geometry and boundary condition in thermal environment. The geometrical parameters of the structure and the temperature difference are given in the table header, and the lamination scheme is [0°/90°/0°]. It can be seen from Table 5 that as the length of the cylindrical shell increases, the stiffness of the structure decreases, and both the forward wave frequency and the backward wave frequency gradually decrease. The rules in Table 6 are the same as those in Figure 4, and thus are not repeated here. These results are valuable to designers and serve as benchmarks for future numerical studies.

Table 5. Dimensionless frequencies of rotating laminated combined conical-cylindrical shell with various length ratio in thermal environment. ($L_1 = 1$, $R_1 = 0.5$ m, $h = 0.05$ m, $m = 1$, $\varphi = 30°$, $\Delta T = 50$ K).

L_2/L_1	Ω, rad/s	n	Forward				Backward			
			CC	SS	CF	FC	CC	SS	CF	FC
0.5	50	1	0.4723	0.4561	0.1127	0.2549	0.4826	0.4664	0.1230	0.2667
		2	0.3809	0.3478	0.0847	0.1157	0.3883	0.3559	0.0928	0.1245
		3	0.3544	0.3153	0.1341	0.1709	0.3595	0.3212	0.1400	0.1776
	100	1	0.4671	0.4509	0.1083	0.2490	0.4877	0.4715	0.1282	0.2726
		2	0.3772	0.3448	0.0818	0.1120	0.3927	0.3595	0.0980	0.1297
		3	0.3522	0.3139	0.1326	0.1702	0.3640	0.3249	0.1459	0.1820
1	50	1	0.3735	0.3618	0.0759	0.1835	0.3846	0.3728	0.0862	0.1945
		2	0.3448	0.3161	0.0361	0.0921	0.3536	0.3249	0.0449	0.1009
		3	0.3404	0.3028	0.0663	0.1687	0.3470	0.3087	0.0729	0.1754
	100	1	0.3676	0.3559	0.0707	0.1776	0.3905	0.3780	0.0914	0.2004
		2	0.3411	0.3124	0.0332	0.0884	0.3581	0.3293	0.0508	0.1061
		3	0.3382	0.3006	0.0670	0.1680	0.3507	0.3131	0.0803	0.1805
1.5	50	1	0.2896	0.2829	0.0553	0.1392	0.3006	0.2947	0.0663	0.1510
		2	0.2505	0.2358	0.0258	0.0781	0.2593	0.2446	0.0346	0.0869
		3	0.2218	0.2019	0.0597	0.1665	0.2291	0.2092	0.0670	0.1724
	100	1	0.2837	0.2778	0.0501	0.1334	0.3065	0.2999	0.0715	0.1562
		2	0.2461	0.2313	0.0236	0.0744	0.2645	0.2498	0.0420	0.0928
		3	0.2203	0.2004	0.0612	0.1650	0.2336	0.2137	0.0752	0.1776
2	50	1	0.2328	0.2284	0.0420	0.1105	0.2439	0.2402	0.0530	0.1223
		2	0.1864	0.1783	0.0214	0.0700	0.1952	0.1871	0.0302	0.0788
		3	0.1606	0.1496	0.0575	0.1540	0.1672	0.1569	0.0641	0.1606
	100	1	0.2269	0.2225	0.0368	0.1046	0.2498	0.2453	0.0589	0.1282
		2	0.1820	0.1739	0.0192	0.0670	0.2004	0.1923	0.0376	0.0847
		3	0.1584	0.1481	0.0589	0.1525	0.1724	0.1621	0.0729	0.1658

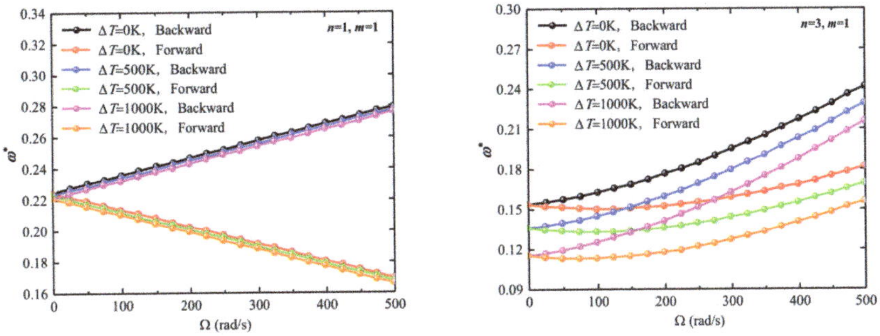

(a) CC

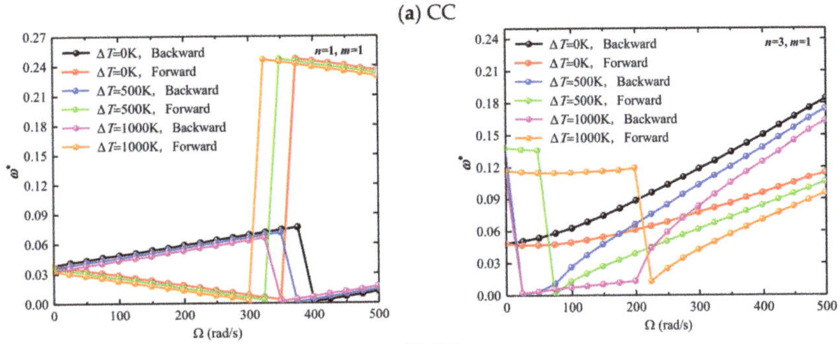

(b) CF

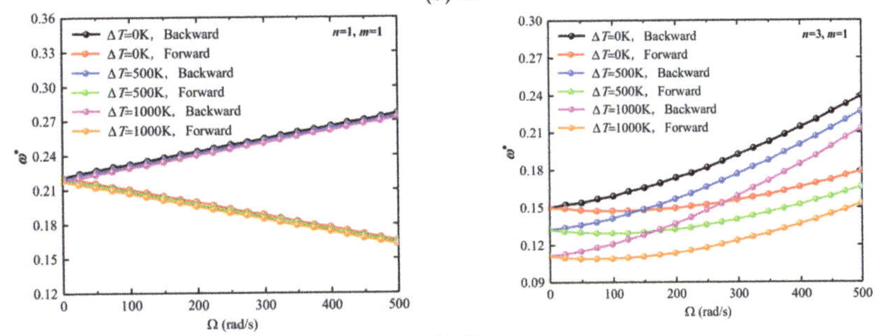

(c) SS

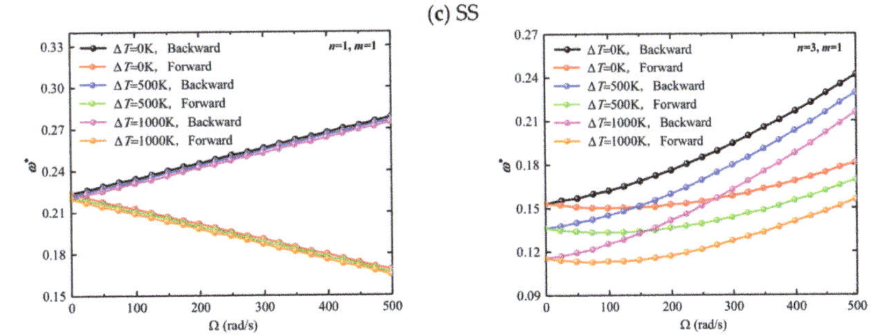

(d) CS

Figure 6. Variation of dimensionless frequencies ω^* of rotating laminated combined conical-cylindrical shell in the thermal environment.

Table 6. Dimensionless frequencies of rotating laminated combined conical-cylindrical shell with various semi-vertex angle in thermal environment. (L_1 = 0.5 m, L_2 = 2 m, R_1 = 0.5 m, h = 0.1 m, m = 1, ΔT = 50 K).

φ	Ω, rad/s	n	Forward				Backward			
			CC	SS	CF	FC	CC	SS	CF	FC
$\pi/6$	50	1	0.2763	0.2645	0.0729	0.1260	0.2881	0.2763	0.0840	0.1378
		2	0.2269	0.2085	0.0766	0.1444	0.2365	0.2173	0.0855	0.1532
		3	0.2807	0.2652	0.2033	0.2763	0.2881	0.2726	0.2107	0.2829
	100	1	0.2704	0.2593	0.0670	0.1201	0.2940	0.2822	0.0899	0.1437
		2	0.2225	0.2041	0.0729	0.1400	0.2417	0.2225	0.0914	0.1584
		3	0.2785	0.2630	0.2019	0.2741	0.2925	0.2770	0.2159	0.2873
$\pi/4$	50	1	0.2859	0.2756	0.0589	0.1297	0.2969	0.2866	0.0700	0.1415
		2	0.2306	0.2115	0.0597	0.1606	0.2402	0.2210	0.0685	0.1695
		3	0.2549	0.2365	0.1599	0.2520	0.2616	0.2431	0.1665	0.2586
	100	1	0.2800	0.2697	0.0538	0.1238	0.3028	0.2925	0.0752	0.1474
		2	0.2269	0.2078	0.0560	0.1562	0.2453	0.2255	0.0744	0.1746
		3	0.2527	0.2343	0.1577	0.2498	0.2667	0.2476	0.1717	0.2638
$\pi/3$	50	1	0.2918	0.2822	0.0494	0.1319	0.3028	0.2940	0.0597	0.1437
		2	0.2343	0.2144	0.0501	0.1857	0.2431	0.2232	0.0597	0.1945
		3	0.2424	0.2218	0.1356	0.2394	0.2498	0.2291	0.1422	0.2468
	100	1	0.2859	0.2770	0.0442	0.1260	0.3087	0.2999	0.0648	0.1496
		2	0.2299	0.2100	0.0472	0.1812	0.2483	0.2284	0.0656	0.1989
		3	0.2402	0.2196	0.1341	0.2372	0.2542	0.2336	0.1481	0.2512
$\pi/2$	50	1	0.2962	0.2873	0.0427	0.1334	0.3080	0.2991	0.0516	0.1451
		2	0.2380	0.2181	0.0457	0.2092	0.2476	0.2269	0.0545	0.2181
		3	0.2372	0.2144	0.1194	0.2321	0.2439	0.2218	0.1267	0.2387
	100	1	0.2903	0.2814	0.0383	0.1275	0.3139	0.3043	0.0560	0.1510
		2	0.2343	0.2137	0.0427	0.2048	0.2520	0.2321	0.0604	0.2232
		3	0.2350	0.2129	0.1186	0.2299	0.2483	0.2262	0.1326	0.2439

4. Conclusions

This paper focuses on the free vibration analysis of laminated combined conical-cylindrical shell rotating in a thermal environment. The equations of motion of the whole system are derived by combining the equations of individual substructures obtained using Hamilton's principle, in which the effects of temperature change, centrifugal and Coriolis forces are taken into account. For numerical calculation of equations of motions, the meshfree strong form method using TRPIM shape function is applied. Through the convergence study, the number of node is determined. The accuracy and reliability of the proposed method are verified through comparison with the results of finite element program. Finally, the effects of parameters such as geometric dimensions, rotating speed and temperature change on the free vibration of combined conical-cylindrical shell are investigated through some numerical examples. The conclusions obtained in this study are as follows:

(1) The meshless Chebyshev-PRIM technique is effective and has relatively high accuracy in the vibration solution of rotating structures. This method has the advantage of fast convergence, and relatively accurate results can be obtained with a smaller number of nodes.

(2) The increase of the half-apex angle of the conical shell reduces the structural rigidity, so the structural frequency decreases. For the combined structure under the CC boundary, after the cone angle increases to a certain extent, the effect of the rotational speed will decrease, and the frequencies corresponding to different rotational speeds will gradually approach.

(3) If the temperature is too high, thermal stress is accumulated inside the structure, the stiffness of the structure is reduced, and the frequency of the combined structure will

also decrease. For the boundary conditions with weakened constraints, such as the CF boundary, thermal buckling also occurs with the increase of the temperature difference.

Author Contributions: Conceptualization, Z.L. and X.Z.; methodology, Z.L. and B.Q.; software, Z.L.; validation, Z.L., S.H. and R.Z.; formal analysis, Z.L.; investigation, Z.L., S.H. and R.Z.; resources, Z.L.; data curation, Z.L.; writing—original draft preparation, Z.L.; writing—review and editing, S.H., R.Z., B.Q. and X.Z.; visualization, Z.L.; supervision, B.Q. and X.Z.; project administration, X.Z.; funding acquisition, X.Z. All authors have read and agreed to the published version of the manuscript.

Funding: This research was funded by National Natural Science Foundation of China, grant number 52105420 and Natural Science Foundation of Hunan Province of China, grant number 2020JJ5760. The authors also gratefully acknowledge the financial support from the State Key Laboratory of High Performance Complex Manufacturing, Central South University, China, grant number ZZYJKT2018-18 and the Fundamental Research Funds for the Central Universities of Central South University, grant number 2022ZZTS0177.

Conflicts of Interest: The authors declare no conflict of interest.

References

1. Haddadpour, H.; Mahmoudkhani, S.; Navazi, H.M. Free vibration analysis of functionally graded cylindrical shells in-cluding thermal effects. *Thin-Walled Struct.* **2007**, *45*, 591–599. [CrossRef]
2. Afshari, H.; Amirabadi, H. Vibration characteristics of rotating truncated conical shells reinforced with agglomerated carbon nanotubes. *J. Vib. Control* **2021**, *28*, 1894–1914. [CrossRef]
3. Jooybar, N.; Malekzadeh, P.; Fiouz, A.; Vaghefi, M. Thermal effect on free vibration of functionally graded truncated conical shell panels. *Thin-Walled Struct.* **2016**, *103*, 45–61. [CrossRef]
4. Shekari, A.; Ghasemi, F.A.; Malekzadehfard, K. Free Damped Vibration of Rotating Truncated Conical Sandwich Shells Using an Improved High-Order Theory. *Lat. Am. J. Solids Struct.* **2017**, *14*, 2291–2323. [CrossRef]
5. Dai, Q.; Cao, Q.; Chen, Y. Frequency analysis of rotating truncated conical shells using the Haar wavelet method. *Appl. Math. Model.* **2018**, *57*, 603–613. [CrossRef]
6. Xiang, X.; Guoyong, J.; Wanyou, L.; Zhigang, L. A numerical solution for vibration analysis of composite laminated conical, cylindrical shell and annular plate structures. *Compos. Struct.* **2014**, *111*, 20–30. [CrossRef]
7. Xue, Y.; Jin, G.; Ma, X.; Chen, H.; Ye, T.; Chen, M.; Zhang, Y. Free vibration analysis of porous plates with porosity distributions in the thickness and in-plane directions using isogeometric approach. *Int. J. Mech. Sci.* **2019**, *152*, 346–362. [CrossRef]
8. Bediz, B. A spectral-Tchebychev solution technique for determining vibrational behavior of thick plates having arbitrary ge-ometry. *J. Sound Vib.* **2018**, *432*, 272–289. [CrossRef]
9. Ye, T.; Jin, G.; Su, Z.; Jia, X. A unified Chebyshev–Ritz formulation for vibration analysis of composite laminated deep open shells with arbi-trary boundary conditions. *Arch. Appl. Mech.* **2014**, *84*, 441–471. [CrossRef]
10. Ma, X.; Jin, G.; Xiong, Y.; Liu, Z. Free and forced vibration analysis of coupled conical–cylindrical shells with arbitrary boundary conditions. *Int. J. Mech. Sci.* **2014**, *88*, 122–137. [CrossRef]
11. Nguyen-Thanh, N.; Rabczuk, T.; Nguyen-Xuan, H.; Bordas, S.P. A smoothed finite element method for shell analysis. *Comput. Methods Appl. Mech. Eng.* **2008**, *198*, 165–177. [CrossRef]
12. Barik, M.; Mukhopadhyay, M. Finite element free flexural vibration analysis of arbitrary plates. *Finite Elements Anal. Des.* **1998**, *29*, 137–151. [CrossRef]
13. Fantuzzi, N.; Tornabene, F.; Viola, E. Four-parameter functionally graded cracked plates of arbitrary shape: A GDQFEM solution for free vibrations. *Mech. Adv. Mater. Struct.* **2016**, *23*, 89–107. [CrossRef]
14. Ye, T.; Jin, G.; Chen, Y.; Shi, S. A unified formulation for vibration analysis of open shells with arbitrary boundary conditions. *Int. J. Mech. Sci.* **2014**, *81*, 42–59. [CrossRef]
15. Caresta, M.; Kessissoglou, N.J. Free vibrational characteristics of isotropic coupled cylindrical–conical shells. *J. Sound Vib.* **2010**, *329*, 733–751. [CrossRef]
16. Tornabene, F.; Fantuzzi, N.; Viola, E.; Reddy, J. Winkler–Pasternak foundation effect on the static and dynamic analyses of laminated doubly-curved and degenerate shells and panels. *Compos. Part B Eng.* **2014**, *57*, 269–296. [CrossRef]
17. Li, H.; Pang, F.; Wang, X.; Du, Y.; Chen, H. Free vibration analysis for composite laminated doubly-curved shells of revolution by a semi analytical method. *Compos. Struct.* **2018**, *201*, 86–111. [CrossRef]
18. Shakouri, M. Free vibration analysis of functionally graded rotating conical shells in thermal environment. *Compos. Part B Eng.* **2019**, *163*, 574–584. [CrossRef]
19. Afshari, H. Effect of graphene nanoplatelet reinforcements on the dynamics of rotating truncated conical shells. *J. Braz. Soc. Mech. Sci. Eng.* **2020**, *42*, 1–22. [CrossRef]
20. Bhangale, R.K.; Ganesan, N.; Padmanabhan, C. Linear thermoelastic buckling and free vibration behavior of functionally graded truncated conical shells. *J. Sound Vib.* **2006**, *292*, 341–371. [CrossRef]

21. Tian, L.; Ye, T.; Jin, G. Vibration analysis of combined conical-cylindrical shells based on the dynamic stiffness method. *Thin-Walled Struct.* **2021**, *159*, 107260. [CrossRef]
22. Qin, Z.; Yang, Z.; Zu, J.; Chu, F. Free vibration analysis of rotating cylindrical shells coupled with moderately thick annular plates. *Int. J. Mech. Sci.* **2018**, *142–143*, 127–139. [CrossRef]
23. Singha, T.D.; Rout, M.; Bandyopadhyay, T.; Karmakar, A. Free vibration of rotating pretwisted FG-GRC sandwich conical shells in thermal environment using HSDT. *Compos. Struct.* **2020**, *257*, 113144. [CrossRef]
24. Talebitooti, M.; Soureshjani, A.H.; Pakravan, I.; Talebitooti, R. Frequency study on panel type of FG-CNTRC joined conical-conical struc-tures. *Compos. Struct.* **2020**, *259*, 113241. [CrossRef]
25. Soureshjani, A.H.; Talebitooti, R.; Talebitooti, M. Thermal effects on the free vibration of joined FG-CNTRC conical-conical shells. *Thin-Walled Struct.* **2020**, *156*, 106960. [CrossRef]
26. Shi, X.; Zuo, P.; Zhong, R.; Guo, C.; Wang, Q. Thermal vibration analysis of functionally graded conical-cylindrical coupled shell based on spectro-geometric method. *Thin-Walled Struct.* **2022**, *175*, 109138. [CrossRef]
27. Ghasemi, A.R.; Mohandes, M.; Dimitri, R.; Tornabene, F. Agglomeration Effects on the Vibrations of CNTS/Fiber/Polymer/Metal Hybrid Laminates Cylindrical Shell. *Compos. Part B Eng.* **2019**, *167*, 700–716. [CrossRef]
28. Liu, S.; Ke, Y.; Davar, A.; Jam, J.E.; Lutfor, M.R.; Sarjadi, M.S. The effects of rotation on the frequencies and critical speed of CNTs/fiber/polymer/metal laminates cy-lindrical shell. *Arab. J. Chem.* **2021**, *15*, 103575. [CrossRef]
29. Dehrouyeh-Semnani, A.M.; Mostafaei, H. Vibration analysis of scale-dependent thin shallow microshells with arbitrary plan-form and boundary conditions. *Int. J. Eng. Sci.* **2021**, *158*, 103413. [CrossRef]
30. Dehrouyeh-Semnani, A.M.; Mostafaei, H. On the mechanics of microshells of revolution. *Int. J. Eng. Sci.* **2021**, *161*, 103450. [CrossRef]
31. Kwak, S.; Kim, K.; Pyon, S.; Li, Y.; Ri, C. A new meshfree approach for three-dimensional free vibration analysis of thick laminated doubly-shell of revolution. *Eng. Anal. Bound. Elements* **2021**, *134*, 199–218. [CrossRef]
32. Kwak, S.; Kim, K.; An, K.; Jong, G.; Yun, J. A novel meshfree method for three-dimensional natural frequency analysis of thick laminated conical, cylindrical shells and annular plates. *Phys. Scr.* **2021**, *96*, 125204. [CrossRef]
33. Liu, G.R.; Gu, Y.T. *An Introduction to Meshfree Methods and Their Programming*; Springer: Dordrecht, The Netherlands, 2005.
34. Zarei, A.; Khosravifard, A. Meshfree investigation of the vibrational behavior of pre-stressed laminated composite plates based on a variationally consistent plate model. *Eng. Anal. Bound. Elements* **2020**, *111*, 118–133. [CrossRef]
35. Mellouli, H.; Jrad, H.; Wali, M.; Dammak, F. Free vibration analysis of FG-CNTRC shell structures using the meshfree radial point interpolation method. *Comput. Math. Appl.* **2020**, *79*, 3160–3178. [CrossRef]
36. Zhang, L.; Lei, Z.; Liew, K. Free vibration analysis of functionally graded carbon nanotube-reinforced composite triangular plates using the FSDT and element-free IMLS-Ritz method. *Compos. Struct.* **2015**, *120*, 189–199. [CrossRef]
37. Fallah, N.; Delzendeh, M. Free vibration analysis of laminated composite plates using meshless finite volume method. *Eng. Anal. Bound. Elem.* **2018**, *88*, 132–144. [CrossRef]
38. Kwak, S.; Kim, K.; Jang, P.; Ri, Y.; Kim, I. A meshfree local weak-form method for free vibration analysis of an open laminated cylindrical shell with elliptical section. *Compos. Struct.* **2021**, *275*, 114484. [CrossRef]
39. Qatu, M.S. *Vibration of Laminated Shells and Plates*; Elsevier: Amsterdam, The Netherlands, 2004; Volume 60, pp. 379–404.
40. Kwak, S.; Kim, K.; Jong, G.; Kim, Y.; Pang, K. Free vibration analysis of laminated elliptic cylindrical panel with varying thickness using a meshfree method. *Eur. Phys. J. Plus* **2022**, *137*, 173. [CrossRef]

materials MDPI

Article

Vibration Characteristics of a Laminated Composite Double-Cylindrical Shell System Coupled with a Variable Number of Annular Plates

Ying Zhang [1,2], Dongyan Shi [1,*] and Dongze He [1]

1 College of Mechanical and Electrical Engineering, Harbin Engineering University, Harbin 150001, China; zhangying2016@hrbeu.edu.cn (Y.Z.); hdz2012071506@126.com (D.H.)
2 College of Engineering, Heilongjiang Bayi Agricultural University, Daqing 163319, China
* Correspondence: shidongyan@hrbeu.edu.com

Abstract: A vibration characteristic analysis model of a laminated composite double cylindrical shell system (LCDCSS) coupled with several annular plates under general boundary conditions is established. Artificial springs are used to simulate the coupling relationship between substructures to ensure the continuity of displacement both at ends of the shells and coupling boundaries. The variable number of annular plates can be distributed unevenly and coupled elastically. Displacement functions of LCDCSS are expressed with improved Fourier series. Based on the principle of energy, obtain the unknown coefficients of the displacement components by using the Rayleigh–Ritz method. The convergence and effectiveness of the proposed method are verified by comparing with the results with literature and FEM, and then carried out parametric investigation to study the free and steady-state response vibration characteristics of LCDCSS. Rapid prediction of free vibration and response vibration of a double-layer cylindrical shell system with various structures and scales is realized by exploiting the model, and some new results of double-layer cylindrical shell system are explored, which can provide reference for further research.

Keywords: laminated composite double-cylindrical shell system; improved Fourier series; Rayleigh–Ritz method; free vibration; steady-state response

Citation: Zhang, Y.; Shi, D.; He, D. Vibration Characteristics of a Laminated Composite Double-Cylindrical Shell System Coupled with a Variable Number of Annular Plates. *Materials* **2022**, *15*, 4246. https://doi.org/10.3390/ma15124246

Academic Editor: Haim Abramovich

Received: 17 May 2022
Accepted: 8 June 2022
Published: 15 June 2022

1. Introduction

As a basic structural member, laminated composite cylindrical shells are used in aviation equipment, ship engineering, construction, transportation, chemical engineering, and other engineering fields. In addition, a double-layer cylindrical shell has been more and more widely used in various fields for excellent physical and chemical properties with the technology progress, such as a typical structure of submarine cabin, seabed oil pipeline, and so on. Studies on vibration characteristics of a double cylindrical shell system have important theoretical significance and practical application value. Scholars continue to explore theoretical and experimental methods for solving the dynamic problems of various kinds of cylindrical shell structures to predict and control the vibration of structures. In recent decades, a number of research studies have been carried out around laminated cylindrical shell structures, which are fully recorded in the literature.

Ming et al. [1] proposed a model to measure the structural characteristics of cylindrical shells, in which the point force excitation is used to replace the circumferential modal force that is difficult to realize in practice. In this method, the transfer function components of different circumferential modes are obtained from the measured data by using the least square method, and the feasibility of this method is verified by the point force excitation experiments of cylindrical shells. Lee and Reddy [2] used the higher-order shear deformation theory to study the vibration characteristics of laminated shell structures and search for a way to control it. Based on the classical shell theory (CST), Zhong et al. [3] proposed a new exact solution to analyze free vibration of cross-ply laminated composite cylindrical shells. Hu et al. [4] studied vibration frequency of a laminated cylindrical thin panel with the Rayleigh–Ritz method, which combined the principle of virtual work with the displacement function of two-dimensional algebraic polynomials, the vibration control equations of laminated cylindrical thin plates with torsion and curvature are established. Based on the first order shear deformation theory, Qu et al. [5] proposed a unified formulation for vibration analysis of composite laminated shell considered both the effects of shear distortion and rotary inertia. Maharjan et al. [6] evaluated the elastic properties of laminated composite cylindrical shells using micro-mechanical approaches. Hafizah et al. [7] analyzed vibration of antisymmetric angle-ply composite annular plates with variable thickness. Civalek [8] presented vibration analysis of laminated composite conical shells based on the shear deformation theory. Zhao et al. [9] proposed a unified analysis model to present the free vibration of laminated composite elliptic cylinders under various boundary conditions.He et al. [10] analyzed the free vibration of composite laminated cylindrical shells with general boundary conditions using a wave-based method (WBM). Kumar [11] explored vibration of laminated composite skew hypar shells by using higher order theory. Jin et al. [12–14] put forward a unified improved Fourier solution for composite laminated structural elements with arbitrary elastic constraints, which can be used to solve the free vibration problems of cylinders, cones, spherical shells and annular plates. Wang et al. [15–17] offered unified solutions for dynamic analysis of circular, annular and sector plate structures of orthotropic materials, laminated composites and functionally graded materials under general boundary conditions. Tornabene et al. [18] completed a general higher-order equivalent single layer theory to study free vibrations of doubly-curved laminated composite shells and panels. Li et al. [19,20] analyzed the vibration characteristics of rotating composite laminated cylindrical shells under subsonic air flow and axial load in humid and hot environments. Zhang et al. [21] analyzed vibration of a composite laminated sector, annular, and circular plate with a simplified plate theory. Zuo et al. [22] combined general shell theory and the wavelet finite element method to present static and vibration characteristic of laminated composite shells. Liu et al. [23] studied the free vibration characteristics of functionally graded cylindrical shells by using the wave-based method. Bisheh et al. [24] carried out free vibration analysis of piezoelectric coupled carbon nanotube (CNT) reinforced composite cylindrical shells, and discussed the influence of boundary conditions on the frequency.

With the development of the research and the expansion of engineering, various complex coupled structures based on cylindrical shells have attracted widespread attention. Some researchers have paid attention to the double shell system. Yamada et al. [25] obtained the vibration control equations of a cylindrical shell using the transfer matrix, and presented the free vibration of a circular cylindrical double-shell system. Yuan et al. [26] established the free vibration characteristic analysis model of a ring cylindrical shell coupling structure system by using the Rayleigh–Ritz method, in which the coupling connection conditions of substructure are simulated by artificial spring technology. Jin et al. [27] explored vibration analysis of circular cylindrical double-shell structures under general coupling and end boundary conditions. Chen et al. [28] derived the dynamic equilibrium equation of the double elastic spherical shell, and obtained the semi analytical solutions of vibration and sound radiation of underwater double spherical shell by using the Dirac-delta function. Dogan et al. [29] integrated the nonlinear spring damper model into the system, established the analytical model of a nonlinear response of a double wall sandwich cylindrical shell system under random excitation, and gave the nonlinear response law. Qing et al. [30] proposed a hybrid state variable technique, which extended the semi-analytical method to the natural frequency and mode analysis of double-layer thick shell structures. Wang et al. [31] carried out a research project on dynamic failure behaviors of steel double-layer latticed cylindrical shell. Zhang et al. [32] presented vibration and sound radiation from submerged double cylindrical shells using the modal superposition method. Huang et al. [33] obtained the wind effects on the double-layer cylindrical latticed shell (DCLS). Zhang et al. [34] studied the free vibration analysis of double cylindrical shells that is rib stiffened, based on a modified Fourier-Ritz method. Xie et al. [35] obtained vibration analysis of double-walled cylindrical shells using a wave-based method. Wali et al. [36] studied free vibration analysis of FGM shell structures by building an efficient 3D-shell model. Choi et al. [37] presented the free vibration of double cylindrical shells based on transfer of an influence coefficient. Mehdi et al. [38] researched vibration of double-bonded micro sandwich cylindrical shells under multi-physical loadings. Chen et al. [39] obtained a vibration frequency model of a ring stiffened cylindrical shell that was stiffened with intermediate ribs by using a wave-based method.

From literature review, researchers have been widely exploring the vibration characteristics of cylindrical shell structures. However, it is regrettable that study of laminated composite double cylindrical shells is seldom done. There may be some limitations to predict vibration character of a double cylindrical shell system with the coupling relationship, scale or boundary conditions of the system changed. The research on predicting the vibration characteristics of composite double cylindrical shell system coupled with variable annular plates under general boundary conditions and coupling constraints has not been reported.

In this research, a unified analytical model for the vibration characteristics of laminated double cylindrical shells coupling with a variable number of annular plates is established. Artificial virtual boundary technique and virtual coupling spring technique are used to simulate the constraint relationship. Based on the first order shear deformation shell theory, effects of tension, bending, and torsion are taken care of. Displacement admissible functions of the system were approached by improved Fourier series. According to the Rayleigh–Ritz method, the unknown coefficients of displacement components were obtained based on the principle of energy. After verifying the convergence and correctness of the proposed method by numerical examples, a series of parametric studies are analyzed to predict the free and state response vibration characteristics of LCDCSS.

2. Analysis Model of the Laminated Composite Double Cylindrical Shell Structure

2.1. Description of the Model

Figure 1 describes the structure of a laminated composite double cylindrical shell system coupled with general boundary conditions. There are two coaxial cylinders with the same length are connected with a variable number of annular plates which distributed non-uniformly along the axial direction. For the convenience of model description, cylindrical coordinates (o, x, θ, r) are set up along the axial, circumferential, and radial direction of the shell structure. The geometric properties of all cylindrical shell structure and annular plates are described based on the cylindrical coordinate system. Length, radius, circumferential angle, and thickness of shells are expressed by L, R_p, h_p, θ_p, in which $p = i,o$ represent the inner and outer shell individually (Cylinder 1 and 2 in Figure 1). In a similar way, the structure of annular plates described with the cylindrical coordinate system (o, r, θ, z), $b = R_o - R_i$ is the width of the annular plate, which is the direction from the internal to outer edge along the radial direction. h_c represents the thickness of annular plate. u_g, v_g, w_g $(g = c, a)$ stand for middle-layer displacements of laminated cylindrical shells and annular plates, respectively. p_a^q denotes position of the qth annular plate along the axial direction of the cylindrical shell.

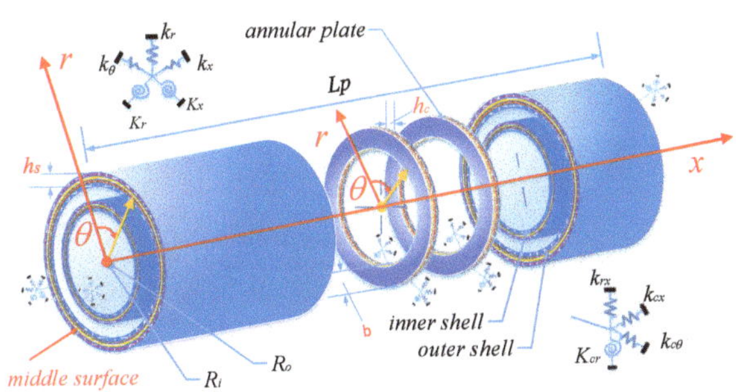

Figure 1. Model of the laminated composite double cylindrical shell system (LCDCSS) with general boundary conditions.

2.2. Kinematic Relations and Stress Resultants

2.2.1. For Cylindrical Shell

Based on shear shell theory (SDST), both in-plane and out-of-plane vibration are considered to describe the stress–strain relationship of a cylindrical shell structure. The displacement and rotation components of any point on the cylindrical shell structure can be represented as follows [12]:

$$\begin{aligned} U^p(x,\theta,z,t) &= u_c^p(x,\theta,t)+z\phi_{cx}^p(x,\theta,t) \\ V^p(x,\theta,z,t) &= v_c^p(x,\theta,t)+z\phi_{c\theta}^p(x,\theta,t) \\ W^p(x,\theta,z,t) &= w_c^p(x,\theta,t) \end{aligned} \tag{1}$$

where $p = i,o$ stand for inner and outer cylindrical shells in the system. t is time variable. U^p, V^p, and W^p represent displacement for an arbitrary point of the cylindrical shell, in axial, circumferential, and radial direction. u_c^p, v_c^p, w_c^p are middle layer displacement of the cylindrical shell, and φ_{cx}^p, $\varphi_{c\theta}^p$, present rotatory displacement component along θz and xz surfaces, respectively. According to the linear elasticity theory [12], strain and displacement relationships are as follows:

$$\begin{aligned} \varepsilon_{xx}^p &= \varepsilon_{xx0}^p+z\chi_{xx}^p,\ \varepsilon_{\theta\theta}^p = \varepsilon_{\theta\theta0}^p+z\chi_{\theta\theta}^p, \gamma_{x\theta}^p = \gamma_{x\theta0}^p+z\chi_{x\theta}^p, \\ \gamma_{xz0}^p &= \frac{\partial w_c^p}{\partial x}+\phi_{cx}^p, \gamma_{\theta z0}^p = \frac{\partial w_c^p}{R_p\partial\theta}-\frac{v_c^p}{R_p}+\phi_{c\theta}^p \end{aligned} \tag{2}$$

where $\varepsilon_{xx0}^p, \varepsilon_{\theta\theta0}^p, \gamma_{x\theta0}^p$ denote the structural strains in the middle surface, and $\chi_{xx}^p, \chi_{\theta\theta}^p, \chi_{x\theta}^p$ are mid-surface changes in curvature. The relationship with displacement can be shown as:

$$\begin{aligned} \varepsilon_{xx0}^p &= \frac{\partial u_c^p}{\partial x},\ \varepsilon_{\theta\theta0}^p = \frac{\partial v_c^p}{R_p\partial\theta}+\frac{w_c^p}{R_p}, \gamma_{x\theta0}^p = \frac{\partial v_c^p}{\partial x}+\frac{\partial u_c^p}{R_p\partial\theta} \\ \chi_{xx}^p &= \frac{\partial\phi_{cx}^p}{\partial x}, \chi_{\theta\theta}^p = \frac{\partial\phi_{c\theta}^p}{R_p\partial\theta}, \chi_{x\theta}^p = \frac{\partial\phi_{cx}^p}{R_p\partial\theta}+\frac{\partial\phi_{c\theta}^p}{\partial x} \end{aligned} \tag{3}$$

The thickness parameter of a fiber layer is between $z_k < z < z_{k+1}$; thus, the corresponding stresses are obtained in terms of the general Hooke's law as:

$$\begin{Bmatrix} \sigma_{cxx}^p \\ \sigma_{c\theta\theta}^p \\ \tau_{cx\theta}^p \end{Bmatrix} = \begin{bmatrix} \overline{Q}_{11}^k & \overline{Q}_{12}^k & \overline{Q}_{16}^k \\ \overline{Q}_{12}^k & \overline{Q}_{22}^k & \overline{Q}_{26}^k \\ \overline{Q}_{16}^k & \overline{Q}_{26}^k & \overline{Q}_{66}^k \end{bmatrix} \begin{Bmatrix} \varepsilon_{xx0}^p \\ \varepsilon_{\theta\theta0}^p \\ \gamma_{x\theta0}^p \end{Bmatrix}, \begin{Bmatrix} \tau_{cxz}^p \\ \tau_{c\theta z}^p \end{Bmatrix} = \begin{bmatrix} \overline{Q}_{55}^k & \overline{Q}_{45}^k \\ \overline{Q}_{45}^k & \overline{Q}_{44}^k \end{bmatrix} \begin{Bmatrix} \gamma_{xz}^p \\ \gamma_{\theta z}^p \end{Bmatrix} \tag{4}$$

where $\sigma_{cxx}^p, \sigma_{c\theta\theta}^p$ are normal stresses, and $\tau_{cx\theta}^p \tau_{cxz}^p$ and $\tau_{c\theta z}^p$ are shear stresses [40]. The kth layer stiffness coefficients of the cylindrical shell are $\overline{Q}_{ij}$ $(i,j = 1,2,4,5,6)$, which represent the elastic properties of the material of the layer, and can be determined as [17]:

$$\begin{gathered} \begin{bmatrix} \overline{Q}_{11}^k & \overline{Q}_{12}^k & \overline{Q}_{16}^k & 0 & 0 \\ \overline{Q}_{12}^k & \overline{Q}_{22}^k & \overline{Q}_{26}^k & 0 & 0 \\ \overline{Q}_{16}^k & \overline{Q}_{26}^k & \overline{Q}_{66}^k & 0 & 0 \\ 0 & 0 & 0 & \overline{Q}_{44}^k & \overline{Q}_{45}^k \\ 0 & 0 & 0 & \overline{Q}_{45}^k & \overline{Q}_{55}^k \end{bmatrix} = T \begin{bmatrix} Q_{11}^k & Q_{12}^k & 0 & 0 & 0 \\ Q_{12}^k & Q_{22}^k & 0 & 0 & 0 \\ 0 & 0 & Q_{44}^k & 0 & 0 \\ 0 & 0 & 0 & Q_{55}^k & 0 \\ 0 & 0 & 0 & 0 & Q_{66}^k \end{bmatrix} T^T \\ T = \begin{bmatrix} c^2 & s^2 & 0 & 0 & -2sc \\ s^2 & c^2 & 0 & 0 & 2sc \\ 0 & 0 & c & s & 0 \\ 0 & 0 & -s & c & 0 \\ sc & -sc & 0 & 0 & c^2-s^2 \end{bmatrix}, s = \sin\alpha_{fiber}^k, c = \cos\alpha_{fiber}^k \\ Q_{11}^k = \frac{E_1^k}{1-\mu_{12}^k\mu_{21}^k}, Q_{12}^k = \mu_{21}^k Q_{11}^k, Q_{22}^k = \frac{E_2^k}{1-\mu_{12}^k\mu_{21}^k}, Q_{44}^k = G_{23}^k \\ Q_{55}^k = G_{13}^k, Q_{66}^k = G_{12}^k \end{gathered} \tag{5}$$

where α^k_{fiber} denotes the included angle between the x-axis and the principal direction of the kth layer. Q^k_{ij} $(i,j = 1,2,4,5,6)$ represents the laminated stiffness coefficients. E^k_1 and E^k_2 express the longitudinal modulus and transverse modulus of the k-th layer, μ^k_{12} and μ^k_{21} are the Poisson's ratios. G^k_{13}, G^k_{23}, and G^k_{12} are the shear moduli and $E^k_1 = E^k_2, G^k_{12} = G^k_{13} = G^k_{23} = E^k_1/2(1+\mu_{12})$.

By integrating the stresses along thickness, the force and moment resultants of laminated thick shells can be obtained as:

$$\begin{Bmatrix} N_{cxx} \\ N_{c\theta\theta} \\ N_{cx\theta} \\ M_{cxx} \\ M_{c\theta\theta} \\ M_{cx\theta} \end{Bmatrix} = \begin{bmatrix} A_{11} & A_{12} & A_{16} & B_{11} & B_{12} & B_{16} \\ A_{12} & A_{22} & A_{26} & B_{12} & B_{22} & B_{26} \\ A_{16} & A_{26} & A_{66} & B_{16} & B_{26} & B_{66} \\ B_{11} & B_{12} & B_{16} & D_{11} & D_{12} & D_{16} \\ B_{12} & B_{22} & B_{26} & D_{12} & D_{22} & D_{26} \\ B_{16} & B_{26} & B_{66} & D_{16} & D_{26} & D_{66} \end{bmatrix} \begin{Bmatrix} \varepsilon^p_{xx0} \\ \varepsilon^p_{\theta\theta0} \\ \gamma^p_{x\theta0} \\ \chi^p_{xx} \\ \chi^p_{\theta\theta} \\ \chi^p_{x\theta} \end{Bmatrix}$$
$$\begin{bmatrix} Q_{cxx} \\ Q_{c\theta\theta} \end{bmatrix} = \kappa_s \begin{bmatrix} A_{44} & A_{45} \\ A_{45} & A_{55} \end{bmatrix} \begin{Bmatrix} \gamma^p_{\theta r0} \\ \gamma^p_{xr0} \end{Bmatrix} \tag{6}$$

where κ_s is the shear correction factor, A_{ij}, B_{ij}, D_{ij} $(i,j = 1,2,4,5,6)$ are tension stiffness, coupled stiffness, and bending stiffness of cylindrical shells, which can be obtained as:

$$\{A_{ij}, B_{ij}, D_{ij}\} = \sum_{k=1}^{N_k} \int_{z_k}^{z_{k+1}} \overline{Q}^k_{ij} \{1, z, z^2\} dz \tag{7}$$

where N_k denotes the number of fiber layers.

2.2.2. For Annular Plate

Using the first order shear deformation theory and taking the cylindrical coordinate system (o, r, x) as a base, $\varepsilon^{0,q}_{\gamma}, \varepsilon^{0,q}_{\theta}, \gamma^{0,q}_{r\theta}, \gamma^{0,q}_{rz}$ and $\gamma^{0,q}_{\theta z}$ denote structural strains in the middle surface, η^q_r, η^q_θ, and $\eta^q_{r\theta}$ are mid-surface change in curvature and twist for the annular plate. They can be written as [17]:

$$\begin{aligned} \varepsilon^{0,q}_r &= \frac{\partial u^q_a}{\partial r}, \ \varepsilon^{0,q}_\theta = \frac{u^q_a}{r} + \frac{1}{r}\frac{\partial v^q_a}{\partial \theta}, \eta^q_r = \frac{\partial \psi^q_r}{\partial r}, \eta^q_\theta = \frac{1}{r}\frac{\partial \psi^q_\theta}{\partial \theta} + \frac{\partial \psi^q_r}{r}, \\ \eta_{r\theta} &= \frac{1}{r}\frac{\partial \psi^q_r}{\partial \theta} + \frac{\partial \psi^q_\theta}{\partial r} - \frac{\psi^q_\theta}{r}, \gamma^{0,q}_{rx} = \psi^q_r + \frac{\partial w^q_a}{\partial r}, \gamma^{0,q}_{\theta x} = \frac{1}{r}\frac{\partial w^q_a}{\partial \theta} + \psi^q_\theta, \\ \gamma^{0,q}_{r\theta} &= \frac{\partial v^q_a}{\partial r} + \frac{1}{r}\frac{\partial u^q_a}{\partial \theta} - \frac{v^q_a}{r} \end{aligned} \tag{8}$$

As mentioned above, the force and moment resultants of a thick annular plate can be obtained as:

$$\begin{Bmatrix} N_r \\ N_\theta \\ N_{r\theta} \end{Bmatrix} = b\int_{-h/2}^{h/2} \begin{bmatrix} \sigma_r \\ \sigma_\theta \\ \tau_{r\theta} \end{bmatrix} dz, \begin{Bmatrix} M_r \\ M_\theta \\ M_{r\theta} \end{Bmatrix} = b\int_{-h/2}^{h/2} \begin{bmatrix} \sigma_r \\ \sigma_\theta \\ \tau_{r\theta} \end{bmatrix} zdz$$
$$\begin{Bmatrix} Q_r \\ Q_\theta \end{Bmatrix} = b\int_{-h/2}^{h/2} \begin{bmatrix} \tau_{rz} \\ \tau_{\theta z} \end{bmatrix} dz \tag{9}$$

2.3. *Energy Expressions*

System Energy includes the energy stored in the cylindrical shells and annular plates, including the corresponding strain energy (U^p_c, U^q_a), which can be written into three parts: tension compression potential energy (U_S), bending potential energy (U_B) and tension compression bending coupling potential energy (U_{BS}); kinetic energy (T^p_c, T^q_a); potential energy (U^p_{cb}) stored by the boundary springs of double cylindrical shells and annular plates; and W_e denotes the work done by the imposed external force or moment.

Energy equations above are obtained as follows:

$$U_c = \frac{1}{2}\int_S \{N_{xx}\varepsilon_{xx0} + N_{\theta\theta}\varepsilon_{\theta\theta0} + N_{x\theta}\gamma_{x\theta0} + M_{xx}\chi_{xx} + M_{\theta\theta}\chi_{\theta\theta} + M_{x\theta}\chi_{x\theta} + Q_\theta\gamma_{\theta z0} + Q_x\gamma_{xz0}\}dS \tag{10}$$

$$U_S = \frac{1}{2}\int_0^L\int_0^{2\pi}\left\{\begin{array}{l} A_{11}\left(\frac{\partial u_c}{\partial x}\right)^2 + \frac{2A_{12}}{R}\left(\frac{\partial v_c}{\partial\theta} + w_c\right)\frac{\partial u_c}{\partial x} + \frac{A_{22}}{R^2}\left(\frac{\partial v_c}{\partial\theta} + w_c\right)^2 \\ +2A_{16}\left(\frac{\partial v_c}{\partial x} + \frac{\partial u_c}{R\partial\theta}\right)\left(\frac{\partial u_c}{\partial x}\right) + \frac{2A_{26}}{R}\left(\frac{\partial u_c}{R\partial\theta} + \frac{\partial v_c}{\partial x}\right)\left(\frac{\partial v_c}{\partial\theta} + w_c\right) \\ +A_{66}\left(\frac{\partial v_c}{\partial x} + \frac{\partial u_c}{R\partial\theta}\right)^2 + \kappa A_{44}\left(\phi_{c\theta} - \frac{v_c}{R} + \frac{\partial w_c}{R\partial\theta}\right)^2 \\ +\kappa A_{55}\left(\phi_{cx} + \frac{\partial w_c}{\partial x}\right)^2 + 2\kappa A_{45}\left(\phi_{c\theta} - \frac{v_c}{R} + \frac{\partial w_c}{R\partial\theta}\right)\left(\phi_{cx} + \frac{\partial w_c}{\partial x}\right) \end{array}\right\}Rdxd\theta \tag{11}$$

$$U_B = \frac{1}{2}\int_0^L\int_0^{2\pi}\left\{\begin{array}{l} D_{11}\left(\frac{\partial\phi_{cx}}{\partial x}\right)^2 + 2\frac{D_{12}}{R}\left(\frac{\partial\phi_{c\theta}}{\partial\theta}\right)\left(\frac{\partial\phi_{cx}}{\partial x}\right) + 2D_{16}\left(\frac{\partial\phi_{cx}}{R\partial\theta} + \frac{\partial\phi_{c\theta}}{\partial x}\right)\left(\frac{\partial\phi_{cx}}{\partial x}\right) \\ +\frac{D_{22}}{R^2}\left(\frac{\partial\phi_{c\theta}}{\partial\theta}\right)^2 + 2\frac{D_{26}}{R}\left(\frac{\partial\phi_{cx}}{R\partial\theta} + \frac{\partial\phi_{c\theta}}{\partial x}\right)\left(\frac{\partial\phi_{c\theta}}{\partial\theta}\right) \\ +D_{66}\left(\frac{\partial\phi_{cx}}{R\partial\theta} + \frac{\partial\phi_{c\theta}}{\partial x}\right)^2 \end{array}\right\}Rdxd\theta \tag{12}$$

$$U_{BS} = \int_0^L\int_0^{2\pi}\left\{\begin{array}{l} B_{11}\frac{\partial u_c}{\partial x}\frac{\partial\phi_{cx}}{\partial x} + \frac{B_{12}}{R}\frac{\partial u_c}{\partial x}\frac{\partial\phi_{c\theta}}{\partial\theta} + B_{16}\frac{\partial u_c}{\partial x}\left(\frac{\partial\phi_{cx}}{R\partial\theta} + \frac{\partial\phi_{c\theta}}{\partial x}\right) \\ +\frac{B_{12}}{R}\left(\frac{\partial v_c}{\partial\theta} + w_c\right)\frac{\partial\phi_{cx}}{\partial x} + \frac{B_{22}}{R^2}\left(\frac{\partial v_c}{\partial\theta} + w_c\right)\frac{\partial\phi_{c\theta}}{\partial\theta} \\ +\frac{B_{26}}{R}\left(\frac{\partial v_c}{\partial\theta} + w_c\right)\left(\frac{\partial\phi_{cx}}{R\partial\theta} + \frac{\partial\phi_{c\theta}}{R\partial\theta}\right) + B_{16}\left(\frac{\partial v_c}{\partial x} + \frac{\partial u_c}{R\partial\theta}\right)\frac{\partial\phi_{cx}}{\partial x} \\ +\frac{B_{26}}{R}\left(\frac{\partial v_c}{\partial x} + \frac{\partial u_c}{R\partial\theta}\right)\frac{\partial\phi_{c\theta}}{\partial\theta} + B_{66}\left(\frac{\partial\phi_{cx}}{R\partial\theta} + \frac{\partial\phi_{c\theta}}{\partial x}\right)\left(\frac{\partial v_c}{\partial x} + \frac{\partial u_c}{R\partial\theta}\right) \end{array}\right\}Rdxd\theta \tag{13}$$

$$T_c = \frac{1}{2}\int_0^L\int_0^{2\pi}\left\{\begin{array}{l} I_0\left(\frac{\partial u_c}{\partial t}\right)^2 + I_0\left(\frac{\partial v_c}{\partial t}\right)^2 + I_0\left(\frac{\partial w_c}{\partial t}\right)^2 + 2I_1\frac{\partial u_c}{\partial t}\frac{\partial\phi_{cx}}{\partial t} \\ +2I_1\frac{\partial v_c}{\partial t}\frac{\partial\phi_{c\theta}}{\partial t} + I_2\left(\frac{\partial\phi_{cx}}{\partial t}\right)^2 + I_2\left(\frac{\partial\phi_{c\theta}}{\partial t}\right)^2 \end{array}\right\}Rdxd\theta \tag{14}$$

$$U_{cb} = \frac{1}{2}\int_0^{2\pi}\left\{\begin{array}{l} \left[k_u u_c^2 + k_v v_c^2 + k_w w_c^2 + k_x\phi_{cx}^2 + k_\theta\phi_{c\theta}^2\right]_{x=0} \\ \left[k_u u_c^2 + k_v v_c^2 + k_w w_c^2 + k_x\phi_{cx}^2 + k_\theta\phi_{c\theta}^2\right]_{x=L} \end{array}\right\}Rd\theta \tag{15}$$

where $(k_u, k_v, k_w, k_x, k_\theta)$ denote the boundary springs related to the boundary constraints of cylindrical shells:

$$U_a = \frac{1}{2}\int\int\int_V\left\{\begin{array}{l} N_r\varepsilon_r^0 + N_\theta\varepsilon_\theta^0 + N_{r\theta}\gamma_{r\theta}^0 + M_r\eta_r \\ +M_\theta\eta_\theta + M_{r\theta}\eta_{r\theta} + Q_r\gamma_{rz}^0 + Q_\theta\gamma_{\theta z}^0 \end{array}\right\}rdrd\theta dz \tag{16}$$

$$T_a = \frac{1}{2}\int_{R_i}^{R_o}\int_0^{2\pi}\left\{\begin{array}{l} I_0\left[\left(\frac{\partial u_a}{\partial t}\right)^2 + \left(\frac{\partial v_a}{\partial t}\right)^2 + \left(\frac{\partial w_a}{\partial t}\right)^2\right] \\ +2I_1\left[\left(\frac{\partial u_a}{\partial t}\right)\left(\frac{\partial\psi_r}{\partial t}\right) + \left(\frac{\partial v_a}{\partial t}\right)\left(\frac{\partial\psi_\theta}{\partial t}\right)\right] \\ +I_2\left[\left(\frac{\partial\psi_\theta}{\partial t}\right)^2 + \left(\frac{\partial\psi_\theta}{\partial t}\right)^2\right] \end{array}\right\}rdrd\theta \tag{17}$$

$$U_{ab} = \frac{1}{2}\int_{-\frac{h}{2}}^{\frac{h}{2}}\int_0^{2\pi}\left\{\begin{array}{l} R_o\left[k_{ru}u_a^2 + k_{rv}v_a^2 + k_{rw}w_a{}^2 + k_{rx}\psi_r^2 + k_{r\theta}\psi_\theta^2\right]_{r=R_o} \\ R_i\left[k_{ru}u_a^2 + k_{rv}v_a^2 + k_{rw}w_a{}^2 + k_{rx}\psi_r^2 + k_{r\theta}\psi_\theta^2\right]_{r=R_i} \end{array}\right\}d\theta dz \tag{18}$$

where $(k_{ru}, k_{rv}, k_{rw}, k_{rx}, k_{r\theta})$ denote the boundary springs related to the boundary constraints of an annular plate.

For the coupling conditions between cylindrical shell and annular plate:

$$u_c|_{x=p_a^q} = -w_a,\ v_c|_{x=p_a^q} = v_a,\ w_c|_{x=p_a^q} = u_a,\ \phi_{c\theta}|_{x=p_a^q} = \psi_\theta \tag{19}$$

In addition, the potential energy (U_{ca}) stored in the coupling springs between cylindrical shell and annular plate is expressed as:

$$U_{ca} = \frac{1}{2}\int_{-\frac{h}{2}}^{\frac{h}{2}}\int_{0}^{2\pi}\left\{\begin{array}{l} R_o\left[k_{cu}(u_c+w_a)^2+k_{cv}(v_c-v_a)^2+k_{cw}(w_c-u_a)^2+k_{cx}(\phi_{c\theta}-\psi_\theta)^2\right]_{r=R_o,x=x_c} \\ R_i\left[k_{cu}(u_c+w_a)^2+k_{cv}(v_c-v_a)^2+k_{cw}(w_c-u_a)^2+k_{cx}(\phi_{c\theta}-\psi_\theta)^2\right]_{r=R_i,x=x_c}\end{array}\right\}d\theta dz \quad (20)$$

where $(k_{cu}, k_{cv}, k_{cw}, k_{cx})$ represent the stiffness coefficients of coupling springs.

In order to simplify the calculation, it is assumed that, in the LCDCSS, only the radial point force is acting on the inner and outer shell. In Equation (21), W_{ex} represents the work done by the external force on the system, δ is the Dirac function, F_0 excitation force amplitude, and the coordinates of the action position:

$$W_{ex} = \frac{1}{2}\int_0^L\int_0^{2\pi} F_0\delta(x-x_F)\delta(\theta-\theta_F)w_c^2(x,\theta)R_p d\theta dx \quad (21)$$

The Lagrangian energy functional (L) of LCDCSS is expressed as:

$$\mathrm{L} = \sum_{p=1}^{2}\left(T_c^p - U_c^p - U_{cb}^p\right) + \sum_{q=1}^{Nq}\left(T_a^q - U_a^q - U_{ab}^q\right) - \sum_{i=1}^{Ni} U_{ca} - W_{ex} \quad (22)$$

N_i is the number of coupling edges.

2.4. Displacement Admissible Functions and Solution Process

The modified Fourier series method is used to describe displacement admissible function, in which several supplementary terms are introduced into the Fourier series expansion to remove any potential discontinuities of the displacement and their derivatives throughout the entire solution domain [12,13,40]. The displacement components of the cylindrical shell and annular plate can be defined as:

$$\begin{array}{c} \mathbf{U} = \sum\limits_{n=-2}^{-1}\sum\limits_{m=-2}^{-1} \mathbf{A}\sin(\lambda_{\partial m}\partial)\sin(n\theta) + \sum\limits_{m=-2}^{-1}\sum\limits_{n=0}^{N} \mathbf{B}\sin(\lambda_{\partial m}\partial)\cos(n\theta) + \\ \sum\limits_{n=-2}^{-1}\sum\limits_{m=0}^{M} \mathbf{C}\sin(\lambda_{\partial m}\partial)\cos(n\theta) + \sum\limits_{m=0}^{M}\sum\limits_{n=0}^{N} \mathbf{D}\cos(\lambda_{\alpha m}\partial)\cos(n\theta) \\ \partial = c, a \end{array} \quad (23)$$

$$\begin{array}{l} \mathbf{U} = \left[u_c^p, v_c^p, w_c^p, \phi_{cx}^p, \phi_{c\theta}^p, u_a^q, v_a^q, w_a^q, \psi_r^q, \psi_\theta^q\right]^T \\ \mathbf{A} = \left[A_{mn}^1, B_{mn}^1, C_{mn}^1, D_{mn}^1, E_{mn}^1, F_{mn}^1, G_{mn}^1, H_{mn}^1, I_{mn}^1, J_{mn}^1\right]^T \\ \mathbf{B} = \left[A_{mn}^2, B_{mn}^2, C_{mn}^2, D_{mn}^2, E_{mn}^2, F_{mn}^2, G_{mn}^2, H_{mn}^2, I_{mn}^2, J_{mn}^2\right]^T \\ \mathbf{C} = \left[A_{mn}^3, B_{mn}^3, C_{mn}^3, D_{mn}^3, E_{mn}^3, F_{mn}^3, G_{mn}^3, H_{mn}^3, I_{mn}^3, J_{mn}^3\right]^T \\ \mathbf{D} = \left[A_{mn}^4, B_{mn}^4, C_{mn}^4, D_{mn}^4, E_{mn}^4, F_{mn}^4, G_{mn}^4, H_{mn}^4, I_{mn}^4, J_{mn}^4\right]^T \end{array} \quad (24)$$

where **U** is the displacement vectors, and subscripts c and a represent the cylindrical shell and annular structures in the system, respectively; $\lambda_{am} = m\pi/L$, and $\lambda_{bm} = m\pi/b$,$b = R_o - R_i$, denoted the auxiliary polynomial functions introduced to in-plane and anti-plate displacement to remove all the discontinuities potentially associated with the second-order derivatives, then ensuring and accelerating the convergence of the series expansion. A. B, C, and D are the expansion coefficients of the trigonometric series in the displacement allowable functions, and M and N are the truncated values of the structural displacement function.

Substitute all the energy formula (Equations (10)–(21)) and the displacement admissible functions (Equations (23) and (24)) of the LCDCSS into the Lagrange energy function

(Equation (22)). By taking variation on these equations based on the Rayleigh–Ritz method, characteristic equations of the LCDCSS can be obtained:

$$\left\{\mathbf{K}-\omega^2\mathbf{M}\right\}\mathbf{\Pi} = \mathbf{F} \tag{25}$$

Therefore, the solution process of the vibration characteristics of the system becomes a simple nonlinear equation solution problem. **K** stands for the stiffness matrix and **M** represents the mass matrix separately. **Π** indicates the Fourier coefficients vector, and **F** is the external force contributions, which can be rewritten as vector form $F_{s2} = \{F_{s2,u},0,0,0,0\}^T$ and $F_{s2} = \{0, F_{s2,v},0,0,0\}^T$ when the LCDCSS receive axial and circumferential excitation load. By solving a standard matrix eigenvalue problem, the frequency parameters are then provided.

3. Numerical Calculation and Analysis

In this section, the simulation calculation will be carried out for the analysis model of the LCDCSS. Firstly, the convergence property and calculation accuracy are verified to show the correctness and effectiveness of the structure model. Then, the parametric investigations of free vibration characteristics of the LCDCSS are carried out, covering the coupled situation and constraint condition of the annular plate, boundary conditions, geometric features, and so on. After that, forced vibration characteristics of the LCDCSS are studied, and some new results and new laws are obtained to enrich the research field.

3.1. Convergence and Validation Study of the LCDCSS

The verification of convergence is of great significance to guarantee the calculation accuracy. In the Table 1 calculation example, three annular plates coupled with cylindrical shells rigidly, which distribute in the axial direction of the cylindrical shell as P_a = [1/12,1/2,11/12] L. The material parameters of the inner and outer cylindrical shells are E_1 = 50 GPa, E_2 = 2 GPa, $G_{12} = G_{13}$ = 1 GPa, G_{23} = 0.4 GPa, μ = 0.25, and the laying scheme is α_{fiber} = [0° 90° 0°]. Furthermore, the physical properties of annular plates are as E_1 = 150 GPa, E_2 = 10 GPa, $G_{12} = G_{13}$ = 6 GPa, G_{23} = 5 GPa, μ = 0.25, α_{fiber} = [0° 90° 0° 90°], ρ = 1500 kg/m^3. Moreover, geometric parameters of the system are as: L = 1.2 m, R_o = 0.5 m, R_i = 0.4 m, $h_c = h_a$ = 0.005 m. For the description of the boundary conditions, the symbols F, C, S, and SD to represent the free boundary, the clamped boundary, the simply supported boundary, and shear diaphragm supported boundary, respectively. For example, the boundary conditions of LCDCSS, written with the form of FF-CC denoting ends with x_p = 0 of the double cylindrical shell, are F, while ends with x_p = L of the double cylindrical shell are C. According to the research on rotary composite structures in reference [10], in the calculation later, the stiffness value of coupling spring remains 1×10^{14} N/m to stimulate a rigid connection effect.

To verify the convergence, the truncated number M and N of substructures are chosen as the same value, which range between 6–48. Table 1 records the lowest five frequency parameters of an LCDCSS under different classical boundary conditions. The numerical comparison shows that natural frequency is obviously changed when M ranges from 6 to 18, and the maximum deviation is 4.82%. The frequency deviation decreases significantly; when M changes from 18 to 26, the maximum deviation decreases to 1.11%, and the tendency flattens out when M over 32 shows that the calculation results tend to converge. The truncated numbers are defined as constant in this regard, which can satisfy the calculation precision.

Next, more examples of the analytical model are analyzed to test the validity and accuracy. Due to vibration characteristics of a laminate composite, a double cylindrical shell structure has not been published. Jin et al. [27] studied the vibration characteristics of an annular plate coupled double-layer cylindrical shell system (DCSS) of isotropic materials, providing reference data as well as the finite element method.

The material parameters of annular plate, inner shell, and outer shell in the system are the same as: $E_1 = E_2$ = 206 GPa, $G_{12} = G_{13} = G_{23} = E_1/2(1+\mu)$, μ = 0.3, ρ = 7850 kg/m^3, and geometric parameters are: $L = 1.2$ m, $R_o = 0.5$ m, $R_i = 0.4$ m, $h_c = h_a = 0.005$ m. The results of the theoretical model under different boundary conditions are compared with those of reference and finite element calculations, which are analyzed by ABAQUS software. The element shell 181 is selected for finite element calculation, and the element edge length is set to 0.01.

Table 1. The lowest five frequency parameters of an LCDCSS with various truncation values.

	$n = 3$												
	m	M = 6	M = 10	M = 14	M = 18	M = 22	M = 26	M = 30	M = 34	M = 38	M = 42	M = 46	M = 50
CC-CC	1	206.2	201.8	198.8	196.2	195.9	195.0	194.5	194.2	193.9	193.8	193.6	193.4
	2	254.8	253.7	251.6	250.1	249.3	248.7	248.5	248.4	248.1	247.9	247.8	247.6
	3	285.1	272.4	267.4	266.4	266.1	265.8	265.7	265.6	265.5	265.4	265.4	265.3
	4	295.7	283.5	278.1	275.1	275.0	274.3	274.0	273.8	273.6	273.6	273.5	273.4
	5	408.2	401.2	397.1	394.4	393.4	392.2	391.7	391.4	390.8	390.5	390.3	390.1
FF-FF	1	190.2	185.0	181.7	180.3	179.4	178.1	177.6	177.2	176.8	176.7	176.5	176.3
	2	220.1	217.6	212.8	210.9	210.8	208.6	208.1	207.9	207.3	207.0	206.9	206.6
	3	270.1	268.4	266.0	265.4	265.4	264.8	264.7	264.7	264.6	264.5	264.5	264.4
	4	285.8	278.4	274.6	272.7	271.9	271.2	270.9	270.7	270.6	270.5	270.4	270.3
	5	333.6	328.6	322.8	320.2	320.1	316.9	316.5	316.1	315.3	315.0	314.7	314.3
SS-SS	1	200.4	195.8	192.7	191.1	190.4	189.2	188.7	188.3	188.0	187.9	187.8	187.6
	2	252.3	249.6	246.9	245.7	245.7	244.4	244.2	243.8	243.6	243.6	243.5	243.3
	3	272.3	270.0	266.9	266.1	266.0	265.4	265.3	265.1	265.1	265.1	265.0	264.9
	4	288.3	280.1	275.7	273.7	272.9	272.0	271.7	271.5	271.3	271.3	271.2	271.1
	5	400.2	392.7	388.3	386.2	386.0	384.0	383.5	382.7	382.4	382.3	382.1	381.8
FC-FC	1	201.6	196.6	196.6	191.9	191.2	190.0	189.5	189.1	188.8	188.7	188.5	188.3
	2	246.3	244.6	244.6	240.7	240.6	239.0	238.8	238.6	238.3	238.1	238.0	237.8
	3	273.2	269.5	269.5	265.7	265.6	265.1	265.0	264.9	264.8	264.8	264.7	264.7
	4	287.9	280.0	280.0	274.1	273.3	272.5	272.2	272.1	271.9	271.8	271.7	271.6
	5	390.4	381.2	381.2	373.3	373.0	370.5	370.1	369.7	369.1	368.8	368.5	368.2

As shown in Table 2, the system with three annular plates coupled between cylindrical shells is investigated. The annular plates are located at p_a = [1/12,1/2,11/12] L, respectively. The boundary conditions are chosen as a different classical boundary, and the annular plates are rigidly coupled with the inner and outer cylindrical shells. It can be seen from the data comparison that the largest error between the present and Ref [27] is lower than 3%. The present results are in good agreement with FEM. On the whole, the calculation results of this method are consistent with Ref and FEM. The error of the two methods does not increase significantly as the order of mode increases. The proposed method maintains reliable calculation accuracy.

Table 2. The first four natural frequencies of DCSS coupled with three annular plates.

BC	m	$n = 1$ Ref [27]	FEM	Present	Error 1 (%)	Error 2 (%)	$n = 2$ Ref [27]	FEM	Present	Error 1 (%)	Error 2 (%)
CC–CC	1	830.67	830.84	830.78	−0.013	0.008	540.78	540.57	541.40	−0.114	−0.153
	2	1382.7	1377.3	1382.6	0.005	−0.387	979.84	975.19	980.62	−0.080	−0.557
	3	1385.5	1382.3	1383.8	0.125	−0.106	995.01	996.17	995.89	−0.089	0.028
	4	1536.2	1534.1	1534.5	0.109	−0.028	1083.9	1083.3	1083.7	0.022	-0.033
SS–SS	1	773.99	774.31	774.33	−0.044	−0.003	428.31	427.85	429.25	−0.220	−0.328
	2	974.97	973.03	974.92	0.005	−0.194	950.53	944.75	951.24	−0.075	−0.687
	3	1218.2	1215.6	1217.6	0.046	−0.168	972.67	974.36	974.13	−0.150	0.024
	4	1380.1	1372.6	1379.6	0.036	−0.510	1076.4	1075.4	1076.0	0.034	−0.059
FF–FF	1	956.58	950.88	957.90	−0.138	−0.738	210.99	208.23	211.88	−0.424	−1.755
	2	1227.3	1221.2	1224.5	0.229	−0.269	236.92	235.29	240.68	−1.589	−2.293
	3	1271.1	1271.1	1272.1	−0.078	−0.078	777.66	777.80	779.93	−0.292	−0.274
	4	1355.3	1346.4	1355.7	−0.028	−0.689	952.73	950.88	946.29	0.676	0.483
BC	m	$n = 3$ Ref [27]	FEM	Present	Error 1 (%)	Error 2 (%)	$n = 4$ Ref [27]	FEM	Present	Error 1 (%)	Error 2 (%)
CC–CC	1	523.93	517.91	521.89	0.389	−0.769	545.06	543.67	546.61	−0.284	−0.540
	2	740.85	742.00	742.18	−0.179	−0.024	588.35	587.24	590.35	−0.339	−0.529
	3	761.54	756.21	759.44	0.275	−0.428	626.64	626.09	625.57	0.171	0.084
	4	805.56	804.71	804.63	0.115	0.010	627.56	626.57	627.94	−0.060	−0.218
SS–SS	1	445.11	443.01	448.10	−0.672	−1.150	502.42	501.33	505.14	−0.542	−0.761
	2	687.93	689.70	690.68	−0.400	−0.142	541.87	545.29	546.62	−0.876	−0.243
	3	722.33	718.60	722.60	−0.038	−0.557	596.94	594.84	596.85	0.015	−0.338
	4	785.97	785.31	785.31	0.083	0.000	601.94	600.24	600.34	0.266	−0.017
FF–FF	1	429.45	425.74	431.62	−0.506	−1.382	500.47	503.97	503.74	−0.653	0.046
	2	515.85	510.74	520.01	−0.807	−1.816	541.04	542.63	544.53	−0.644	−0.349
	3	720.87	717.15	721.64	−0.107	−0.626	590.51	588.05	589.81	0.119	−0.299
	4	729.39	723.6	733.88	−0.615	−1.420	599.67	597.89	597.77	0.317	0.020

Figure 2 shows some modes in Table 2. It can be seen from the figure that, under three different classical boundary conditions, the vibration modes of the inner and outer cylindrical shell in DCSS system are always consistent, indicating that the energy transfer between substructures is relatively stable, and the system has high stability. It is verified that the proposed method can predict the natural vibration characteristics of DCSS accurately. Then, parametrical study on free vibration of the LCDCSS is carried out.

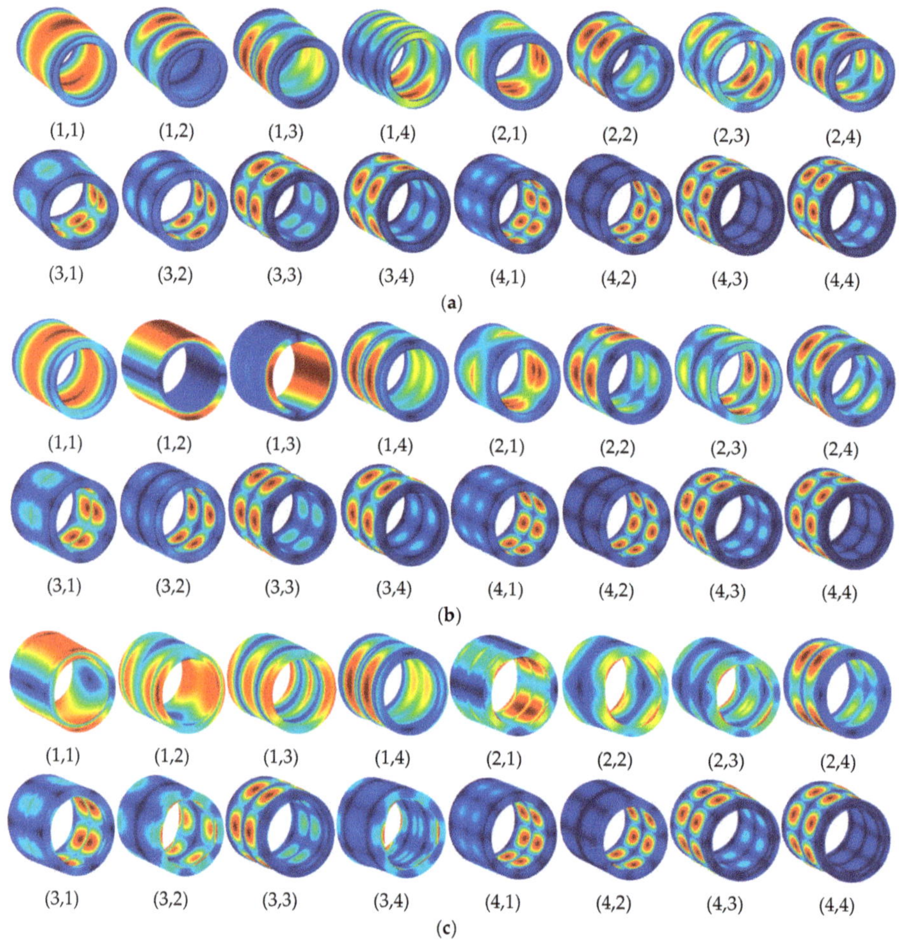

Figure 2. Mode shapes of DCSS with different boundary conditions. (**a**) CC-CC; (**b**) SS-SS; (**c**) FF-FF.

3.2. Free Vibration Analysis of the LCDSS

Firstly, the influence of the coupling relationship between the shells and the annular plates on the vibration characteristics is investigated conveniently. In the next numerical example, there are two annular plates between double cylindrical shells, which are located at p_a = [1/3 2/3] L. One of four sets spring on coupled boundaries is set to 10^5 maintaining others' rigidity. The physical parameters of cylindrical shell and annular plates are still consistent with Figure 2, and the geometric parameters of the LCDCSS are defined as: α_{fiber}= [0° 90° 90° 0°], [45° −45° 45° −45°], L = 3, R_o = 0.5 m, R_i = 0.3 m, $1/4h_a = h_c$ = 0.005 m.

From calculations in Table 3, it can be seen that the radial spring stiffness k_{cu} has an obvious influence on the natural frequency of the structure, which shows that the out-plane vibration of an annular plate has a great influence on the natural frequency of the coupling system, especially on the lower order frequency. Moreover, according to the comparison of the calculations with the two laying schemes, the natural vibration frequency of α_{fiber} = [45° −45° 45° −45°] is higher than that of [0° 90° 90° 0°].

Table 3. Influence of coupling relationship on vibration frequency of LCDCSS.

B.C.	[0° 90° 90° 0°] CC–CC				[45° −45° 45° −45°] CC–CC			
m	$k_{cu} = 10^5$	$k_{cv} = 10^5$	$k_{cw} = 10^5$	$k_{cx} = 10^5$	$k_{cu} = 10^5$	$k_{cv} = 10^5$	$k_{cw} = 10^5$	$k_{cx} = 10^5$
1	44.967	189.840	191.196	190.676	57.105	282.229	305.843	309.305
2	44.971	360.911	372.063	371.453	57.213	482.443	657.440	682.661
3	190.583	553.187	671.764	674.137	311.144	494.711	1029.792	999.627
4	371.525	629.980	679.502	678.936	689.441	677.832	1099.645	1053.731
5	674.644	687.826	688.259	684.037	745.276	832.207	1132.119	1144.327
B.C.	SS–SS				SS–SS			
m	$k_{cu} = 10^5$	$k_{cv} = 10^5$	$k_{cw} = 10^5$	$k_{cx} = 10^5$	$k_{cu} = 10^5$	$k_{cv} = 10^5$	$k_{cw} = 10^5$	$k_{cx} = 10^5$
1	44.96	165.65	166.72	166.196	57.08	163.09	161.20	166.102
2	44.97	357.85	368.91	368.460	57.20	374.63	543.35	565.734
3	166.14	538.79	607.20	591.192	166.77	431.90	969.53	953.477
4	368.32	606.6	660.2	667.569	571.7	658.95	1093.0	1052.548
5	636.8	625.42	673.8	674.025	744.76	698.5	1125.9	1141.095
B.C.	FF–FF				FF–FF			
m	$k_{cu} = 10^5$	$k_{cv} = 10^5$	$k_{cw} = 10^5$	$k_{cx} = 10^5$	$k_{cu} = 10^5$	$k_{cv} = 10^5$	$k_{cw} = 10^5$	$k_{cx} = 10^5$
1	44.968	202.37	222.30	200.59	57.079	178.34	245.07	246.23
2	45.515	282.73	297.49	294.36	57.888	238.50	282.33	289.75
3	195.49	378.91	385.95	384.23	242.777	331.24	355.58	367.23
4	294.88	452.46	472.03	470.0	287.619	424.28	685.21	718.05
5	384.34	573.59	659.19	646.0	373.156	516.01	1046.3	1015.0

Then, the influence of coupled position of annular plate on structural vibration is investigated. There is annular plate coupling in the system, and the position moves from $x_c = 0$ to $x_c = l$ end along the cylindrical shell axis. In the calculation, the annular plate and cylindrical shells couple rigidly. The dimensionless frequency is $\Omega = \omega L^2/h\,(\rho/E_2)^{1/2}$. Figure 3 shows the free vibration frequency variation curves of different orders when the annular plate moves along the axial direction of the cylindrical shell under the fixed and simply supported boundary conditions. Frequency parameters in Figure 3 are distributed symmetrically on the left and right, which reflect the axial symmetry of double cylindrical shell structure.

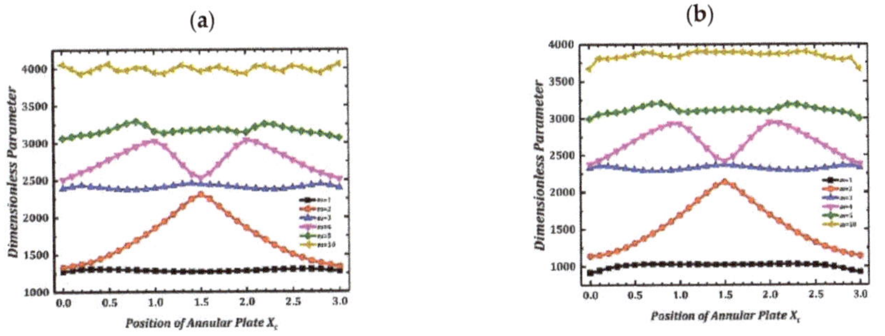

Figure 3. Variation of frequency parameters of LCDCSS with various coupling positions. (**a**) CC-CC; (**b**) SS-SS.

Figure 4 investigates change of LCDCSS free vibration frequency caused by different fiber angles under classical and elastic boundary conditions. Coupling position and material parameters of the system are consistent with Table 3, and the geometric parameters are as: $L = 3$ m, $R_o = 0.5$ m, $R_i = 0.3$ m, $h_{c\frac{1}{4}} = h_a = 0.005$ m. The laying scheme of laminated material is [90° α_{fiber} 90°], in which α_{fiber} changes from 0 to 180. It can be found in Figure 5 that curves of natural frequencies are symmetrically distributed, and there is an extremum value at the midpoint where $\alpha_{\text{fiber}} = 90°$. It can be concluded from the results that, when the angle between adjacent fibers is 0°, the natural vibration frequency of the structure is the minimum.

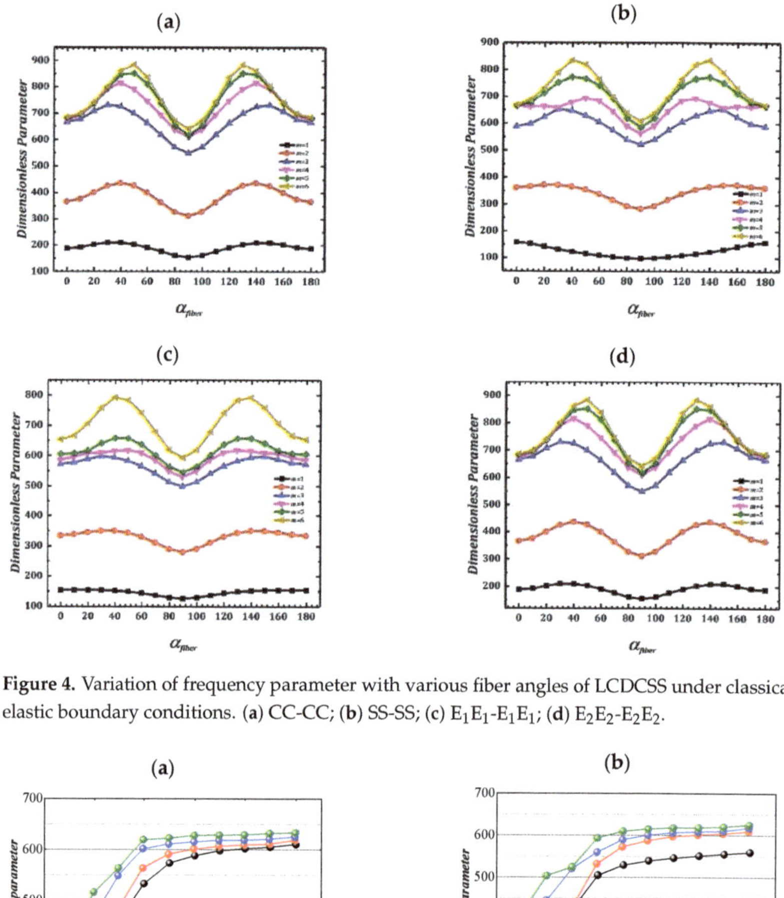

Figure 4. Variation of frequency parameter with various fiber angles of LCDCSS under classical and elastic boundary conditions. (**a**) CC-CC; (**b**) SS-SS; (**c**) E_1E_1-E_1E_1; (**d**) E_2E_2-E_2E_2.

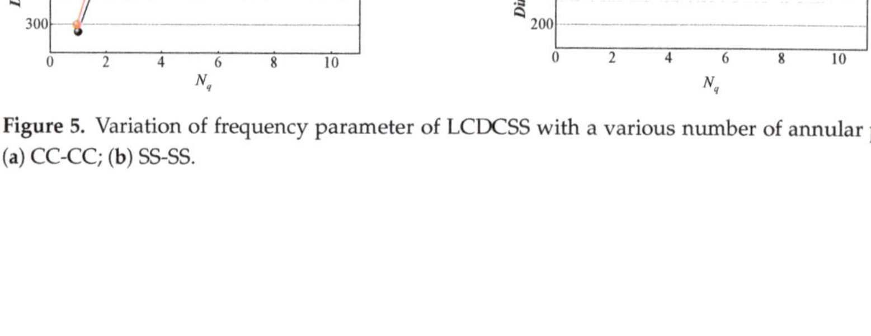

Figure 5. Variation of frequency parameter of LCDCSS with a various number of annular plates. (**a**) CC-CC; (**b**) SS-SS.

The model we have established is abstracted from the actual structure of the project, which needs to consider the specific requirements of the actual working conditions for the structural dynamic characteristics. Theoretically, the greater the number of annular plates as the inner and outer shell connection structure, the higher the stability of the system. However, it can be seen from the calculation results given in the figure that the length of the cylindrical shell is l = 0.5 m, and other geometric and material parameters are the same as those in Table 1. As shown in Figure 5, for a double cylindrical shell with finite length, when the number of annular plates increases from 1 to 10 and distributes along the x-axis direction of the cylindrical shell, the structural frequency parameters first increase significantly. However, when the number of annular plates exceeds 6, the structural frequency parameters tend to be stable, and the impact of continuous increase of annular plates on the structural frequency is reduced. It can be estimated that the minimum number of annular plates can be used to ensure the stability of the LCDCSS.

In the example shown in Figure 6, the influence of the thickness coefficient h_o/h_i of inner and outer cylindrical shells on the vibration frequency of LCDCSS is studied. The system parameters are set as: L = 1.5 m, h_i = L/500, h_o/h_i = 1~20, R_i = 0.3 m, R_o = 0.5 m, E_1 = 740 GPa, E_2 = 18.5 GPa, μ_{12} = 0.25, G_{12} = G_{13} = 111 GPa, G_{23} = 92.5 GPa, α_{fiber} = [30° 0° −30° 0° 30°], ρ = 1600 kg/m^3. As showed in Figure 6, with the increase of h_o/h_i, the variation extent of low-order vibration frequency is small, and the high-order frequency increases significantly. When the thickness coefficient h_o/h_i increases to a certain threshold, the variable trend of the system with classical boundary conditions listed in the figure decreases obviously.

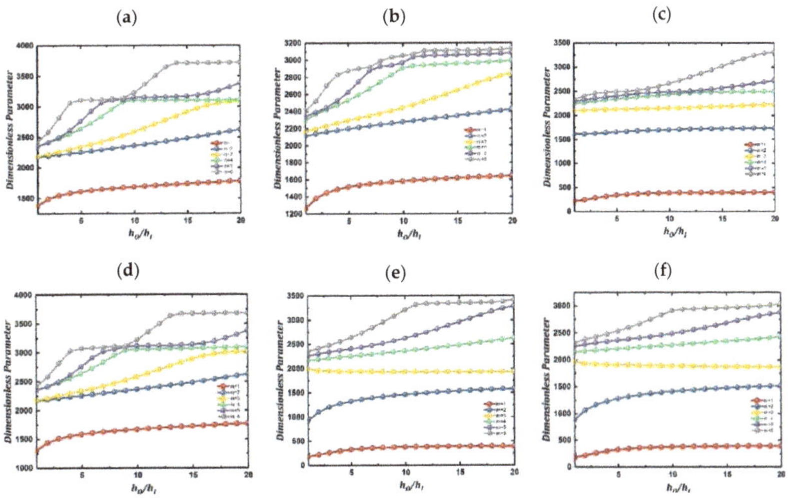

Figure 6. Variation of frequency parameter Ω versus thickness for an LCDCSS with classical boundary conditions. (**a**) CC-CC; (**b**) SS-SS; (**c**) FF-FF; (**d**) CS-CS; (**e**) CF-CF; (**f**) SF-SF.

Figure 7 shows the variation of vibration characteristics of LCDCSS when the thickness coefficient of both cylindrical shell and annular plates changes simultaneously. The system parameters are as: α_{fiber} = [0° 90° 0° 90°], L = 2 m, h_a = 0.02~0.4, h_i = 0.004 m, h_o/h_i = 1~20. From Figure 7, it can be found that the frequencies continue to rise with the increase of structural thickness, which shows that the larger the thickness, the system is more stable. It shows that the geometric parameters affect the vibration characteristic stiffness matrix of the LCDCSS in a certain range, and the vibration characteristics of the system can be adjusted by changing the parameters, which has important practical significance for the performance control of double shell structure.

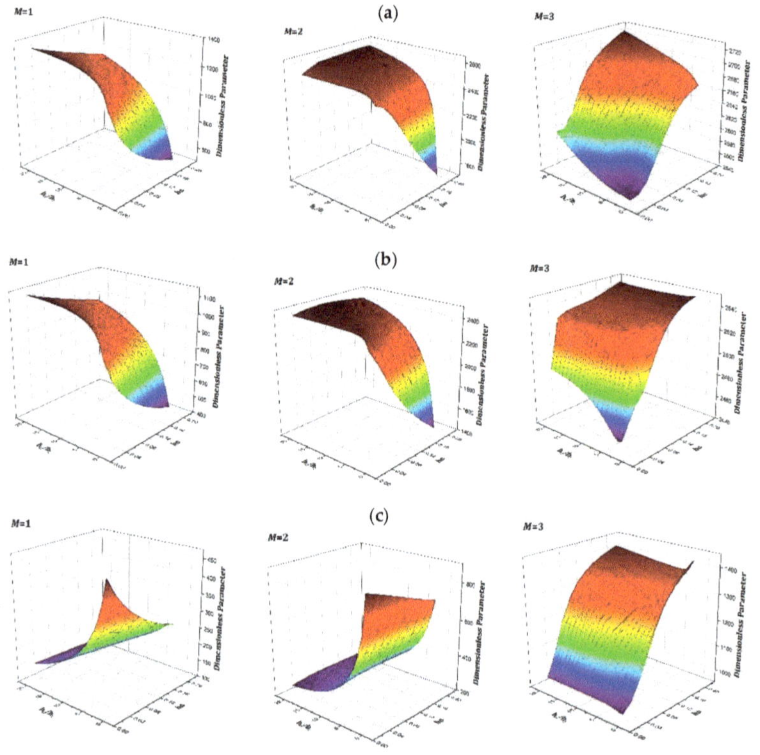

Figure 7. Variation of frequency parameter with various thickness of cylindrical shell and annular plate in LCDCSS. (**a**) CC-CC; (**b**) SS-SS; (**c**) E_1E_1-E_1E_1.

3.3. Steady State Response Analysis of the LCDCSS

This section attempts to verify the effectiveness of the method proposed to predict forced vibration characteristics of the LCDCSS. Firstly, take the isotropic material double-cylindrical shell (DCSS) as an example to test the effectiveness of the model for the steady state response analysis. When the boundary condition of FF-CC is considered, there are two annular plates coupled in DCSS located at $P_a = [0\ 1/2]L$. The physical and geometric properties of the DCSS are defined as: $E = 206$ GPa, $G_{12} = G_{13} = G_{23} = E/2\,(1 + \mu)$, $\mu = 0.3$, $\rho = 7850\ \mathrm{kg/m^3}$, $L = 1.2$ m,$R_o = 0.5$ m, $R_i = 0.4$ m, $h_c = h_a = 0.005$ m. It is assumed that the outer cylindrical shell of the double-layer system is subjected to a point force $F = 1$ N located at ($x = 0.6$ m, $\theta = 60°$) in cylindrical coordinate (o, x, θ, r), which is opposite to the normal direction. Results of measure point 1# located at ($x = 0.3$ m, $\theta = 60°$) on the outer shell, point 2# located at ($x = 0.3$ m, $\theta = 60°$) on the inner shell, point 3# located at ($r = 0.05$, $\theta = 60°$) on the annular plate on the left end of the cylindrical shell. As shown in Figure 8, the out-plane response curves of the measure points with this method are compared with the FEM results. In the finite element calculation, 181 shell elements are selected as calculation units, mesh size is 0.01×0.01 mm, and its results coincide quite well with the results mentioned above, which proves that a parametrical study on the forced vibration of the system will be carried out accurately and reliably.

As Figure 9 shows, variation of response vibration frequency of LCDCSS with various layer scheming is investigated. The material properties of the cylindrical shells are as follows: $E_1 = 50$ GPa, $E_2 = 2$ GPa, $\mu_{12} = 0.25$, $G_{12} = G_{13} = 1$ GPa, $G_{23} = 0.4$ GPa, $\rho = 1600\ \mathrm{kg/m^3}$, and those of annular plates are: $E_1 = 150$ GPa, $E_2 = 10$ GPa, $\mu_{12} = 0.25$, $G_{12} = G_{13} = 6$ GPa, $G_{23} = 5$ GPa, $\rho = 1600\ \mathrm{kg/m^3}$. The laying scheme of cylindrical shell and annular plate is

α_{fiber} = [0° 90° 0° 90°]$_n$ (n = 2, 3, 4). In addition, the geometric parameters, boundary conditions, position and size of external force excitation of LCDCSS are consistent with Figure 8. Results of measure points are: point 1# located at (x = 0.3 m, θ = 60°) on the outer shell, point 2# located at (x = 0.3 m, θ = 60°) in the outer shell, point 3# located at (r = 0.05, θ = 60°) on the annular plate coupled on the left end of cylindrical shell, point 4# located at (r = 0.05, θ = 90°) coupled in the middle of the cylindrical shell. The curve comparison shows that, with the increase of the number of fiber layers, the response amplitude changes slightly, but the change amplitude is small. The formant of the response displacement curve moves to the right, which is more obvious in the higher frequency region. Such characterization shows that the increase of the fiber layer can improve the structural stability.

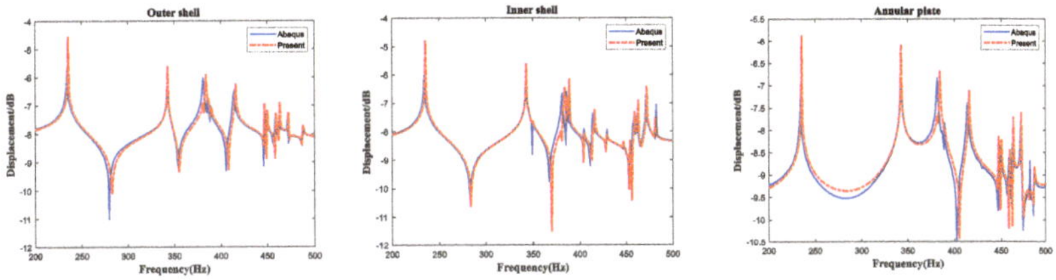

Figure 8. Comparison of steady state response of DCSS.

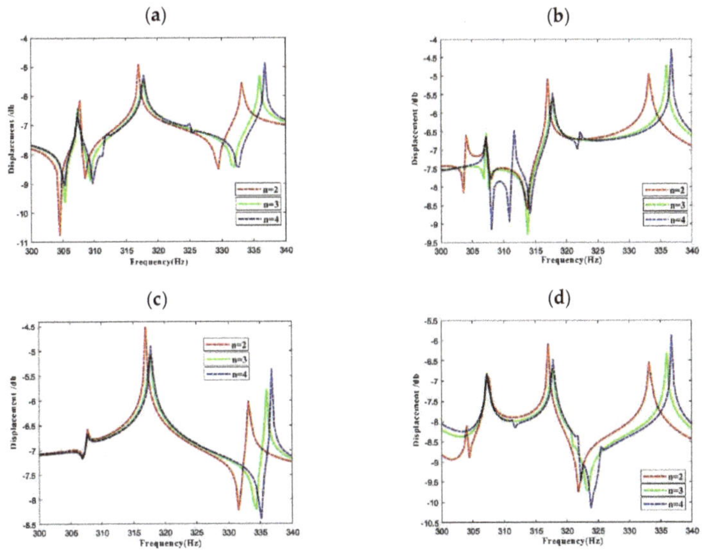

Figure 9. Variation of response vibration frequency for LCDCSS with various material layer scheming. (**a**) Point 1#; (**b**) Point 2#; (**c**) Point 3#; (**d**) Point 4#.

In Figure 10, the influence of excitation amplitude on response vibration is investigated. In the LCDCSS system, an annular plate couples in the axial middle of the cylindrical shell. The boundary condition and laying scheme is CC-CC with other parameters including external force and measure points are consistent with Figure 8. From the comparison of the response curves of F = 1N, 2N, 3N in Figure 10, it can be seen that the peak response increases in proportion to the increase of the corresponding excitation amplitude. However, waveform of the response curve of the system will not change, and the displacement curve will not shift within the frequency range, which is consistent with the general cognition.

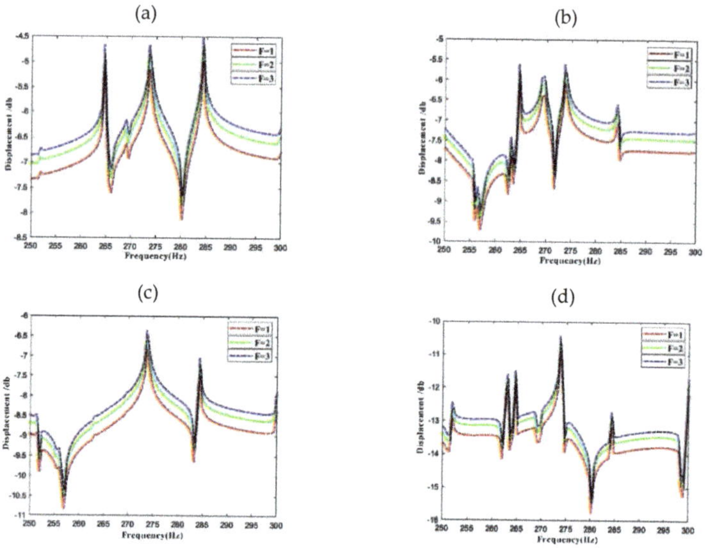

Figure 10. Variation of response vibration frequency for LCDCSS with various external excitation amplitude. (**a**) Point 1#; (**b**) Point 2#; (**c**) Point 3#; (**d**) Point 4#.

Figure 11 shows the axial vibration response frequency of LCDCSS at four response points with thickness of the annular plate is $h_a = 0.007$ m, 0.008 m, and 0.009 m, respectively. From the change of the response curves in Figure 11, it can be seen that the formant in the frequency region moves to the right, and the frequency amplitude also decreases, which means that the structural stiffness is improved. The greater the thickness of annular plate, the stronger the structural constraints.

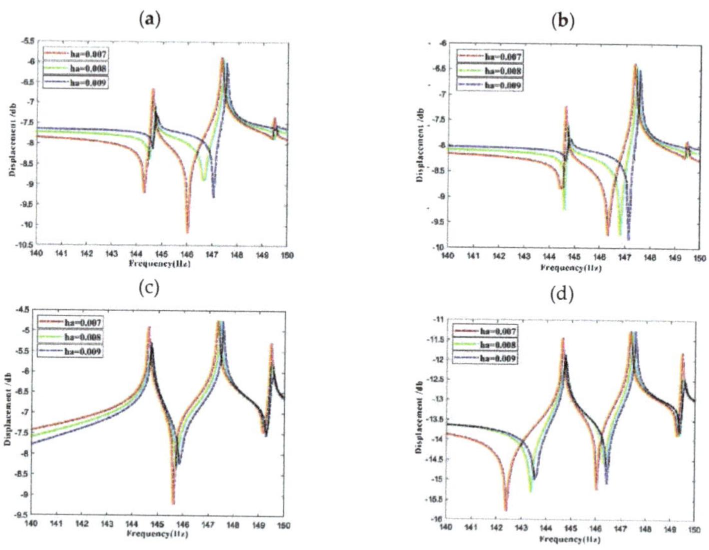

Figure 11. Variation of axial dynamic response vibration frequency for LCDCSS with various thicknesses of annular plates. (**a**) Point 1#; (**b**) Point 2#; (**c**) Point 3#; (**d**) Point 4#.

4. Conclusions

Based on the Rayleigh–Ritz energy method, a unified analysis model for the vibration characteristics of the LCDCSS with general boundary condition is established in this paper. The accuracy and reliability of the method are verified by simulation calculation of different examples and compared with the results of literature and finite element method. Through the parametric study, free vibration and forced vibration of double cylindrical shell system are predicted, which enriches the relevant data in the research field and provides reference for the structure design. The research results show that the method can effectively solve the problems of free vibration and forced vibration of LCDCSS with an arbitrary annular plate. This method can be further extended to the study of vibration characteristics of multilayer plates, multilayer shells, and more complex coupled structural systems.

Author Contributions: Conceptualization, D.S. and Y.Z.; methodology, D.S.; software, D.H.; data curation, Y.Z. writing—original draft preparation, Y.Z.;visualization, D.H. All authors have read and agreed to the published version of the manuscript.

Funding: This research was funded by the Natural Science Foundation of Heilongjiang Province of China (grant number [E2016024]).

Data Availability Statement: All data, models, and code generated or used during the study appear in the submitted article.

Acknowledgments: The authors gratefully acknowledge the Natural Science Foundation of Heilongjiang Province of China (Grant No. E2016024).

Conflicts of Interest: The authors declare no conflict of interest. The funders had no role in the design of the study; in the collection, analyses, or interpretation of data; in the writing of the manuscript, or in the decision to publish the results.

References

1. Ming, P.; Norton, M. The measurement of structural mobilities of a circular cylindrical shell. *J. Acoust. Soc. Am.* **2000**, *107*, 1374–1382. [CrossRef] [PubMed]
2. Lee, S.J.; Reddy, J. Vibration suppression of laminated shell structures investigated using higher order shear deformation theory. *Smart Mater. Struct.* **2004**, *13*, 1176–1194. [CrossRef]
3. Zhong, R.; Tang, J.; Wang, A.; Shuai, C.; Wang, Q. An exact solution for free vibration of cross-ply laminated composite cylindrical shells with elastic restraint ends. *Comput. Math. Appl.* **2019**, *77*, 641–661. [CrossRef]
4. Hu, X.; Tsuiji, T. Vibration analysis of laminated cylindrical thin panels with twist and curvature. *Int. J. Solids Struct.* **2001**, *38*, 2713–2736. [CrossRef]
5. Qu, Y.; Long, X.; Wu, S.; Meng, G. A unified formulation for vibration analysis of composite laminated shells of revolution including shear deformation and rotary inertia. *Compos. Struct.* **2013**, *98*, 169–191. [CrossRef]
6. Maharjan, A.; Sang-Youl, L. Free Vibration Characteristics of Laminated Composite Cylindrical Shells Reinforced by SWCNT with a Central Cutout. *J. Korean Soc. Adv. Compos. Struct.* **2019**, *10*, 15–20. [CrossRef]
7. Hafizah, A.K.N.; Viswanathan, K.K.; Aziz, Z.A.; Lee, J.H. Vibration of antisymmetric angle-ply composite annular plates of variable thickness. *J. Mech. Sci. Technol.* **2018**, *32*, 2155–2162. [CrossRef]
8. Civalek, O. Vibration analysis of laminated composite conical shells by the method of discrete singular convolution based on the shear deformation theory. *Compos. Part B Eng.* **2013**, *45*, 1001–1009. [CrossRef]
9. Zhao, J.; Choe, K.; Shuai, C.; Wang, A.; Wang, Q. Free vibration analysis of laminated composite elliptic cylinders with general boundary conditions. *Compos. Part B Eng.* **2019**, *158*, 55–66. [CrossRef]
10. He, D.; Shi, D.; Wang, Q.; Shuai, C. Wave based method (WBM) for free vibration analysis of cross-ply composite laminated cylindrical shells with arbitrary boundaries. *Compos. Struct.* **2019**, *213*, 284–298. [CrossRef]
11. Kumar, A.; Bhargava, P.; Chakrabarti, A. Vibration of laminated composite skew hypar shells using higher order theory. *Thin-Walled Struct.* **2012**, *63*, 82–90. [CrossRef]
12. Jin, G.; Ye, T.; Chen, Y.; Su, Z.; Yan, Y. An exact solution for the free vibration analysis of laminated composite cylindrical shells with general elastic boundary conditions. *Compos. Struct.* **2013**, *106*, 114–127. [CrossRef]
13. Jin, G.; Ye, T.; Jia, X.; Gao, S. A general Fourier solution for the vibration analysis of composite laminated structure elements of revolution with general elastic restraints. *Compos. Struct.* **2014**, *109*, 150–168. [CrossRef]
14. Jin, G.; Ye, T.; Ma, X.; Chen, Y.; Su, Z.; Xie, X. A unified approach for the vibration analysis of moderately thick composite laminated cylindrical shells with arbitrary boundary conditions. *Int. J. Mech. Sci.* **2013**, *75*, 357–376. [CrossRef]
15. Wang, Q.; Shi, D.; Liang, Q.; Pang, F. Free vibrations of composite laminated doubly-curved shells and panels of revolution with general elastic restraints. *Appl. Math. Model.* **2017**, *46*, 227–262. [CrossRef]

16. Wang, Q.; Shi, D.; Pang, F.; Liang, Q. Vibrations of Composite Laminated Circular Panels and Shells of Revolution with General Elastic Boundary Conditions via Fourier-Ritz Method. *Curved Layer. Struct.* **2016**, *3*, 105–136. [CrossRef]
17. Wang, Q.; Shi, D.; Liang, Q.; Ahad, F. A unified solution for free in-plane vibration of orthotropic circular, annular and sector plates with general boundary conditions. *Appl. Math. Model.* **2016**, *40*, 9228–9253. [CrossRef]
18. Tornabene, F.; Viola, E.; Fantuzzi, N. General higher-order equivalent single layer theory for free vibrations of doubly-curved laminated composite shells and panels. *Compos. Struct.* **2013**, *104*, 94–117. [CrossRef]
19. Li, X.; Li, Y.; Xie, T. Vibration characteristics of a rotating composite laminated cylindrical shell in subsonic air flow and hygrothermal environment. *Int. J. Mech. Sci.* **2019**, *150*, 356–368. [CrossRef]
20. Li, X. Parametric resonances of rotating composite laminated nonlinear cylindrical shells under periodic axial loads and hygrothermal environment. *Compos. Struct.* **2020**, *255*, 112887. [CrossRef]
21. Zhang, H.; Zhu, R.; Shi, D.; Wang, Q. A simplified plate theory for vibration analysis of composite laminated sector, annular and circular plate. *Thin-Walled Struct.* **2019**, *143*, 106252. [CrossRef]
22. Zuo, H.; Chen, Y.; Jia, F.; Yang, Z. Unified wavelet finite element formulation for static and vibration analysis of laminated composite shells. *Compos. Struct.* **2021**, *272*, 114207. [CrossRef]
23. Liu, T.; Wang, A.; Wang, Q.; Qin, B. Wave based method for free vibration characteristics of functionally graded cylindrical shells with arbitrary boundary conditions. *Thin-Walled Struct.* **2019**, *148*, 19. [CrossRef]
24. Bisheh, H.; Wu, N.; Rabczuk, T. Free vibration analysis of smart laminated carbon nanotube-reinforced composite cylindrical shells with various boundary conditions in hygrothermal environments. *Thin-Walled Struct.* **2020**, *149*, 106500. [CrossRef]
25. Yamada, G.; Irie, T.; Tamiya, T. Free vibration of a circular cylindrical double-shell system closed by end plates. *J. Sound* **1986**, *108*, 297–304. [CrossRef]
26. Yuan, J.; Dickinson, S.M. The Free Vibration of Circularly Cylindrical Shell and Plate Systems. *J. Sound Vib.* **1994**, *175*, 241–263. [CrossRef]
27. Zhang, C.; Jin, G.; Ma, X.; Ye, T. Vibration analysis of circular cylindrical double-shell structures under general coupling and end boundary conditions. *Appl. Acoust.* **2016**, *110*, 176–193. [CrossRef]
28. Chen, J.M.; Huang, Y.Y.; Chen, Y.B. Vibration and acoustic radiation from submerged spherical double-shell. *China Ocean Eng.* **2003**, *17*, 341–354.
29. Dogan, V.; Vaicaitis, R. Nonlinear Response of Double-Wall Cylindrical Shells Subjected to Random Excitation. In Proceedings of the 9th Biennial International Conference on Engineering, Construction and Operations in Challenging Environment, League City, Houston, TX, USA, 7–10 March 2004; pp. 46–54.
30. Qing, G.-H.; Liu, Y.-H.; Qiu, J.-J.; Meng, X.-J. A semi-analytical method for the free vibration analysis of thick double-shell systems. *Finite Elements Anal. Des.* **2006**, *42*, 837–845. [CrossRef]
31. Wang, J.G.; Li, H.W. Analysis on Dynamic Failure Behaviors of Steel Double-Layer Latticed Cylindrical Shells with Three Circle Centers Used in a Gymnasium. *Appl. Mech. Mater.* **2013**, *256–259*, 706–709.
32. Zhang, C.; Shang, D.J.; Li, Q. Effect of Drive Location on Vibro-Acoustic Characteristics of Submerged Double Cylindrical Shells with Damping Layers. *Appl. Mech. Mater.* **2013**, *387*, 59–63.
33. Huang, Y.-Q.; Zhong, L.; Fu, J.-Y. Wind-induced vibration and equivalent wind load of double-layer cylindrical latticed shells. *J. Vibroeng.* **2014**, *16*, 1063–1078.
34. Zhang, C.; Jin, G.; Ma, X.; Su, Z. Free Vibration of Rib Stiffened Double Cylindrical Shells with General Boundary Condition. In Proceedings of the 22nd International Congress on Sound and Vibration (ICSV), Florence, Italy, 12–16 July 2015.
35. Xie, K.; Chen, M. Wave based method for vibration analysis of double-walled cylindrical shells. *Appl. Acoust.* **2018**, *139*, 293–306. [CrossRef]
36. Wali, M.; Hentati, T.; Jarraya, A.; Dammak, F. Free vibration analysis of FGM shell structures with a discrete double directors shell element. *Compos. Struct.* **2015**, *125*, 295–303. [CrossRef]
37. Choi, M.-S.; Yeo, D.-J. Free Vibration Analysis of Double Cylindrical Shells Using Transfer of Influence Coefficent. *J. Power System Eng.* **2017**, *21*, 48–54. [CrossRef]
38. Mohammadimehr, M.; Zarei H., B.; Parakandeh, A.; Arani, A.G. Vibration analysis of double-bonded sandwich microplates with nanocomposite facesheets reinforced by symmetric and un-symmetric distributions of nanotubes under multi physical fields. *Struct. Eng. Mech.* **2017**, *64*, 361–379.
39. Chen, M.; Wei, J.; Xie, K.; Deng, N.; Hou, G. Wave Based Method for Free Vibration Analysis of Ring Stiffened Cylindrical Shell with Intermediate Large Frame Ribs. *Shock Vib.* **2013**, *20*, 459–479. [CrossRef]
40. Chen, Y.; Jin, G.; Liu, Z. Free vibration analysis of circular cylindrical shell with non-uniform elastic boundary constraints. *Int. J. Mech. Sci.* **2013**, *74*, 120–132. [CrossRef]

materials

MDPI

Article

Algorithmic-Aided Approach for the Design and Evaluation of Curvilinear Steel Bar Structures of Unit Roofs

Jolanta Dzwierzynska * and **Patrycja Lechwar**

Faculty of Civil and Environmental Engineering and Architecture, Rzeszow University of Technology, Al. Powstancow Warszawy 12, 35-959 Rzeszow, Poland; p.lechwar@prz.edu.pl

* Correspondence: joladz@prz.edu.pl

Abstract: Rationalization in structural design in the field of steel structures mostly consists inreducing structural material. The aim of this work was to develop an algorithmic-aided, original and practical approach to shaping curvilinear steel bar structures of modular roofs, enabling their optimization. The first stage of shaping consists in creating algorithms that define the structures of shelters made of four roof units. Algorithmic definitions of the structures made it possible to obtain many variants of the roof structures with the adopted preliminary criteria. In order to evaluate the effectiveness of the individual variants, the genetic optimizations of the structures' forms were carried out. Assuming that the structures were loaded with self-weights, the cross-sections of the structures' members were optimized with the permissible deflections, while the structures' weights were the optimization criteria. This allowed us to eliminate the design variants unfavorable in terms of shape and weight. In contrast, the structures with the most advantageous properties were then optimized for weight under snow and wind loads. The research allowed us to notice how the shapes of the structures influenced their efficiency. The dual approach proposed for shaping, which takes advantage of the generative design and consistent flow of information during shaping, allowed us to achieve better solutions compared to the traditional approach.

Keywords: steel bar structures; parametric design; genetic algorithms; structural optimization; lightweight structures; manufacturing and processing

Citation: Dzwierzynska, J.; Lechwar, P. Algorithmic-Aided Approach for the Design and Evaluation of Curvilinear Steel Bar Structures of Unit Roofs. *Materials* **2022**, *15*, 3656. https://doi.org/10.3390/ma15103656

Academic Editors: Lihong Su, Xing Zhao, Peitang Wei and Hui Wang

Received: 21 April 2022
Accepted: 18 May 2022
Published: 20 May 2022

1. Introduction

At present, thanks to technological changes over the years, the production of steel is definitely better in terms of quality than the steel produced in the past. As a result of the development of material engineering, designers have at their disposal the steel of new grades not only with higher strength, but also steel that meets special requirements. The properties of steel and its advantages make it applicable in the design of various building structures, which encompasses bar, shell and tension structures [1]. However, the advantages of steel such as its high tensile and compressive strength as well as its ability to prefabricate are the reasons why it is widely used in the building industry [1]. Due to this fact, the rational shaping of steel structures deserves special attention.

The shaping of steel structures is defined as the progressive optimization of their shapes due to assumed criteria of strength. These criteria constitute a set of related rules concerning reliability [2], including bearing capacity, serviceability as well as resistance to exceptional environmental influences, which are mostly defined by standards [3–6]. Another important aspect of shaping steel bar structures is the calculation and design of joints [7]. Over the years, many different methodologies for shaping steel structures have been developed. Among them, the most popular are minimization of the energy used for producing structural material, minimization of material consumption, as well as material reuse [8].

However, the behavior of any structure depends not only on the kind of material used, acting loads and technological solutions, but also on the geometric form of it. Due to this fact, the so-called geometric shaping criteria of steel structures are also very important. The geometric shaping criteria mainly deal with geometric parameters of the module, the structure's height and its span, as well as the characteristics of the structural members [9]. Some of these criteria have been determined in long-term practice as well as profound studies in the field of structural optimization. Nowadays, they are considered rational due to a present state of knowledge and the technology advancement level, as well as sustainable design [10].

In the old pre-computer age, the optimal shapes of the structures have been found by experiments such as, for example, inverted hanging models or soap film methods. The structures achieved by these experiments were of minimal surface shapes [11,12]. They were characterized by high stiffness with a minimal amount of bending, which was guaranteed by their forms. In general, shape optimization deals with the form or the contour of the structure; however, for the discrete bar structures it means the position of the nodes in the whole structure's form.

Curvilinear steel bar structures are particularly demanding in terms of both shaping the geometric forms of them and establishing their topologies. However, the positions of bars in structures can be determined using both the numerical and geometric method. In the first method, in order to establish structure's nodes, the coordinates of the grid points and the so-called directional vectors can be applied, whereas in the second method, the so-called base solids can be used, which divide a given space, or the so-called base surface, which is divided in order to obtain grids of bars [8].

Curvilinear steel bar structures require a more sophisticated approach to structural shaping as they can be realized with the minimal use of materials. In this case, the form and topology significantly influence efficiency of the structure and the flow of forces in it [13]. As it has been shown in previous publications, the geometry of the base surface, which is divided during shaping, has a significant impact on the properties of the structure formed on its basis [14]. Ruled surfaces, which are the surfaces composed of straight lines, are especially advantageous due to their discretization [15]. However, as far as ruled surfaces are concerned, the subgroup of them, the so-called Catalan surfaces, are the most popular for shaping roof structures [16]. However, considering the Catalan surfaces, the hyperbolic paraboloids have found the greatest application for shaping single or complex roof structures. This work deals with shaping curvilinear steel bar structures composed of the modules of Catalan surfaces as the units of a cylindroid or a conoid. It is a sequel to the authors' previous study on the parametric design of roof structures composed of the repetitive modules of Catalan surfaces [16]. However, contrary to the previous publication, where reinforced concrete structures are concerned, here, steel bar structures are taken into consideration.

The way of shaping structures is also dependent on the shaping tools used, which change according to the development of technology. CAD and BIM software currently available on the market, apart from their basic purpose, i.e., digital recording of the structure, modeling and coding the entire information about it, are integrated with Computer Integrated Manufacturing [17,18]. However, modern tools for parametrically shaping structures such as ones working in Rhinoceros 3D environment, created by Robert McNeel and associates, present new possibilities for shaping [19]. Commonly, when new tools become available, the aspects of their proper usage arise. Noteworthy are the modern tools that allow algorithmic aided design, such as Revit (Dynamo) and Rhinoceros (Grasshopper). Thanks to this software, a structural model can be described using a script with variable parameters and then generated.

Algorithmic-aided structural shaping, which means the process when both the geometric model and structural analysis are realized by means of algorithms, is rather a new field of research. However, some studies explore the concept of algorithmic-aided structural shaping in architecture, engineering and construction [20–24]. The possibility of using genetic algorithms for the optimization of the structures being designed is explored in [25,26]. A genetic algorithm approach to designing non-linear steel frames with semi-rigid connections is presented in [27,28]. It deals with searching for a frame with the minimum weight by selecting the most suitable steel cross sections from a standard set. However, a generative design strategy for the minimum-weight design of diagrid tall buildings is proposed in [23].

Appropriate digital tools can facilitate the work of the constructor-designer and lead to the best solutions. In this context, the study aims at developing an original approach to shaping curvilinear steel bar structures of the modular roofs created from the repeating cylindroid and conoid surface units. This approach uses genetic algorithms as a tool to search for optimal solutions of the structures and for their evaluation. It also allows us to notice how the shapes of the structures influence their efficiency.

2. Shaping Approach

2.1. Possible Surface Discretization

The surface discretization method is very important. As mentioned previously, the load bearing capacity of each steel bar structure is directly related to the topology of the grid it constitutes. The discretization methods vary in the number of face edges, node complexity as well as face's curvature. Another aspect regarding the discretization of the base surface for shaping a steel bar structure is the anticipated way of presenting this structure in the future, due to the fact that the steel bar structure can constitute an invisible supported system or be a visible steel-panel mesh. In the second case, the mesh directly determines the aesthetics of the shaped object, so its pattern should be of a high-quality. Triangulation of the surfaces is a common way to discretize surfaces. In general, triangular meshes are more rigid and stronger than the meshes of other types. However, quadrilateral or hexagonal meshes also have some structural advantages as they can be torsion free structures. On the other hand, in the case of the structures based on triangular meshes, six beams typically meet in one node. This means a significantly higher number of bar elements and node complexities compared to other types of meshes. Often, these disadvantages also result in low structural transparency, which can be an important feature of a see-through covering material such as glass or plastic polycarbonate sheets for roofing. Moreover, the per-area cost of triangular glass panels is higher than that of quadrilateral panels due to the fact that the weight of the structure is smaller when less steel is used. From this point of view, quadrilateral meshes are more optimal. On the other hand, triangulated grid structures can easily approximate free form shapes, which is not so easy in the case of the other meshes. The most commonly used method of curved surfaces discretization is so-called tessellation. This method involves the process of filling the surface or space with a repeated geometric motif that is usually a regular polygon. Various methods of surface tessellation are presented in [29].

Taking the above pros and cons into consideration, quadrilateral meshes will be applied on the surfaces constituting the roofs' units. Additionally, a light polycarbonate material will be applied for roofing. Additionally, the planarity of the mashes' cells will be tested in order to check the possibility of using planar panels.

2.2. The Shaping Tools Used

The most intensive development of various optimization methods in structure design began in the 1970s. It was related with the application of computer technology for solving various engineering problems regarding structural analysis, as well as material properties. Optimal shaping of steel bar structures has always been the necessary goal of structural engineering. Therefore, this problem has been the focus of scientists and practicing engineers and is reflected in numerous publications [24,30,31].

The optimization method applied often depends on the design tools used. In our research, Rhinoceros 3D software, was used, which is software equipped with visual scripting language. The software user needs to build an algorithm that is a graphical code, which is a fully functioning program. The idea of visual programming language is to replace the textual syntax with graphic objects in which are coded individual commands and data, as well as to create links between them. The Rhino scripts of the considered structures were created by means of Grasshopper and Karamba 3D [26].

Next, it has been applied so-called evolutionary optimization EO (genetic optimisation), which is based on a genetic algorithm inspired by the Darwinian law of natural selection. This process of optimization mimics the natural process for genetic coding and selection, as well as reproduction. The reason of it is to iteratively improve the individuals in each generation, according to a given optimization criteria associated with fitness functions. This is because EO deals with searching for optimal solutions in a population of possible variables, contrary to other optimization methods, which deal with improving one variable.

Nowadays, there is more and more interest in the application of EO in structural engineering. The genetic algorithm approach to structural topology optimization is applied in [32–35]. However, some attempts to design and optimize steel bar structures using visual scripts were undertaken in [24,36,37].

In our case study, visual scripts were created in order to determine digital models of curvilinear steel bar structures as canopies composed of four modules of cylindroid shape or four modules of conoid shape. The structures have been defined by several geometric parameters. Next, the scripts describing structures' geometries were extended in order to obtain structural models and connection properties, as well as loads applied. Then, the structures under self-loads were optimized due their masses with the application of EO. The results of the simulations were evaluated. After evaluation and the selection of the proper structures, their further optimization was carried out with the application of Robot Structural Analysis Professional software, assuming additional environmental loads acting on the structures [38]. A general structure shaping diagram is shown in Figure 1.

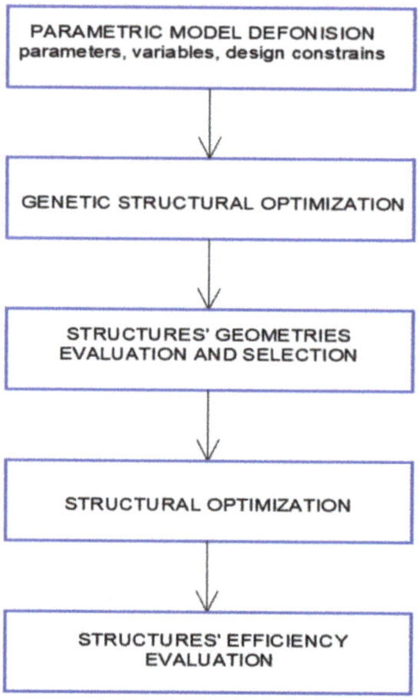

Figure 1. Basic scheme of the integrated shaping process.

3. Results

As has been noted previously, due to their geometric properties, Catalan surfaces are applied as geometrical shapes for the formation of roofs [16,39]. However, the hyperbolic paraboloid, which is the surface of second order, is the most widely used. Due to their double curvature, hyperbolic paraboloids are exceptionally stiff and this is the reason for their common application. They are also widely discussed in the literature [24,37]. The cylindroid and the conoid are not as popular; however, they deserve attention as they can constitute interesting modular forms for the composition of compound roofs. Due to this fact, finite models of cylindroids and conoids have been considered as the roof's units.

From a geometrical point of view, both the cylindroid and the conoid as the ruling surfaces can be defined as the surfaces composed of straight lines, the so-called surface rulings. Moreover, they can be determined by two generatrix lines and the so-called director plane to which all rulings are parallel. However, in the case of the cylindroid, both directrix lines are curved, whereas in the case of the conoid, one of the directrix lines is straight. Dependently on the shapes of the curved lines being surfaces' directrix lines, various forms of the surfaces' units and, thus, various forms of the composite roofs, can be obtained.

3.1. Geometric Modeling of Compound Unit Roofs by Grasshopper

Each surface as a three-dimensional object can be described mathematically by a single equation with three space variables. In our case, in order to create a parametric surface by means of Grasshopper, each surface needs to be described by two parameters (u, v) measured along directrices and rulings. Parametric modelling began by establishing a series of points on two lines included in various planes and being the surfaces' directrix lines. In the case of the cylindroid, both lines were curved and in the case of conoid, one of them was curved, whereas the second one was straight. The same number of points on each line were established, whereas the position of each point was described by parameter u. In this way, the surfaces' rulings joined points of the lines, which corresponded to the same parameter value along u direction. As a result, the surface of the rulings was parallel to the same vertical director plane. The example modules of a cylindroid shape are presented in Figure 2, and the example modules of conoid shape are presented in Figure 3.

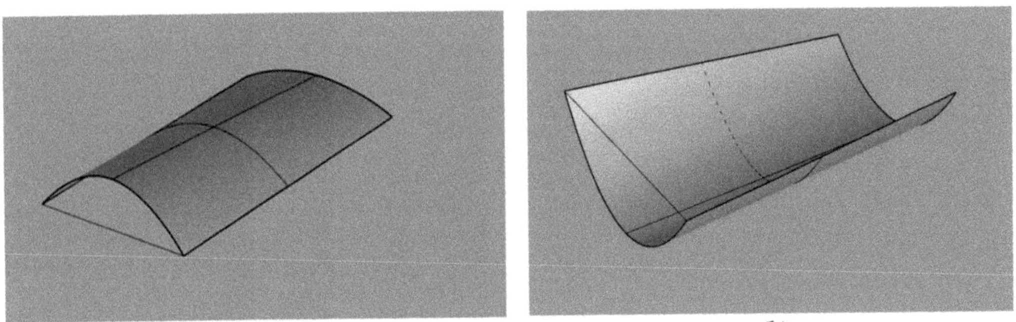

(a) (b)

Figure 2. Example modules of cylindroid: (**a**) witch positive arch; (**b**) witch negative arch.

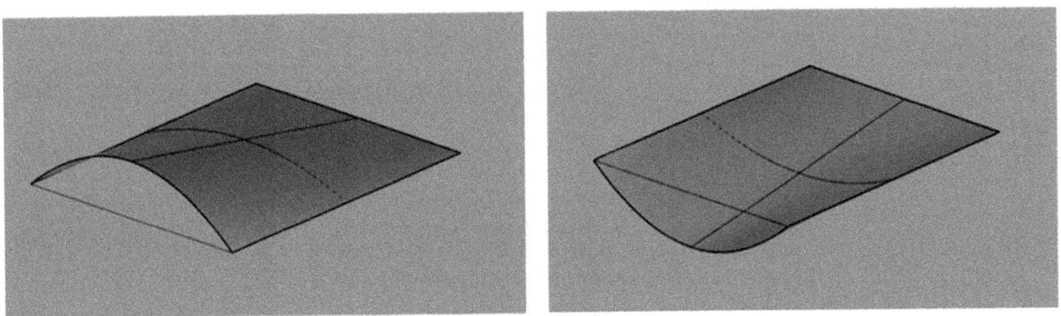

(a) (b)

Figure 3. Example modules of conoid: (**a**) witch positive arch; (**b**) witch negative arch.

The shapes obtained constituted a base for creation of the four-module roofs supported by four columns. In the analyzed case, a sample square (9.6 m × 9.6 m) was covered with a roof consisting of four modules. However, due to the fact that the effective structures are sought, it has been assumed that each structure has two axes of symmetry, and it is composed of homogeneous modules of cylindroid grids or conoid grids. The geometry of the structure was described in an algorithmic way using Grasshopper's visual script components. The example of a structure with markings of individual lengths is presented in Figure 4.

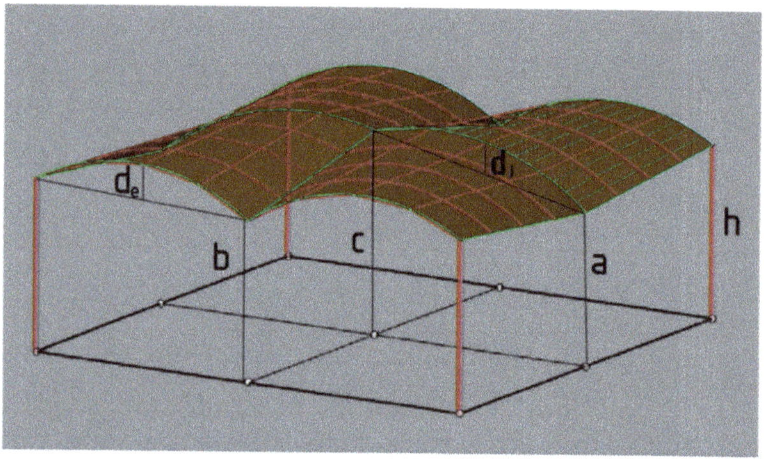

Figure 4. An example of the four-unit structure.

3.2. Geometric Modeling of Compound Unit Roofs by Grasshopper

Shaping of any steel bar structure can be treated as solving a problem according to the needed requirements and constrains. However, algorithmic-aided shaping is an action aimed at the logical description of a problem using a script and choosing the values of some structural parameters that meet the shaping criteria. The shaping criteria can be different; however, in the case of the steel bar structures, the main goal of shaping is structures' masses reduction, unification of the structures' elements, as well as using uncomplicated joints. They all influence a structural cost minimization.

The general scheme is shown in Figure 1 and applied to shape the considered curvilinear steel bar structures in order to reduce structural mass, encompassing two optimization processes. The first process was as follows: geometric modeling of the structures and establishing their structural models, which encompassed defining the shaping variables, identifying constrains, identifying the optimization criteria, optimizing the structures under dead loads and selecting the best solutions. The second process consisted of performing the structural analysis for the best roof forms, taking into account various load combinations, including environmental loads, optimization of structures due to their masses, selection and evaluation of the best results.

In order to perform the first optimization of the considered structures, the scripts determining the geometrical characteristics of the structures were developed by means of the Karamba 3D plug-in. For that reason, the bar grids' vertices were transformed into structural nodes, the supports were defined as rigid and the structural joints were established: rigid for grid and hint joints connected the lattices with columns. However, in order to obtain a bar grid with a repeatable assortment of bars, an even division into four or five parts was assumed when discretizing the surface, which constituted one of the design variables. As a cover, it has been applied polycarbonate sheets attached to the bars with screws.

The simulations were performed with five different assumptions for both the roofs with cylindroid modules and the roofs with conoid modules, as listed below and according to the dimensional variables a, b, c, d_i, d_e presented in Figure 4. However, the starting point for each simulation was the structure defined by the smallest parameters: a, b, c, d_i, d_e.

- **Case 1**—Cylindroid structure 1

  Columns' heights h = 4 m, the height of internal arches d_i = 0.7 m,
  intervals of variables: a = 3–4 m, b = 3–4 m, c = 3–4 m, the height of the external arches d_e = −0.7–0.7 m,

- **Case 2**—Cylindroid structure 2

  Columns' heights h = 4 m, the height of the external arches d_e = 0.7 m, the height of internal arches d_i = −0.7 m,
  intervals of variables: a = 3–4 m, b = 3–4 m, c = 3–4 m, the height of the internal arches d_i = −0.7–0.7 m

- **Case 3**—Cylindroid structure 3

  Columns' heights h = 3 m, the height of internal arches d_i = 0.7 m,
  intervals of variables: a = 3–4 m, b = 3–4 m, c = 3–4 m, the height of the external arches d_e = −0.7–0.7 m,

- **Case 4**—Cylindroid structure 4

  Columns' heights h = 3 m, the height of the external arches d_e = 0.7 m,
  intervals of variables: a = 3–4 m, b = 3–4 m, c = 3–4 m, the height of internal arches d_i = −0.7 m,

- **Case 5**—Cylindroid structure 5

  Columns' heights h = 3 m, the height of the external arches d_e = 0.7 m, the height of internal arches d_i = 0.7 m,
  intervals of variables: a = 3–4 m, b = 3–4 m, c = 3–4 m,

- **Case 6**—Conoid structure 1

  Columns' heights h = 4 m, the height of the internal arches d_i = 0.7 m,
  intervals of variables: a = 3–4 m, b = 3–4 m, c = 3–4 m,

- **Case 7**—Conoid structure 2

  Columns' heights h = 4 m, the height of the external arches d_e = −0.7 m,
  intervals of variables: a = 3–4 m, b = 3–4 m, c = 3–4 m,

- **Case 8**—Conoid structure 3

  Columns' heights h = 3 m, the height of the internal arches d_i = 0.7 m,
  intervals of variables: a = 3–4 m, b = 3–4 m, c = 3–4 m,

- **Case 9**—Conoid structure 4

  Columns' heights h = 3 m, the height of the internal arches d_e = −0.7 m,
  intervals of variables: a = 3–4 m, b = 3–4 m, c = 3–4 m,

- **Case 10**—conoid structure 5

  Columns' heights h = 3 m, the height of the external arches d_i = 0.7 m,
  intervals of variables: a = 3–4 m, b = 3–4 m, c = 3–4 m,

According to [1], maximum deflection for all structures was established equal to 3.8 cm. During the simulation, each structure was assumed to be loaded with its own weight and the structures' masses were optimized taking into account the maximum allowable deflection. As the results of the performed simulations, five geometries of the roof structures composed of cylindroid units with the minimal masses, presented in Figure 5, and five roof structures with the conoid units with the minimal masses, presented in Figure 6, were established.

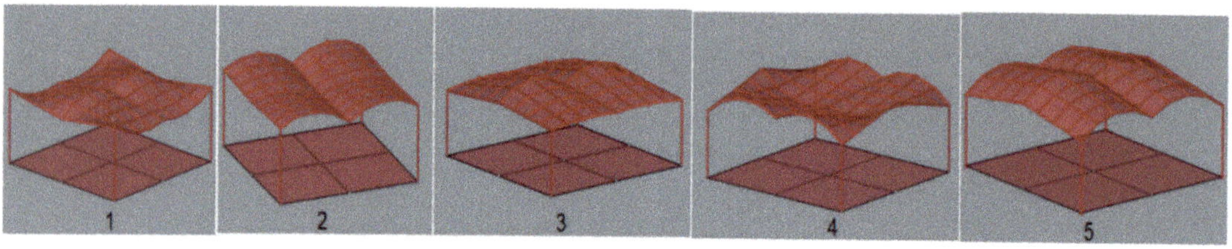

Figure 5. Various models of the roof structures composed of cylindroid units obtained due to simulations.

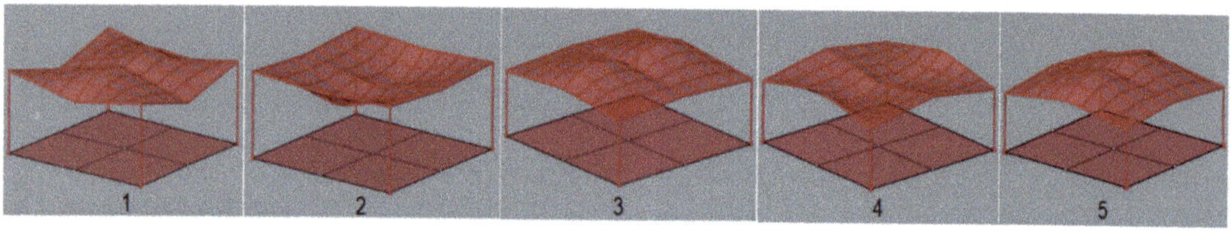

Figure 6. Various models of the roof structures composed of conoid units obtained due to simulations.

The dimensions which characterize the optimized roof structures with cylindroid units and with conoid units are given, respectively in Tables 1 and 2.

Table 1. Characteristic dimensions of various roof structures with cylindroid units.

Kind of the Structure	a [m]	b [m]	c [m]	d_e [m]	d_i [m]	h [m]
Cylindroid structure 1	3.28	3.39	3.89	−0.45	0.70	4.0
Cylindroid structure 2	3.7	3.86	3.66	0.70	0.36	4.0
Cylindroid structure 3	3.14	3.24	3.52	0.12	0.70	3.0
Cylindroid structure 4	3.47	4.00	3.08	0.70	−0.54	3.0
Cylindroid structure 5	3.0	3.60	3.30	0.70	0.70	3.0

Table 2. Characteristic dimensions of various roof structures with conoid units.

Kind of the Structure	a [m]	b [m]	c [m]	d_e [m]	d_i [m]	h [m]
Conoid structure 1	3.02	3.29	3.89	0.70	-	4.0
Conoid structure 2	3.74	3.81	3.10	−0.70	-	4.0
Conoid structure 3	3.61	3.63	3.85	-	0.70	3.0
Conoid structure 4	3.62	3.75	3.06	-	−0.70	3.0
Conoid structure 5	3.35	3.55	4.00	-	0.7	3.0

The structures obtained through simulations have been evaluated. The roofs that were unfavorable in terms of shape due to the possibility of snow or rainwater accumulation such as: the cylindroid structure 2, the cylindroid structure 4, the conoid structure 2, the conoid structure 4 were rejected. Thus, the following unit structures of favorable cylindroid shape were analyzed further: the cylindroid structure 1, the cylindroid structure 3, and the cylindroid structure 5, presented in Figure 7, as well as the following unit structures of favorable conoid shape: the conoid structure 1, the conoid structure 3, the conoid structure 5, are presented in Figure 8.

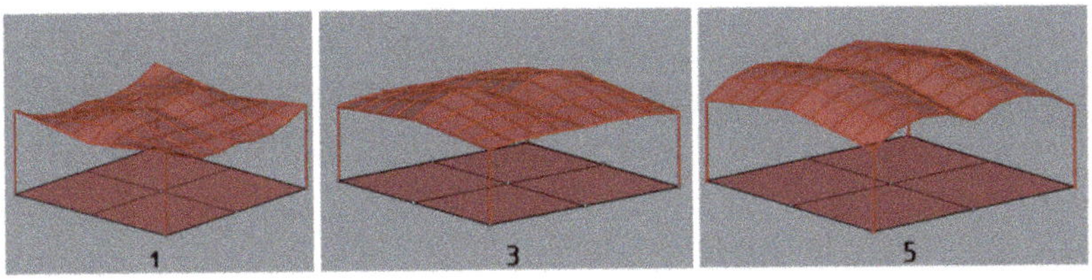

Figure 7. The models of the roof structures composed of cylindroid units favorable in terms of shape.

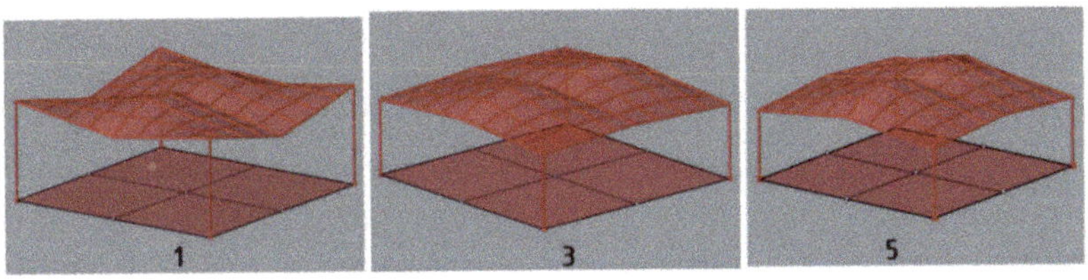

Figure 8. The models of the roof structures composed of conoid units favorable in terms of shape.

3.3. Structural Analysis by Means of Robot Structural Analysis Professional

Further analysis, optimization and dimensioning of the considered structures was carried out with the application of Robot structural Analysis Professional software. According to European Standards included in Eurocodes [3–6], both Ultimate Limit States (ULS) and Serviceability Limit States (SLS) have been verified during the process of shaping structures.

The selected structures were subjected to FEM analysis taking into account both permanent and environmental loads. Due to this fact, it has been assumed that each structure was loaded by permanent loads (self–weight), as well as environmental loads: snow load and wind load. The permanent load constituted cladding loads and structures' loads. For the roof cladding, polycarbonate plastic sheets with a density of 0.02 kg/m^3 and a panel thickness of 1.0 cm were used. The structural material steel of S 235 grade was used and the structural members with circular hollow cross-sections and wall thicknesses no less than 3.2 mm were also used. As far as environmental loads are concerned, the boundary conditions regarding snow and wind loads were assumed for the objects located in Rzeszow, Poland. Taking into account the third snow load zone location and the structures' characteristics, it was assumed:

- characteristic value of snow load on the ground s_k = 1.2 kN/m^2,
- the thermal coefficient C_t =1.0,
- the exposure coefficient C_e = 1.0.

Based on the standard guidelines and the symmetry of the structures, some simplifications have been proposed assuming the loads acting on the structures. However, in the performed analysis, two snow load cases have been considered: even snow load and the possibility of snowdrifts, whereas the roof's shape coefficients have been assumed as for cylindrical roofs [5]. As far as the wind loads are concerned, it was assumed that base wind velocity pressure was equal to 0.3 kN/m^2 and the loads were generated automatically, assuming several various wind directions [6].

The structures were optimized according to mass, taking into account the most unfavorable combination of loads. The analysis showed uneven stress distribution in individual elements of the structure. Therefore, in order to minimize the masses of the structures, their elements were divided into several groups for the purposes of dimensioning. The results of the structural analysis and the optimizations performed are presented in Tables 3 and 4.

Table 3. The results of the structural analysis and optimization of the roof structures composed of cylindroid units.

Kind of the Structure	Max. Axial Force [kN]	Max. Bending Moment [kNm]	Max. Deformation [cm]	Structure's Mass [kg]
Cylindroid structure 1	13.17	4.23	3.8	2308
Cylindroid structure 3	15.36	4.00	3.8	2223
Cylindroid structure 5	12.50	4.93	3.7	2272

Table 4. The results of the structural analysis and optimization of the roof structures composed of conoid units.

Kind of the Structure	Max. Axial Force [kN]	Max. Bending Moment [kNm]	Max. Deformation [cm]	Structure's Mass [kg]
Conoid structure 1	14.55	5.07	3.4	2322
Conoid structure 3	26.32	5.00	3.7	2924
Conoid structure 5	12.76	7.98	3.7	3869

3.4. Analysis of the Possibility of Using Flat Panels on the Structure

Another aspect which is very important in the case of the structures being analyzed is the possibility of using glass panels instead of the polycarbonate ones. The considered structures were analyzed with the use of lightweight polycarbonate panels, which minimized the total masses. Due to the easy adaptation of polycarbonate to any curved surface, the curvature of the panels used was not significant. When glass panels are used as planes, it is important for them to be flat quadrilaterals. Therefore, in order to check the planarity of individual panels of the structure, a verification algorithm was developed. The individual structures were verified in terms of the planarity of the polygons delineating the bar mesh. The most favorable shapes were found among the roofs of cylindroid shapes. Namely, in the case of the cylindroid structure 5, all flat panels can be applied, and in the case of the cylindroid structure 1, four flat panels can be applied in each roof unite, Figure 9.

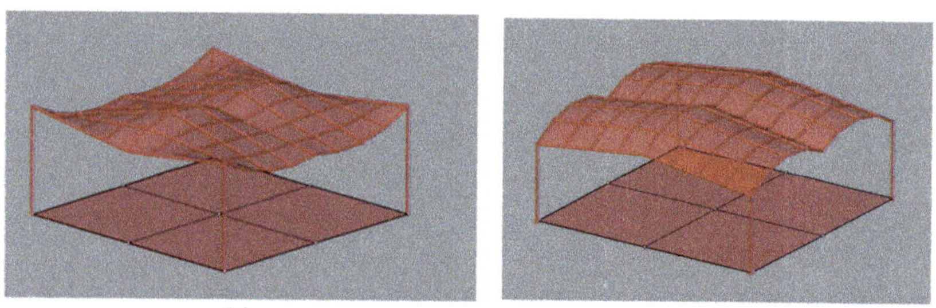

(a) (b)

Figure 9. The structures with the possibility of using flat panels: (**a**) a cylindroid structure 1; (**b**) a cylindroid structure 5.

It is worth mentioning that in the case of glass panels applied as roofing material, it is recommended to use stainless steel spider connections. However, depending on the node position, a one-, two- or four-point connector should be used.

4. Importance of Research

An algorithmic-aided method of efficiently shaping steel bar shelters with roofs composed of four units of cylindroid and conoid surfaces has been proposed.

The developed visual scripts made it possible to test a large number of structures in the assumed specific dimensional ranges. The first stage of shaping, consisting of a algorithmically aided geometric design and structural design of the structures made it possible to distinguish several representative structures in terms of shape and mass. For this purpose, the optimization of each structure in terms of weight was carried out with a self-weight load, maintaining the permissible deflections. Based on the optimization results, five different geometries of the shelters with the roofs made of cylindroid units and five different geometries with the roofs made of conoid units were distinguished. The geometric characteristics of these structures are presented in Tables 1 and 2. The shapes of the structures were evaluated and the structures with shapes unfavorable due to the possibility of snowfall accumulation were eliminated. The three most favorable structural geometries of the roofs with the cylindroid units and three with the conoid units were subjected to structural analysis assuming environmental loads and optimized due to their masses. The structures' members were dimensioned. The summary of the results of these analyses are presented in Tables 3 and 4 and show that, in general, the canopies with roofs made of cylindroid modules are lighter. On the other hand, among the structures with conoidal roofs, the structure called the conoid structure 1 is characterized by the smallest mass. The maps on bars for the most efficient structures with cylindroid modules: the cylindroid structure 1, structure 3, structure 5 are shown in Figure 10, whereas maps on bars on the most efficient structure 1 with conoid module are presented in Figure 11.

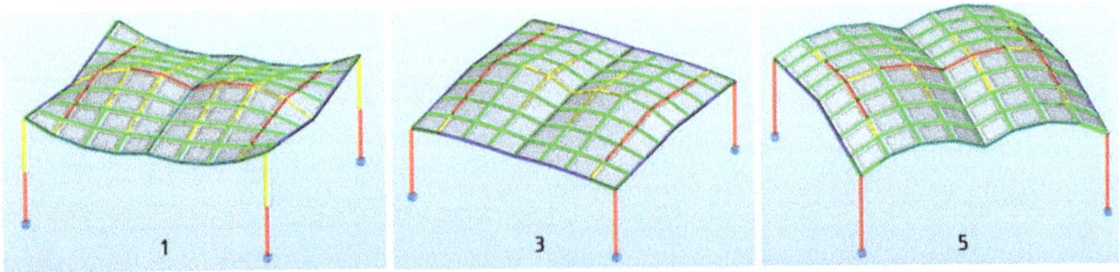

Figure 10. Maps on bars of the most efficient structures with cylindroid roofs.

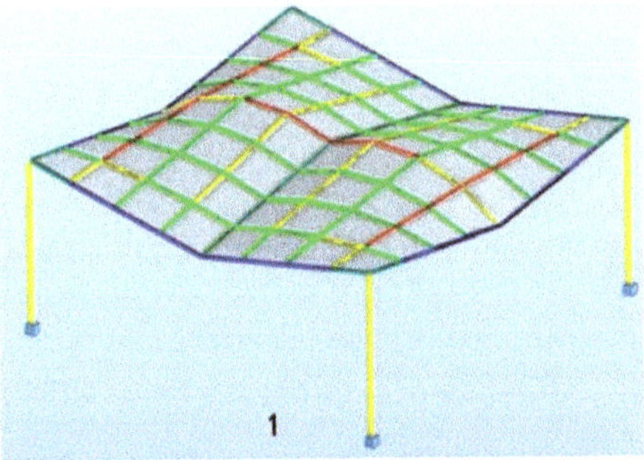

Figure 11. Maps on bars of the most efficient structure with conoid roof.

The above structures were designed to be applied as light material for roofing that is polycarbonate and easily adapts to curvilinear shapes. However, in the case of steel grid structures, it is convenient to use flat panels. In general, the grid structures created based on screw ruled surfaces as Catalan surfaces are characterized by no planar grid cells being spatial polygons. However, the grid roofs of the considered structures were tested in terms of using flat panels, e.g., glass. Thanks to the algorithm modification, the planarity of the grid cells were tested. The results of the analysis showed that among the structures tested, the polygons constituted the grid cells of the structures with cylindroid modules; that is, cylindroids 1 and cylindroid 5, are characterized by planarity. However, in the case of the structure called cylindroid 1, only four polygons of the bar grid of each module are flat. However, these modules are symmetrically located in the structure so the interesting solutions can be obtained. In contrast, in the case of the structure cylindroid 5, all of the lattice polygons are planar, so the entire roof can be covered with flat panels. Due to this fact, this structure seems to be the most efficient and universal.

The research showed that although the obtained structures due to optimization have two planes of symmetry, individual modules do not have such a property, therefore the obtained solutions are more effective than the initial assumptions of the simulation when symmetrical modules have been taken. The most efficient structural models both of cylindroid and conoid shape obtained as the result of the conducted research are economical, so they can be proposed as effective steel bar structures and they can find practical applications. Therefore, as the next step, laboratory tests are planned in order to verify the obtained results.

However, the research has shown that a holistic approach to the issue of shaping the steel bar structures, which is based on algorithmic aid, is appropriate, as it allows us to obtain optimal solutions within the assumed ranges of the design variables. Due to this fact, the research is worth continuing and extending in order to shape more complex structures, both single and modular ones.

In a broader sense, this research has shown how to apply generative tools to integrate structures' geometries and structural efficiency as a single goal, in order to obtain optimal steel bar structures.

5. Conclusions

An algorithmic-aided method for shaping curvilinear steel bar structures of shelters with roofs composed of four grid shell units was proposed. The bar grids were created on the basis of the Catalan surface shapes, in particular the cylindroid and the conoid. The novelty of this method consists in:

- the use of algorithmically-aided structural analysis for the preliminary bars' cross sections optimization and evaluation of the received solutions,
- a dual approach in shaping method which allows us to obtain optimal shapes of the steel bar structures in order to guarantee their effectiveness.

The applied method allowed us to:

- achieve the effective structures of both cylindroid and conoid shape in terms of load transfer and optimize them using mass as an optimization criterion,
- obtain the curvilinear steel bar structures with flat quadrilateral panels, which significantly influences the structural cost and permits the application of glass panels.

The authors plan to develop this method further. However, the achieved solutions and the developed method of shaping structures turned out to be promising, not only in terms of the possibility of structures' optimization, but also as a design tool for searching for the efficient structures in the initial design phase.

Author Contributions: Conceptualization, J.D.; methodology, J.D.; software, J.D. and P.L.; validation, J.D. and P.L.; formal analysis, J.D.; investigation, J.D.; resources, J.D.; data curation, J.D.; writing—original draft preparation, J.D.; writing—review and editing, J.D.; visualization, J.D.; supervision, J.D.; project administration, J.D.; funding acquisition, J.D. All authors have read and agreed to the published version of the manuscript.

Funding: This research was supported by the Polish Ministry of Science and Higher Education grant to maintain research potential.

Institutional Review Board Statement: Not applicable.

Informed Consent Statement: Not applicable.

Data Availability Statement: Not applicable.

Conflicts of Interest: The authors declare no conflict of interest.

References

1. Bródka, J.; Broniewicz, M. Structure safety. In *Design of Steel Structures According to Eurocodes*, 1st ed.; Polskie Wydawnictwo Techniczne PWT: Warsaw, Poland, 2013; pp. 101–155.
2. Woliński, S. On the criteria of shaping structures. *Sci. Pap. Rzesz. Univ. Technol.* **2011**, *276*, 399–408.
3. *PN—EN 1991-1-1:2004 Eurocode 1*; Actions on Structures. Part 1-1: General Actions—Densities, Self-Weight, Imposed Loads for Buildings. PKN: Warsaw, Poland, 2004.
4. *PN—EN 1993-1-1:2006 Eurocode 3*; Design of Steel Structures. Part 1-1: General Rules and Rules for Buildings. PKN: Warsaw, Poland, 2006.
5. *PN—EN 1991-1-1:2004 Eurocode 1*; Actions on Structures—Part 1–3: General Actions—Snow Loads. PKN: Warsaw, Poland, 2004.
6. *PN—EN 1991-1-1:2004 Eurocode 1*; Actions on Structures—Part 1–4: General Actions—Wind Actions. PKN: Warsaw, Poland, 2004.
7. *PN-EN 1993-1-8:2006 Eurocode 3*; Ddesign of Joints—Part 1–8. PKN: Warsaw, Poland, 2006.
8. Dzwierzynska, J.; Labuda, I. Modeling of Curvilinear Steel Rod Structures Based on Minimal Surfaces. *Materials* **2021**, *14*, 6826. [CrossRef] [PubMed]
9. Rebielak, J. Bar space structures—Rules of Shaping. In Proceedings of the Symmetry: Natural and Artificial. In Proceedings of the 3rd Interdisciplinary Symmetry Symposium and Exhibition, Washington, DC, USA, 14–20 August 1995.
10. Hussain, A. Sustainable Structural Design. *Int. J. Res. Eng. Appl. Sci.* **2012**, *2*, 19–30.
11. Richardson, J.N.; Adriaenssens, S.; Filomeno Coelho, R.; Bouillard, P. Coupled form-finding and grid optimization approach for single layer grid shells. *Eng. Struct.* **2013**, *52*, 230–239. [CrossRef]
12. Li, N.; Lu, J.; Zong, W.; Fan, W. Form-finding optimization methods for free-form reticulated shells: Reverse realization and numerical simulation. *Adv. Mech. Eng.* **2017**, *9*. [CrossRef]
13. Zalewski, W.; Allen, E. *Shaping Structures Statics*, 1st ed.; John Wiley & Sons: New York, NY, USA, 1998; ISBN 9780471169680.
14. Dzwierzynska, J. Shaping of Curvilinear Steel Bar Structures for Variable Environmental Conditions Using Genetic Algorithms—Moving towards Sustainability. *Materials* **2021**, *14*, 1167. [CrossRef]
15. Grey, A. *Modern Differential Geometry of Curves and Surfaces with Mathematica*; CRC Press LCC: Boca Raton, FL, USA, 1999.
16. Dzwierzynska, J.; Prokopska, A. Pre-Rationalized Parametric Designing of Roof Shells Formed by Repetitive Modules of Catalan Surfaces. *Symmetry* **2018**, *10*, 105. [CrossRef]
17. Dzwierzynska, J. Single-image-based Modelling Architecture from a Historical Photograph. *IOP Conf. Ser. Mater. Sci. Eng.* **2017**, *245*, 062015. [CrossRef]
18. Dzwierzynska, J. Reconstructing Architectural Environment from a Panoramic Image. *IOP Conf. Ser. Earth Environ. Sci.* **2016**, *44*, 042028. [CrossRef]
19. Rhino 7. Available online: https://www.rhino3d.com/ (accessed on 18 May 2022).
20. Januszkiewicz, K.; Banachowicz, M. Nonlinear Shaping Architecture Designed with Using Evolutionary Structural Optimization Tools. *IOP Conf. Ser. Mater. Sci. Eng.* **2017**, *245*, 082042. [CrossRef]
21. Dzwierzynska, J. Integrated Parametric Shaping of Curvilinear Steel Bar Structures of Canopy Roofs. *Buildings* **2019**, *9*, 72. [CrossRef]
22. Dzwierzynska, J. Rationalized Algorithmic-Aided Shaping a Responsive Curvilinear Steel Bar Structure. *Buildings* **2019**, *9*, 61. [CrossRef]
23. Cascone, F.; Faiella, D.; Tomei, V.; Mele, V. A Structural Grammar Approach for the Generative Design of Diagrid-Like Structures. *Buildings* **2021**, *11*, 90. [CrossRef]
24. Dzwierzynska, J. Multi-Objective Optimizing Curvilinear Steel Bar Structures of Hyperbolic Paraboloid Canopy Roofs. *Buildings* **2020**, *10*, 39. [CrossRef]
25. Preisinger, C. Linking Structure and Parametric Geometry. *Archit. Des.* **2013**, *83*, 110–113. [CrossRef]
26. Wang, S.Y.; Tai, K.; Wang, M.Y. An enhanced genetic algorithm for structural topology optimization. *Int. J. Numer. Methods Eng.* **2006**, *65*, 18–44. [CrossRef]
27. Hayalioglu, M.S.; Degertekin, S.O. Genetic algorithm based optimum design of non-linear steel frames with semi-rigid connections. *Steel Compos. Struct.* **2004**, *4*, 453–469. [CrossRef]
28. Saka, M.P.; Kameshki, E.S. Optimum design of nonlinear steel frames with semi-rigid connections using a genetic algorithm. *Comput. Struct.* **2001**, *79*, 1593–1604.
29. Tarczewski, R. *Topologia form Strukturalnych. Naturalne i Tworzone Przez Człowieka Prototypy form Konstrukcyjnych w Architekturze*; Oficyna Wydawnicza Politechniki Wrocławskiej: Wrocław, Poland, 2011.
30. Łukowicz, A.; Urbańska-Galewska, E.; Deniziak, P.; Gordziej-Zagórowska, M. Classification of restraints in the optimization problem of the cold-formed profil. *Adv. Sci. Technol. Res. J.* **2015**, *9*, 61–67. [CrossRef]
31. Maute, K.; Ramm, E. Adaptive Topology Optimization of Shell Structures. *AIAA J.* **1997**, *35*, 1767–1773. [CrossRef]
32. Lee, I.; Zuo, W.; Bai, J.; Li, B. A hybrid OC–GA approach for fast and global truss optimization with frequency constraints. *Appl. Soft Comput.* **2014**, *14*, 528–535.
33. Qu, D.Y.; Huang, Y.Y.; Song, L.Y. Structural topology optimization based on improved genetic algorithm. In Proceedings of the 4th International Conference on Mechanical Engineering and Green Manufacturing (MEGM2015), Guilin, China, 30–31 August 2015.

34. Kanga, P.; Younb, S.-K. Isogeometric topology optimization of shell structures using trimmed NURBS surfaces. *Finite Elem. Anal. Des.* **2016**, *120*, 61. [CrossRef]
35. Lee, I.; Bae, E.; Hinton, E. Shell topology optimization using the layered artificial material model. *Int. J. Numer. Methods Eng.* **2000**, *47*, 843–867. [CrossRef]
36. Pugnale, A.; Sassone, M. Morphogenesis and Structural Optimization of Shell Structures with the aid of a Genetic Algorithm. *J. Int. Assoc. Shell Spat. Struct.* **2007**, *48*, 161–166.
37. Dzwierzynska, J. Shaping of Spatial Steel Rod Structures Based on a Hyperbolic Paraboloid. *Arch. Civ. Eng.* **2018**, *64*, 309–320. [CrossRef]
38. BIM Structural Analysis Software for Engineers. Available online: https://www.autodesk.com/products/robot-structural-analysis/overview (accessed on 18 May 2022).
39. Tofil, J. Application of Catalan surface in designing roof structures—An important issue in the education of the future architect engineer. In Proceedings of the International Conference on Engineering Education (ICEE 2007), Coimbra, Portugal, 3–7 September 2007.

materials

MDPI

Article

Investigation of a Novel Hydrogen Depressurization Structure Constituted by an Orifice Plate with Tesla-Type Channels

Bei Li [1], Yu Liu [1,2,*], Jiaqing Li [3,*], Bin Liu [4], Xingxing Wang [1] and Guanyu Deng [2,*]

1 School of Mechanical Engineering, Nantong University, Nantong 226019, China; 2010110241d@stmail.ntu.edu.cn (B.L.); wangxx@ntu.edu.cn (X.W.)
2 School of Mechanical, Materials, Mechatronic and Biomedical Engineering, University of Wollongong, Wollongong, NSW 2522, Australia
3 College of Chemical Engineering, Fuzhou University, Fuzhou 350116, China
4 School of Mechanical Engineering, Shijiazhuang Tiedao University, Shijiazhuang 050043, China; liubin@stdu.edu.cn
* Correspondence: liuyu_me@ntu.edu.cn (Y.L.); jiaqing@fzu.edu.cn (J.L.); gd577@uowmail.edu.au (G.D.)

Abstract: A hydrogen depressurization system is required to supply the hydrogen to the fuel cell stack from the storage. In this study, a Tesla-type depressurization construction is proposed. Parallel Tesla-type channels are integrated with the traditional orifice plate structure. A computational fluid dynamics (CFD) model is applied to simulate high-pressure hydrogen flow through the proposed structure, using a commercial software package, ANSYS-Fluent (version 19.2, ANSYS, Inc. Southpointe, Canonsburg, PA, USA). The Peng–Robinson (PR) equation of state (EoS) is incorporated into the CFD model to provide an accurate thermophysical property estimation. The construction is optimized by the parametric analysis. The results show that the pressure reduction performance is improved greatly without a significant increase in size. The flow impeding effect of the Tesla-type orifice structure is primarily responsible for the pressure reduction improvement. To enhance the flow impeding effect, modifications are introduced to the Tesla-type channel and the pressure reduction performance has been further improved. Compared to a standard orifice plate, the Tesla-type orifice structure can improve the pressure reduction by 237%. Under low inlet mass flow rates, introduction of a secondary Tesla-type orifice construction can achieve better performance of pressure reduction. Additionally, increasing parallel Tesla-type channels can effectively reduce the maximum Mach number. To further improve the pressure reduction performance, a second set of Tesla-type channels can be introduced to form a two-stage Tesla-type orifice structure. The study provides a feasible structure design to achieve high-efficiency hydrogen depressurization in hydrogen fuel cell vehicles (HFCVs).

Keywords: hydrogen; hydrogen fuel cell; depressurization; orifice plate structure; computational fluid dynamics; numerical model

Citation: Li, B.; Liu, Y.; Li, J.; Liu, B.; Wang, X.; Deng, G. Investigation of a Novel Hydrogen Depressurization Structure Constituted by an Orifice Plate with Tesla-Type Channels. *Materials* **2022**, *15*, 4918. https://doi.org/10.3390/ma15144918

Academic Editor: Shinichi Tashiro

Received: 13 June 2022
Accepted: 12 July 2022
Published: 14 July 2022

1. Introduction

Contemporarily, the transportation sector represents more than one-quarter of carbon gas emissions [1]. An increasing application of hydrogen is considered a potential strategy to gradually fulfill net-zero carbon emissions in the transportation sector [2]. Hydrogen is also considered the ultimate conventional energy source of the 21st century due to its cleanness and sustainability [3]. Therefore, the application of hydrogen fuel cell vehicles (HFCVs) has attracted significant interest [4]. A practical and economical method for HFCV applications is by using pressurized hydrogen storage tanks, as the volumetric energy density of gaseous hydrogen is extremely low [5]. To improve the travelling distance of HFCVs, the tank pressure has been continuously increased during the development of HFCVs. However, the optimal working pressure of the fuel cell tends to be low, leading to the growing demand on the performance of depressurization system [6–8].

In recent years, a number of studies on high-pressure gas depressurization have been undertaken. Luo and his colleagues [9] developed a pressure reducing valve that has a fixed pressure ratio. The pressure and leakage characteristics were theoretically analyzed through simulations. The results show that as the operating pressure increases, the pressure ratio reduces to the designed value. Ulanicki et al. [10] investigated the oscillation of pressure reducing valves (PRVs) at low flow rates. The study was motivated by an industrial case analysis. The purpose of this study is eliminating pressure fluctuation. The results show that the PRV is less stable for small valve openings. Binod and his group [11] utilized a computational fluid dynamic (CFD) model to investigate the transient process in pressure regulation and shut-off valves. Okhotnikov et al. [12] studied pressure drops and steady flow torques of the valve at various flow rates and orifice openings by the CFD method, and relative information, such as the discharge co-efficient and flow jet angles dependencies on the orifice opening, was obtained from this study. Jin and his group [13,14] designed a high-level multi-stage PRV (HMPRV) for hydrogen depressurization in hydrogen refueling stations. It was found that the HMPRV can successfully control the gas pressure and working temperature and is less prone to block flow. In their previous work, the mechanisms of pressure reduction and energy conversion was investigated based on a novel PRV with an orifice plate. In order to optimize valve performance, a parametric study on the throttling portion of a HMPRV was undertaken by Hou and his team [15]. It was found that larger hydrogen kinetic energy causes a stronger turbulent vortex, higher energy consumption, larger multistage injection casing diameter, injection-plate diameter, and pressure ratio. Chen et al. [16] investigated the effects of valve openings on flow characteristics in detail. It was found that larger pressure and velocity gradients mainly appeared at the throttling components for all valve openings. A larger valve opening resulted in more energy consumption. Chen et al. [17,18] simulated the compressible turbulent flow in an HMPRV using CFD software ANSYS-Fluent to analyze the noise and energy consumption. Liu and his group [19] studied the hydrogen flow through a perforated plate in a pressure-reducing system based on a CFD model. The thermodynamic properties of hydrogen were described using a real fluid equation of state (EoS). In addition, the effect of the types of perforated plate was investigated. The results show that the size of the perforated plate has a significant effect on the hydrogen flow.

The above-mentioned PRVs comprise rotating parts with complex structures which will cause excessive turbulence and noise; the complexity of the structures will also result in manufacturing difficulties. In recent years, the Tesla valve [20] has attracted growing attention in relation to pressure depressurization, as it can cause a significant pressure drop when the flow of fluids is reversed. Tesla valves have a fixed geometry with no moving parts, therefore, they may have a longer lifetime and can facilitate mass production. A large number of investigations have been undertaken on using Tesla valve for pressure reduction, mainly focusing on structure optimization. The Tesla valve shape is optimized through two-dimensional (2D) CFD simulations combined with an optimization procedure [21]. A three-dimensional (3D) parametric model is proposed for the Tesla valve by Zhang et al., and his group optimized the geometric relationships of Tesla valve [22]. De Vries et al. [23] designed a new Tesla valve and symmetrically integrated it into a single rotating pulsating heat pipe (PHP). They then investigated the flow characteristics and thermal performance of the PHP. Bao et al. [24] designed a novel Tesla valve with a special tapering/widening structure, analyzed and compared it with other types of Tesla valve, to find which showed a superior absolute pressure drop ratio. Monika et al. [25] developed a multi-stage Tesla valve configuration to enhance heat transfer. Zhang and his colleagues [26] designed a multistage pressure-reducing valve; the valve combined a Tesla-type orifice valve and a sleeve pressure structure valve. In this study, the influences of working parameters on fluid pressure and velocity distributions were analyzed. Qian et al. [27,28] performed simulations for hydrogen reverse flow in a multi-stage Tesla valve. They summarized the power–law relationship in the flow rate, the number of stages, and the pressure ratio, and evaluated them using Mach number, turbulent dissipation rate, and blown-barrel loss as criterions. Jin and his team [29] studied the influence of different structural parameters of a single-stage Tesla valve on the hydrogen pressure reduction. The results show that a smaller hydraulic diameter, a smaller inner curve radius, and a larger valve angle could provide a higher pressure drop at a larger inlet velocity. Qian and Jin et al. [27,29] predicted the physical properties of the ideal gas EoS during simulation. However, since the ideal gas EoS does not take into account the effect of intermolecular potential energy, it will produce large errors under high-pressure conditions.

As mentioned above, several studies were conducted on Tesla valves as well as multi-stage pressure-reducing structures. However, studies connecting the Tesla valve to the traditional perforated plate structure are seldom found. The actual gas EoS refers to the mathematical expression of the functional relationship between the state parameters when a certain amount of gas reaches equilibrium state. The ideal gas completely ignores the interaction between gas molecules and cannot explain phenomena such as gas–liquid change and throttling in which molecular forces play an important role. However, the hydrogen depressurization system operates at very high pressures under which the ideal gas EoS may produce large errors. Additionally, Peng–Robinson (PR) EoS, a real gas EoS, is simple to calculate and accurate to the physical property of pure gas. Some studies on flow and heat transfer under complex conditions offer a great help to this paper's investigation of the flow through the Tesla-type orifice structure. Rezaei et al. [30] studied electro-osmotic flow of an aqueous solution of NaCl using the molecular dynamics simulation to investigate the effects of the electric field and temperature on the flow properties. Toghraie et al. [31] conducted a simulation to study boiling heat transfer through a volume fraction (VOF) method, and they also studied the quench phenomena through a fluid jet on a hot horizontal surface. Li and his co-workers [32] investigated the fluid flow and heat transfer using two-phase approach mixed convection of a non-Newtonian nanofluid in a porous H-shaped cavity. These studies showed that the simulation technology was able to simulate the fluid flow through complex geometric conditions.

In this paper, a 3D CFD model integrated with the PR real gas EoS was proposed to investigate the pressure decrease in a new Tesla-type orifice structure. The traditional perforated plate structure is displaced by a flow channel with Tesla valves in the hydrogen pressure reduction system. The structure (Tesla-type orifice plate structure) consists of multiple Tesla valves in parallel to achieve higher pressure reduction. In addition, the optimization for the Tesla valve is undertaken to improve the characteristics of the backflow impact in its flow channel. Furthermore, a two-stage Tesla-type orifice plate structure, which comprises two Tesla in series, is introduced. The effects of structural parameters on the flow characteristics are investigated to obtain better depressurization performance. This research offers technical support for HFCVs.

2. Numerical Methods

2.1. Governing Equations

The CFD software ANSYS-Fluent is employed for the numerical solution. ANSYS-Fluent uses the finite volume method to discretize the governing differential equations of fluid flow based on the Navier–Stokes (N–S) equation, which involves the solution of mass, momentum, and energy conservation equations [33], as expressed in Equations (1)–(3).

$$\frac{\partial \rho}{\partial t} + \nabla \cdot (\rho v) = 0 \tag{1}$$

$$\frac{\partial}{\partial t}(\rho v) + \nabla \cdot (\rho v) = -\nabla \cdot p + \nabla \cdot \tau + \rho g \tag{2}$$

$$\frac{\partial}{\partial t}(\rho E) + \nabla \cdot [v(\rho E + p)] = \nabla \cdot (k_{eff} \Delta T - \tau v) \tag{3}$$

where ρ is the density, t the time, $\boldsymbol{v}$ the velocity vector, $\boldsymbol{p}$ the pressure vector, τ the viscous stress tensor, $\boldsymbol{g}$ the gravitational acceleration, E the total energy per unit control body, and k_{eff} the effective thermal conductivity.

2.2. Turbulence Model

An appropriate turbulence model is crucial to simulate hydrogen flow with high compressible pressure gradient. As the influence of the compressibility on turbulence dissipation cannot be explained by the standard k-ε model [34], the realizable k-ε model is applied in this work. In practice, the realizable k-ε turbulence model [35] has been successfully used in various flows, such as separated flows, channel and boundary layer flows, and rotating homogeneous shear flows. Particularly, the realizable k-ε model can better predict the diffusion rates of axisymmetric and planar jets. The realizable k-ε model is described as:

$$\frac{\partial}{\partial t}(\rho k)+\frac{\partial}{\partial x_i}(\rho k u_i)=\frac{\partial}{\partial x_j}\left[\left(\mu+\frac{\mu_t}{\sigma_k}\right)\frac{\partial k}{\partial x_j}\right]+G_k+G_b-\rho\varepsilon-Y_M+S_k \quad (4)$$

$$\frac{\partial}{\partial t}(\rho \varepsilon)+\frac{\partial}{\partial x_j}(\rho \varepsilon u_j)=\frac{\partial}{\partial x_j}\left[\left(\mu+\frac{\mu_t}{\sigma_\varepsilon}\right)\frac{\partial \varepsilon}{\partial x_j}\right]+\rho C_1 S\varepsilon-\rho C_2\frac{\varepsilon^2}{k+\sqrt{v\varepsilon}}+C_{1\varepsilon}\frac{\varepsilon}{k}C_{3\varepsilon}G_b+S_\varepsilon \quad (5)$$

where G_k denotes the generation of turbulence kinetic energy owing to the average velocity gradients; G_b represents the generation of turbulence kinetic energy owing to buoyancy; Y_M is the contribution of the fluctuating dilatation incompressible turbulence to the overall dissipation rate; C_2 and $C_{1\varepsilon}$ are constants; σ_k and σ_ε are the turbulent Prandtl numbers for k and ε, respectively; and S_K and S_ε are user-defined source terms.

2.3. PR EoS

An accurate prediction of the thermodynamic properties of the fluid is essential to achieve satisfactory accuracy in the CFD model. In this study, a real gas Eos, i.e., the PR EoS, is applied to predict better thermodynamic properties of high-pressure hydrogen. The PR EoS [36] is illustrated by:

$$P=\frac{RT}{v-b}-\frac{a(T)}{v^2+2bv-b^2} \quad (6)$$

$$b=0.0778\frac{RT_c}{P_c} \quad (7)$$

$$a(T)=0.45724\frac{R^2T_c^2}{P_c}[1+n(1-(T/T_c)^{0.5}]^2 \quad (8)$$

$$n=0.37464+1.54226\omega-0.26993\omega^2 \quad (9)$$

where R represents the universal gas constant, v the molar volume, P_c the critical pressure, T_c the critical temperature, and ω the eccentricity factor of the gas.

Hydrogen densities at various conditions, which are adopted from the experimental data of Michels et al. [37] are utilized to evaluate the accuracy of the PR EOS. Figure 1 shows the comparison between experimental data and EoS predictions. It is obvious that the ideal gas EoS will cause large discrepancies at high pressures. The densities calculated by PR EOS are consistent with the measurements and the maximum relative error is 3.8%. GERG-2008 (Groupe Européen de Recherches Gazières) EoS [38] performs slightly better than the PR EoS at higher pressures. However, comparing to the PR EoS, it is much more time-consuming to solve the GERG-2008 EoS at runtime during the 3D CFD simulations; therefore, it is an adequate choice to employ PR EoS in the simulations to ensure acceptable accuracy.

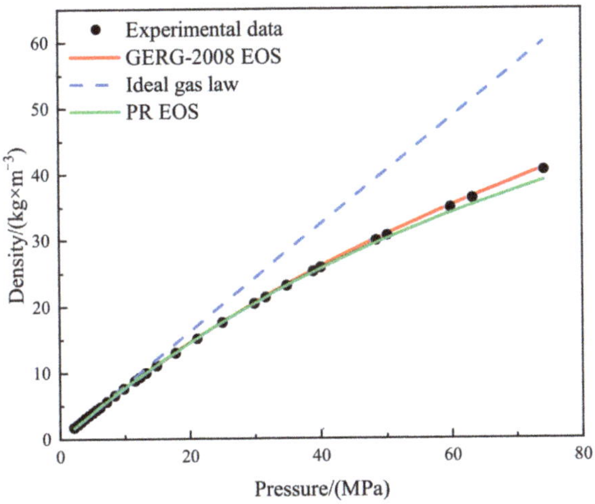

Figure 1. Comparison between different equations of state and experimental hydrogen thermophysical parameters.

2.4. Verification of the Numerical Methods

As measured data of hydrogen through the Tesla-type channel is scarce, the experiment performed by Liu et al. [39] for water flowing through Tesla-type channel is used for model validation. Figure 2a shows a structured hexahedral mesh generated for the single stage Tesla-type channel used in the experiment. The mesh independence was verified using meshes with different numbers of cells. Figure 2b shows the predicted pressure reduction with increasing cell numbers. It was found that the pressure reduction did not change much when the number of cells increased to 1.379 million; therefore, the grid with 1.379 million cells was used for the validation simulation. To ensure the accuracy of simulation results, grid-independent verification has been carried out for all subsequent simulations.

The experimental and simulated pressure reduction for different inlet flow rates are shown in Figure 3. It displaces a great consistency between prediction and measurement. The CFD model somewhat over-predicted the pressure reduction. The largest relative error between the results from simulation and observation is 4.48%, indicating that the CFD model can produce satisfactory prediction of fluid flow in a Tesla-type channel.

2.5. Computational Domain and Boundary Conditions

The benchmark structure of the traditional orifice plate valve was introduced and analyzed by Chen et al. [16,17]. For a better pressure reduction effect, Tesla-type channels are integrated into a conventional orifice plate structure to form a novel hydrogen depressurization structure. Figure 4a shows the structure of the traditional orifice plate valve. The central flow domain is a 200 mm diameter circular channel with a 50 mm long inlet section, a 450 mm long outlet section, and a 25 mm thick orifice plate. There are 37 holes on the plate which are staggered in equilateral triangles (see Figure 4b). In Figure 4a, Point A is where the orifice plate connected to the inlet section, while Point B is where the orifice plate connected to the outlet section. A Tesla-type channel usually has a good effect on pressure reduction. The pressure reduction performance can be further optimized when Tesla-type channels are integrated into conventional orifice plate structures. Figure 4c shows a modified structure which replaces the straight orifice flow channel with a Tesla-type channel in a traditional orifice plate. This Tesla-type orifice structure uses a circular channel with a diameter of 5 mm, so as to better couple with the main flow channel. Other structural parameters of the Tesla-type channel are: inlet length: L_1 = 5 mm; outlet length: L_2 = 5 mm; side straight channel length: L = 10 mm; the angle between side channel and main channel: α = 45°; the angle between bending channel and main channel:

$\beta = 130°$; and the radius of the curve in the circular section: $R = 2.5$ mm. In Figure 4c, Point C is where the Tesla-type channel connected to the inlet section, and Point D is where the Tesla-type channel connected to the outlet section. Due to the symmetrical geometry, the computational domain uses half of the Tesla-type orifice structure.

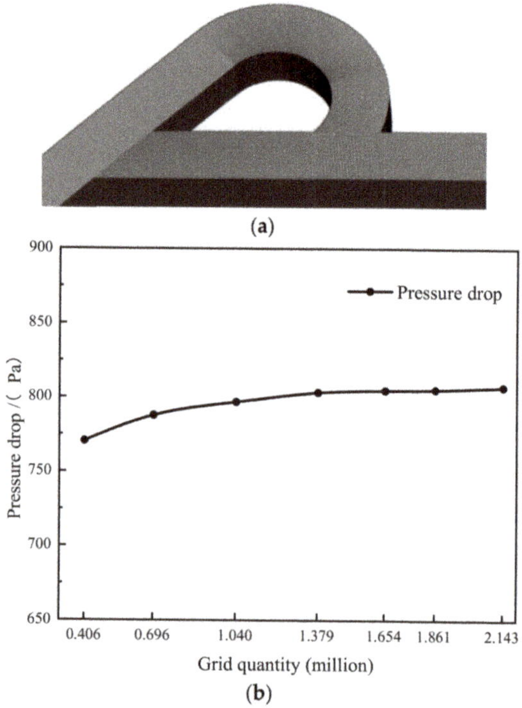

Figure 2. Computational mesh and grid-independence study. (**a**) Mesh for the single stage Tesla-type channel. (**b**) Pressure reduction between inlet and outlet under different grid densities.

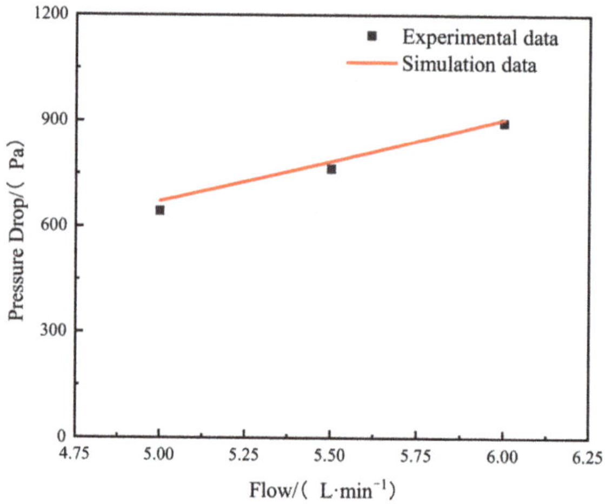

Figure 3. Pressure reduction: predicted vs. measured.

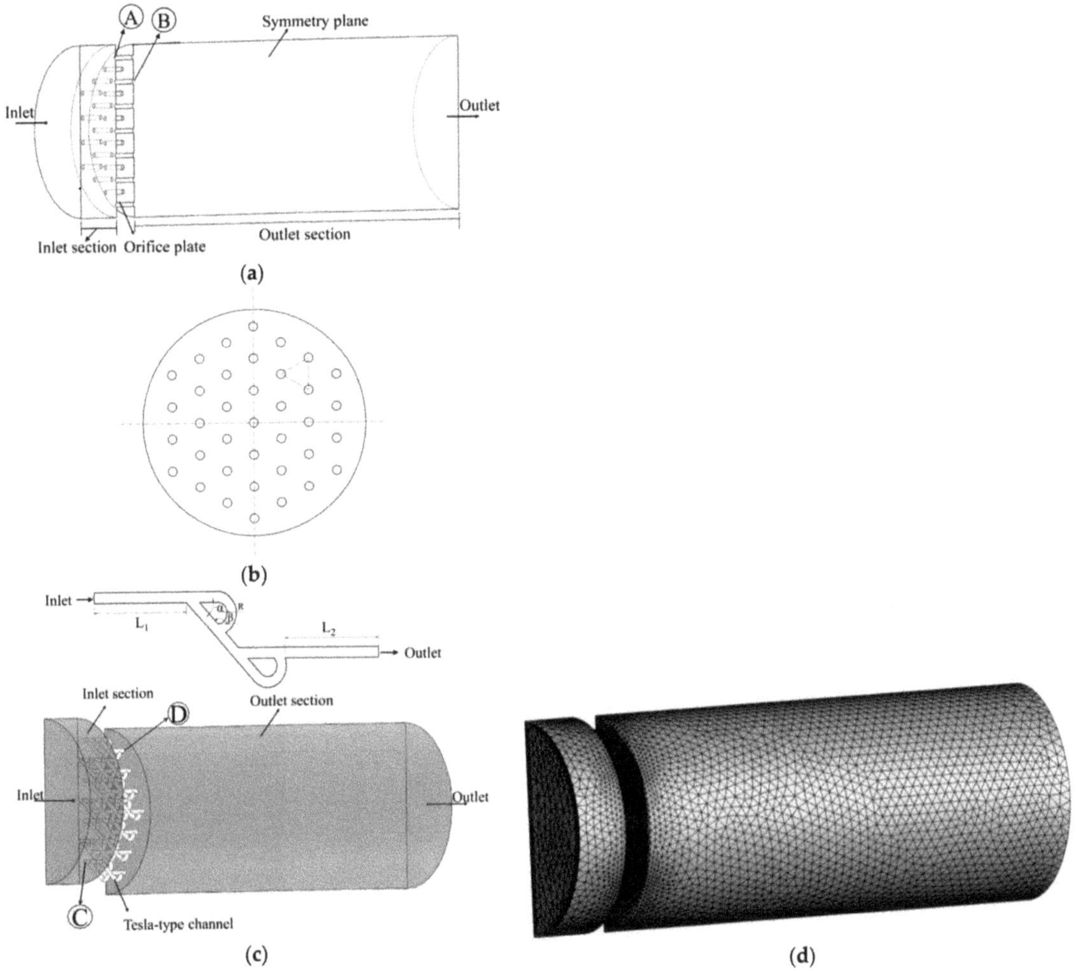

Figure 4. Computational domain and meshing. (**a**) Schematic of the fluid domain of conventional orifice plate structure. (**b**) The distribution of holes on the plate. (**c**) Schematic of fluid domain of the Tesla-type orifice structure. (**d**) Mesh division of Tesla-type orifice structure.

The boundary conditions for conventional orifice plate structure and Tesla-type orifice structure are similar, which are defined as (see Figure 4): (a) inlet: mass flow inlet (flow rate Q_m) with a constant temperature (300 K); (b) outlet: pressure outlet with a 0.2 MPa constant pressure; (c) symmetry plane: symmetrical impermeable boundary conditions with zero gradients of all variables; and (d) adiabatic wall boundary conditions specified to other boundary surfaces.

As the orifice plate has a relatively complex structure, a non-structural tetrahedral mesh was generated for the computational domain, as shown in Figure 4d. Mesh refinement was applied around the Tesla-type channel. Additionally, the energy residuals are 10 to the minus 6, and everything else is 10 to the minus 3. Key information of the model implementation is shown in Table 1.

Table 1. Key information of the model implementation.

Computational Time	Number of Iterations	Convergence Criteria
40–80 h/case	20	$1 \times 10^{-6}/1 \times 10^{-3}$

3. Results and Discussion

3.1. Distribution of Pressure and Density

The pressure distributions for traditional orifice plate structure and Tesla-type orifice structure on the symmetry plane under different inlet mass flow rates are shown in Figures 5 and 6, respectively. It is found that the pressure reduction varies with inlet mass flow rate Q_m for both structures. As shown in Figure 5a for the conventional orifice plate structure at Q_m = 0.02 kg s^{-1}, the maximum pressure gradient mainly occurs at Point A (refer to Figure 4a). At Q_m = 0.1 kg s^{-1} (Figure 5b), the pressure begins to change dramatically at the location of the plate orifice. In general, there is a continuous pressure distribution between the orifice plate and outlet section when Q_m is less than 0.1 kg s^{-1}. When Q_m increases to 0.5 kg s^{-1} or 1 kg s^{-1} (Figure 5c,d), step change in pressure gradient occurs at the connection between the orifice plate and outlet section. When Q_m = 0.5 kg s^{-1} (Figure 5c), a small annular region with sudden pressure reduction is observed at Point B (refer to Figure 4a). The conventional orifice plate structure reduces the size of flow channel to throttle the hydrogen to achieve the pressure reduction. As shown in Figure 6, the Tesla-type orifice structure can achieve higher pressure reduction due to the increased resistance in the structure. Figure 6 shows that the pressure reductions are not only observed at connections between the Tesla channel and the inlet and outlet sections (Points C and D in Figure 4c), but also great pressure reduction can be seen between two stages of the Tesla channel. It is worth noting that the low-pressure zone at Point D shown in Figure 6a,b disappears with the rise of mass flow rate.

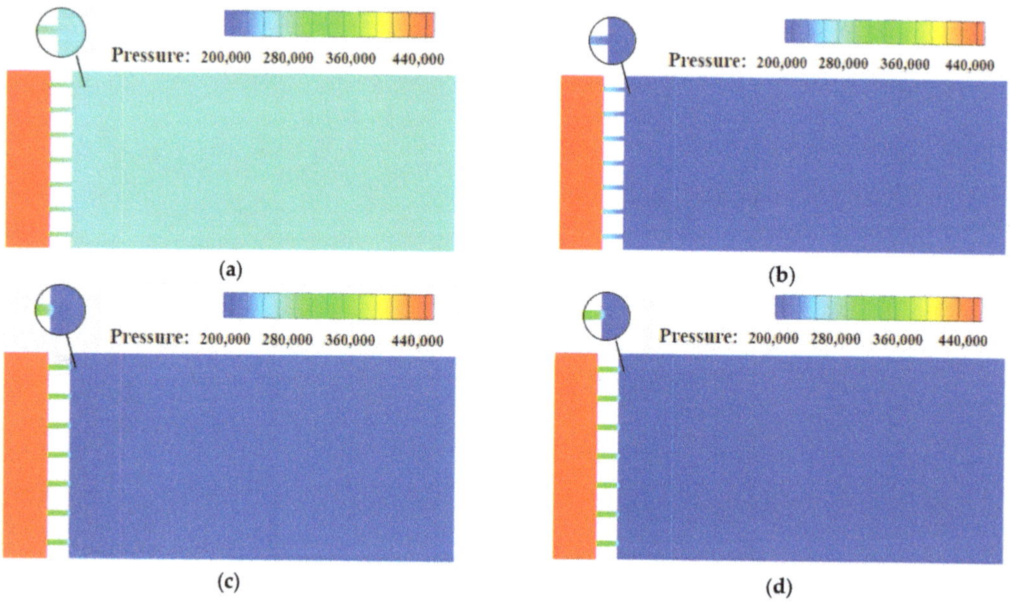

Figure 5. Pressure distribution on the symmetry plane for conventional orifice plate under different mass flow rates (pressure in Pascal). (**a**) Q_m = 0.02 kg s^{-1}. (**b**) Q_m = 0.1 kg s^{-1}. (**c**) Q_m = 0.5 kg s^{-1}. (**d**) Q_m = 1 kg s^{-1}.

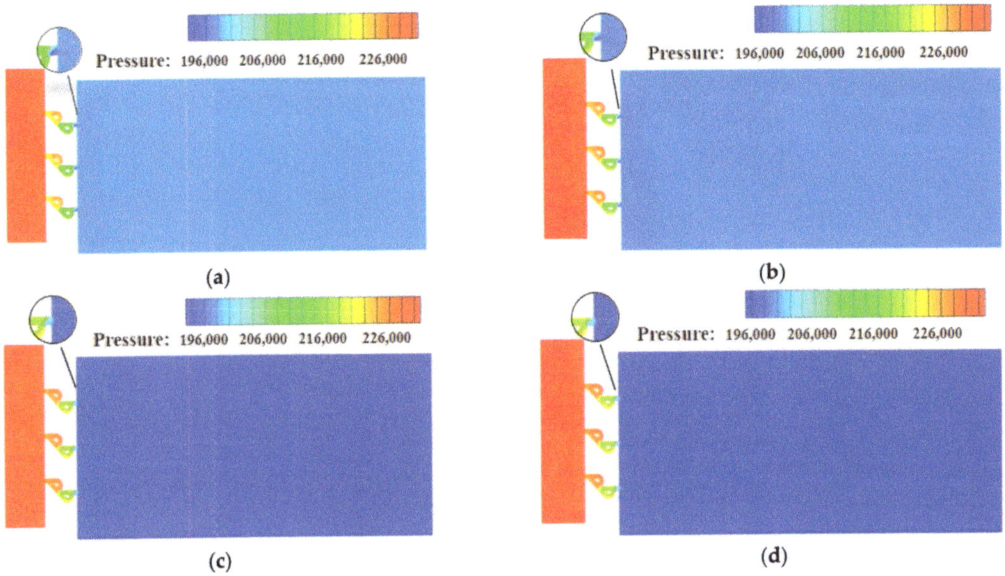

Figure 6. Pressure distribution on the symmetry plane for the Tesla-type orifice structure under different mass flow rates (pressure in Pascal). (**a**) $Q_m = 0.02$ kg s^{-1}. (**b**) $Q_m = 0.1$ kg s^{-1}. (**c**) $Q_m = 0.5$ kg s^{-1}. (**d**) $Q_m = 1$ kg s^{-1}.

Figures 7 and 8 are the density distributions for the traditional orifice plate structure, as well as the Tesla-type orifice structure. The reduction in hydrogen pressure leads to the reduction in density. When the flow rate increases, the pressure and density gradients in the flow field increase. It is seen that the pressure greatly influences hydrogen density. This highlights the necessity of using real gas EoS in the simulation. In the Tesla-type orifice structure, the pressure varies in the channel due to the impact of the bending section. This is also reflected in the variation in the density.

3.2. Analysis of Mach Number and Turbulence Intensity

Figure 9 shows the Mach number distribution in the conventional orifice plate structure under different inlet mass flow rate conditions. Similar Mach number distributions were observed when the inlet mass flow rates were $Q_m = 0.02$ and 0.1 kg s^{-1}. Under both mass flow rates, the Mach number is less than 1 throughout the flow domain. When hydrogen passes Point A, after adiabatic expansion, the pressure energy is converted into kinetic energy. This is reflected in the sudden decrease in the pressure and sharp rise in the velocity/Mach number. After entering the outlet section, the jet boundary is restricted by decreasing kinetic energy and velocity. The Mach number is distributed in such a way that the area near the wall is small and the area in the middle flow domain is large. As the mass flow rate rises to 1 kg s^{-1}, the downstream jet flow affected area increases as well. When $Q_m = 0.02$ kg s^{-1}, Mach numbers in the traditional orifice plate structure and Tesla-type orifice structure are less than 0.3 while hydrogen behaves as a subsonic flow. When Q_m rises to 0.5 or 1 kg s^{-1}, the traditional orifice plate structure and the Tesla-type orifice structure have a supersonic flow at Point B and Point D. The hydrogen flows from the inlet section into the channel with abruptly decreasing area and then flows into the outlet section with much larger area. The flow is similar to that in a Laval nozzle [40].

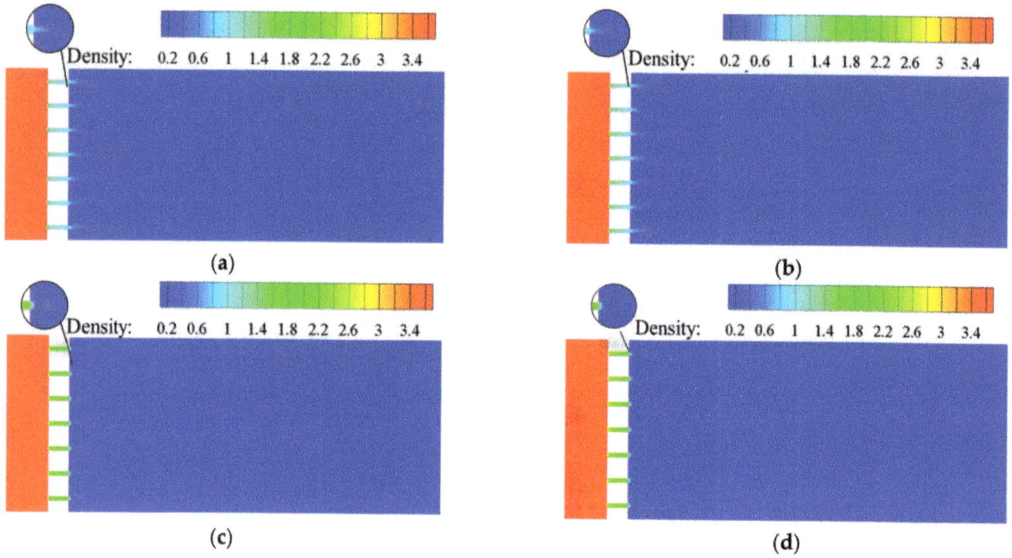

Figure 7. Density distribution on the symmetry plane for conventional orifice plate structure under different mass flow rates (density in kg m^{-3}). (**a**) Q_m = 0.02 kg s^{-1}. (**b**) Q_m = 0.1 kg s^{-1}. (**c**) Q_m = 0.5 kg s^{-1}. (**d**) Q_m = 1 kg s^{-1}.

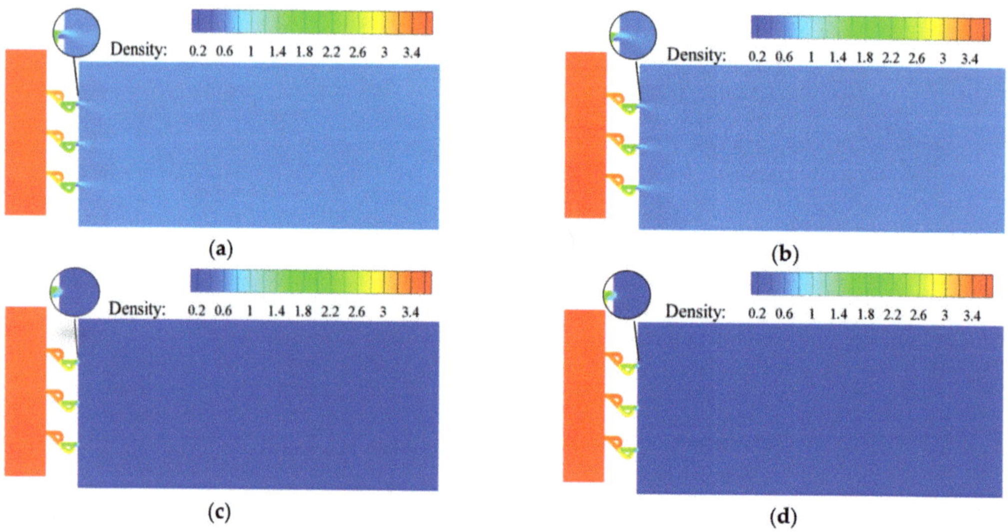

Figure 8. Density distribution on the symmetry plane for Tesla-type orifice structure under different mass flow rates (density in kg m^{-3}). (**a**) Q_m = 0.02 kg s^{-1}. (**b**) Q_m = 0.1 kg s^{-1}. (**c**) Q_m = 0.5 kg s^{-1}. (**d**) Q_m = 1 kg s^{-1}.

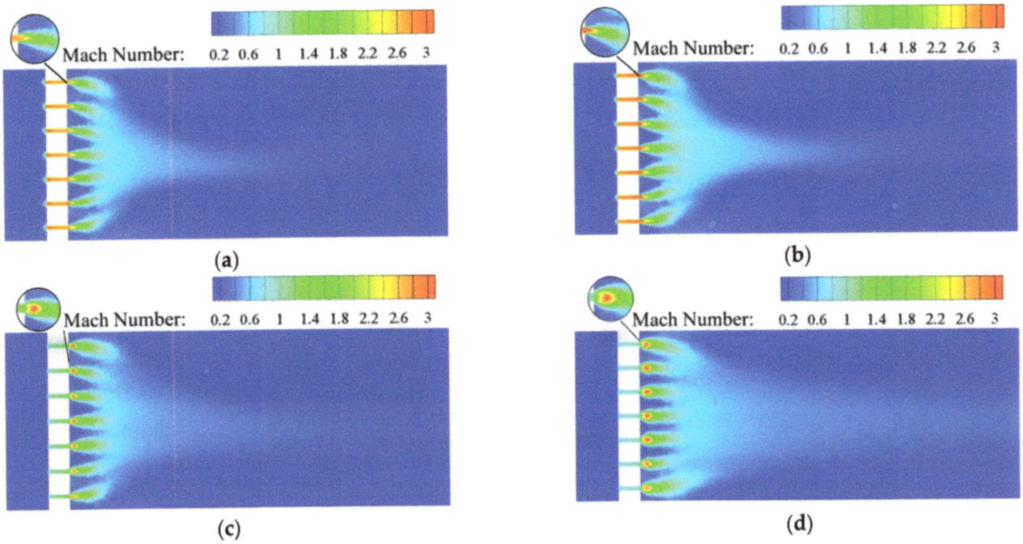

Figure 9. Mach number distribution on the symmetry plane for the conventional orifice plate structure under different mass flow rates. (**a**) Q_m = 0.02 kg s^{-1}. (**b**) Q_m = 0.1 kg s^{-1}. (**c**) Q_m = 0.5 kg s^{-1}. (**d**) Q_m = 1 kg s^{-1}.

As shown in Figures 9 and 10, hydrogen is accelerated to the speed of sound in both the conventional orifice plate and the Tesla-type channels. Eventually, supersonic speed is achieved at the outlet section with an expanded cross-section. For both structures, it can be found that when Q_m = 0.5 and 1 kg s^{-1}, there is an area at Point B and Point D with low pressure and large Mach number. It is evident that expansion waves are generated here. In the Tesla-type orifice structure, the larger Mach number in the bending channel indicates that the velocity in the bending channel is higher than that in the straight channel, as the hydrogen flows more easily in the bending channel. When the hydrogen with high velocity flows out from the bending channel, it will impede the hydrogen in the straight channel, reducing the flow rate of hydrogen in the straight track and lowering the Mach number. The Mach number distribution in the outlet section of the Tesla-type orifice structure is clearly different from that of the conventional orifice plate structure. The Mach number near the lower part is larger than that in the upper part. The comparison between Figures 9 and 10 demonstrates the area of the Tesla-type orifice structure with a Mach number greater than 1 is smaller than that of the traditional orifice plate structure. This indicates that the Tesla-type orifice structure reduces the area of fluid with high velocity, which can also help achieve better pressure reduction.

Figures 11 and 12 show the turbulence intensity on the symmetry plane of the conventional orifice plate structure and the Tesla-type orifice structure, respectively. We can see that the maximum turbulence intensity increases with the inlet mass flow rates. As Q_m reaches 0.02 and 0.1 kg s^{-1}, the maximum turbulence intensity of the conventional orifice plate structure appears at Point A. The turbulence intensity at Point B is much higher than that at the end of outlet section. As Q_m = 0.02 kg s^{-1}, the jet of each plate orifice at the exit has less influence on each other. The turbulence intensity between the two plates is smaller and the velocity of the hydrogen jet is lower. Additionally, the velocity gradient in this area is low because the diversion effect between the two jets of the plate orifice is small. When Q_m = 0.1 kg s^{-1}, the area of low turbulence intensity zone between the plate orifice exits decreases as the plate orifice exit velocity becomes larger. As Q_m reaches 0.5 and 1 kg s^{-1}, the maximum turbulence intensity of the conventional orifice plate structure appears at Point B and near the wall.

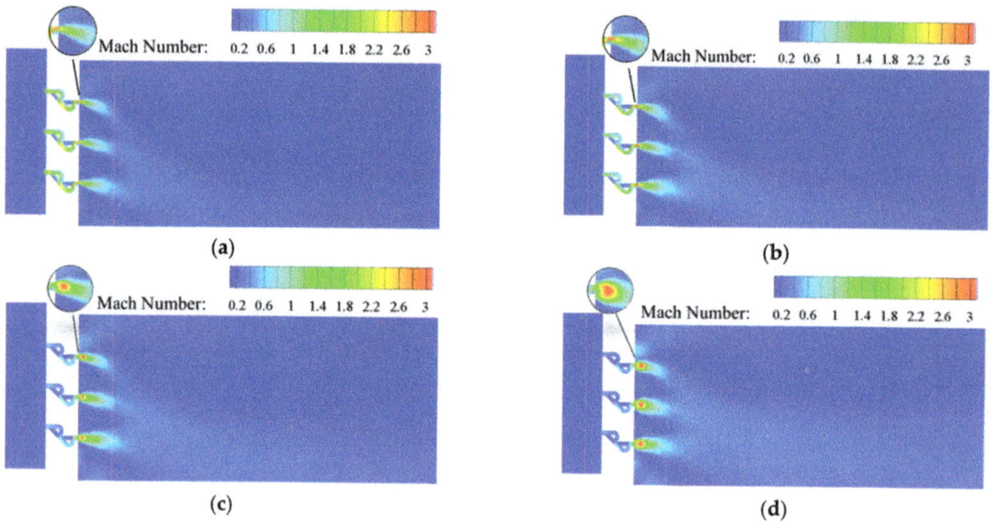

Figure 10. Mach number distribution on the symmetry plane for Tesla-type orifice structure under different mass flow rates. (**a**) Q_m = 0.02 kg s^{-1}. (**b**) Q_m = 0.1 kg s^{-1}. (**c**) Q_m = 0.5 kg s^{-1}. (**d**) Q_m = 1 kg s^{-1}.

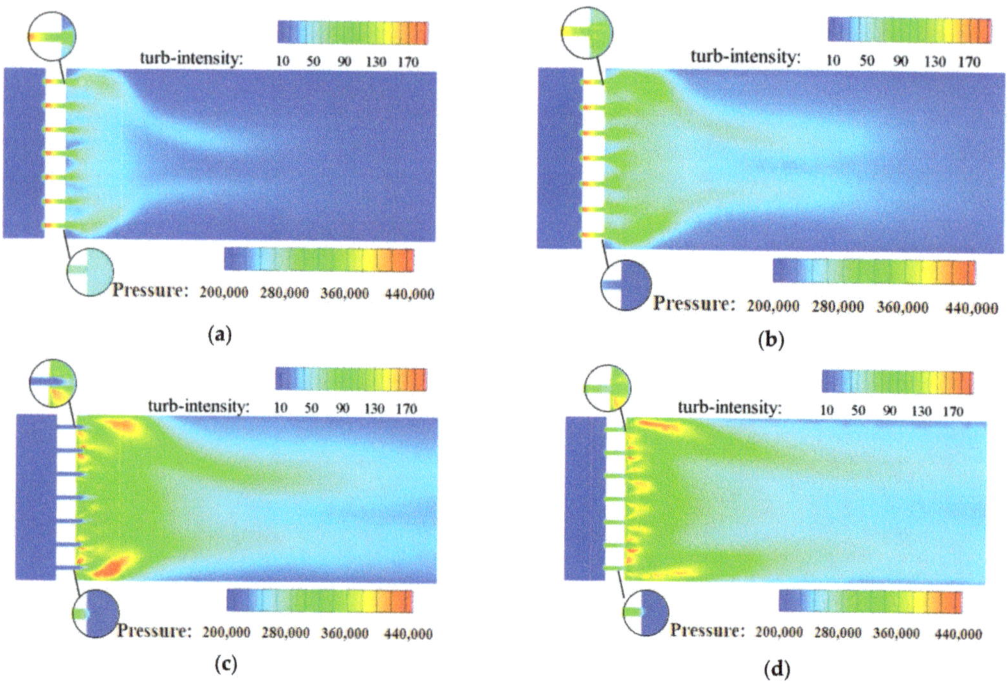

Figure 11. Turbulence intensity distribution on the symmetry plane for the conventional orifice plate structure under different mass flow rates (%). (**a**) Q_m = 0.02 kg s^{-1}. (**b**) Q_m = 0.1 kg s^{-1}. (**c**) Q_m = 0.5 kg s^{-1}. (**d**) Q_m = 1 kg s^{-1}.

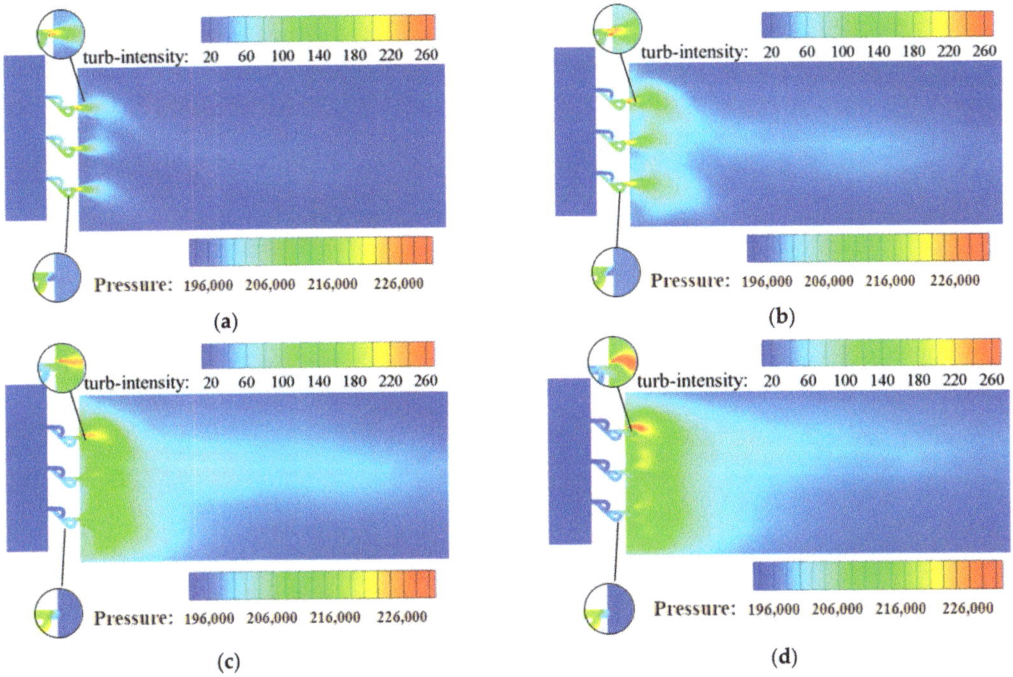

Figure 12. Turbulence intensity distribution on the symmetry plane for the Tesla-type orifice structure under different mass flow rates (%). (**a**) Q_m = 0.02 kg s^{-1}. (**b**) Q_m = 0.1 kg s^{-1}. (**c**) Q_m = 0.5 kg s^{-1}. (**d**) Q_m = 1 kg s^{-1}.

As shown in Figure 12, the maximum turbulence intensity in the Tesla-type orifice structure appears at Point D. The maximum turbulence intensity is observed at the uppermost plate orifice exit when a supersonic flow occurs there. The turbulence intensity is more significant in the exit section of the Tesla-type orifice structure near the upper wall. It is found that changes in hydrogen flow rate, as well as changes in the structure, affect turbulence intensity. Comparison of Figures 11 and 12 shows that the Tesla-type orifice structure is easier to enable the formation of vortices due to higher turbulent intensity induced; this is mainly because the Tesla-type orifice structure makes it easier for the fluid to enter the bending channel. The interaction between the fluid in straight and the bending channel results in the increase in turbulent intensity and also leads to the abrupt pressure drop.

4. Optimization of Tesla-Type Orifice Structure

4.1. Optimization Methods

The above study shows that a higher pressure-reduction performance can be achieved by replacing the conventional orifice plate structure with a simple orifice plate integrated with a Tesla-type orifice structure flow channel. In order to achieve a better performance on pressure reduction, the structure of a Tesla-type orifice can be further optimized by improving the Tesla-type channel. Figure 13a shows the improved Tesla-type orifice structure flow path. The pressure reduction performance of the Tesla-type orifice structure is investigated. The above study illustrates the main reason that a Tesla valve can reduce pressure is that its bending channel has an impeding effect on the flow in the straight channel. To enhance the flow impeding effect, a novel construction was introduced at the junction of the bending and straight channels. In the new structure, the flow in the bending channel was brought to interact with the flow in the straight channel earlier (see

Figure 13a). The angle β between the main channel and the bent channel was increased. These modifications increase the impeding effect of the return flow in bending channel. In the new structure, L_3 = 8.6 mm (Figure 13a).

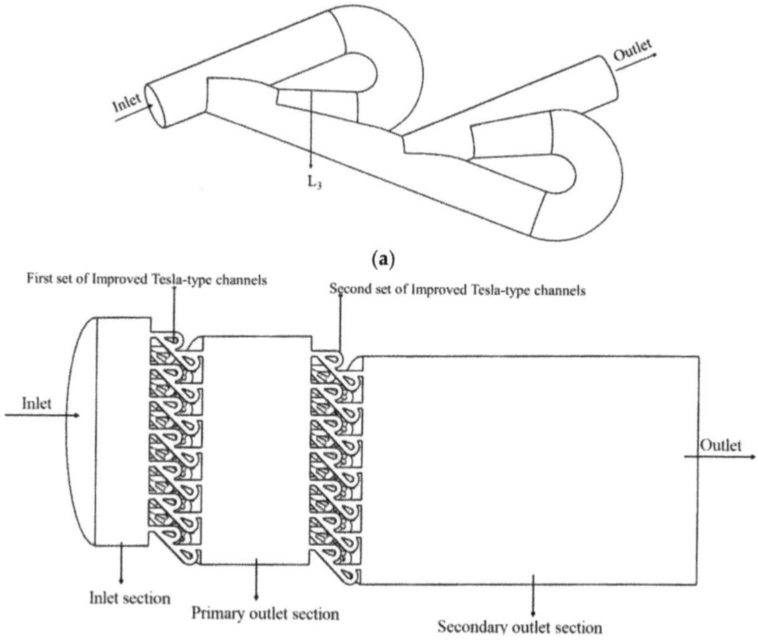

Figure 13. Improved Tesla-type channel and two-stage Tesla-type orifice structure. (**a**) Improved Tesla-type channel. (**b**) Two-stage Tesla-type orifice structure.

In addition to the improvement in the Tesla-type channel, another set of parallel Tesla-type channels was introduced to the Tesla-type orifice structure, which forms a two-stage Tesla-type orifice structure, as shown in Figure 13b. The lengths of the inlet section, the primary outlet section, and the secondary outlet section are 50, 100, and 200 mm, respectively. The main flow channel is also a circular channel with a diameter of 200 mm. Additionally, half of the proposed structure is used as the computational domain (Figure 13b).

To evaluate the pressure reduction performance of the improved Tesla-type orifice structure, the same mass flow rates of 0.01, 0.1, 0.5, and 1 kg s^{-1} and outlet pressure of 0.2 MPa are used for the CFD simulations.

4.2. Flow Field Analysis of the Two-Stage Tesla-Type Orifice Structure

Figure 14 shows the pressure distribution on the symmetry plane of the two-stage Tesla-type orifice structure under different inlet mass flow rates. It is found that the pressure in the orifice structure reduces several times. In addition to the pressure reduction when entering the Tesla-type flow channel orifice in the inlet section, a significant pressure reduction can be seen each time when hydrogen flows through the junction of the bending and straight channels. In each outlet section of the two-stage Tesla-type orifice structure, the pressure of hydrogen is evenly distributed with minimal variation. As the mass flow rate increases, the pressure reduction in the orifice structure increases continuously. Compared to Figure 6, it is obvious that the improved Tesla-type orifice structure achieved better pressure reduction performance. Additionally, the pressure in the improved Tesla-type orifice structure was reduced much more smoothly.

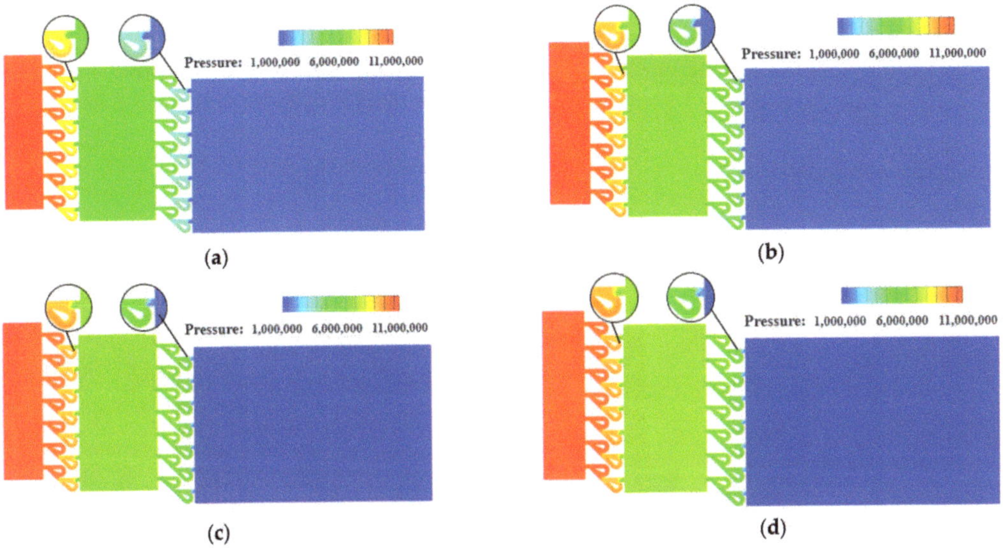

Figure 14. Pressure distribution on the symmetry plane for the two-stage Tesla-type orifice structure under different mass flow rates (Pressure in Pascal). (**a**) Q_m = 0.02 kg s^{-1}. (**b**) Q_m = 0.1 kg s^{-1}. (**c**) Q_m = 0.5 kg s^{-1}. (**d**) Q_m = 1 kg s^{-1}.

The Mach number distributions are shown in Figure 15. As the inlet mass flow rate increases, the maximum Mach number in the two-stage Tesla-type orifice structure gradually increases. When Q_m = 0.02 kg s^{-1}, the largest Mach number appears behind the bending channel of the second stage. This is caused by the sharp decrease in pressure and the rise in velocity at this location because of the hydrogen in the bending channel joining the hydrogen from the straight channel. There is limited difference in Mach numbers between the two stages of Tesla-type channels. When Q_m = 0.1 kg s^{-1}, the maximum Mach number appears at the rear position of the bending channel of the second stage of the new orifice channels. This indicates that the velocity is higher at the second stage of Tesla-type orifice structure, implying more pressure reduction. When Q_m rises to 0.5 and 1 kg s^{-1}, a supersonic flow occurs in the two-stage Tesla-type orifice structures. The hydrogen gas flows out of the second stage Tesla-type channels with a sudden increase in cross-section area. An expansion wave is generated, causing a sudden increase in flow velocity, with maximum Mach numbers observed near the channel exits.

Figure 16 shows the relative magnitude of maximum Mach number for four types of valves under different mass flow rates. In Figure 16, F-1 represents the conventional orifice plate structure, F-2 the original Tesla-type orifice structure, F-3 the one-stage improved Tesla-type orifice structure, and F-4 the two-stage optimized Tesla-type orifice structure. Figure 16a shows that, when the traditional orifice plate structure is replaced by the Tesla-type orifice structure, up to 0.5 kg s^{-1} mass flow rate, there is a relatively large growth in Mach number; however, a further increase in the mass flow rate causes very limited improvement in the Mach number. When Q_m = 1 kg s^{-1}, the Mach number of Tesla-type orifice structure becomes smaller than that of the conventional orifice plate structure. A similar trend is demonstrated in Figure 16c, but the growth in mass flow rate will result in a greater reduction in Mach number for the one-stage improved Tesla-type orifice structure. Figure 16b demonstrates that the one-stage improved Tesla-type orifice structure initially shows a larger Mach number than the original Tesla-type orifice structure; however, the growth in the mass flow rate will lead to a Mach number smaller than that of the original Tesla-type orifice structure. Figure 16d shows that, when the mass flow rate is less than 0.1 kg s^{-1}, the two-stage Tesla-type orifice structure has a slightly larger Mach number than the one-stage one. However, the increase in the mass flow rate will soon cause a smaller Mach number for the two-stage Tesla-type orifice structure. Overall, for most of the mass flow rates, the two-stage Tesla-type orifice structure shows a smaller maximum Mach number. This proves that the two-stage Tesla-type orifice structure is an effective way to obtain the same pressure reduction with lower Mach number.

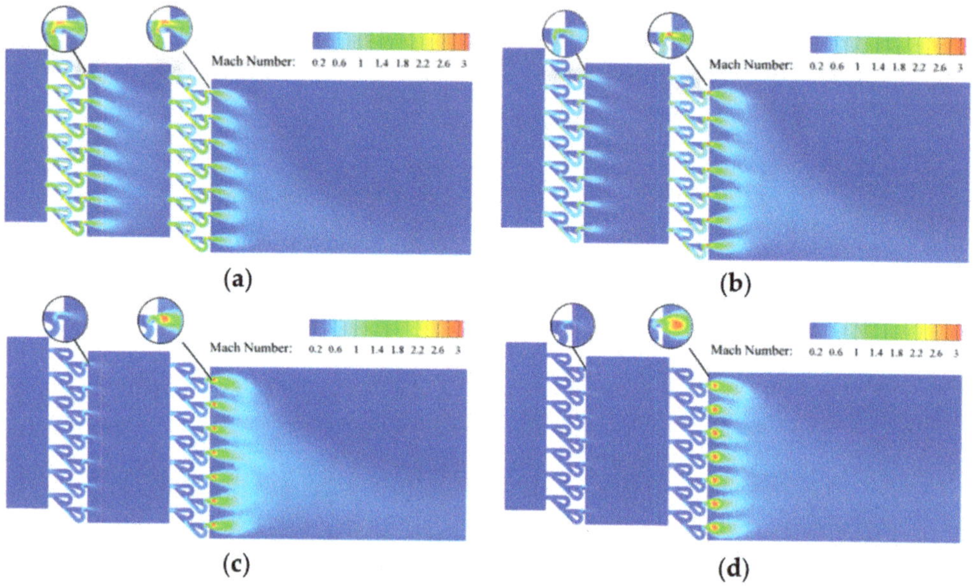

Figure 15. Mach number distribution on the symmetry plane for the two-stage Tesla-type orifice structure under different mass flow rates. (**a**) $Q_m = 0.02$ kg s^{-1}. (**b**) $Q_m = 0.1$ kg s^{-1}. (**c**) $Q_m = 0.5$ kg s^{-1}. (**d**) $Q_m = 1$ kg s^{-1}.

Figure 17 demonstrates the pressure on the centerline of the symmetry plane of the four structures. The change in hydrogen pressure due to the difference in the structure is well reflected. In Figure 17, X = 0 corresponds to where the inlet section is connected to the plate orifice and positive X corresponds to the flow direction of hydrogen. It is shown in Figure 17 that, for all four types of pressure reduction valves, higher inlet mass flow rate leads to higher pressure reduction. Among these valves, the conventional orifice plate structure has the worst pressure reduction performance. Comparison between the original and the optimized Tesla-type orifice structures shows that the first stage pressure reduction is formed when the hydrogen enters the plate orifice, and the subsequent two stages of pressure reduction is due to the flow channel characteristics of Tesla valve structure. The investigation reveals that the optimized Tesla-type orifice structure does improve the effect of impeding flow and obtained better pressure reduction performance.

It can also be seen from the figure that, for the Tesla-type valves, as Q_m reaches 0.02 and 0.1 kg s^{-1}, the hydrogen pressure experiences a recovery before the hydrogen enters the outlet section. This is different from the pressure recovery for $Q_m = 0.5$ and $Q_m = 1$ kg s^{-1}. When $Q_m = 0.5$ and $Q_m = 1$ kg s^{-1}, low-pressure sectors are formed due to the expansion when hydrogen enters the outlet section, and the pressure recovery curve is smoother. In contrast, when $Q_m = 0.02$ and 0.1 kg s^{-1}, the pressure recovery is due to the vortex formed at the intersection of the bending and straight channels, and the pressure recovery curve is sharper than that for higher inlet mass flow rates.

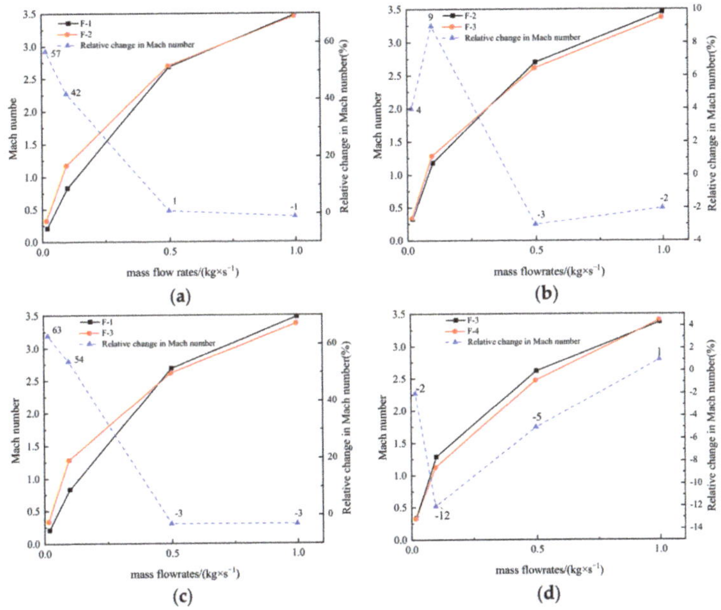

Figure 16. Relative change in Mach number of four structures under different mass flow rates. (**a**) Relative change in Mach number between F-1 and F-2. (**b**) Relative change in Mach number between F-2 and F-3. (**c**) Relative change in Mach number between F-1 and F-3. (**d**) Relative change in Mach number between F-3 and F-4.

Figure 18 shows the comparison of the pressure reduction performance of the four structures at different inlet mass flow rates. The pressure reduction is enhanced when the straight channel in the traditional orifice plate structure is replaced by the Tesla valve flow channel (Figure 18a). When Q_m reaches 0.02 kg s^{-1}, there is an up to 170% increase in the magnitude of pressure reduced. As the mass flow rate increases, the relative increase in the magnitude of pressure reduction across the Tesla-type orifice structure decreases. At Q_m = 0.5 kg s^{-1}, the downward trend slows down after the appearance of the supersonic flow. Similarly, the improved Tesla-type orifice structure achieved better pressure reduction performance than the conventional orifice plate structure. However, at Q_m = 0.02 kg s^{-1}, it achieved a 237% increase in the magnitude of pressure reduced compared to the conventional orifice plate structure.

In Figure 18b, the pressure reduction performance between the original and improved Tesla-type orifice structures is compared. It indicates that the one-stage improved Tesla-type orifice structure achieved further pressure reduction. For Q_m < 0.1 kg s^{-1}, the growth rate of the relative pressure reduction in the one-stage improved Tesla-type orifice structure becomes more significant than that of the original Tesla-type orifice structure. When supersonic flow presents in the orifice structure, the relative increase in the magnitude of pressure reduction in the one-stage improved Tesla-type orifice structure shrinks. As shown in Figure 18d, the two-stage improved Tesla-type orifice structure shows obvious improvement in the pressure reduction performance compared to the one-stage one. However, the relative increase in the pressure reduction magnitude reduces with the increase in mass flow rate. When supersonic flow forms in the valve, the relative increase in the pressure reduction magnitude reduces to its minimum value. Under this situation, increase in the mass flow rate will result in the increase in pressure reduction magnitude. Overall, it is found that the improved Tesla valve is able to achieve better pressure reduction performance than the original Tesla valve. Under a low inlet mass flow rate, a second stage of Tesla valve can be introduced to further improve the pressure reduction performance.

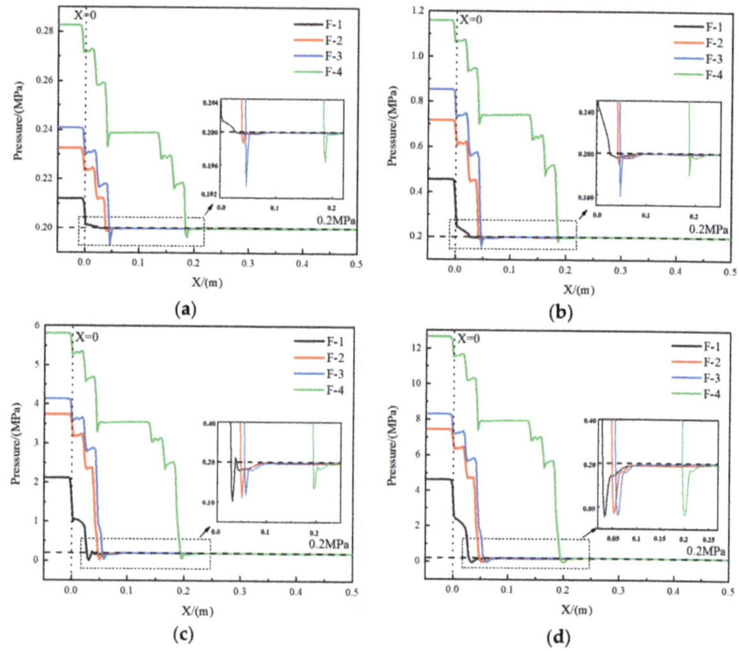

Figure 17. Pressure distribution on the centerline of the symmetry plane for the four structures under different mass flow rates. (**a**) $Q_m = 0.02$ kg s^{-1}. (**b**) $Q_m = 0.1$ kg s^{-1}. (**c**) $Q_m = 0.5$ kg s^{-1}. (**d**) $Q_m = 1$ kg s^{-1}.

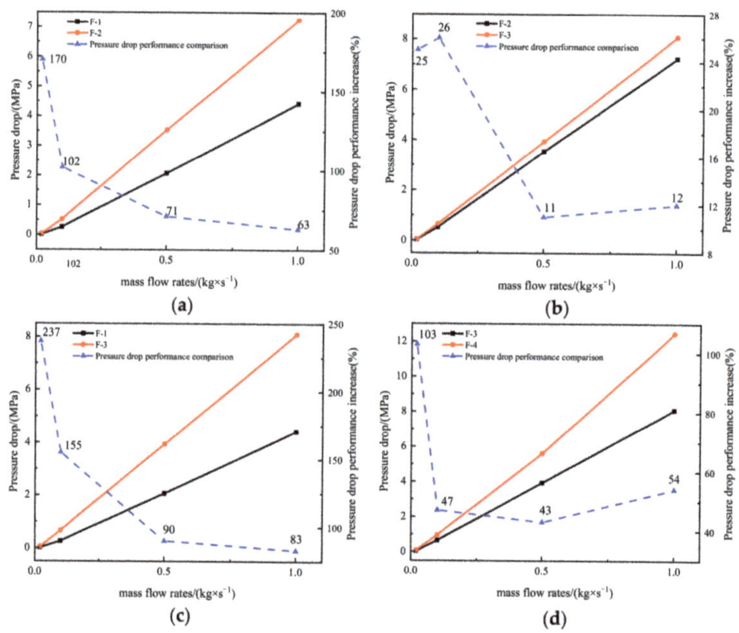

Figure 18. Comparison of pressure reduction performance of four structures. (**a**) F-1 and F-2 pressure reduction. (**b**) F-2 and F-3 pressure reduction. (**c**) F-1 and F-3 pressure reduction. (**d**) F-3 and F-4 pressure reduction.

5. Conclusions

In this work, a novel Tesla-type orifice structure used for high-pressure hydrogen depressurization in HFCVs is proposed. The cylindrical channel in a traditional orifice plate structure is replaced by a Tesla valve flow channel. It is found that the pressure reduction performance could be improved significantly without a significant increase in size. The flow impeding effect of the Tesla-type orifice structure is primarily responsible for the pressure reduction improvement. To enhance the flow impeding effect, modifications are introduced to the Tesla-type channel and the pressure reduction performance has been further improved. It can be concluded that:

(1) In contrast to the conventional orifice structure, the Tesla-type orifice structure has a better performance on pressure reduction. Modifications introduced to the Tesla channel can further improve the pressure reduction performance. Under an inlet mass flow rate of 0.02 kg s^{-1}, the pressure reduction can be increased by 237% compared to the conventional orifice structure;

(2) To further improve the pressure reduction performance, a second set of Tesla-type channels can be introduced to form a two-stage Tesla-type orifice structure. Additionally, the angle β between the bent channel and the main channel increased by more than 130° and L_3 reduced to 8.6 mm in the two-stage Tesla-type orifice structure;

(3) Under the same mass flow rate, the maximum Mach number in the Tesla-type orifice structure is greater than that in the conventional orifice plate structure before the occurrence of supersonic flow. A lower Mach number can alleviate the start-up noise of fluid flow and save energy. When the supersonic flow is formed, the Tesla-type orifice structure shows a similar or smaller maximum Mach number. The two-stage Tesla-type orifice structure can effectively reduce the maximum Mach number with the same pressure reduction;

(4) Due to the asymmetry of the Tesla-type orifice structure, hydrogen flows towards the lower wall when entering the outlet section, producing a wall-fitting effect on the lower wall surface. The vortex can lead to mechanical energy consumption because it generally aggravates the turbulence of the hydrogen flow. A large vortex is formed in the upper area of the outlet chamber with a low turbulence intensity. In contrast to the traditional orifice plate structure, the Tesla-type orifice structure shows less vortices in the high turbulence intensity region, reducing energy consumption.

Author Contributions: Conceptualization, Y.L. and G.D.; Data curation, B.L. (Bin Liu); Formal analysis, B.L. (Bei Li), J.L., X.W. and G.D.; Investigation, B.L. (Bei Li), Y.L., J.L. and G.D.; Methodology, Y.L., J.L., B.L. (Bin Liu), X.W. and G.D.; Project administration, Y.L. and B.L. (Bin Liu); Resources, Y.L. and B.L. (Bin Liu); Software, J.L. and B.L. (Bin Liu); Supervision, Y.L. and G.D.; Validation, X.W.; Visualization, J.L. and X.W.; Writing—original draft, B.L. (Bei Li), Y.L. and B.L. (Bin Liu); Writing—review & editing, J.L., X.W. and G.D. All authors have read and agreed to the published version of the manuscript.

Funding: This research received no external funding.

Institutional Review Board Statement: Not applicable.

Informed Consent Statement: Not applicable.

Conflicts of Interest: The authors declare no conflict of interest.

Nomenclature

$C_2C_{1\varepsilon}$	Constants
E	Total energy per unit control body
g	Gravitational acceleration
G_k	Generation of turbulence kinetic energy due to the mean velocity gradients
G_b	Generation of turbulence kinetic energy due to buoyancy
k_{eff}	Effective thermal conductivity

L_2	Outlet length
$\boldsymbol{p}$	Pressure vector
P_c	Critical pressure
Q_m	Flow rate
R	Universal gas constant
t	Time
T_c	Critical temperature
$\boldsymbol{v}$	Velocity vector
v	Molar volume
Y_M	Contribution of the fluctuating dilatation incompressible turbulence to the overall dissipation rate
Greek symbols	
α	Angle between side channel and main channel
β	Angle between bending channel and main channel
ρ	Gas density
τ	Viscous stress tensor
σ_k σ_ε	Turbulent Prandtl numbers for k and ε
ω	Eccentricity factor of the gas
Abbreviations list	
EoS	Equation of State
PHP	Pulsating Heat Pipe
2D/3D	two-dimensional/three-dimensional
PR	Peng–Robinson
GERG-2008	Groupe Européen de Recherches Gazières
HMPRV	High-level Multistage PRV
CFD	Computational Fluid Dynamic
PRVs	Pressure Reducing Valves
HFCVs	Hydrogen Fuel Cell Vehicles

References

1. IEA. *Global Energy & CO2 Status Report 2019*; International Energy Agency: Paris, France, 2019.
2. Gyamfi, B.A.; Adedoyin, F.F.; Bein, M.A.; Bekun, F.V.; Agozie, D.Q. The anthropogenic consequences of energy consumption in E7 economies: Juxtaposing roles of renewable, coal, nuclear, oil and gas energy: Evidence from panel quantile method. *J Clean. Prod.* **2021**, *295*, 126373. [CrossRef]
3. Dong, X.; Wu, J.; Xu, Z.; Liu, K.; Guan, X. Optimal coordination of hydrogen-based integrated energy systems with combination of hydrogen and water storage. *Appl. Energy* **2021**, *308*, 118274. [CrossRef]
4. Wilberforce, T.; Alaswad, A.; Palumbo, A.; Dassisti, M.; Olabi, A.G. Advances in stationary and portable fuel cell applications. *Int. J. Hydrog. Energy* **2016**, *41*, 16509–16522. [CrossRef]
5. Pearson, R.J.; Turner, J.W.G. Renewable Fuels. In *Comprehensive Renewable Energy*; Elsevier: Amsterdam, The Netherlands, 2012; pp. 305–342.
6. Lin, Z.; Ou, S.; Elgowainy, A.; Reddi, K.; Veenstra, M.; Verduzco, L. A method for determining the optimal delivered hydrogen pressure for fuel cell electric vehicles. *Appl. Energy* **2018**, *216*, 183–194. [CrossRef]
7. Mori, D.; Hirose, K. Recent challenges of hydrogen storage technologies for fuel cell vehicles. *Int. J. Hydrog. Energy* **2009**, *34*, 4569–4574. [CrossRef]
8. Barthélémy, H.; Weber, M.; Barbier, F. Hydrogen storage: Recent improvements and industrial perspectives. *Int. J. Hydrog. Energy* **2016**, *42*, 7254–7262. [CrossRef]
9. Liang, L.; Xiaofeng, H.; Ben, D.; Xing, H. Theoretical and Experimental Research on a Pressure-Reducing Valve for a Water Hydraulic Vane Pump. *J. Press. Vessel Technol.* **2014**, *136*, 021601.
10. Ulanicki, B.; Skworcow, P. Why PRVs Tends to Oscillate at Low Flows. *Proc. Eng.* **2014**, *89*, 378–385. [CrossRef]
11. Saha, B.K.; Chattopadhyay, H.; Mandal, P.B.; Gangopadhyay, T. Dynamic simulation of a pressure regulating and shut-off valve. *Comput. Fluids* **2014**, *101*, 233–240. [CrossRef]
12. Okhotnikov, I.; Noroozi, S.; Sewell, P.; Godfrey, P. Evaluation of steady flow torques and pressure losses in a rotary flow control valve by means of computational fluid dynamics. *Int. J. Heat Fluid Flow* **2017**, *64*, 89–102. [CrossRef]
13. Jin, Z.-J.; Chen, F.-Q.; Qian, J.-Y.; Zhang, M.; Chen, L.-L.; Wang, F.; Fei, Y. Numerical analysis of flow and temperature characteristics in a high multi-stage pressure reducing valve for hydrogen refueling station. *Int. J. Hydrog. Energy* **2016**, *41*, 5559–5570. [CrossRef]
14. Jin, Z.-J.; Wei, L.; Chen, L.-L.; Qian, J.-Y.; Zhang, M. Numerical simulation and structure improvement of double throttling in a high parameter pressure reducing valve. *J. Zhejiang Univ. A* **2013**, *14*, 137–146. [CrossRef]

15. Hou, C.-W.; Qian, J.-Y.; Chen, F.-Q.; Jiang, W.-K.; Jin, Z.-J. Parametric analysis on throttling components of multi-stage high pressure reducing valve. *Appl. Therm. Eng.* **2018**, *128*, 1238–1248. [CrossRef]
16. Chen, F.-Q.; Gao, Z.-X.; Qian, J.-Y.; Jin, Z.-J. Numerical Study on Flow Characteristics in High Multi-Stage Pressure Reducing Valve. In Proceedings of the ASME 2017 Fluids Engineering Division Summer Meeting, Waikoloa, HI, USA, 30 July–3 August 2017. [CrossRef]
17. Chen, F.; Ren, X.; Hu, B.; Li, X.; Gu, C.; Jin, Z. Parametric analysis on multi-stage high pressure reducing valve for hydrogen decompression. *Int. J. Hydrog. Energy* **2019**, *44*, 31263–31274. [CrossRef]
18. Chen, F.; Zhang, M.; Qian, J.; Chen, L.; Jin, Z. Pressure analysis on two-step high pressure reducing system for hydrogen fuel cell electric vehicle. *Int. J. Hydrog. Energy* **2017**, *42*, 11541–11552. [CrossRef]
19. Liu, B.; Li, S. Computational Fluid Dynamics Analysis on a Perforated Plate in a High Pressure Reducing System for Hydrogen Fuel Cell Electric Vehicle Using Real Gas Equation of State. In Proceedings of the 2019 IEEE 28th International Symposium on Industrial Electronics (ISIE), Vancouver, BC, Canada, 12–14 June 2019; pp. 1067–1071. [CrossRef]
20. Tesla, N. Valvular Conduit. U.S.Patent 1329559A, 8 July 1920.
21. Gamboa, A.R.; Morris, C.J.; Forster, F.K. Improvements in Fixed-Valve Micropump Performance Through Shape Optimization of Valves. *J. Fluids Eng.* **2004**, *127*, 339–346. [CrossRef]
22. Zhang, S.; Winoto, S.; Low, H. Performance simulations of Tesla microfluidic valves. *Int. Conf. Integr. Commer. Micro Nanosyst.* **2007**, *42657*, 15–19.
23. De Vries, S.; Florea, D.; Homburg, F.; Frijns, A. Design and operation of a Tesla-type valve for pulsating heat pipes. *Int. J. Heat Mass Transf.* **2017**, *105*, 1–11. [CrossRef]
24. Bao, Y.; Wang, H. Numerical study on flow and heat transfer characteristics of a novel Tesla valve with improved evaluation method. *Int. J. Heat Mass Transf.* **2022**, *187*, 122540. [CrossRef]
25. Monika, K.; Chakraborty, C.; Roy, S.; Sujith, R.; Datta, S.P. A numerical analysis on multi-stage Tesla valve based cold plate for cooling of pouch type Li-ion batteries. *Int. J. Heat Mass Transf.* **2021**, *177*, 121560. [CrossRef]
26. Zhang, Y.; Liu, B.; She, X.; Luo, Y.; Sun, Q.; Teng, L. Numerical study on the behavior and design of a novel multistage hydrogen pressure-reducing valve. *Int. J. Hydrog. Energy* **2022**, *47*, 14646–14657. [CrossRef]
27. Qian, J.; Chen, M.; Gao, Z.; Jin, Z. Mach number and energy loss analysis inside multi-stage Tesla valves for hydrogen decompression. *Energy* **2019**, *179*, 647–654. [CrossRef]
28. Qian, J.-Y.; Wu, J.-Y.; Gao, Z.-X.; Wu, A.; Jin, Z.-J. Hydrogen decompression analysis by multi-stage Tesla valves for hydrogen fuel cell. *Int. J. Hydrog. Energy* **2019**, *44*, 13666–13674. [CrossRef]
29. Jin, Z.; Gao, Z.; Chen, M.; Qian, J. Parametric study on Tesla valve with reverse flow for hydrogen decompression. *Int. J. Hydrog. Energy* **2018**, *43*, 8888–8896. [CrossRef]
30. Rezaei, M.; Azimian, A.R.; Toghraie, D. Molecular dynamics study of an electro-kinetic fluid transport in a charged nanochannel based on the role of the stern layer. *Physica A* **2015**, *426*, 25–34. [CrossRef]
31. Toghraie, D. Numerical thermal analysis of water's boiling heat transfer based on a turbulent jet impingement on heated surface. *Physica E* **2016**, *84*, 454–465. [CrossRef]
32. Li, Z.; Barnoon, P.; Toghraie, D.; Dehkordi, R.B.; Afrand, M. Mixed convection of non-Newtonian nanofluid in an H-shaped cavity with cooler and heater cylinders filled by a porous material: Two phase approach. *Adv. Powder Technol.* **2019**, *30*, 2666–2685. [CrossRef]
33. Fluent, A. *Ansys Fluent Theory Guide*; ANSYS Inc.: Canonsburg, PA, USA, 2011.
34. Liu, X.; Godbole, A.; Lu, C.; Michal, G.; Venton, P. Source strength and dispersion of CO2 releases from high-pressure pipelines: CFD model using real gas equation of state. *Appl. Energy* **2014**, *126*, 56–68. [CrossRef]
35. Shih, T.-H.; Liou, W.W.; Shabbir, A.; Yang, Z.; Zhu, J. A new k-epsilon eddy viscosity model for high Reynolds number turbulent flows: Model development and validation. *NASA STI Recon Tech. Rep. N* **1994**, *95*, 11442.
36. Peng, D.; Robinson, D.B. A new two-constant equation of state. *Ind. Eng. Chem. Fundam.* **1976**, *15*, 59–64. [CrossRef]
37. Michels, A.; De Graaff, W.; Wassenaar, T.; Levelt, J.; Louwerse, P. Compressibility isotherms of hydrogen and deuterium at temperatures between-175C and + 150C (at densities up to 960 amagat). *Physica* **1959**, *25*, 25–42. [CrossRef]
38. Kunz, O.; Wagner, W. The GERG-2008 Wide-Range Equation of State for Natural Gases and Other Mixtures: An Expansion of GERG-2004. *J. Chem. Eng. Data* **2012**, *57*, 3032–3091. [CrossRef]
39. Liu, H.; Ma, J.; Wang, S.; Zheng, R. Research on the one-direction flow characteristic of tesla valve. *Phys. Eng.* **2020**, *30*, 120–124. (In Chinese)
40. Tan, D.H.; Herman, P.K.; Janakiraman, A.; Bates, F.S.; Kumar, S.; Macosko, C.W. Influence of Laval nozzles on the air flow field in melt blowing apparatus. *Chem. Eng. Sci.* **2012**, *80*, 342–348. [CrossRef]

materials

MDPI

Article

A Measurement Solution of Face Gears with 3D Optical Scanning

Xinxin Lu [1,2], Xing Zhao [1,2,*], Bo Hu [1,2], Yuansheng Zhou [1,2], Zhezhen Cao [1,2] and Jinyuan Tang [1,2]

1 State Key Laboratory of High Performance Complex Manufacturing, Central South University, Changsha 410083, China
2 College of Mechanical and Electrical Engineering, Central South University, Changsha 410083, China
* Correspondence: xingzhao@csu.edu.cn

Abstract: Gears are usually measured by the contact metrology method in gear measuring centers or coordinate measuring machines. Recently, three-dimensional (3D) optical scanning, a non-contact metrology method, has been applied in the industry as an advanced measurement technology mainly due to its high efficiency. However, its applications to gears with complicated geometry, such as face gears, are still limited due to its relatively low accuracy and the void of related measurement solutions. In this work, an accurate measurement solution with 3D optical scanning is proposed for the tooth surface deviations of orthogonal face gears. First, point cloud collection is carried out by the 3D scanner. Furthermore, the measurement solution is implemented with a three-stage algorithm by aligning point clouds with the design model. Subsequently, 3D modeling is studied by numbering the points and reconstructing the real tooth surfaces. An example with a measurement experiment and loaded tooth contact analysis is given to show the validity of the proposed method.

Keywords: face gears; 3D optical scanning; point clouds; tooth surface deviations; loaded tooth contact analysis

Citation: Lu, X.; Zhao, X.; Hu, B.; Zhou, Y.; Cao, Z.; Tang, J. A Measurement Solution of Face Gears with 3D Optical Scanning. *Materials* **2022**, *15*, 6069. https://doi.org/10.3390/ma15176069

Academic Editor: Alessandro P. Fantilli

Received: 21 July 2022
Accepted: 30 August 2022
Published: 1 September 2022

1. Introduction

Face gears [1–4] have been applied in some transmission systems due to the advantages of smooth transmission, large coincidence, and small footprint. Compared with the bevel gear drives, face gear drives have smaller axial forces and lower sensitivity to axial installation errors. Moreover, the support apparatus of face gear drives can be designed in a relatively more compact way than spiral bevel gear drives, and it can significantly save weight. The above strengths make face gears receive attention and research. Some car companies, such as Audi, have developed the advanced technology of face gears and applied it to automobile differentials. In the aerospace industry, face gears have been successfully used in helicopter transmission systems. More types of face gears are studied nowadays. According to the macroscopic form of the tooth surface, the types of face gears can be divided into spur face gears, helical face gears, conical face gears, worm face gears, asymmetric helical face gears [5,6], and so on. The performance of face gear drives is constantly improved, and the application scenarios are wider.

In terms of processing, face gears can be machined by grinding, CNC plunge milling, hobbing, and other processing methods [7–9]. The spur face gears, a simple type of face gears, can reach the desired accuracy level in the aerospace industry by grinding with "grinding-measurement-correction" [10,11]. The precision grade is evaluated by the measurement, so the reliable measurement of face gears is the basis of the correction of the machine tool settings in the processing. The measurement method of face gears generally adopts the contact measuring machines [12–14], such as coordinate measuring machines (CMMs) and gear measuring centers (GMCs). CMMs are universal measuring tools and can measure various workpieces. GMCs are specialized machines for measuring gears, which are made by adding rotary tables to CMMs. They both utilize the touch signal emitted by a probe when the tip ball comes into contact with the surface of a workpiece and records the probe coordinates at that time. The recorded coordinates are not valid points on the surface of the workpiece due to the signal delay, the inaccurate measurement coordinate system, and the radius of the probe, so the measured data need to be compensated further by the numerical algorithm.

The real face gears are the workpieces after processing, and they usually contain machining errors, which will affect the meshing performance. During the meshing process, the tooth surfaces of face gears and pinions will contact at a point at one moment, and the contact path will generate by connecting all contact points at every moment. Tooth surface deviations will alter the original tooth surface morphology, which will change the original contact path and further has a significant impact on the meshing performance of gear drives. In the industry, the meshing performance of real face gears can usually be analyzed by roll testing experiment, but this method is costly and timely. Therefore, some scholars have conducted tooth contact analysis for the real tooth surfaces, reconstructed with the measured data using CMMs or GMCs. Wang et al. [15] used the compensated measured data to reconstruct the tooth surface of face gears and conducted the digital tooth contact analysis using the proposed robust algorithm. Lin et al. [16] presented a least rotation angle method and the improved quad-tree search algorithm to compute the contact pattern and transmission error of a single flank fitted by the data measured on GMCs. Din et al. [17] proposed a collaborative manufacturing system for hypoid gears based on data-driven programming that the data are collected using CMMs, and the machine setting is modified by the tooth geometric and meshing performance.

The above researchers all adopted the contact metrology method. Although the contact metrology method, with the aid of compensation technology, can reach high accuracy, it has two disadvantages stated as follows.

(1) The border area of the tooth surface cannot be measured accurately since other locations would be touched. Moreover, the accuracy of the extrapolation method is insufficient when reconstructing the tooth surfaces. Hence, it is impossible to accurately determine whether edge contact occurs during the meshing of the real face gear drives. Edge contact refers to the meshing of the border areas of the tooth surface between face gears and pinions, which leads to stress concentration and meshing impact [18,19]. It will cause obvious vibration and noise during the meshing and reduces the service life of gears.
(2) The efficiency is low when measuring multiple teeth since every point needs to be contacted.

3D optical scanning, as a non-contact metrology method, has become a new trend in measuring the macroscopic morphology of surfaces with the emergence and development of computer vision [20–22]. Recently, this technique has been studied to inspect and analyze gear quality. Chen et al. [23] proposed an optical inspection method for tooth surfaces using the projection moire technique and measured the deviation based on the reference gear tooth profiles measured by CMM. Gonzalez-Perez et al. [24] performed tooth contact analysis of cylindrical gears reconstructed from point clouds. Urbas et al. [25,26] studied improved methods for optical measurement accuracy of spur gears. Three-dimensional (3D) optical scanning can be a concern by scholars due to the advantages as follows.

(1) As a non-contact metrology method, it is more efficient than the contact metrology method when measuring multiple teeth.
(2) The collected data using this method obtain the complete geometry of a tooth, including the top land surface, the bottom land surface, and both sides of the holistic tooth surface.

However, 3D optical scanning for measurement has not been widely used in the gear industry due to the low measurement accuracy. Its accuracy typically varies from 10 μm up to 100 μm [27–30]. The measurement errors come from two aspects. One is the systematic errors, and the other is the alignment errors between the measured workpiece and the design model. The former can only be reduced by upgrading the hardware specifications, while the latter can be reduced by alignment using the numerical algorithm. Thus, it is possible to achieve an accuracy of several microns with some numerical algorithms for compensation. Moreover, the previous research objects are cylindrical gears, while geometric characteristics of face gears are different, and the macroscopic morphology of tooth surfaces is extremely complex.

In order to solve the problem of contact and non-contact metrology method and meet the increasing need for measuring holistic geometric information, a 3D optical scanning method is adopted for measuring the tooth surface of orthogonal face gears in this work, and the measurement solution is presented to ensure the measurement accuracy. In Section 2, the method of point cloud collection is introduced, and point clouds are processed by filtering and down-sampling. In Section 3, the measurement solution for tooth surfaces of face gears with 3D optical scanning is proposed to obtain the tooth surface deviations based on the alignment, and three-stage algorithm (3SA) is presented to align three important features between point clouds and the design model of face gears. In Section 4, the 3D modeling method is further investigated. An example is given to show the validity of the measurement solution with 3D optical scanning, and loaded tooth contact analysis (LTCA) for the real tooth surfaces of face gears by finite element method (FEM) is conducted to show the influence of the deviations on meshing performance in Section 5.

2. Point Cloud Collection

Point clouds containing the holistic geometry information of face gears can be collected at a rate of millions of points per second by 3D optical scanning. A way to implement 3D optical scanning is by binocular fringe projection method [31–34], as shown in Figure 1. The main principle is to project the fringes generated by the system onto the surface of objects in the corresponding order. Binocular cameras are applied to take images, and the original fringes will be found to be deformed due to the height change of the surface of objects. Then, the phase information of the surface of objects can be solved by the phase-shifting algorithm with the collected data of the deformed fringes. According to the phase information, the coordinate value of each point in the space can be calculated by triangulation, and then the point cloud data of the surface can be obtained. The binocular fringe projection method has been widely studied, and the detailed implementation methods will not be further described in the paper.

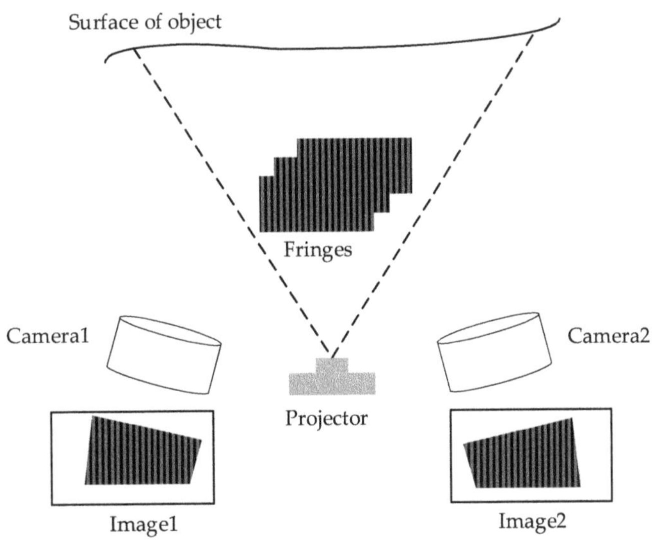

Figure 1. Fringe projection.

Currently, the available 3D scanning technology has been integrated into commercial equipment by some companies. Choosing the proper scanner equipment is a requirement for obtaining high-quality point clouds. Generally, the higher the spatial resolution and measurement accuracy of the equipment, the higher the quality of the collected point clouds of face gears. This work takes ATOS 3D Scanner to collect point clouds, as shown in Figure 2a. ATOS 3D Scanner is an industrial optical system with a charge-coupled device camera at both ends. The measured data by it can be transmitted to the computer, and the coordinate information of point clouds can be shown in the special software. The main system configuration is shown in Table 1.

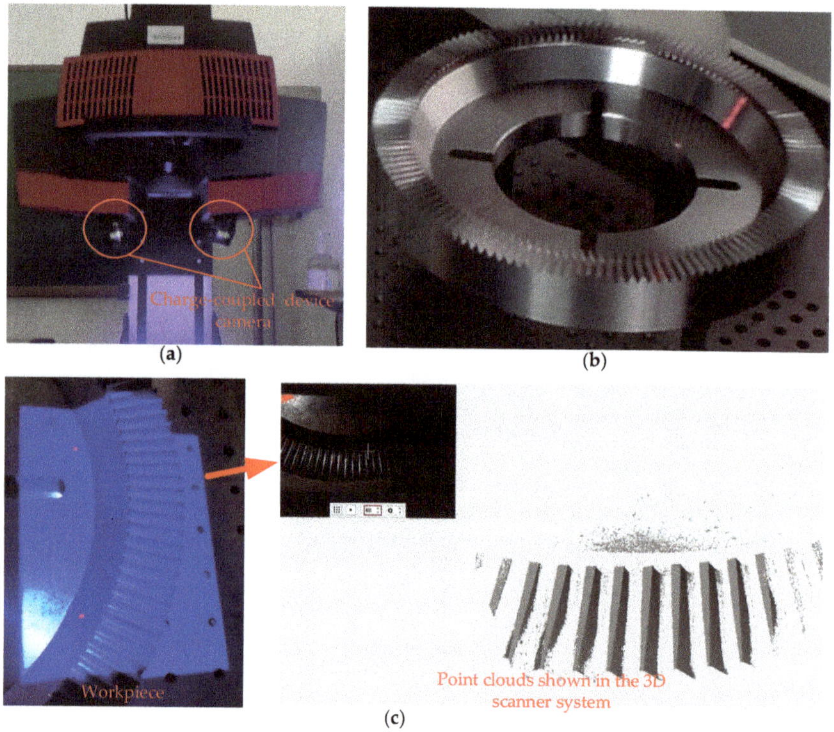

Figure 2. (**a**) ATOS 3D Scanner, (**b**) a face gear, and (**c**) the point cloud collection.

Table 1. The configuration of ATOS 3D Scanner.

System	ATOS 3D Scanner
Camera pixels	2 × 3296 × 2472 pixels
Measuring volumes	38 × 29 × 15 – 2000 × 1500 × 1500 mm^3
Point spacing	0.01–0.61 mm
Working distance	490–2000 mm
Projected light source	Structured blue light
Operating Temperature	5–40 °C

When performing the operation, first ensure that the experiments are in a stable working environment, such as a stable light and warm-up environment, in which the measurement results of the 3D scanner are not affected by external factors. Then, the scanner is set up on the stationary support, and the settings of equipment are adjusted, such as lens position and aperture size. Subsequently, the scanner needs to be calibrated with the calibration plates. Through calibration, the system can accurately calculate the position of the scanning equipment relative to the scanning objects. Therefore, accurate calibration is a prerequisite for measurement accuracy. Moreover, the white developers need to be sprayed onto the face gear tooth surfaces to make them diffuse reflection for a better quality of point clouds. Since the bottom surface of face gears is the processing reference with high accuracy, face gears can be placed flat on the workbench when collecting point clouds, as shown in Figure 2b. Finally, the scanner system can be operated to take pictures of the workpieces, and the system would automatically obtain the coordinate data of the workpiece so that it obtains the point cloud data. In Figure 2c, a workpiece of face gear is being scanned by the 3D scanner, and the point clouds can be seen in the scanner system.

Subsequently, point clouds would require data processing. As shown in Figure 3a, the unprocessed point clouds will contain unnecessary point data. Only the points on the tooth surface are needed, and the other points should be removed. Here take the conditional filters to remove the outliers. Since the table is horizontal, the high area is the point cloud of tooth surfaces, and the low area is the point cloud of the conical surface and upper surface. When setting the appropriate z-value, the two different point clouds could be distinguished. The needed point cloud of tooth surfaces is shown in Figure 3b. If multiple point clouds need to be stitched together, the noise points will be generated, and further noise point reduction is needed with filter algorithms, such as the Gaussian filter [35]. Subsequently, the data volume of the point cloud can be reduced by the down-sampling method [36], as shown in Figure 3c. The different grid boxes in the space are divided, and a specific density of points in the grid is taken.

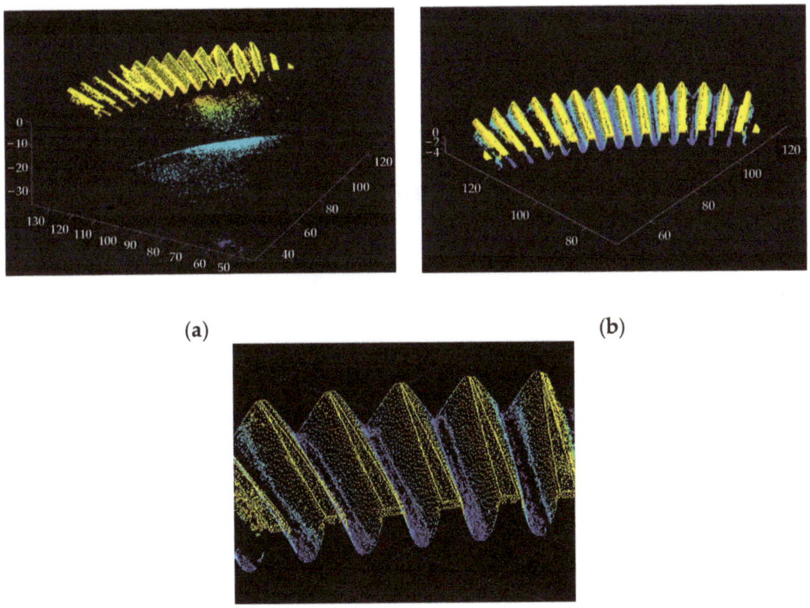

(a) (b) (c)

Figure 3. (**a**) Unprocessed point cloud, (**b**) filtering, and (**c**) down-sampling.

3. The Measurement Solution

The point cloud collected in Section 2 is in the measurement coordinate system (MCS), which is different from the design coordinate system (DCS). DCS is a fixed coordinate system determined in the design process of face gears, while MCS is the coordinate system set randomly in the measurement process. In order to obtain the tooth surface deviations of face gears, the point cloud containing the geometry information must be compared to the design model in the same coordinate system. The tooth surface deviation of one point can be represented as

$$dev_i = (\boldsymbol{pm}_i - \boldsymbol{pt}_i) \cdot \boldsymbol{n}_i \tag{1}$$

where $\boldsymbol{pt}_i$ and $\boldsymbol{n}_i$ are the coordinates and the unit normal of one point on the theoretical tooth surface in DCS, respectively. $\boldsymbol{pm}_i$ is the measured point which is in the direction of $\boldsymbol{n}_i$. Before computing the tooth surface deviations, alignment is required, and the geometric features between the point cloud and the design model are aligned, which is mathematically equivalent to transforming the coordinate system. The alignment is taken into three stages: the alignment for the top land surface, the center of circles, and the tooth surface, as shown in Figure 4, and 3SA is presented as follows.

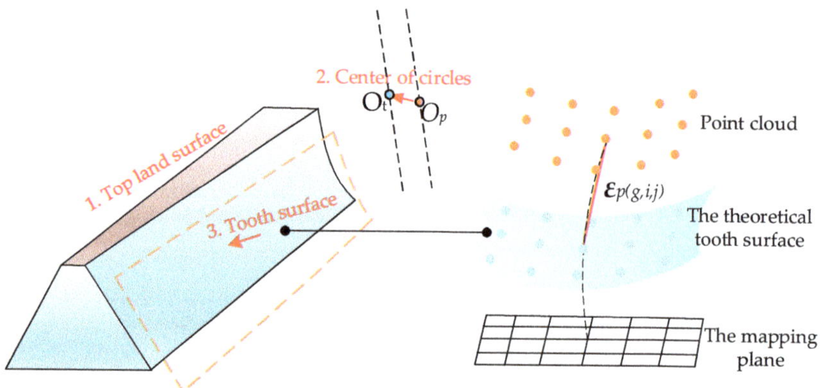

Figure 4. The three-stage alignment for top land surface, center of circles, and tooth surface.

(1) The first stage

The top land surface of the designed orthogonal face gears is a plane, which is parallel to the xoy plane in DCS, and the points on it are the same z-value. The collected points on the top land surface will not be the same z-value in MCS for the measurement and machining errors. Hence, the plane of the measured top land surface is fitted by the following equation as

$$z_{ls} = ax_{ls} + by_{ls} + c \tag{2}$$

The subscript ls represents the top land surface of the workpiece, and a, b, c are the coefficients. The solution can be obtained by solving the overdetermined systems of equations as

$$\mathbf{A} \cdot \mathbf{X} = \mathbf{b} \Leftrightarrow \mathbf{X} = \left(\mathbf{A}\mathbf{A}^{\mathrm{T}}\right)^{-1}\mathbf{A}^{\mathrm{T}}\mathbf{b} \tag{3}$$

where

$$\mathbf{X} = \begin{pmatrix} a \\ b \\ c \end{pmatrix}, \mathbf{A} = \begin{bmatrix} x_1 & y_1 & 1 \\ x_2 & y_2 & 1 \\ x_3 & y_3 & 1 \\ \vdots & \vdots & \vdots \\ x_n & y_n & 1 \end{bmatrix}, \mathbf{b} = \begin{pmatrix} z_1 \\ z_2 \\ z_3 \\ \vdots \\ z_n \end{pmatrix} \tag{4}$$

(x_1, y_1, z_1), (x_2, y_2, z_2), (x_3, y_3, z_3), $\cdots$, (x_n, y_n, z_n) are the coordinates of the points on the top land surface of the workpiece. Then, the point cloud is translated along the z-axis as

$$\Delta z = z_{ls} - z_t \tag{5}$$

where z_t is the value of the theoretical top land surface along the z-axis in DCS.

(2) The second stage

The center of circles O_t of the designed face gears is zero point, while the center of circles O_p of the point cloud is set roughly. Hence, they are aligned by the pattern search technique [37,38]. The maximum and the minimum radius in the point cloud on the xoy plane are respectively $R_{r\max}$ and $R_{r\min}$, and their coordinates are respectively ($x_{r\max}$, $y_{r\max}$, $z_{r\max}$) and ($x_{r\min}$, $y_{r\min}$, $z_{r\min}$). The maximum and the minimum radius of the designed face gears are $R_{t\max}$ and $R_{t\min}$, respectively. The zero point in MCS is set on the inner side of the point cloud, so the initial value for the translation vector of the point cloud is set by the following equation as

$$\boldsymbol{Ini} = \left(\frac{(R_{r\max} - R_{t\max})}{R_{r\max}} x_{r\max}, \frac{(R_{r\max} - R_{t\max})}{R_{r\max}} y_{r\max}, 0 \right) \tag{6}$$

The following optimization objective is established as

$$\min f = \text{norm}(R_{r\max} - R_{t\max}, R_{r\min} - R_{t\min}) \tag{7}$$

Then the optimal solution is searched as follows.

Step 1. Set the initial translation vector $\boldsymbol{q}_0 = \boldsymbol{Ini}$, the initial step $\delta_0 = 1$, the contraction factor $\alpha = 0.6$, and the tolerance error $\varepsilon = 10^{-6}$.

Step 2. Let $\boldsymbol{p} = \boldsymbol{q}_k$ and $j = 1$.

Step 3. If $f(\boldsymbol{p} + \delta_k \boldsymbol{e}_j) < f(\boldsymbol{p}), \boldsymbol{p} = \boldsymbol{p} + \delta_k \boldsymbol{e}_j$ and turn to Step 5. Otherwise, turn to Step 4.

Step 4. If $f(\boldsymbol{p} - \delta_k \boldsymbol{e}_j) < f(\boldsymbol{p}), \boldsymbol{p} = \boldsymbol{p} - \delta_k \boldsymbol{e}_j$ and turn to Step 5. Otherwise, turn to Step 5.

Step 5. If $j < n$, $j = j + 1$ and turn to Step 3. Otherwise, $\boldsymbol{q}_{k+1} = \boldsymbol{p}$ and turn to Step 6.

Step 6. If $f(\boldsymbol{q}_{k+1}) < f(\boldsymbol{q}_k)$, let $\boldsymbol{p} = \boldsymbol{q}_{k+1} + (\boldsymbol{q}_{k+1} - \boldsymbol{q}_k)$, $\delta_{k+1} = \alpha\delta_k$, $k = k + 1$, $j = 1$ and turn to Step 3. Otherwise, turn to Step 7.

Step 7. If $f(\boldsymbol{q}_k) < \varepsilon$, stop computation, and $\boldsymbol{q}_k$ is the optimal solution. Otherwise, turn to Step 8.

Step 8. If $\boldsymbol{q}_{k+1} = \boldsymbol{q}_k$, let $\delta_{k+1} = \alpha\delta_k$, $k = k + 1$ and turn to Step 2. Otherwise, let $\boldsymbol{q}_{k+1} = \boldsymbol{q}_k$, $\delta_{k+1} = \delta_k$, $k = k + 1$ and turn to Step 2.

The optimal solution is obtained through the above steps, as

$$\boldsymbol{q}^* = (q_1{}^*, q_2{}^*, 0) \tag{8}$$

where the superscript * represents the optimal solution. Then, the point cloud is translated according to the optimal solution as follows.

$$\begin{cases} \Delta x = q_1{}^* \\ \Delta y = q_2{}^* \end{cases} \tag{9}$$

(3) The third stage

In the third stage, the points after being processed in the second stage will be aligned with the theoretical tooth surface. The 5 × 9 grid on each side of the measured tooth surface of a tooth is selected randomly. The grids are translated and rotated around the *z*-axis. The transformed grids are represented as

$$\begin{bmatrix} \boldsymbol{p}_{\text{tr}} \\ 1 \end{bmatrix} = \boldsymbol{M}_{tr} \cdot \begin{bmatrix} \mathbf{p}_{\text{mg}} \\ 1 \end{bmatrix} \tag{10}$$

where

$$\boldsymbol{p}_{tr} = \begin{bmatrix} x_{\text{tr}} \\ y_{\text{tr}} \\ z_{\text{tr}} \end{bmatrix}, \mathbf{p}_{\text{mg}} = \begin{bmatrix} \text{x}_{\text{mg}} \\ \text{y}_{\text{mg}} \\ \text{z}_{\text{mg}} \end{bmatrix}, \boldsymbol{M}_{\text{tr}} = \begin{bmatrix} \cos\alpha & -\sin\alpha & 0 & dx \\ \sin\alpha & \cos\alpha & 0 & dy \\ 0 & 0 & 1 & dz \\ 0 & 0 & 0 & 1 \end{bmatrix} \tag{11}$$

$\boldsymbol{p}_{\mathrm{mg}}$ represents the points on the grids before the transformation, $\boldsymbol{M}_{\mathrm{tr}}$ represents the transformation matrix, and $\boldsymbol{p}_{\mathrm{tr}}$ represents the point on the transformed grids. α is the rotation angle around the z-axis, and dx, dy, and dz are the translation values along the x-, y-, and z-axis, respectively. They are all chosen as the optimization variables. The reason for the second translation here is that the data processed in the first two stages will contain systematic errors from the top land surface and center of circles, so the tooth surfaces need to be further aligned to reduce the systematic errors.

Then, the rotation radius r and height h of each point of grids can be obtained after rotating the grids around the z-axis, and the points are mapped to the mapping plane. Every point on the theoretical tooth surface with the same rotation radius and height is calculated according to the tooth surface equation, as follows.

$$\begin{cases} \mathrm{r}_{(g,i,j)} = \sqrt{x_{\mathrm{tr}\,(g,i,j)}(\boldsymbol{var})^2 + y_{\mathrm{tr}\,(g,i,j)}(\boldsymbol{var})^2} \\ \mathrm{h}_{(g,i,j)} = z_{\mathrm{tr}\,(g,i,j)}(\boldsymbol{var}) \end{cases}, g = 1,2; i = 1,2,\cdots,5; j = 1,2,\cdots,9 \quad (12)$$

$$\begin{cases} \sqrt{px_{(g,i,j)}(uv_1, uv_2)^2 + py_{(g,i,j)}(uv_1, uv_2)^2} = \mathrm{r}_{(g,i,j)} \\ pz_{(g,i,j)}(uv_1, uv_2) = \mathrm{h}_{(g,i,j)} \end{cases}, g = 1,2; i = 1,2,\cdots,5; j = 1,2,\cdots,9 \quad (13)$$

where the different values of the subscript g represent the grid on the different tooth surfaces of a tooth. The subscript i and j represent the point number on the grid. $\left(px_{(g,i,j)}, py_{(g,i,j)}, pz_{(g,i,j)}\right)$ represents the coordinate of the point on the theoretical tooth surface. $\boldsymbol{var}$ represents the unknown variables (α,dx,dy,dz). (uv_1,uv_2) represents two unknown variables of the tooth surface equation. The tooth surface equation of face gears has been derived in our previous literature [39], and here it is not stated in detail. The vector between the point $\boldsymbol{p}_{\mathrm{tr}}$ on the transformed grid and the corresponding theoretical point $\boldsymbol{p}_{\mathrm{tp}}$ can be yielded as

$$\boldsymbol{\varepsilon}_{\mathrm{p}(g,i,j)} = \boldsymbol{p}_{\mathrm{tr}(g,i,j)}(\boldsymbol{var}) - \boldsymbol{p}_{\mathrm{tp}(g,i,j)}(\boldsymbol{var}) \quad (14)$$

In order to achieve the best alignment between the transformed grid and the theoretical tooth surface, a nonlinear least square minimization optimization is established, and it is represented as

$$\min f_{\varepsilon} = \frac{1}{2}\sum_{g,i,j} \boldsymbol{\varepsilon}_{\mathrm{p}(g,i,j)}(\boldsymbol{var})^{\mathrm{T}} \boldsymbol{\varepsilon}_{\mathrm{p}(g,i,j)}(\boldsymbol{var}) \quad (15)$$

The above over-determined nonlinear equation is expanded by second-order Taylor to transform it into a trust region problem, as

$$\min O_{\varepsilon k}(\boldsymbol{d}) = f_{\varepsilon k} + \nabla f_{\varepsilon k}{}^{\mathrm{T}}\boldsymbol{d} + \frac{1}{2}\boldsymbol{d}^{\mathrm{T}}\nabla^2 f_{\varepsilon k}\boldsymbol{d} \quad (16)$$

where $\boldsymbol{d}$ is the optimal step of the iteration. $f_{\varepsilon k}$ is the value of f_{ε} in the kth iteration. $\nabla f_{\varepsilon k}$ and $\nabla^2 f_{\varepsilon k}$ are the first and second gradient, respectively. For convenience, $\nabla f_{\varepsilon k}$ is marked as g_k. Since the calculation for $\nabla^2 f_{\varepsilon k}$ is too complex, the approximate Hessian matrix $\boldsymbol{B}_k$ is used to replace, as

$$\boldsymbol{B}_k = J_k{}^{\mathrm{T}} J_k \quad (17)$$

where $\boldsymbol{J}_k$ is the Jacobi matrix. Subsequently, the trust region dogleg method is used to solve the optimization function. The optimization problem is represented as

$$\begin{aligned} &\boldsymbol{d} = \mathrm{argmin} f_{\varepsilon k} + g_k{}^{\mathrm{T}}\boldsymbol{d} + \tfrac{1}{2}\boldsymbol{d}^{\mathrm{T}}\boldsymbol{B}_k\boldsymbol{d} \\ &s.t. \|\boldsymbol{d}\| \le \Delta_k \end{aligned} \quad (18)$$

where Δ_k is the trust region radius. The solution to Equation (18) can be given as

$$\boldsymbol{d}^* = \begin{cases} \tau \cdot \boldsymbol{C}, 0 \le \tau \le 1 \\ -\boldsymbol{C} + (\tau - 1)\cdot(\boldsymbol{GN} + \boldsymbol{C}), 1 < \tau \le 2 \end{cases} \quad (19)$$

where

$$C = -\frac{{g_k}^{\mathrm{T}} g_k}{{g_k}^{\mathrm{T}} \boldsymbol{B}_k g_k} g_k \tag{20}$$

$$GN = -(\boldsymbol{B}_k)^{-1} g_k \tag{21}$$

C is the Cauchy step, and GN is the Gauss-Newton step. τ can be solved by the following scalar quadratic equation as

$$\|-C + (\tau - 1) \cdot (GN + C)\|^2 = \Delta_k \tag{22}$$

The ratio ρ_k is used to measure the approximation between Equations (15) and (18), as follows.

$$\rho_k = \frac{f_\varepsilon(\boldsymbol{var}_k) - f_\varepsilon(\boldsymbol{var}_k + \boldsymbol{d})}{O_{\varepsilon k}(\boldsymbol{0}) - O_{\varepsilon k}(\boldsymbol{d})} \tag{23}$$

where $\boldsymbol{var}_k$ is the value of optimization variables in the kth iteration. When $0.9 < \rho_k \leq 1$, the approximation is well and Δ_k is increased. When $0.1 < \rho_k \leq 0.9$, Δ_k remains the same. When $\rho_k \leq 0.1$, Δ_k is reduced.

When the optimal solution is solved, the point cloud can be transformed with

$$\boldsymbol{M}_{\mathrm{tr}}{}^{*} = \begin{bmatrix} \cos\alpha^* & -\sin\alpha^* & 0 & dx^* \\ \sin\alpha^* & \cos\alpha^* & 0 & dy^* \\ 0 & 0 & 1 & dz^* \\ 0 & 0 & 0 & 1 \end{bmatrix} \tag{24}$$

4. 3D Modelling

To better visualize the tooth surface deviations using 3D software and further carry out contact simulation by FEM [40–43], the 3D model of real five-pairs-of-teeth face gears needs to be modeled. For complex surface reconstruction, non-uniform rational b-spline (NURBS) [44–46] is the common modeling method. However, the disorderly scattered points make it difficult to reconstruct the face gear tooth surface accurately. Therefore, it is necessary to determine the regularly arranged points in the range of the tooth surface and label the points with serial numbers.

As shown in Figure 5a, the theoretical points arranged equidistantly within the range of tooth surface are calculated first according to the equation of face gear tooth surface [39], and (a,b) is used to record the sequence number of points. The nearest points $\boldsymbol{cp}$ from the point cloud are found by calculating the minim Euler distance E from the theoretical points $\mathbf{tp}_{(a,b)}$, and $\boldsymbol{cp}$ will be numbered as same as $\mathbf{tp}_{(a,b)}$, as follows.

$$\mathrm{minE} = \|\boldsymbol{cp}_{(a,b)} - \mathbf{tp}_{(a,b)}\|_2 \tag{25}$$

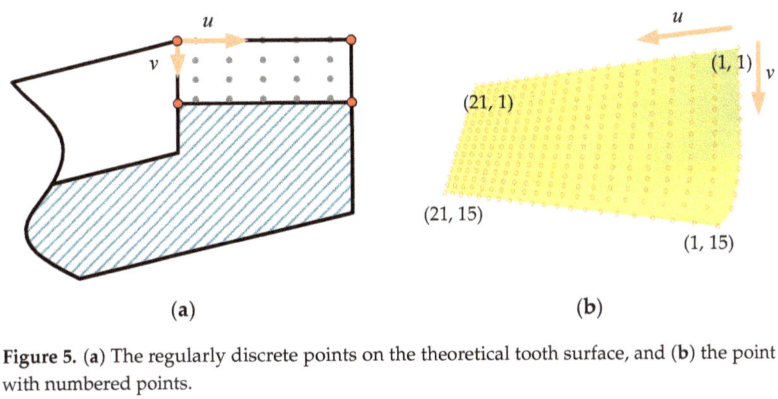

Figure 5. (**a**) The regularly discrete points on the theoretical tooth surface, and (**b**) the point cloud with numbered points.

The point cloud with numbered points is obtained, as shown in Figure 5b. Subsequently, the tooth surface is represented as a bicubic NURBS surface, as

$$S(u,v) = \frac{\sum_{i=0}^{m}\sum_{j=0}^{n} N_{i,3}(u)N_{j,3}(v)\omega_{i,j}d_{i,j}}{\sum_{i=0}^{m}\sum_{j=0}^{n} N_{i,3}(u)N_{j,3}(v)\omega_{i,j}} \tag{26}$$

where

$$\begin{cases} N_{i,0}(u) = \begin{cases} 1, u_i \le u < u_{i+1} \\ 0, \text{otherwise} \end{cases} \\ N_{i,p}(u) = \frac{u-u_i}{u_{i+p}-u_i}N_{i,p-1}(u) + \frac{u_{i+p+1}-u}{u_{i+p+1}-u_{i+1}}N_{i+1,p-1}(u), p = 1,2,3 \end{cases} \tag{27}$$

$$\begin{cases} N_{j,0}(v) = \begin{cases} 1, v_j \le v < v_{j+1} \\ 0, \text{otherwise} \end{cases} \\ N_{j,p}(v) = \frac{v-v_j}{v_{j+p}-v_j}N_{j,p-1}(v) + \frac{v_{j+p+1}-v}{v_{j+p+1}-v_{j+1}}N_{j+1,p-1}(v), p = 1,2,3 \end{cases} \tag{28}$$

u is along the radius of the face gear, and v is along the depth of the tooth. m and n are the number of control vertex of u and v, respectively. Set $m = 21$ and $n = 15$. $\omega_{i,j}$ is the weight factor. $N_{i,3}$ and $N_{j,3}$ are the basis functions of u and v, respectively. $d_{i,j}$ is the control vertex.

According to Equation (26), a tooth surface can be reconstructed, and a tooth space surface is composed of two tooth surfaces. Another tooth surface can be reconstructed by the same method. Then, the tooth space model can be obtained by Boolean operation between the tooth space and the face gear blank. Further, the 3D model of real five-pairs-of-teeth face gears can be obtained by segmenting the blank with the other tooth space surfaces, as shown in Figure 6.

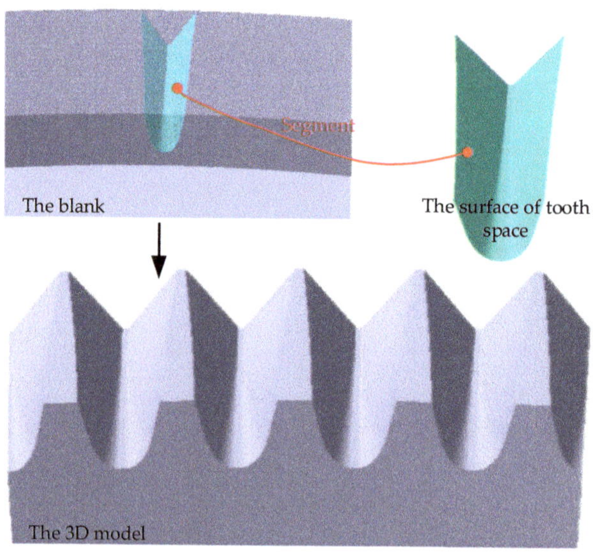

Figure 6. A 3D model of real five-pairs-of-teeth face gear.

5. Example and Discussion

Experiment with point cloud collection on the workpiece of face gear machined by grinding is conducted as stated in Section 2, and the parameters of a face gear are shown in Table 2. After being downsampled, the point cloud with 40,377 points is obtained. Subsequently, the point cloud is aligned using the measurement solution proposed in Section 3. In Figure 7a, the curve of objective value versus iteration number shows that the objective value of Equation (7) can reach the set accuracy at the 775th iteration. Then, the approximation of the solution at each iteration to the solution at the 775th iteration is defined by the 2-norm distance between them in the parametric space. The curve in Figure 7b shows that the solution tends to converge. Figure 8a shows that the change of the objective value of Equation (15) is extremely little after the 1000th iteration, reaching below 10^{-4} mm^2. Figure 8b shows that the approximation between the solution at each iteration and the solution at the 1000th iteration is convergent. The convergence accuracy achieved at the 1000th iteration has almost no effect on the measurement results requiring micron-level accuracy. The above shows that 3SA has convergence characteristics.

Table 2. The parameters of a face gear.

Parameters	Values
Shaper teeth number N_s	22
Pinion teeth number N_1	19
Face gear teeth number N_2	142
Module m	1.95 mm
Pressure angle of the rack cutter α	25°
Shaft angle γ_m	90°
Inner radius L_1	128 mm
Outer radius L_2	152 mm

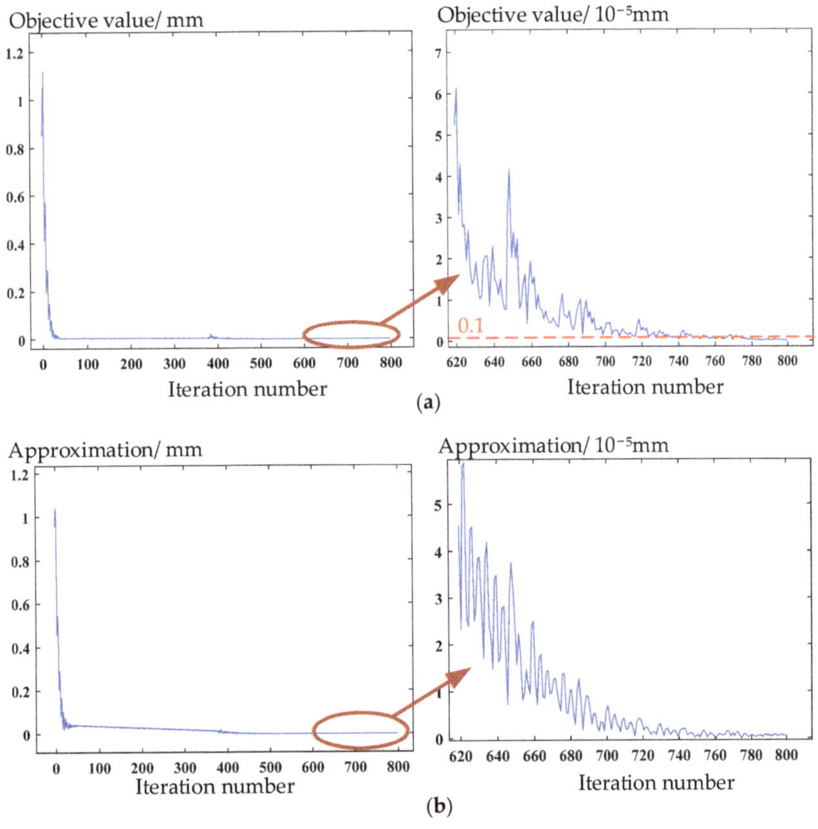

Figure 7. Convergence curves of (**a**) objective value and (**b**) iterative solution in the second stage of the algorithm.

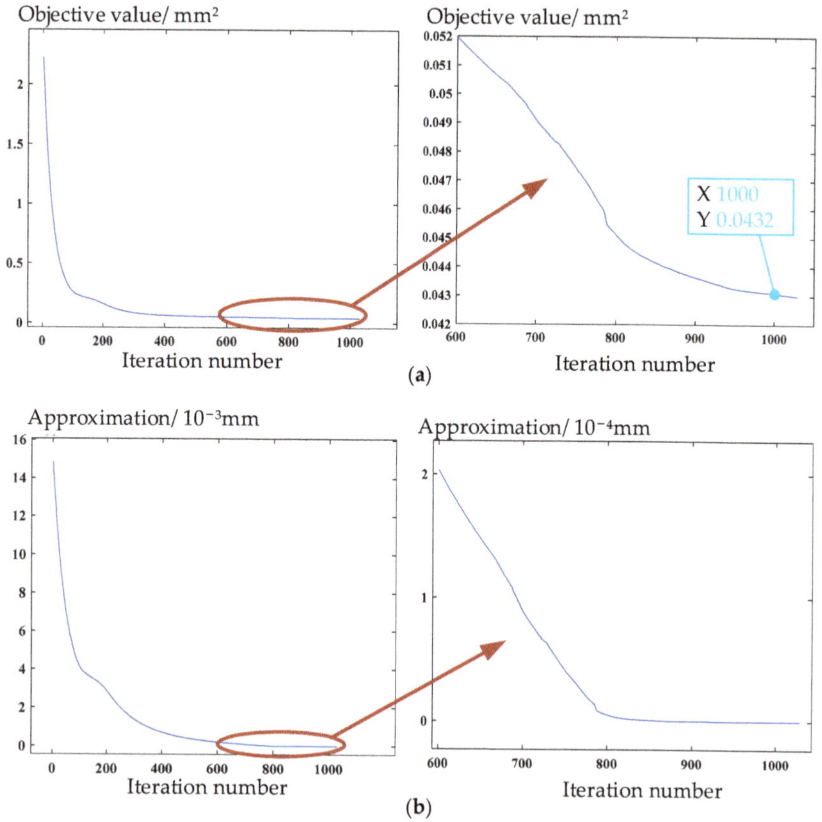

Figure 8. Convergence curves of (**a**) objective value and (**b**) iterative solution in the third stage of the algorithm.

As shown in Table 3, the objective value after the second stage of 3SA reaches 9.537×10^{-7} mm, which is a great improvement compared with 0.121 mm at the beginning, and the objective value of the tooth surface in the third stage increases from 2.23 mm^2 to 0.0457 mm^2.

Table 3. The comparison of objective value before and after computing in 3SA.

		Second Stage	Third Stage
Objective value	Before	0.868 mm	2.23 mm^2
	After	9.537×10^{-7} mm	0.0043 mm^2

The tooth surface reconstructed from the point cloud is compared with the theoretical tooth surface in the CATIA, as shown in Figure 9a. When viewing from top to bottom, the left side of the tooth is defined as the left tooth surface, and the right side is the right tooth surface. The result is stated as follows.

(1) Maximum deviation, not more than 0.030 mm, is on the border areas. The deviations above 0.022 mm are extremely few, and most are below 0.022 mm.

(2) The tooth surface deviations are inclined. The deviations of the inner side on the left tooth surface are negative, the deviations of the middle area trend to 0, and the deviations of the outer side are positive. It indicates that the inner side is overcut, and the outer side is undercut. While the deviations of the inner side on the right tooth surface are

positive, and the deviations of the outer side are negative, which indicates that the inner side is undercut, and the outer side is overcut.

(3) The deviation trends of five teeth are almost the same.

The probable reason for the above result of deviation is that there is an angle error in the installation of the tool. Compared the measurement result of the proposed method with that of GMC shown in Figure 9b, the trend of deviation distribution is consistent. Although the maximum deviation is larger, it is distributed in the border areas that cannot be measured by GMC. The error between tooth surface deviations obtained by GMC and the proposed method is less than 7 μm, which meets the measurement requirements for most face gears. It demonstrates that the measurement solution with 3D optical scanning is valid.

Moreover, the measurement solution proposed in the paper can show deviations in the border areas and more details of deviations. For example, the deviations increase significantly in some local areas shown in Figure 9a, which may be caused by the profile error of the worm wheel.

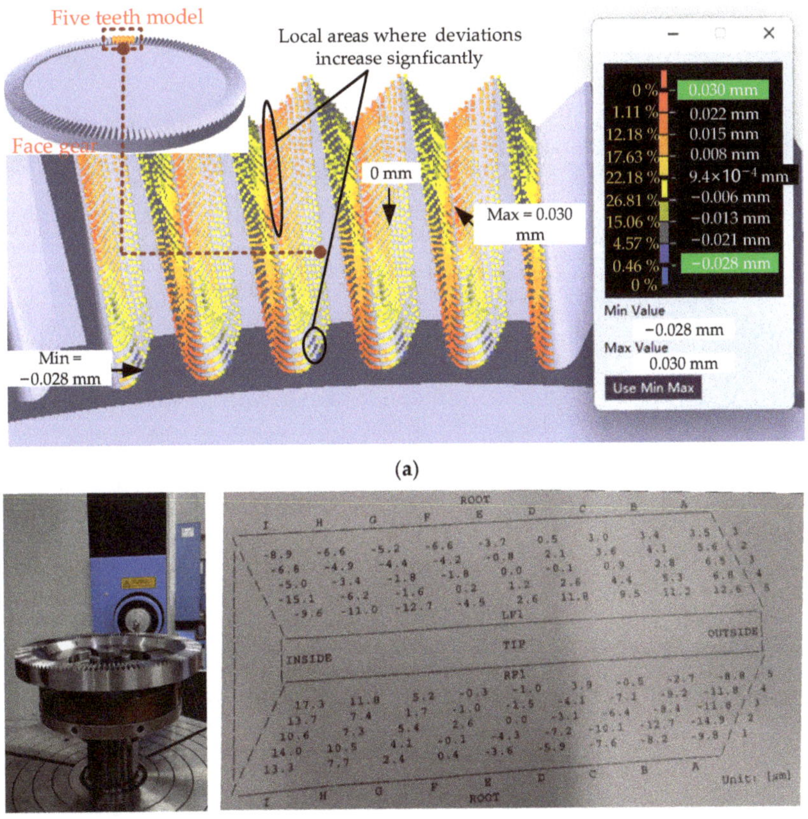

Figure 9. (**a**) Holistic deviation of the tooth surfaces obtained by the proposed method and (**b**) the measurement report of GMC.

In order to further study the effect of the deviations on the contact performance of the face gear drives in this example, the 3D model of real five-pairs-of-teeth face gear is obtained according to Section 4, and LTCA is conducted further by FEM. It should be noted that the model here contains the tooth surface deviations, and the deviations of different teeth are a little different, so the previous method cannot be used to generate the mesh. In the previous mesh generation method, the finite element meshes of a tooth model are generated first, and the finite element meshes of other teeth models are obtained by rotation. While the five teeth model contains tooth surface deviations, the finite element meshes of every tooth needs to be generated, and all finite element meshes are combined, as shown in Figure 10a. Due to the proven processing technology of the pinions, the designed model of the pinion without deviation in this example is applied, and the face gear drive is assembled, as shown in Figure 10b. A torque load of 160,000 N·mm is applied to the pinion to conduct the simulated meshing process of the real face gear drive.

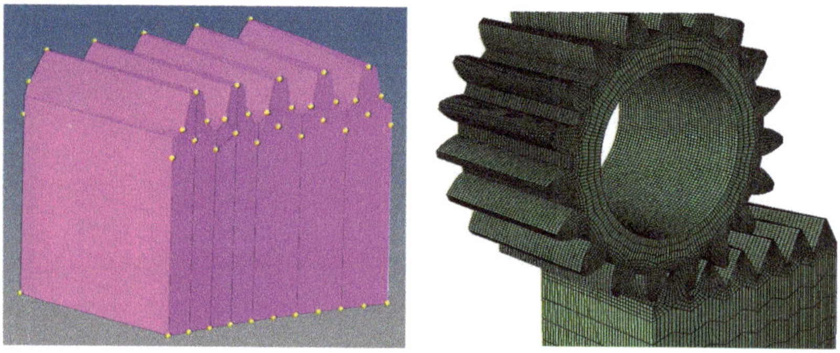

(**a**) (**b**)

Figure 10. FEM: (**a**) Mesh generation, (**b**) the finite element model of real five-pairs-of-teeth face gear drive.

The result of combining all the simulation frames is shown in Figure 11a,b and Figure 12a,b. The first and fifth teeth are the incoming and outgoing teeth of the model, respectively, so they are excluded for comparison. Figure 11 shows the contact pattern of real face gear tends to the area with the maximum tooth surface deviation and the outer side, which will cause edge contact. The maximum contact stress of real face gear is about 200 Mpa, larger than that of designed face gear. Moreover, the maximum contact stress of real face gear occurs in the border areas, which is the area with the maximum deviation meanwhile. The reason is that the "bulge" formed by the positive deviations of the tooth surface in this example will make contact earlier than the original contact points during the meshing of face gear drives. Further, it destroys the original contact path, and the "bulge" bears more contact load so that the contact stress is more concentrated in the border areas.

From Figure 12, the root bending stress of real face gear is 330.9 Mpa, while that of designed face gear is about 200 Mpa. The root bending stress increased by about 0.65 times, which will greatly affect the service life of face gear. The deviation difference between the different teeth has almost no influence on the meshing of face gear drive in this example.

The result of LTCA shows that the deviations in the border areas cause the change of contact pattern, contact stress, and root bending stress. The specific influence in this example is stated as follows.

(1) Edge contact occurs due to the deviations. The real face gear drive cannot make smooth contact as the designed model, and the tooth surfaces of the face gear and pinion collide with each other, which will cause vibration and noise. Hence, the stability of the transmission of the face gear drive will be reduced.

(2) Edge contact causes the contact stress concentrated in the border areas, and contact stress is increased by approximately 0.4, resulting in more easily occurring tooth surface pitting failure. The root bending stress is increased by approximately 0.65, resulting in more easily occurring tooth root breakage failure. The increase in contact stress and root bending stress will reduce the service life of the face gear.

Therefore, further processing should be conducted for the areas with large positive deviations, especially the border areas of the tooth surface. A simple method is to use an end mill cutter to mill the border areas of the tooth surface. The more flexible and practical method is to correct the settings of the grinding machine tool to reduce the machining errors or to modify the profile of the grinding tool to process the higher-performance tooth surface according to the measurement result of the proposed method.

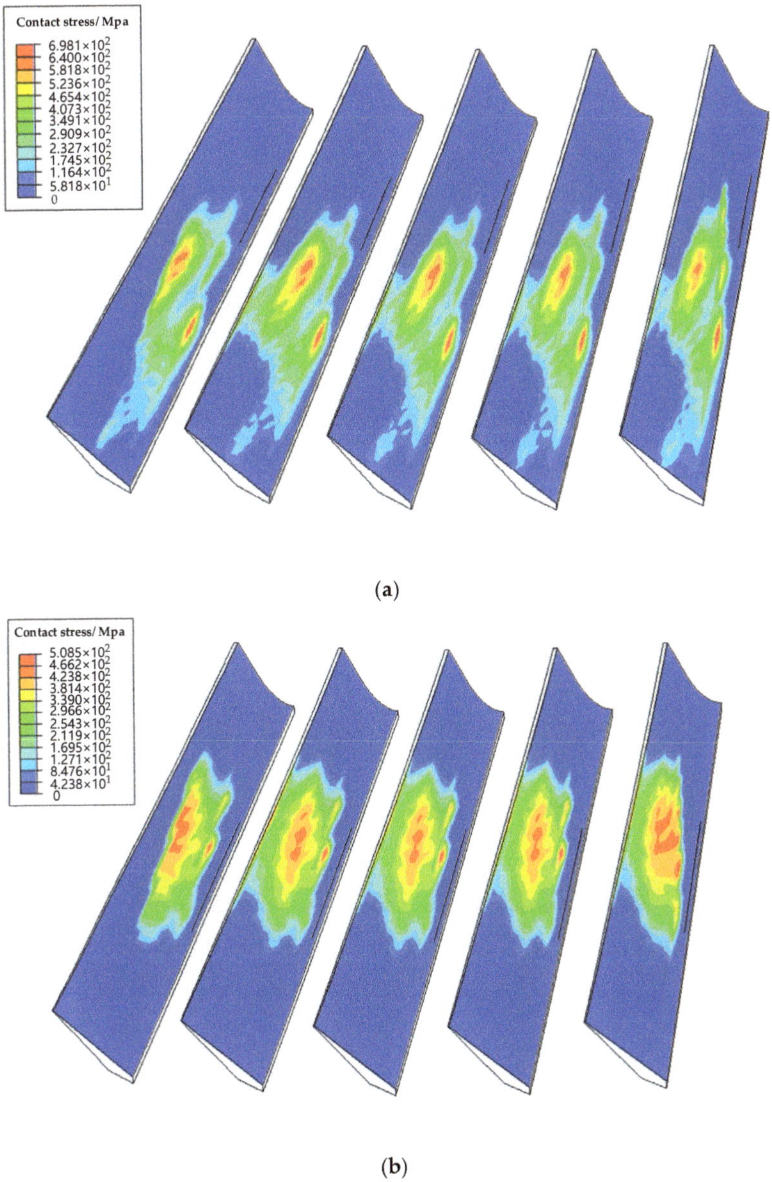

Figure 11. The simulation results for the contact pattern and contact stress of (**a**) the real face gear drive and (**b**) the designed face gear drive.

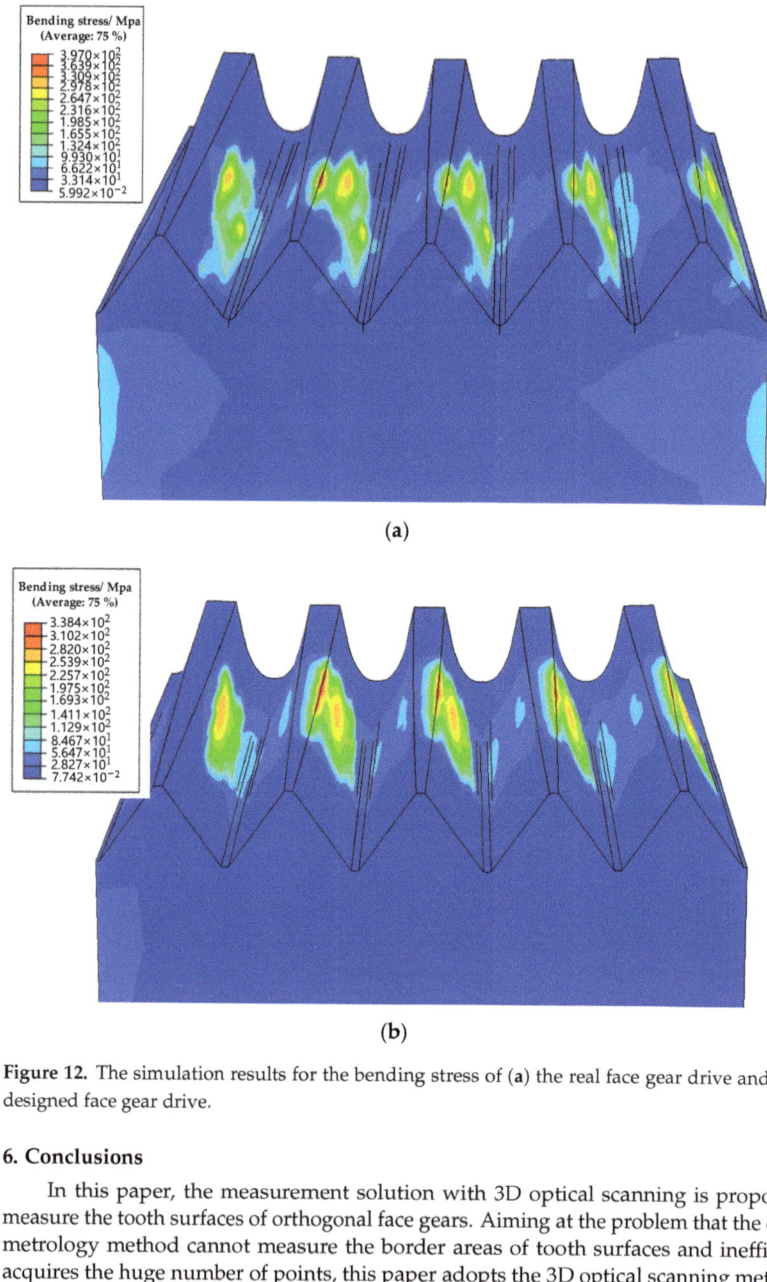

Figure 12. The simulation results for the bending stress of (**a**) the real face gear drive and (**b**) the designed face gear drive.

6. Conclusions

In this paper, the measurement solution with 3D optical scanning is proposed to measure the tooth surfaces of orthogonal face gears. Aiming at the problem that the contact metrology method cannot measure the border areas of tooth surfaces and inefficiently acquires the huge number of points, this paper adopts the 3D optical scanning method to acquire point clouds, and the measurement solution is proposed to ensure measurement accuracy for the tooth surfaces. A 3SA is presented to align the features between point clouds and the designed face gears, including top land surfaces, centers of circles, and tooth surfaces. Moreover, the 3D modeling method is studied by reconstructing the real tooth surface with the points after being numbered. An example is given to show that the measurement solution with 3D optical scanning is valid. With the comparison, the proposed measurement solution can show the deviations in the border areas and more details that the previous measurement method could not reveal. LTCA is further conducted to study the meshing performance of the real tooth surface. The results of LTAC show that the deviations in the border areas cause edge contact, which will significantly reduce the contact strength and bending strength of real face gear. Summarily, the proposed measurement solution can provide a more reliable base for correction processing.

The proposed measurement solution can meet the measurement required for the tooth surfaces of face gears applied in automobile and other industries. In the future, the numerical algorithm of precise multi-point cloud splicing will be studied to measure face gears with all teeth, and the measurements of more precision face gears that can be applied to the aerospace industry will be conducted.

Author Contributions: Conceptualization, X.L. and Y.Z.; methodology, X.L.; software, X.L. and Z.C.; validation, X.Z. and B.H.; formal analysis, X.L.; investigation, X.L., X.Z. and B.H.; resources, X.Z., Y.Z. and J.T.; data curation, X.Z., B.H. and Y.Z.; writing—original draft preparation, X.L.; writing—review and editing, X.Z. and B.H.; visualization, X.L. and X.Z.; supervision, Y.Z. and J.T.; project administration, Y.Z. and J.T.; funding acquisition, Y.Z. and J.T. All authors have read and agreed to the published version of the manuscript.

Funding: This research was funded by the National Science and Technology Major Project of China through Grant No. 2019-VII-0017-0158, the National Natural Science Foundation of China through Grant No. 51705541, Industrial Technology Development Program No. JCKY2020213B006, Hunan Provincial Natural Science Foundation of China through Grant No. 2021JJ20071, and State Key Laboratory of High Performance Complex Manufacturing through Grant No. Kfkt2019-08 and No. ZZYJKT2020-06.

Institutional Review Board Statement: Not applicable.

Informed Consent Statement: Not applicable.

Data Availability Statement: Not applicable.

Acknowledgments: The authors gratefully acknowledge the support from the National Science and Technology Major Project of China, National Natural Science Foundation of China, Industrial Technology Development Program, Hunan Provincial Natural Science Foundation of China, State Key Laboratory of High Performance Complex Manufacturing, and the assistance of Zhimin Guo in the revised version of the manuscript.

Conflicts of Interest: The authors declare no conflict of interest.

References

1. Litvin, F.L.; Fuentes, A.; Howkins, M. Design, generation and TCA of new type of asymmetric face-gear drive with modified geometry. *Comput. Methods Appl. Mech. Eng.* **2001**, *190*, 5837–5865. [CrossRef]
2. Litvin, F.L.; Wang, J.-C.; Bossler, R.B., Jr.; Chen, Y.-J.D.; Heath, G.; Lewicki, D.G. Application of face-gear drives in helicopter transmissions. *J. Mech. Des.* **1994**, *116*, 672–676. [CrossRef]
3. Lu, X.; Zhou, Y.; He, D.; Zheng, F.; Tang, K.; Tang, J. A novel two-variable optimization algorithm of TCA for the design of face gear drives. *Mech. Mach. Theory* **2022**, *175*, 104960. [CrossRef]
4. Mo, S.; Song, W.; Zhu, S.; Feng, Z.; Tang, W.; Gao, H. Complex geometric modeling and tooth contact analysis of a helical face gear pair with arc-tooth. *J. Cent. South Univ.* **2022**, *29*, 1213–1225. [CrossRef]
5. Mo, S.; Zhang, Y.; Luo, B.; Bao, H.; Cen, G.; Huang, Y. The global behavior evolution of non-orthogonal face gear-bearing transmission system. *Mech. Mach. Theory* **2022**, *175*, 104969. [CrossRef]
6. Mo, S.; Luo, B.; Song, W.; Zhang, Y.; Cen, G.; Bao, H. Geometry design and tooth contact analysis of non-orthogonal asymmetric helical face gear drives. *Mech. Mach. Theory* **2022**, *173*, 104831. [CrossRef]
7. Tang, J.; Cui, W.; Zhou, H.; Yin, F. Integrity of grinding face-gear with worm wheel. *J. Cent. South Univ.* **2016**, *23*, 77–85. [CrossRef]
8. Wang, Y.; Chu, X.; Zhao, W.; Wang, Z.; Su, G.; Huang, Y. A precision generating hobbing method for face gear with assembly spherical hob. *J. Cent. South Univ.* **2019**, *26*, 2704–2716. [CrossRef]
9. Zschippang, H.A.; Weikert, S.; Wegener, K. Face-gear drive: Simulation of shaping as manufacturing process of face-gears. *Mech. Mach. Theory* **2022**, *172*, 104791. [CrossRef]
10. Zhou, Y.; Tang, Z.; Shi, X.; Tang, J.; Li, Z. Efficient and accurate worm grinding of spur face gears according to an advanced geometrical analysis and a closed-loop manufacturing process. *J. Cent. South Univ.* **2022**, *29*, 1–13. [CrossRef]
11. Tang, Z.; Zhou, Y.; Wang, S.; Zhu, J.; Tang, J. An innovative geometric error compensation of the multi-axis CNC machine tools with non-rotary cutters to the accurate worm grinding of spur face gears. *Mech. Mach. Theory* **2022**, *169*, 104664. [CrossRef]

12. Xu, H.; Zhou, Y.; He, Y.; Tang, J. An efficient approach to the five-axis flank milling of non-ferrous spiral bevel gears. *Materials* **2021**, *14*, 4848. [CrossRef] [PubMed]
13. Litvin, F.L.; Fuentes, A. *Gear Geometry and Applied Theory*; Cambridge University Press: Cambridge, UK, 2004.
14. Zhou, Z.; Zhang, Y.; Tang, K. Sweep scan path planning for efficient freeform surface inspection on five-axis CMM. *Comput.-Aided Des.* **2016**, *77*, 1–17. [CrossRef]
15. Wang, S.; Zhou, Y.; Tang, J.; Tang, K.; Li, Z. Digital tooth contact analysis of face gear drives with an accurate measurement model of face gear tooth surface inspected by CMMs. *Mech. Mach. Theory* **2022**, *167*, 104498. [CrossRef]
16. Lin, C.-H.; Fong, Z.-H. Numerical tooth contact analysis of a bevel gear set by using measured tooth geometry data. *Mech. Mach. Theory* **2015**, *84*, 1–24. [CrossRef]
17. Ding, H.; Wan, Z.; Zhou, Y.; Tang, J. A data-driven programming of the human-computer interactions for modeling a collaborative manufacturing system of hypoid gears by considering both geometric and physical performances. *Robot. Comput.-Integr. Manuf.* **2018**, *51*, 121–138. [CrossRef]
18. Stadtfeld, H.J. *Handbook of Bevel and Hypoid Gears: Calculation, Manufacturing, Optimization*; Rochester Institute of Technology: Rochester, NY, USA, 1993.
19. Rong, K.; Ding, H.; Chen, S.; Wang, Y.; Zhou, Z.; Lu, R. Top-Rem grinding tool modification considering loaded edge contact for spiral bevel gears. *Adv. Eng. Inform.* **2022**, *53*, 101697. [CrossRef]
20. Kuş, A. Implementation of 3D optical scanning technology for automotive applications. *Sensors* **2009**, *9*, 1967–1979. [CrossRef]
21. Hawryluk, M.; Ziemba, J. Application of the 3D reverse scanning method in the analysis of tool wear and forging defects. *Measurement* **2018**, *128*, 204–213. [CrossRef]
22. Valigi, M.C.; Logozzo, S.; Affatato, S. New challenges in tribology: Wear assessment using 3D optical scanners. *Materials* **2017**, *10*, 548. [CrossRef]
23. Chen, Y.-C.; Chen, J.-Y. Optical inspection system for gear tooth surfaces using a projection moiré method. *Sensors* **2019**, *19*, 1450. [CrossRef]
24. Gonzalez-Perez, I.; Fuentes-Aznar, A. Tooth contact analysis of cylindrical gears reconstructed from point clouds. In *New Approaches to Gear Design and Production*; Goldfarb, V., Trubachev, E., Barmina, N., Eds.; Springer: Cham, Switzerland, 2020; pp. 219–237.
25. Urbas, U.; Zorko, D.; Černe, B.; Tavčar, J.; Vukašinović, N. A method for enhanced polymer spur gear inspection based on 3D optical metrology. *Measurement* **2021**, *169*, 108584. [CrossRef]
26. Urbas, U.; Hrga, T.; Povh, J.; Vukašinović, N. Novel alignment method for optical 3D gear metrology of spur gears with a plain borehole. *Measurement* **2022**, *192*, 110839. [CrossRef]
27. Brajlih, T.; Valentan, B.; Drstevenser, I.; Balic, J.; Pogacar, V. Testing the accuracy of ATOS[TM] 3d optical scanner measuring volumes. *Ann. DAAAM Proc.* **2007**, *2007*, 111–113.
28. Hwang, J.; Yu, S.; Kim, J.; Lee, S. 3D face modeling using the multi-deformable method. *Sensors* **2012**, *12*, 12870–12889. [CrossRef]
29. Vagovský, J.; Buranský, I.; Görög, A. Evaluation of measuring capability of the optical 3D scanner. *Procedia Eng.* **2015**, *100*, 1198–1206. [CrossRef]
30. Hawryluk, M.; Ziemba, J.; Zwierzchowski, M.; Janik, M. Analysis of a forging die wear by 3D reverse scanning combined with SEM and hardness tests. *Wear* **2021**, *476*, 203749. [CrossRef]
31. Zhong, K.; Li, Z.; Li, R.; Shi, Y.; Wang, C. Pre-calibration-free 3D shape measurement method based on fringe projection. *Opt. Express* **2016**, *24*, 14196–14207. [CrossRef]
32. Huang, J.; Wu, Q. A new reconstruction method based on fringe projection of three-dimensional measuring system. *Opt. Lasers Eng.* **2014**, *52*, 115–122. [CrossRef]
33. He, W.; Zhong, K.; Li, Z.; Meng, X.; Cheng, X.; Liu, X.; Shi, Y. Accurate calibration method for blade 3D shape metrology system integrated by fringe projection profilometry and conoscopic holography. *Opt. Lasers Eng.* **2018**, *110*, 253–261. [CrossRef]
34. Valagiannopoulos, C.A. On smoothening the singular field developed in the vicinity of metallic edges. *Int. J. Appl. Electromagn. Mech.* **2009**, *31*, 67–77. [CrossRef]
35. Young, I.T.; van Vliet, L.J. Recursive implementation of the Gaussian filter. *Signal Process.* **1995**, *44*, 139–151. [CrossRef]
36. Al-Rawabdeh, A.; He, F.; Habib, A. Automated feature-based down-sampling approaches for fine registration of irregular point clouds. *Remote Sens.* **2020**, *12*, 1224. [CrossRef]
37. Rizk-Allah, R.M.; Hassanien, A.E. A hybrid equilibrium algorithm and pattern search technique for wind farm layout optimization problem. *ISA Trans.* 2022, *in press*. [CrossRef]
38. Melo, I.D.; Antunes, M.P. Bad data correction in harmonic state estimation for power distribution systems: An approach based on generalised pattern search algorithm. *Electr. Power Syst. Res.* **2022**, *204*, 107684. [CrossRef]
39. Zhou, Y.; Wu, Y.; Wang, L.; Tang, J.; Ouyang, H. A new closed-form calculation of envelope surface for modeling face gears. *Mech. Mach. Theory* **2019**, *137*, 211–226. [CrossRef]
40. Barone, S.; Borgianni, L.; Forte, P. Evaluation of the effect of misalignment and profile modification in face gear drive by a finite element meshing simulation. *J. Mech. Des.* **2004**, *126*, 916–924. [CrossRef]
41. Pietrusiak, D.; Wróbel, J.; Czechowski, M.; Fiebig, W. Dynamic NVH numerical analysis of power steering in the presence of lubricant in the system. *Materials* **2022**, *15*, 2406. [CrossRef]

42. Fu, X.; Wang, B.; Wang, T. Experimental and numerical investigations on the deformation defects by gear rolling process with local induction heating. *Int. J. Adv. Manuf. Technol.* **2021**, *117*, 835–844. [CrossRef]
43. Ge, Y.; Liu, Z.; Zhou, K.; Yang, J.; Liu, D. Investigation of the electromagnetic characteristics and operating performance of a bidirectional PM excited machine. *Sci. Rep.* **2022**, *12*, 8408. [CrossRef]
44. Piegl, L.; Tiller, W. *The NURBS Book*; Springer: Berlin/Heidelberg, Germany, 1996.
45. Oesterle, B.; Geiger, F.; Forster, D.; Fröhlich, M.; Bischoff, M. A study on the approximation power of NURBS and the significance of exact geometry in isogeometric pre-buckling analyses of shells. *Comput. Methods Appl. Mech. Eng.* **2022**, *397*, 115144. [CrossRef]
46. Pham-Tan, H.; Thai, C.H.; Phung-Van, P. NURBS-based refined plate theory for metal foam plates with porosities. *Thin-Walled Struct.* **2022**, *175*, 109246. [CrossRef]

materials

MDPI

Article

Rough Surface Characterization Parameter Set and Redundant Parameter Set for Surface Modeling and Performance Research

Duo Yang [1,2], Jinyuan Tang [1,2,*], Fujia Xia [1,2] and Wei Zhou [3]

1 State Key Laboratory of High Performance Complex Manufacturing, Central South University, Changsha 410017, China
2 College of Mechanical and Electrical Engineering, Central South University, Changsha 410017, China
3 Hunan Provincial Key Laboratory of High Efficiency and Precision Machining of Difficult-to-Cut Material, Hunan University of Science and Technology, Xiangtan 411199, China
* Correspondence: jytangcsu_312@163.com or jytang@csu.edu.cn

Abstract: Among the 26 roughness parameters described in ISO 25178 standard, the parameters used to characterize surface performance in characterization parameter set (**CPS**) lack scientificity and unity, resulting in application confusion. The current **CPS** comes from empirical selection or small sample experiments, thus featuring low generality. A new method for constructing **CPS** in rough surfaces is proposed to solve the above issues. Based on a data mining method, statistical theory, and roughness parameters definitions, the 26 roughness parameters are divided into **CPS** and redundant parameter sets (**RPS**) with the help of reconstructed surfaces and machining experiments, and the mapping relationships between **CPS** and **RPS** are established. The research shows that **RPS** accounts for 50%, and **CPS**, of great significance for surface performance, and has the ability to fully cover surface topography information. The birth of **CPS** provides an accurate parameter set for the subsequent study of different surface performance, and it provides more effective parameters for evaluating the workpiece surface performance from the same batch.

Keywords: roughness surface; morphological characterization; characterization parameter set; redundant parameter set; ISO25178

Citation: Yang, D.; Tang, J.; Xia, F.; Zhou, W. Rough Surface Characterization Parameter Set and Redundant Parameter Set for Surface Modeling and Performance Research. *Materials* **2022**, *15*, 5971. https://doi.org/10.3390/ma15175971

Academic Editors: Pawel Pawlus and Ştefan Ţălu

Received: 24 July 2022
Accepted: 26 August 2022
Published: 29 August 2022

1. Introduction

With the rapid development of the modern precision manufacturing industry, products have shifted from macroscopic shape control to microscopic shape collaborative design and manufacturing. Surface roughness, as a part of the micro-structural information, can significantly affect surface sealing, wear, fatigue, etc. [1–3]. In this sense, it is of great significance to carry out surface roughness characterization research [4]. Due to the complex information covered by surface roughness, researchers have tried to formulate some roughness parameters to describe different surface morphology characteristics. Therefore, surface roughness characterization is transformed into a correlation study among roughness parameters.

In 1929, G. Schmalz in Germany was the first to propose the surface roughness measurement baseline and evaluation coefficients, which provided an effective index to open the era of quantitative evaluation in surface roughness [5]. Subsequently, by constructing the curve between the surface peak-valley depth and the bearing length rate, Abbott formulated some roughness parameters to display the surface roughness information from different angles [6]. After that, other countries began to define their national surface roughness standards on the basis of their industrial application scenarios and actual production needs. The symbols and definitions of the parameters are of great difference, thus resulting in bad situations such as small application range, low reliability and versatility, the phenomenon of "parameter big bang" [7]. The number of parameters reached more than one hundred at one time [8] in standards.

In order to unify roughness parameters standards, the technical committee of the International Organization for Standardization (ISO) carried out a revision of the surface topography standard specification in 1996, and divided roughness parameters into seven categories based on the definitions [9]. However, the definitions of two-dimensional (2D) roughness parameters were based on the root mean square of the profile section height and could only cover the surface characteristic information in the X and Y directions. It inevitably led to the loss of topographic information and could not achieve a good characterization of three-dimensional (3D) roughness surface.

With the development of surface topography measurement and computer digital technology, the lossy stylus measurement in surface roughness was gradually replaced by the non-destructive optical measurement. This great improvement provided reliable technical support for the birth of 3D surface roughness parameters based on topography mid-surface. The University of Birmingham took the lead in defining 3D roughness parameters, later called "Birmingham 14 parameters" [10], to lay the foundation for subsequent surface roughness standards. In 2010, the ISO extended the "Birmingham 14 parameters" to ISO 25178 3D roughness parameters standards with the help of two-dimensional standards and topography measurement technology, and the categories were upgraded to six: height parameters, spatial parameters, hybrid parameters, functional parameters, volume parameters, and feature parameters [11,12].

However, the increasement and extension from 2D to 3D led to complex internal correlations and a low degree of matching between their definitions and categories. Some parameters even own repeated characterization information. This means that some of 3D roughness parameters in ISO 25178 are not developed through rigorous research and do not actually contain valid information to characterize surface performance. As roughness parameters are widely used in the study of surface performance and once there is some deviation in the selection of roughness parameters or the roughness parameters themselves do not have valid information, the results will be unreliable. It is key to establish a reasonable characterization method to find the roughness parameters suitable for the subsequent research of various surface performance.

In recent years, researchers have carried out internal correlation analysis in 3D roughness parameters and explored the correlation between 3D roughness parameters and surface performance. For example, Franco studied the correlation of S-series and V-series parameters in ISO 25178 standard, and believed that there was information redundancy between ***Spk*** and ***Vmp*** [13]. Pawlus carried out the study on the correlation of two-dimensional roughness height parameters, and elucidated that there was a strong positive correlation between ***Ra*** and ***Rq*** [14]. Qi et al. used the Spearman rank correlation coefficient to judge the internal correlation of the six categories in 3D roughness parameters, and established the parameter rank tree to distinguish the correlation strength [15]. However, the above work either involves an incomplete number of parameters, or the correlation analysis method just fits in dividing the roughness parameters according to the correlation level and cannot accurately distinguish the correlation strength in the same level.

Therefore, some researchers tried to select reasonable 3D **CPS** based on surface performance and parameter application frequency. M. Sedlaček et al. studied the correlation of ***Sq***, ***Ssk*** and ***Sku*** in the height parameters with the surface friction coefficient and clarified the influence of these three roughness parameters on the friction performance [16–18]. B. He et al. found the relationships between the critical load and 3D roughness parameters ***Spc***, ***Sq***, ***Vvc***, ***Sdq*** in the micro-connected structure, and gave application range of ***Spc***, ***Sq***, ***Vvc***, ***Sdq*** [19]. Zeng et al. combined the surface height probability density function, autocorrelation function and height parameters to evaluate the correlation of 2D and 3D roughness parameters to surface bearing capacity, friction and lubrication performance [20]. Todhunter analyzed the industry application of different roughness parameters by collecting the utilization frequency of the parameters from a total of 179 industrial users in 34 countries [21]. However, the method of selecting target roughness parameters only by engineering experience or application frequency lacks effective theoretical support. Besides, due to the large number of application scenarios and various surface properties (friction, wear, fatigue, lubrication, sealing, etc.), the performance screening method still results in "parameter big bang" to characterize surfaces. Therefore, it is urgent to design an effective characterization method to unify the selection range of target roughness parameters under different properties and ensure that there is no information redundancy in selection to avoid the deviation of research objectives.

It is generally acknowledged that each surface performance parameter is closely influenced by the surface geometry. Therefore, taking the surface geometry characterization research as the start, the roughness parameter characterization set will be established to characterize different performance parameters with full potential. Although there exist different definitions and expressions in ISO 25178, the data trend analysis shows that some roughness parameters fluctuate with others, and thus the information of these parameters is actually expressed by other parameters. If roughness parameters expressed by others are regarded as the "redundant parameter sets (**RPS**)" and the rest are classified as the "characterization parameter sets (**CPS**)", the comprehensive characteristics description of the surface topography by **CPS** will be realized with the removal of redundant information.

Combined with the data feature, Pearson correlation analysis is used to roughly delineate **CPS** and **RPS** selection. With the principle of statistical non-strong and non-weak introduced [22], the core roughness parameters (**CRP**) used to characterize different redundant parameters are selected from **CPS**. Based on the current situation that the polynomial regression model is prone to lead to redundant items, a new method of item number reduction is proposed to construct the optimal explicit expressions between **CRP** and **RPS**, which realizes the information coverage from **CRP** to **RPS**. The optimal explicit expression can automatically find roughness parameters with strong characterization and furtherly determine **RPS** and **CPS**. Finally, the research verifies the reliability of the optimal explicit expression by the reconstruction and the measured surfaces. The designed characterization method can provide the guidance and basis for selecting reasonable target **CPS** for industrial applications.

2. Basic Concepts and Research Methods

The overall method technical route is in Appendix A. The followings are the detailed introductions.

(1) Due to the large number and complex correlation of roughness parameters, and the instability under small samples, the paper uses the stochastic process theory and surface reconstruction technology to set the value interval of seven reconstruction coefficients, and randomly combines them in their respective intervals. Finally, 1000 sets of reconstructed surfaces are generated for the research data and it ensures the reliability (Section 2.3 for details);

(2) 1000 groups of surface roughness parameters are obtained by ISO25178 definitions, and six types of roughness parameters are initially divided into **CPS** and **RPS** by combining Pearson correlation analysis and non-strong and non-weak statistical principles. Considering that a single redundant parameter should not be characterized by all the parameters in **CPS**, **CRS** characterizing different redundant parameters are selected from **CPS** and the correctness of parameter sets is analyzed based on the parameters definitions (Sections 2.2 and 3.1 for details);

(3) The process in step 2 still cannot clear the quantitative characterization relationships between **CPS** and **RPS**. By means of polynomial regression model with strongly nonlinear characterization ability, a standard deviation of automatic pruning method is born. The method can automatically find roughness parameters of the surface with strong characterization ability and eliminate irrelevant factors. It realizes the explicit formula expression from **CPS** to **RPS** and uses the experiment to prove that **CPS** can cover the rough surface information (Sections 2.4, 3.2 and 3.3 for details);

(4) To illustrate the engineering significance of **CPS** on surface performance, the paper describes the application of **CPS** in surface performance research by means of roughness

parameter definition, neural network, sensitivity analysis, optimization algorithm, finite element calculation, etc. The reliability of these applications is discussed by the existing research, which provides the direction for the **CPS** application, and clarifies the practical significance (Section 3.4 for details).

2.1. 3D Roughness Parameter

Due to the complexity of 3D roughness surface information, ISO 25178 has defined a total of 26 main roughness parameters in six categories to characterize the surface roughness and describe the different topography features. The detailed definitions and descriptions can be found in literature [6,10]. The paper only briefly introduces their relevant symbols and definitions in Table 1.

Table 1. Roughness Parameters.

Category	Symbol	Definition
Height Parameters	***Sa***	Arithmetical mean height
	Sz	Maximum height
	Sq	Root mean square height
	Ssk	Skewness
	Sku	Kurtosis
	Sp	Maximum peak height
	Sv	Maximum pit depth
Hybrid parameters	***Sdq***	Root mean square gradient
	Sdr	Developed interfacial area ratio
Feature parameters	***Spd***	Density of peaks
	Spc	Arithmetic mean peak curvature
	S5p	Five-point peak height
	S5v	Five-point pit height
	S10z	Ten-point height of surface
Functions parameters	***Sk***	Core height
	Spk	Reduced peak height
	Svk	Reduced valley height
	Smr1	Material ratio in peak
	Smr2	Material ratio in valley
	Sxp	Peak extreme height
Volume parameters	***Vmp***	Peak material volume
	Vmc	Core material volume
	Vvc	Core void volume
	Vvv	Dale void volume
Space parameters	***Sal***	Autocorrelation length
	Str	Texture aspect ratio

2.2. Definition of Parameter Set

As the definitions and expressions of 3D roughness parameters vary greatly and some of them are defined based on experience, it is often difficult to analyze their internal correlation from definitions or formulas, which becomes the key point that has puzzled researchers for many years [22]. Differences always lie in some parameter formulas, but there exists an obvious data fluctuation trend between them. For example, there are great differences between the formulas of ***Sa*** and ***Sq***, but the research shows that they follow a linear trend with nearly 0.8 times between the two [23,24]. It reflects the fact that the information of some redundant parameters is actually represented by other parameters among the ISO 25178 standard, thus causing the application confusion in industrial production.

Therefore, for the sake of clarifying the quantitative correlation, the research tries to use data mining method and statistical theory to illustrate the correlation of different roughness parameters and find the roughness parameter set that truly characterizes the surface topography based on the idea that some roughness parameters follow the fluctuations of others. In order to facilitate the development of the work, the concepts of "redundant parameter sets (**RPS**)", "characterization parameter sets (**CPS**)", and "core roughness parameters (**CRP**)" are proposed and introduced in the following.

(1) Redundant parameter sets

The redundant parameter set is defined as the parameter set characterized by other roughness parameters. Specifically, it includes the parameters that can be determined by explicit expressions from others, so as to complete the information coverage and remove redundant information;

(2) Characterization parameter sets

The characterization parameter set is the parameter set used to predict **RPS**, and **CPS** is regarded as the set that truly characterizes the surface topography without redundant information, and is the research object with high correlation in different surface performance studies. Figure 1 shows the brief construction of **CPS**.

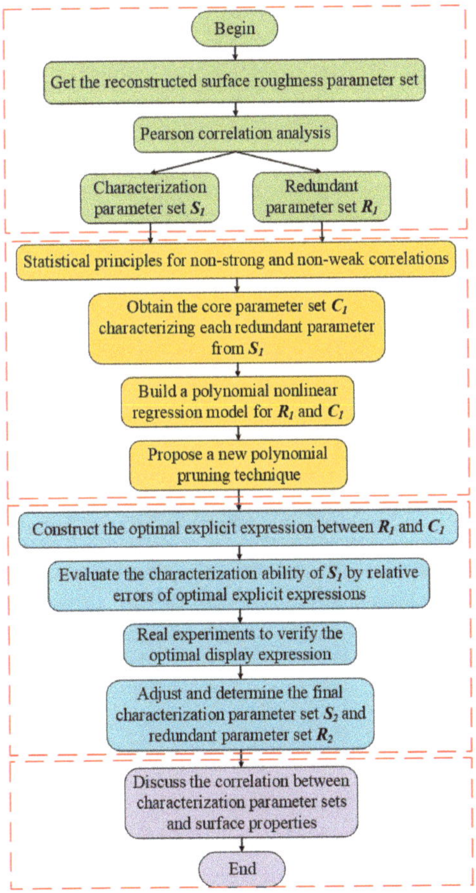

Figure 1. Construction of **CPS**.

a. Based on the unclear correlation dilemma, Pearson correlation analysis is used to roughly distinguish the correlation between different roughness parameters. The all parameters are divided into **CPS** and **RPS**;
b. Owing to the large number of characteristic parameters in the initial division, it is not conducive to constructing the optimal explicit expression between **CPS** and **RPS**. Therefore, the statistics principle of non-strong and non-weak is introduced to screen **CRP** from **CPS** for the sake of characterizing each redundant parameter and then the parameter in **CRP** is regarded as the independent variable in the optimal explicit expression;
c. A new polynomial pruning method is proposed to establish the optimal explicit expressions between **CRP** and **RPS**. The relative error of the expression is used to evaluate the information reflection ability from **CRP** to **RPS**, and finally the reliability of **RPS** and **CPS** is verified and adjusted based on real experiments;
d. After the reliability of **CPS** is verified based on theory and experiments, the correlation between the **CPS** and surface performance parameters is extended and analyzed to point out the practical significance of **CPS** in engineering research.

(3) Core roughness parameters

The core roughness parameters are parameters from **CPS**, furtherly used as the independent variables of the optimal explicit expression.

2.3. Surface Reconstruction

As a result of the complex internal correlation among surface roughness parameters, it is difficult to obtain accurate correlation if the number of research samples is small in characterization. A reliable parameter characterization method needs to be built on the analysis with a large number of surface samples to ensure its generality and stability. The large number of real surfaces experiments, especially at the early stage of the research, will result in a high cost of trial and error for researchers. And if the research is just based on experiments, it will easily lead to a sudden increase in time and money costs.

The utilization of numerical method to generate reconstructed surfaces with different roughness characteristics can avoid the above problems and provide a large number of surface samples in a short time. The grinding surface is the common representative one among non-Gaussian surfaces, so the paper takes the surface with grinding characteristics as the object. Its surface reconstruction is generally based on the surface height probability density function and the autocorrelation function. The following is the brief introduction to the reconstruction surface method.

According to the Johnson transformation method [25], the height probability density function can be obtained by three characteristic coefficients, while the autocorrelation function is controlled by four characteristic coefficients. These coefficients are defined as follows:

(1) Characteristic coefficient of height probability density function

The three characteristic coefficients k_1, k_2 and k_3 [26] of the height probability density function are the second, third and fourth order center distances of the reconstructed surface, respectively:

$$k_1 = \frac{1}{N}\sum(Z - \overline{Z})^2 \tag{1}$$

$$k_2 = \frac{1}{N}\sum(Z - \overline{Z})^3 \tag{2}$$

$$k_3 = \frac{1}{N}\sum(Z - \overline{Z})^4 \tag{3}$$

Here, Z is the surface height matrix; $\overline{Z}$ is the mean value of Z; N is the number of elements in the matrix.

(2) Characteristic coefficient of autocorrelation function

The four coefficients a_1, a_2, a_3 and a_4 of the autocorrelation function are shown in Formula (4) [27]:

$$C = \left[a_1 e^{-a_2 \tau_x} + (1 - a_1)\cos(a_3 \tau_x)\right] e^{-a_4 \tau_y} \tag{4}$$

Here, C is the surface autocorrelation function.

After the coefficients of the surface height probability density function and the autocorrelation function have been obtained, the random matrix $\boldsymbol{R}$ can be generated by combining the random process theory [28] and the fast Fourier transform method.

$$R = \text{ifft2}\left(e^{i2\pi\phi}\right) \tag{5}$$

Here, ifft2() represents the inverse Fourier transform of the matrix; ϕ is the characteristic function of the fast Fourier transform.

After that, C will be expanded into the autocorrelation function matrix C' through the symmetry assumption with the random matrix $\boldsymbol{R}$ [29]. By further applying the transformation to the random matrix R and the autocorrelation function matrix C' with the following formula, the height matrix Z_f of the grinding surface can be obtained.

$$Z_f = \frac{\text{fft2}(C')^{\circ 1/2}\text{fft2}(R)}{|\text{fft2}(R)|} \tag{6}$$

Here, fft2() represents the Fourier transform of the matrix; $^\circ$ means that each element of the matrix is operated separately.

Figure 2 displays the schematic diagram of the real measurement grinding surface and reconstructed grinding surface. The reconstructed surface has obvious grinding texture characteristics.

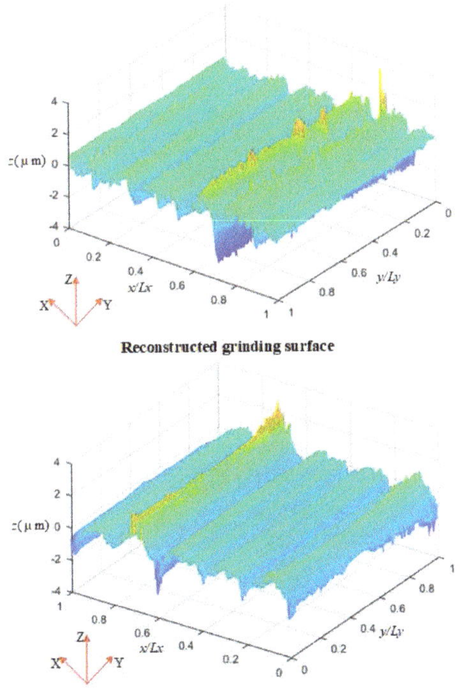

Figure 2. Measured and reconstructed grinding surface topography.

2.4. Principle of Optimal Explicit Expression

Since the correlation of roughness parameters is difficult to explore theoretically from the perspective of definition, the correlation model between **RPS** and **CPS** often stands on BP neural network or nonlinear regression model. However, BP neural network is not fit to provide a simple and explicit expression with less ability to realize the practical application in production. Therefore, the paper establishes the optimal explicit expressions between **RPS** and **CPS** by means of polynomial nonlinear regression model. **RPS** and **CPS** will be roughly distinguished by Pearson correlation analysis at first.

(1) Pearson correlation analysis

As roughness parameters belong to continuous data, the correlation among them should be studied through Pearson correlation analysis [30], and the calculation is as follows:

$$r_{ij} = \frac{s_{ij}}{\sqrt{s_{ii}s_{jj}}} \tag{7}$$

r_{ij} is the correlation coefficient between variable $\boldsymbol{i}$ and variable $\boldsymbol{j}$;
s_{ij} is the covariance between variable $\boldsymbol{i}$ and variable $\boldsymbol{j}$;
s_{ii} and s_{jj} are the variances of variable $\boldsymbol{i}$ and variable $\boldsymbol{j}$, respectively.

(2) Polynomial nonlinear regression model

The polynomial nonlinear regression model is the data method with strong nonlinear fitting ability. It is applicable to the situation where the correlation between variables is not clear, so as to explore the influence of variables and realize the explicit expression of independent variables to dependent variables [31,32]. Here explains how it works:

Regard the dependent variable as a complete polynomial combination of $\boldsymbol{n}$ independent variables:

$$\begin{aligned} f(x) = \; & a_0 + a_1x_1 + \cdots + a_nx_n + a_{n+1}x_1^2 \\ & + a_{n+2}x_1x_2 + \cdots + a_{N-1}x_k^m = \sum_{i=0}^{N-1} a_iu_i(x) \end{aligned} \tag{8}$$

Here,

$u_i(x)$ is the complete polynomial under the m power of independent variables $x = (x_1, x_2, \ldots, x_n)$;

a_i is the undetermined coefficient corresponding to the complete polynomial;

The total number of model terms is $N = (n+m)!/(n!m!)$;

However, when the polynomial nonlinear regression model is directly used for the research, based on the total number of items in the model, the increase of the independent variables will lead to an explosive growth of the items, and it is easy to produce many redundant terms. These redundant terms are meaningless to improve the accuracy of the model, so the number of terms should be reasonably pruned [32,33] to establish the optimal explicit expression.

(3) Item number pruning method

Aiming at the situation that the polynomial nonlinear regression model is prone to generating redundant items, the research designs a new method of reducing items and establishes the optimal explicit expressions between **RPS** and **CPS**. Its specific calculation is shown in Figure 3.

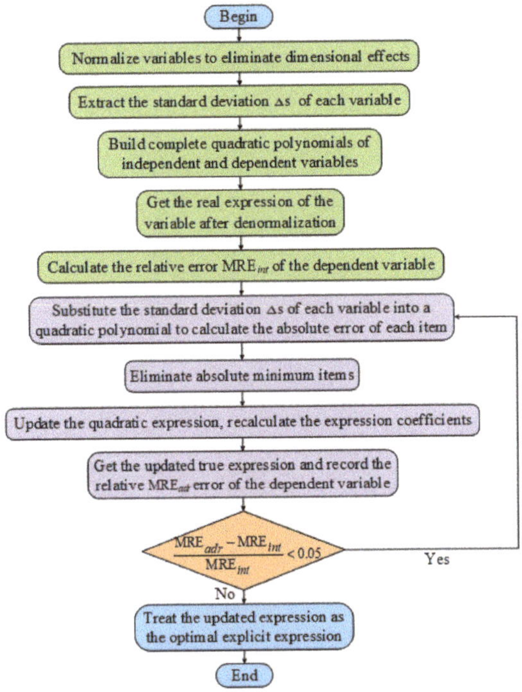

Figure 3. Optimal explicit expression screening process.

I. In order to remove the errors caused by different variable dimensions, the variables are normalized, and the standard deviation Δs of each variable after normalization will be calculated;

II. Establish the complete quadratic polynomial expression between the independent variables and the dependent variables, obtain the true expression after denormalization of the variables, and then calculate the relative error of the dependent variable MRE_{int};

III. Substitute the standard deviation Δs of the independent variable into the quadratic polynomial expression, calculate the absolute value of each item, and then remove the item with the smallest absolute value. Take the remaining terms as the updated quadratic expression, and solve for the updated expression coefficients;

IV. Calculate the updated denormalized quadratic expression and record the relative error MRE_{adr} of the dependent variable at this time. If

$$\frac{MRE_{adr} - MRE_{int}}{MRE_{int}} < 0.05,$$

go back to step **III** to solve again and update the expression. Otherwise, take the final expression as the optimal explicit expression.

3. Results and Analysis

Based on Li [27] and Yang [22] et al.'s research on the correlation judgement and surface reconstruction, the paper firstly sets the coefficients of the height probability density function and the autocorrelation function, shown in Table 2. After a random selection of each coefficient, 1000 groups of reconstructed grinding surfaces are generated for subsequent research.

Table 2. Reconstructed coefficients.

	Reconstructed Coefficients						
	k_1	k_2	k_3	a_1	a_2	a_3	a_4
Minimum	0.4	−5.5	0.6	0.6	0.08	0.08	0.0005
Maximum	3.4	2.5	50	0.9	0.12	0.12	0.0012

3.1. Classification for CPS and RPS

As mentioned above, the correlation among 26 roughness parameters is complex with hugely different definitions and formulas, thus causing the troubles. Therefore, the paper tries to preliminarily judge the correlation between the roughness parameters from data analysis and determines the correctness and error of the results with the help of parameters definitions. Since the data type belongs to continuous data, Pearson correlation analysis is used to distinguish the correlation between the roughness parameters.

When the Pearson correlation coefficient of the bivariate is large, there is a strong correlation between the two parameters. Variable **A** will closely follow the change of the variable **B**. At this time, it can be considered that variable **A** is controlled by variable **B**. When the correlation coefficient is extremely small, variable **A** either has a curve trend with variable **B**, or there is no actual correlation between the two. At the moment, it is generally necessary to use the trend distribution diagram and definitions to assist in judgement.

The criterion for determining the correlation strength is shown in Table 3 [14,15,30].

Table 3. Correlation judgment criteria.

Range	Conclusion
$0 \leq \left\|r_{ij}\right\| < 0.1$	Very weakly correlated
$0 \leq \left\|r_{ij}\right\| < 0.3$	Weakly correlated
$0.3 \leq \left\|r_{ij}\right\| < 0.7$	Moderately correlated
$0.7 \leq \left\|r_{ij}\right\| < 0.9$	Strongly correlated
$0.9 \leq \left\|r_{ij}\right\| < 1$	Very strongly correlated

As is known, ***Sa*** is widely used to characterize the surface quality in the application of surface characterization and ISO 25178. The research intends to use ***Sa*** as the first benchmark in **CPS**. The specific method is as follows:

(1) Take ***Sa*** as the benchmark (the first selected to **CPS**), and evaluate the correlation between the remaining parameters and ***Sa***. The parameters whose correlation coefficient with ***Sa*** is greater than 0.9 are put into **RPS**, and the parameter with the smallest coefficient is selected as the next one into **CPS**. The remaining are classified as the undetermined set;

(2) Treat the second selected parameter as the next benchmark, calculate the correlation coefficient between the second and each parameter in the undetermined set. Parameters with the coefficient greater than 0.9 are put into **RPS**, and the one with the smallest coefficient is selected as the third one into **CPS**. The remaining are still used as the updated undetermined set;

(3) With similar operation as step 2, **CPS** and **RPS**, two different rough parameter sets, are finally obtained until the item number in the undetermined set is 0.

The trend of each parameter in **CPS** during the screening process is shown in Figure 4. There is no obvious curve distribution of each parameter. Although ***Spc*** and ***Spk*** seem to present certain trend at the end of the screening, a careful observation reveals that in the same ***Spc***, the fluctuation of ***Spk*** accounts for 60% in the overall range, making it impossible to meet the standard that the correlation coefficient is greater than 0.9. Therefore, ***Spk*** enters into **CPS**. The specific screening process can be seen in Figure 5. **CPS** consists of ***Sa, Ssk, Sku, Sp, Sv, Str, Spk, Smr1, Sxp, Vvv, Spd*** and ***Spc***. The following is an introduction to the parameters in **CPS** to help determine the rationality of selection.

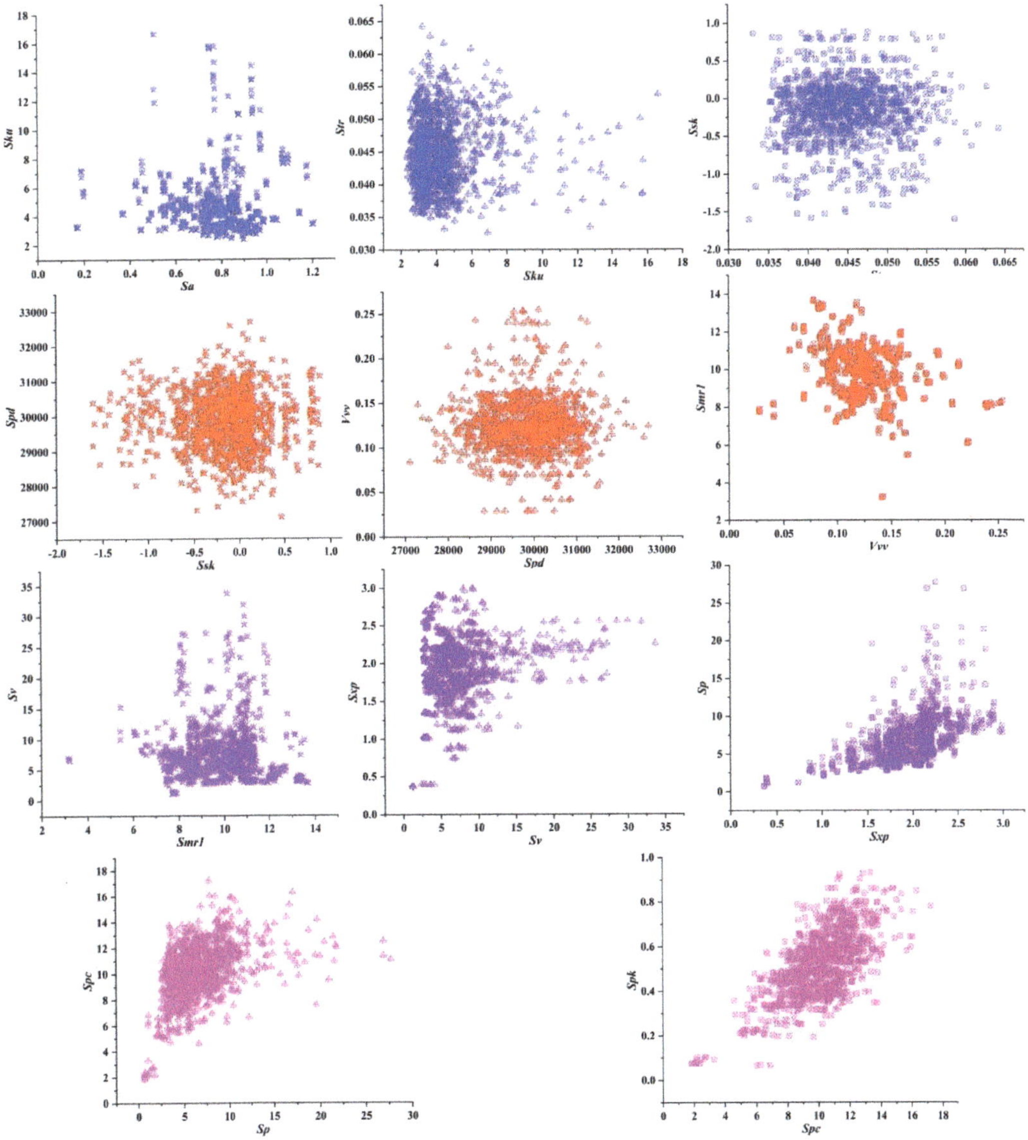

Figure 4. CPS scatter trend.

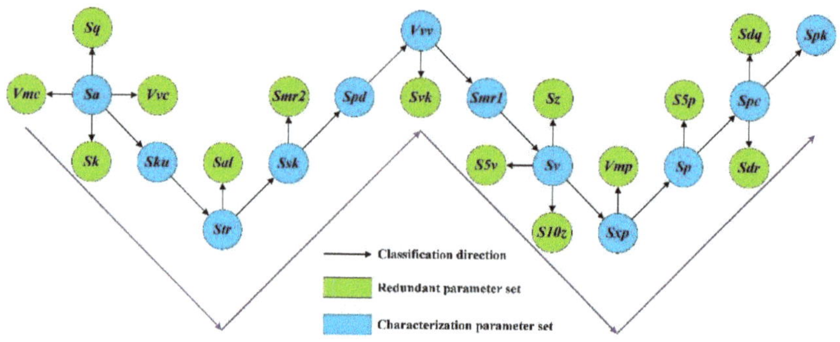

Figure 5. CPS and RPS.

The above analysis shows that the method of distinguishing **CPS** and **RPS** with strong correlation not only illustrates the classification rationality from the perspective of data, but also proves the reliability of the discriminating method in combination with the physical meaning and definition of roughness parameters.

Sa is the most widely used parameter, characterizing the average height difference in the surface. ***Sq*** which is highly related to ***Sa*** is classified into **RPS**. It is consistent with the work of Pawlus et al. [14]. While ***Ssk*** and ***Sku*** stand for surface skewness and kurtosis respectively, they are used to evaluate the symmetry and steepness of the surface height distribution. ***Ssk*** and ***Sku*** belong to the key factors of surface reconstruction theory [26] and are indispensable for surface characterization [34]. ***Sp*** and ***Sv*** represent the surface extremum features. Since $\boldsymbol{Sp} + \boldsymbol{Sv} = \boldsymbol{Sz}$, the information of ***Sz*** is actually expressed by ***Sp*** and ***Sv***. Therefore, ***Sz*** is kicked into **RPS**. ***Str***, reflecting surface anisotropy and defined by the surface autocorrelation function, is usually used to characterize the surface texture direction and spatial information. According to the definitions, ***Sal*** and ***Str*** can iterate over each other, so information intersection exists in the two parameters. As for ***Spk*** and ***Vmp***, both of them represent peak features above the core surface so that it is reasonable to put one of them into **CPS** [13]. ***Vvv*** and ***Svk*** characterize surface valley. ***Spd*** and ***Spc*** comprehensively describe the average surface peak density and peak curvature radius. They are the characteristic descriptions of different asperity peaks on the surface. ***Sxp*** and ***Smr1*** can be used to help define other roughness parameters, such as ***Sk***, ***Svk***, etc. [35].

3.2. Establishment of Optimal Explicit Expression

Although the separation of **CPS** and **RPS** is achieved through the classification method, it is still unknown how **CPS** represents the information in **RPS**. Besides, it has not been verified whether **CPS** can predict all parameters and thus achieve comprehensive characterization of surface topography. Therefore, only through the mapping relationship between **CPS** and **RPS** and the establishment of a quantitative model between them, **CPS** has the capability of topography characterization under the control of all roughness parameters.

3.2.1. Core Roughness Parameters in CPS

In order to meet the needs of industrial production, the paper sets out from the polynomial regression model with strong nonlinear ability and designs a new method of items reduction to establish the optimal explicit expression between **CPS** and **RPS**. After selection, there are 12 parameters in **CPS** and 14 parameters in **RPS**. If the parameters in **CPS** are regarded as independent variables and each parameter in **RPS** is treated as the dependent variable, the number of polynomial terms will explode, and it is easy to overfit. Even if the pruning method is introduced at this time, the efficiency of solving the optimal expression will be greatly reduced.

In addition, combined with the definitions, not every redundant parameter needs all parameters in **CPS** to characterize their information.

According to the statistical regression theory and the interpretation of the independent variables by Friend et al. [36], the independent variables in the regression model should minimize the collinearity. Collinear variables do not increase the fit of the model. Then Yang et al. [22] put forward the principle of "non-strong and non-weak" to find the core independent variables with low collinearity, which have strong correlation with the dependent variable, but are relatively independent. In this study, the principle of "non-strong and non-weak" is introduced to screen the **CRP** and establish the optimal explicit expression. Figure 6 shows **CRP** corresponding to **RPS**. The number and types of **CRP** are extremely different, which is consistent with the actual cognition.

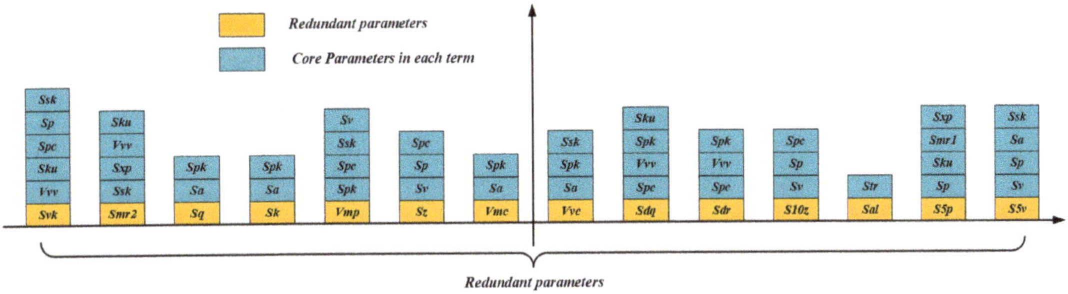

Figure 6. CRP.

3.2.2. Optimal Explicit Expression

Although the core roughness parameters to characterize **RPS** are further determined with the principle of "non-strong and non-weak", it can be seen from Figure 5 that the definition and formula of $\boldsymbol{Sp}$ is $\boldsymbol{Sp} + \boldsymbol{Sv} = \boldsymbol{Sz}$, but the **CRP** of $\boldsymbol{Sz}$ are $\boldsymbol{Sp}$, $\boldsymbol{Sv}$ and $\boldsymbol{Spc}$. The **CRP** of $\boldsymbol{Sq}$ are $\boldsymbol{Sa}$ and $\boldsymbol{Spk}$. It is different from other works. The core parameter of $\boldsymbol{Sal}$ is $\boldsymbol{Str}$, and they are consistent with the definitions. This reveals that the judgement method only by the principle of "non-strong and non-weak" will lead to the situation that some parameters still have information confusion and that the results are not reliable enough. Therefore, the paper designs a new redundant item pruning method to automatically identify and find roughness parameters with strong characterization ability with reliability.

Since the roughness parameter sets with clear equality are $\boldsymbol{Sz}$, $\boldsymbol{Sp}$ and $\boldsymbol{Sv}$ ($\boldsymbol{Sp} + \boldsymbol{Sv} = \boldsymbol{Sz}$), the research will discuss these three parameters in detail to verify the reliability and stability of the redundancy information reduction technique.

At the beginning, $\boldsymbol{Sz}$ is introduced to the dependent variable, while the core parameters of $\boldsymbol{Sp}$, $\boldsymbol{Sv}$ and $\boldsymbol{Spc}$ in Figure 6 are regarded as independent variables. With the coefficients in model (9) solved, the relative error MRE will be obtained. After that, the polynomial of $\boldsymbol{Sz}$ will be addressed by the designed redundant item pruning method. Table 4 records the deleted item and the relative error change in the whole process, where 0 means the item is eliminated and 1 means retained.

Table 4. Polynomial pruning of $\boldsymbol{Sz}$.

Dependent Variable	Num	Polynomial terms										MRE
		Sp^2	Sv^2	Spc^2	$Sp*Sv$	$Sp*Spc$	$Sv*Spc$	Sp	Sv	Spc	Const	
	01	1	1	1	1	1	1	1	1	1	1	2.1205×10^{-7}
	02	1	1	1	1	1	1	1	1	0	1	2.1259×10^{-7}
	03	1	1	0	1	1	1	1	1	0	1	2.1243×10^{-7}
	04	1	1	0	1	1	0	1	1	0	1	2.1147×10^{-7}
Sz	05	1	1	0	1	0	0	1	1	0	1	2.1036×10^{-7}
	06	0	1	0	1	0	0	1	1	0	1	2.0950×10^{-7}

Table 4. *Cont.*

Dependent Variable	Num	Polynomial terms										MRE
		Sp^2	Sv^2	Spc^2	$Sp * Sv$	$Sp * Spc$	$Sv * Spc$	Sp	Sv	Spc	Const	
	07	0	0	0	1	0	0	1	1	0	1	2.0807×10^{-7}
	08	0	0	0	0	0	0	1	1	0	1	2.0543×10^{-7}
	09	0	0	0	0	0	0	1	1	0	0	1.4×10^{-3}

In the pruning process of ***Sz***, the first item to be eliminated is ***Spc***, then ***Spc***^{2}, ***Sv*** * ***Spc*** and ***Sp*** * ***Spc***. All of them contain ***Spc***. Combined with the parameter definition, the pruning process shows that the method can preferentially identify and eliminate items with low correlation with ***Sz***. The method is highly reliable and can actively discriminate the roughness parameters with strong characterization ability. In addition, on account of the error jump in 09, the 08 is selected as the optimal explicit expression at the end: $Sz = Sp + Sv + 6.2654 \times 10^{-8}$, and the relative error is 2.0543×10^{-7}.

Compared with the real formula $Sp + Sv = Sz$, the optimal explicit expression is consistent with the real one. The pruning method ensures the reliability of the accuracy. At the same time, the influence of ***Spc*** is removed through continuous deletion. It verifies the ability of the proposed pruning method to remove the chaotic representation of parameters and greatly reduce the number of terms. The method is suitable for the optimal explicit expression.

After the accuracy of ***Sz*** is verified, each parameter in **RPS** gets a similar treatment. The rank of the core parameters is shown in Table 5(a). Due to the large number and types in **CRP**, it is bound to cause confusion in the model. Therefore, the quadratic square terms are sorted in order $x_1^2, x_2^2, \ldots x_n^2$, and then the cross terms $x_1x_2, x_1x_3, \ldots, x_1x_n, x_2x_3, \ldots, x_2x_n, \ldots x_{n-1}x_n$ are arranged as above. Finally, sort the first order items and add the constant item at the end. The optimal term for each redundant parameter expression is selected in Table 5(b).

As ***Sz*** has been analyzed in the previous section, only other parameters are further elaborated here. For ***Sq***, the largest coefficient term in the optimal expression is the first-order ***Sa***, and the coefficients about ***Sv*** and ***Spk*** are all relatively smaller. It shows that ***Sq*** is mainly regulated by the primary term ***Sa***, while the surface peak-valley extreme features controlled by ***Sv*** and ***Spk*** account for a relatively low proportion in the information characterization of ***Sq***.

This also explains why ***Sa*** and ***Sq*** are strongly correlated, but not completely correlated. ***Sal*** and ***Str*** present a completely linear expression, conforming to the definitions of these two parameters. In ***Sk***, it is expressed by the first term of ***Sa*** and ***Spk***, indicating that the surface core height is actually characterized by the surface arithmetic mean height and the protruding peak height, and the influence of ***Sa*** is greater.

For *Svk*, it still retains quite a number of terms. The largest coefficient is the first-order term ***Vvv***, illustrating that ***Vvv*** has a good ability to characterize the surface valley features defined by ***Svk***, but at the same time it needs other parameters to realize the additional information of ***Svk***. In the optimal explicit expression of ***Smr2***, the quadratic cross term and the first term have larger coefficients. It indicates that ***Smr2*** is mainly affected by the coupling effect of surface valley void volume, surface skewness, kurtosis, and peak maximum height. These four reflect the surface bearing ability from different angles.

However, ***Vvc*** and ***Vmc*** are mainly controlled by ***Sa***, verifying the highest frequency of ***Sa*** in industrial applications. ***Vmp*** is mainly affected by the independent characterization of ***Spk*** and the coupled characterization of surface protruding peak height and skewness. The independent characterization of the first order ***Spk*** is stronger than the coupled characterization of the surface protruding peak height and skewness. It explains why ***Vmp*** and ***Spk*** are strongly correlated but with a little redundant information.

Table 5. Part (a) Rank of core parameters in **RPS**; Part (b) Optimal explicit expression of **RPS**.

Core Parameters in RPS													
(a)													
Sz	*Sq*	*Sal*	*Sk*	*Svk*	*Smr2*	*Vvc*	*Vmp*	*Vmc*	*Sdq*	*Sdr*	*S10z*	*S5p*	*S5v*
$x_1 = Sp$ $x_2 = Sv$ $x_3 = Spc$	$x_1 = Sa$ $x_2 = Sv$ $x_3 = Spk$	$x_1 = Str$	$x_1 = Sa$ $x_2 = Spk$	$x_1 = Ssk$ $x_2 = Sku$ $x_3 = Sp$ $x_4 = Vvv$ $x_5 = Spc$	$x_1 = Ssk$ $x_2 = Sku$ $x_3 = Sxp$ $x_4 = Vvv$	$x_1 = Sa$ $x_2 = Ssk$ $x_3 = Spk$	$x_1 = Ssk$ $x_2 = Sv$ $x_3 = Spk$ $x_4 = Spc$	$x_1 = Sa$ $x_2 = Spk$	$x_1 = Sku$ $x_2 = Spk$ $x_3 = Vvv$ $x_4 = Spc$	$x_1 = Spk$ $x_2 = Vvv$ $x_3 = Spc$	$x_1 = Sp$ $x_2 = Sv$ $x_3 = Spc$	$x_1 = Sku$ $x_2 = Sp$ $x_3 = Smr1$ $x_4 = Sxp$	$x_1 = Sa$ $x_2 = Ssk$ $x_3 = Sp$ $x_4 = Sv$
(b)													
Sz	*Sq*	*Sal*	*Sk*	*Svk*	*Smr2*	*Vvc*	*Vmp*	*Vmc*	*Sdq*	*Sdr*	*S10z*	*S5p*	*S5v*
0	0	0	0	-3.86×10^{-2}	6.58×10^{-1}	-1.01×10^{-1}	-2.15×10^{-3}	0	-1.97×10^{-4}	0	-1.49×10^{-2}	-1.47×10^{-2}	0
0	-1.66×10^{-4}	4.08×10^{2}	0	-1.17×10^{-3}	2.07×10^{-2}	5.00×10^{-2}	-1.20×10^{-5}	0	0	0	-1.07×10^{-2}	-1.29×10^{-2}	0
0	1.97×10^{-1}	2.05	0	-1.40×10^{-4}	-1.13×10^{-1}	-2.07×10^{-1}	0	0	0	1.61×10^{-2}	-1.44×10^{-2}	-1.53×10^{-2}	-2.64×10^{-3}
0	6.64×10^{-3}		3.43	0	0	3.47×10^{-1}	1.97×10^{-5}	1.22	-2.54×10^{-4}	0	2.30×10^{-3}	-4.31×10^{-1}	-1.55×10^{-2}
0	-1.90×10^{-1}		-4.67×10^{-1}	4.03×10^{-5}	1.06×10^{-1}	3.16×10^{-1}	-2.24×10^{-4}	-1.69×10^{-1}	2.64×10^{-3}	0	1.64×10^{-2}	-8.62×10^{-4}	0
0	-1.31×10^{-3}		-1.15×10^{-2}	2.38×10^{-3}	−2.10	-2.24×10^{-1}	1.57×10^{-2}	-1.81×10^{-3}	4.94×10^{-2}	1.62	1.57×10^{-2}	7.35×10^{-2}	0
1	1.20			2.26×10^{-3}	-1.78×10^{1}	1.62	-5.85×10^{-4}		-4.70×10^{-4}	0	7.13×10^{-1}	1.05×10^{-1}	2.08×10^{-1}
1	5.78×10^{-3}			0	2.41×10^{-1}	4.43×10^{-2}	1.46×10^{-3}		0	−3.04	7.30×10^{-1}	-2.21×10^{-2}	-5.04×10^{-2}
0	5.35×10^{-2}			-4.78×10^{-3}	−2.40	-1.85×10^{-1}	-4.40×10^{-5}		0	2.96×10^{-2}	2.14×10^{-1}	1.55×10^{-1}	-1.35×10^{-1}
6.27×10^{-8}	-4.31×10^{-5}			-2.74×10^{-4}	0	1.29×10^{-3}	-8.31×10^{-4}		3.68×10^{-2}	1.54×10^{-2}	2.09×10^{-2}	1.54×10^{-1}	3.84×10^{-3}
				0	2.69		-1.47×10^{-3}		4.30×10^{-3}			-4.10×10^{-1}	-1.07×10^{-1}

Table 5. *Cont.*

Core Parameters in RPS													
(a)													
Sz	*Sq*	*Sal*	*Sk*	*Svk*	*Smr2*	*Vvc*	*Vmp*	*Vmc*	*Sdq*	*Sdr*	*S10z*	*S5p*	*S5v*
				5.41×10^{-4}	-6.05×10^{-1}		-1.12×10^{-4}		-6.40×10^{-3}			5.39×10^{-1}	1.98×10^{-1}
				0	4.87		9.30×10^{-2}		-1.88×10^{-1}			-6.40×10^{-2}	1.54×10^{-1}
				1.48×10^{-4}	-8.12×10^{1}		5.21×10^{-4}		2.61×10^{-2}			1.12×10^{-2}	6.11×10^{-1}
				0	9.14×10^{1}		5.93×10^{-5}		-4.46×10^{-3}			9.52×10^{-1}	2.34×10^{-1}
				-4.83×10^{-2}									
				2.82×10^{-2}									
				6.56×10^{-3}									
				4.71									
				-7.66×10^{-3}									
				-1.01×10^{-1}									

For the optimal explicit expression of ***Sdq***, the final retained coefficients are relatively evenly distributed, indicating that ***Sdq*** covers a wide range of information and is defined by different features of the surface. In ***Sdr***, all items of ***Spk*** are removed, indicating that the height of the surface protruding peak has little effect on it. The surface extremum features covered by ***S10z***, ***S5p*** and ***S5v*** can be characterized with the coupling of ***Sp***, ***Sv*** and other parameters. In order to better distinguish the characterization ability of the optimal explicit expression, Figure 7 reveals the sample error distribution and average relative error of each redundant parameter in 1000 reconstructed surfaces. In the 1000 surfaces, the relative error of **RPS** is less than 0.1 and the proportion of samples is more than 90% except for ***Sdr***. Some parameters even reach 100%. For the surface proportion with a relative error less than 0.05, the remaining are basically above 80%, but ***Sdq***, ***Sdr***, ***S10z***, ***S5p*** and ***S5v*** drop to a large extent. In addition, the maximum average relative error of the parameters is 0.07 and the rest of the parameters are basically within 0.05 from the broken line distribution. It indicates that **CPS** can achieve full coverage of the surface topography information and remove the redundancy through the optimal expression.

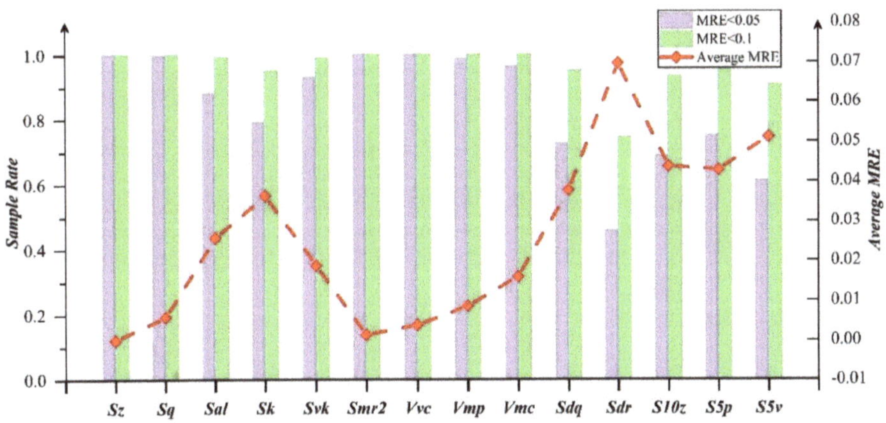

Figure 7. Parameter error distribution of optimal explicit expression.

3.3. Experiment Verification

Although **RPS** and **CPS** have been distinguished with the help of 1000 reconstructed surfaces and the optimal explicit expression of **RPS** has been established, the reconstructed surfaces are generated by the mathematical model and since factors such as tool runout and measurement errors cannot be avoided in the real machining process, the real surface has more randomness in the height distribution. Therefore, after the preliminary theoretical exploration is carried out with the reconstructed surfaces, it is necessary to verify the reliability of the method and analysis with the real grinding surfaces. The experiment conditions are shown in Table 6.

Table 6. Experiment conditions.

Machining Parameters	Parameter Values
Grinding wheel	CBN grinding wheel
Wheel radius	100 mm
Wheel mesh	120
Wheel speed	500–3000 r/min
Cutting speed	200 mm/min
Cutting depth	5–30 μm

The component is 12Cr2Ni4A steel, the surface roughness morphology gets measured with the white light interferometer Wyko NT9100, the sampling area is 0.48 mm × 0.64 mm

= 0.3072 mm^2, and the sampling interval is 1 μm. The machining process and 3D roughness topography measurement are shown in Figure 8.

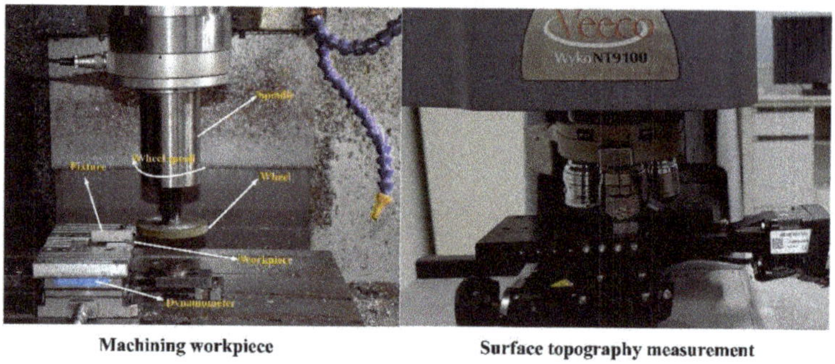

Figure 8. Machining and topography measurement.

A total of 44 surfaces were collected through the grinding experiment, and 26 surface roughness parameters were substituted into the optimal explicit expression of **RPS** to verify its generalization and reliability.

It can be seen from Figure 9 that parameters ***Svk***, ***S10z***, ***S5p*** and ***S5v*** with the relative errors less than 0.1 mainly account for 60–70%. However, with the relative errors less than 0.2, the proportion of these parameters has increased significantly, basically reaching about 0.9. Besides, the average error of parameters other than ***Sdq*** and ***Sdr*** is basically within 10%, indicating that the optimal explicit expressions of the remaining parameters are still reliable.

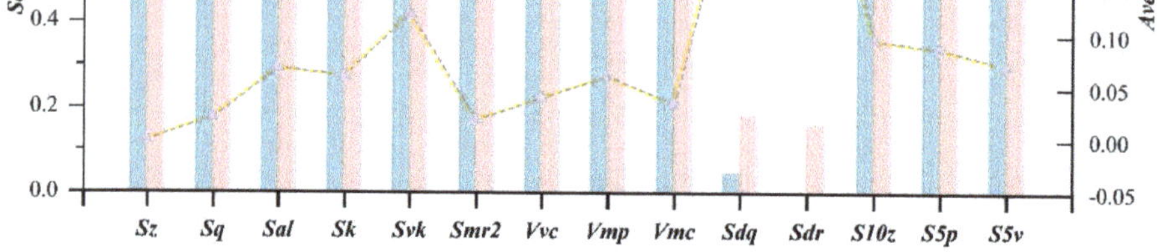

Figure 9. Parameter error distribution of experiment surfaces.

However, as ***Sdq*** and ***Sdr*** present such a large deviation, further research must be considered. The two are significantly different from others in terms of the relative error proportional distribution and average relative error. ***Sdr*** of the reconstructed surface in Figure 7 also belongs to the maximum error term. Considering that there are some slight differences between the reconstructed surface and the measured surface, the mathematical model cannot realize the topography feature control defined by all roughness parameters. The accumulation of these factors further exacerbates the degradation of the optimal expression prediction ability about ***Sdq*** and ***Sdr***.

In addition, the correlation coefficient of ***Sdq*** and ***Sdr*** is as high as 0.9843, and the two show a strong linear correlation. One should be selected into **CPS** and the other gets into **RPS**. Taking ***Sa*** as the benchmark, the correlation coefficients between ***Sdq*** and ***Sdr*** to ***Sa*** are 0.829 and 0.806, respectively. ***Sdr*** with small correlation with ***Sa*** (less repetition of topographic information with ***Sa***) is selected into **CPS**, and ***Sdq*** belongs to **RPS**.

Then the "non-strong and non-weak" principle and "polynomial pruning method" are used to find the core parameters and the optimal explicit expression is constructed. The core parameters of ***Sdq*** are ***Sdr***, ***Sxp***, ***Sv*** and ***Sp***, and the optimal expression constructed is as follows:

$$\begin{aligned}\boldsymbol{Sdq} = -4.5972 \times 10^{-5}\boldsymbol{Sp}^2 - 1.5786 \times 10^{-5}\boldsymbol{Sv}^2 - 0.0141\boldsymbol{Sxp}^2 - 0.0012\boldsymbol{Sdr}^2 + \\ 9.1066 \times 10^{-5}\boldsymbol{SpSv} + 0.002\boldsymbol{SpSxp} - 0.0027\boldsymbol{SvSxp} + 1.9366 \times 10^{-4}\boldsymbol{SvSdr} - \\ 0.0019\boldsymbol{SxpSdr} - 0.0035\boldsymbol{Sp} + 0.005\boldsymbol{Sv} + 0.0673\boldsymbol{Sxp} + 0.0517\boldsymbol{Sdr} + 0.0346\end{aligned}$$

The adjusted optimal expression of ***Sdq*** can achieve the accuracy of approximately 0.9 on the measured surface with high reliability. So far, combined with theoretical and experiment analysis, the number of parameters in **CPS** to truly characterize and control surface roughness morphology is 13. The method achieves 50% reduction in the overall roughness parameters and builds a rational characterization model with fewer parameters to describe the integrity of the surface features.

3.4. Significance of CPS for Surface Performance

The establishment of **RPS** and **CPS** and the optimal explicit expressions clarify the correlation between roughness parameters and realize the complete characterization of surface features with fewer parameters. The method finds the parameters truly controlling the surface roughness and provides guidance for researchers to apply, but these analyses are still limited to the internal characterization of surface geometry and do not discuss about the relationships between **CPS** and surface performance. The roughness parameter that can well characterize surface performance is the focus of industry and research.

Different performance parameters are always closely related to the surface geometry topography characteristics described by roughness parameters. Owing to the large number of performance parameters and the complex correlation between the 26 roughness parameters, the existing performance screening and characterization parameter method [22] will result in the explosion of the final characterization parameters and unreliability. For example, the selected parameter is actually regulated by other roughness parameters, which will lead to deviations from the expected target, so that the better characterization effect is more likely to lose. **CPS** and **RPS** proposed in this paper can solve the above issues.

Since **CPS** can realize the complete description of the surface topography information, when the correlation between surface performance and all roughness parameters is studied, more attention should be paid to observe the correlation between the surface performance and **CPS**. The topographic features, mainly influencing the surface performance, can be identified by selecting the parameters from **CPS**. Therefore, a method is designed to study the correlation between **CPS** and surface performance and to explain the significance of **CPS** in Figure 10.

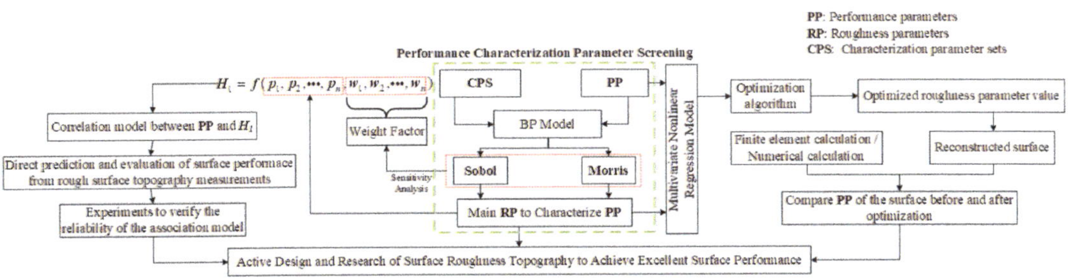

Figure 10. The method of correlation between **CPS** and surface performance.

The research focuses on the geometric characterization of the rough surface and clarifies the correlation among roughness parameters, so the correlation between **CPS** and surface performance belongs to the further expansion of the research. The method of correlation between **CPS** and surface performance will help prove the basic significance and applicability of this research. It will enable researchers to better understand the engineering significance of **CPS** for the realization in co-design and manufacture of rough surfaces. Therefore, here only provides a feasibility assessment and a rough introduction to the correlation method between **CPS** and surface performance.

(1) Regarding the performance characterization screening method described in the green box at the core position in Figure 10, this part belongs to the improvement and expansion of the method proposed in the literature [22] to screen the main roughness parameters based on the contact performance. Literature [22] mentions that since there is no direct physical model for roughness parameters and contact stress, the BP neural network model is introduced to establish the mapping relationship between the two, and then the main roughness parameters affecting the contact stress are identified by the quantitative **Sobol** and qualitative **Morris** analysis in the sensitivity analysis. The BP neural network has universality in fitting continuous data and different performance parameters belong to continuous data, so the extension of its contact performance to different performance parameters in this section will not affect the reliability of the method. In addition, the method replaces 26 roughness parameters with **CPS**, which will make the final parameters used to characterize the surface performance more accurate. Then by substituting different performance parameters and counting the frequency of different selected parameters in **CPS**, the ability of roughness parameters to characterize various performance characteristics will be distinguished;

(2) On the right side of the green box in Figure 10, the work of establishing the multivariate nonlinear regression model between the performance parameters and the main roughness parameters in **CPS** is easy to be completed with the help of the polynomial nonlinear regression and pruning techniques proposed in the paper. Regarding the inverse optimization of the explicit regression model, it is not difficult to find the suitable optimization algorithm to get the best parameter range. Although it is difficult to control the roughness parameters in the actual surface machining, the surface reconstruction algorithm will be an ideal way to realize it.

For the generation of reconstructed surfaces with specified roughness parameters, the literature [10] proposed a reliable method. The contact fatigue calculation model of rough surfaces, such as KE finite element model and Wen's numerical calculation model [37], can complete the performance prediction in the reconstructed surface.

This part belongs to theoretical research based on mathematical models and contact theory. It has the advantages of low experimental cost and high efficiency and can provide paths and basic guidance for finding and designing the suitable rough surface with excellent performance;

(3) On the left side of the green box in Figure 10, it focuses on experiment research and verification. Even if the same batch of workpieces (surface residual stress, hardness and other material properties are considered to be nearly the same) are under the same loading conditions, the influence of other surface features in addition to the morphology features defined by ***Sa*** will still lead to great differences in contact properties, friction and wear, fatigue and other properties. However, the producers cannot judge the quality of the same batch, which will greatly reduce the service performance and increase the production cost of the enterprise.

This section will solve the above problems. At first, the weight factors of **CPS** are extracted, and the function $H_1 = f(p_1, p_2, \ldots, p_n, w_1, w_2, \ldots, w_n)$ is constructed by combining the selected roughness parameters in **CPS**. The purpose of this step is to facilitate the observation and analysis about the comprehensive influence of the selected parameters on performance parameters. If only based on a single parameter, it will inevitably lead to the incompleteness of information. Secondly, the way to establish a correlation model between performance parameters and $\boldsymbol{H_1}$ is more convenient for experiment verification.

Due to the huge experiment cost, if the multiple regression analysis on the right side is carried out, a large amount of experiment is required to get an accurate and reliable model. However, the purpose of the experiment research in this part is to rank the surface quality of different workpieces from the selected roughness parameters in **CPS**, so a particularly accurate model is not a must. Therefore, the feasibility of this part based on experiments is extremely high.

(4) Whether it is to screen the performance characterization parameters, or to establish an accurate nonlinear regression model from the theoretical view, or to evaluate the characterization parameters through experimental research, the establishment of the initial **CPS** is indispensable. Due to the continuous accumulation of errors, if the correct **CPS** cannot be obtained or they are selected only by experience, the correlation analysis between **CPS** and the surface performance will inevitably be unreliable.

4. Conclusions

(1) Based on 1000 reconstructed surfaces, 26 roughness parameters are roughly classified into **CPS** and **RPS** by Pearson correlation analysis. The principle of "non-strong and non-weak" helps **CPS** extract key factors from **CRP** to facilitate the establishment of subsequent expressions. The results demonstrate that the **RPS** information can be covered by **CPS**;
(2) The optimal explicit expressions of **CPS** and **RPS** get established, and the accuracy is basically above 90%. Then a polynomial pruning method is designed to find roughness parameters with strong ability to characterize surface information. The correlation between **CPS** and **RPS** is quantified to clarify the cause of application confusion. The results show **RPS** is independently affected and coupled by several different core parameters;
(3) The experiment verifies the reliability of the optimal explicit expression of **RPS** and surface characterization method and helps fix the number in **CPS** at 13. They are ***Sa, Ssk, Sku, Sp, Sv, Str, Spk, Smr1, Sxp, Vvv, Spd, Spc*** and ***Sdq***. **RPS** accounts for 50% of the overall roughness parameter set, and the method realizes the comprehensive description of surface features with a smaller number of parameters, which has been well verified by the theoretical and experimental analysis;
(4) A surface characterization method for screening **CPS** is designed to find the key factors that really control the surface morphology. It also solves the dilemma of blindly or empirically selecting roughness parameters in industrial production. The reliability of the method to explore the correlation between **CPS** and different surface performance parameters is analyzed in detail. It proves the engineering significance of **CPS** for realizing co-design and manufacture in rough surfaces.

Author Contributions: Formal analysis, D.Y.; resources.; data curation, F.X.; writing—original draft preparation, D.Y. and W.Z.; writing—review and editing, D.Y. and W.Z.; project administration, J.T.; funding acquisition, J.T. and W.Z. All authors have read and agreed to the published version of the manuscript.

Funding: This research was funded by National Key R&D Program of China, grant number 2020YFB2008200, National Science and Technology Major Project, grant number 2019-VII-0017-0158, National Natural Science Foundation of China, grant number 51705142 and the APC was funded by National Key R&D Program of China, grant number 2020YFB2008200.

Institutional Review Board Statement: Not applicable.

Informed Consent Statement: Not applicable.

Data Availability Statement: All data that support the findings of this study are included within the article.

Acknowledgments: The authors gratefully acknowledge the support of the National Key R&D Program of China (Grant No. 2020YFB2008200) and National Science and Technology Major Project (2019-VII-0017-0158), National Natural Science Foundation of China (NSFC) through Grant No. 51705142.

Conflicts of Interest: The authors declare no conflict of interest.

Appendix A. The Overall Technical Route

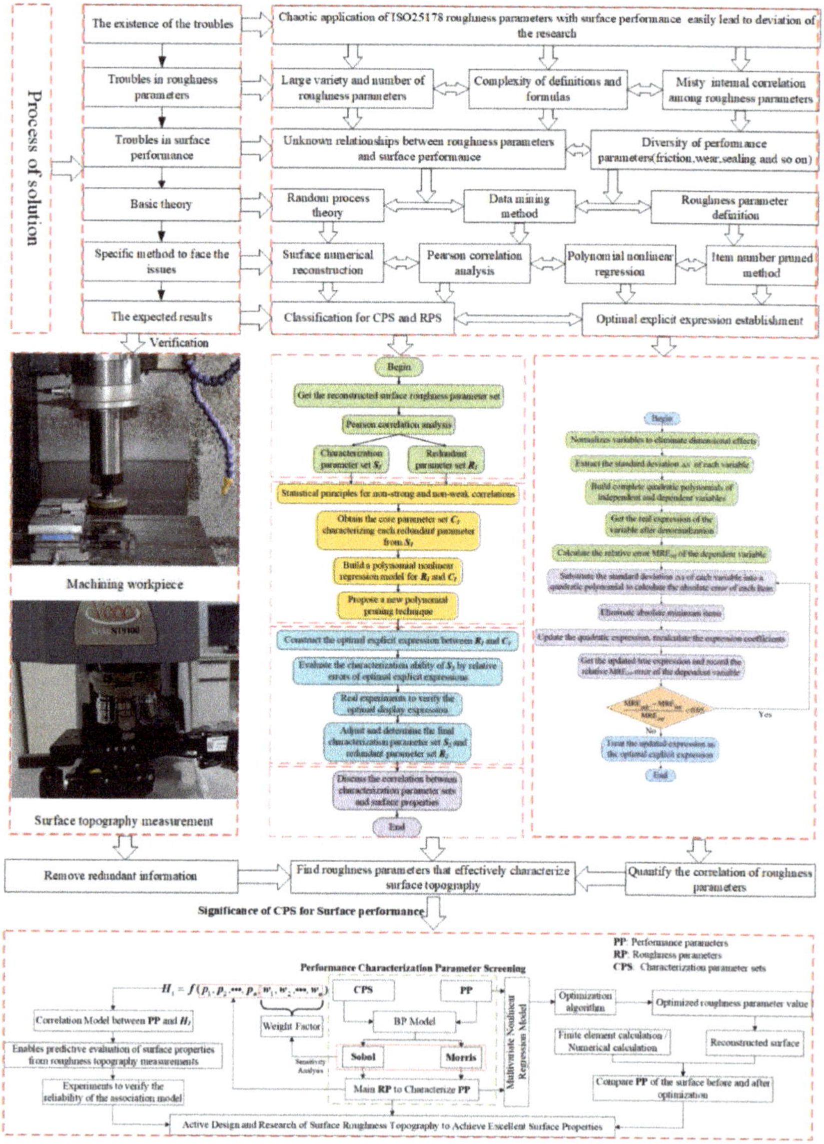

References

1. Yánez, A.; Fiorucci, M.P.; Cuadrado, A.; Martel, O.; Monopoli, D. Surface roughness effects on the fatigue behaviour of gyroid cellular structures obtained by additive manufacturing. *Int. J. Fatigue* **2020**, *138*, 105702. [CrossRef]
2. Guo, L.; Li, G.; Gan, Z. Effects of surface roughness on CMAS corrosion behavior for thermal barrier coating applications. *J. Adv. Ceram.* **2021**, *10*, 472–481. [CrossRef]
3. Dinh, T.D.; Han, S.; Yaghoubi, V.; Xiang, H.; Erdelyi, H.; Craeghs, T.; Segers, J.; Van Paepegem, W. Modeling detrimental effects of high surface roughness on the fatigue behavior of additively manufactured Ti-6Al-4V alloys. *Int. J. Fatigue* **2021**, *144*, 106034. [CrossRef]
4. Ţălu, Ş. *Micro and Nanoscale Characterization of Three Dimensional Surfaces. Basics and Applications*; Napoca Star Publishing House: Cluj-Napoca, Romania, 2015; ISBN 978-606-690-349-3.
5. Kumar, V.; Singh, H. Regression analysis of surface roughness and micro-structural study in rotary ultrasonic drilling of BK7. *Ceram. Int.* **2018**, *44*, 16819–16827. [CrossRef]
6. Lou, S.; Zhu, Z.; Zeng, W.; Majewski, C.; Scott, P.; Jiang, X. Material ratio curve of 3D surface topography of additively manufactured parts: An attempt to characterise open surface pores. *Surf. Topogr. Metrol. Prop.* **2021**, *9*, 1–13. [CrossRef]
7. He, B.; Wei, C.; Liu, B.; Ding, S.; Shi, Z.; Wei, C. Three-dimensional surface roughness characterization and application. *Opt. Precis. Eng.* **2018**, *26*, 1994–2011. [CrossRef]
8. Whitehouse, D.J. Surface metrology. *Meas. Sci. Technol.* **1997**, *8*, 955. [CrossRef]
9. *ISO 4287*; Geometrical Product Specifications (GPS)–Surface Texture: Profile Method–Terms, Definitions and Surface Texture Parameters. International Organization for Standardization: Geneva, Switzerland, 1997.
10. Manesh, K.; Ramamoorthy, B.; Singaperumal, M. Numerical generation of anisotropic 3D non-Gaussian engineering surfaces with specified 3D surface roughness parameters. *Wear* **2010**, *268*, 1371–1379. [CrossRef]
11. Marinello, F.; Pezzuolo, A. Application of ISO 25178 standard for multiscale 3D parametric assessment of surface topographies. *IOP Conf. Ser. Earth Environ. Sci.* **2019**, *275*, 012011. [CrossRef]
12. Aver'Yanova, I.; Bogomolov, D.Y.; Poroshin, V. ISO 25178 standard for three-dimensional parametric assessment of surface texture. *Russ. Eng. Res.* **2017**, *37*, 513–516. [CrossRef]
13. Franco, L.A.; Sinatora, A. 3D surface parameters (ISO 25178-2): Actual meaning of Spk and its relationship to Vmp. *Precis. Eng.* **2015**, *40*, 106–111. [CrossRef]
14. Pawlus, P.; Reizer, R.; Wieczorowski, M. Characterization of the shape of height distribution of two-process profile. *Measurement* **2020**, *153*, 1–9. [CrossRef]
15. Qi, Q.; Li, T.; Scott, P.J.; Jiang, X. A correlational study of areal surface texture parameters on some typical machined surfaces. *Procedia Cirp.* **2015**, *27*, 149–154. [CrossRef]
16. Sedlaček, M.; Podgornik, B.; Vižintin, J. Planning surface texturing for reduced friction in lubricated sliding using surface roughness parameters skewness and kurtosis. *Proc. Inst. Mech. Eng. Part J J. Eng. Tribol.* **2012**, *226*, 661–667. [CrossRef]
17. Sedlaček, M.; Podgornik, B.; Vižintin, J. Correlation between standard roughness parameters skewness and kurtosis and tribological behaviour of contact surfaces. *Tribol. Int.* **2012**, *48*, 102–112. [CrossRef]
18. Skela, B.; Sedlaček, M.; Kafexhiu, F.; Podgornik, B. Wear behaviour and correlations to the microstructural characteristics of heat treated hot work tool steel. *Wear* **2019**, *426*, 1–11. [CrossRef]
19. He, B.; Petzing, J.; Webb, P.; Leach, R. Improving copper plating adhesion on glass using laser machining techniques and areal surface texture parameters. *Opt. Lasers Eng.* **2015**, *75*, 39–47. [CrossRef]
20. Zeng, Q.; Qin, Y.; Chang, W.; Luo, X. Correlating and evaluating the functionality-related properties with surface texture parameters and specific characteristics of machined components. *Int. J. Mech. Sci.* **2018**, *149*, 62–72. [CrossRef]
21. Todhunter, L.D.; Leach, R.; Lawes, S.D.; Blateyron, F. Industrial survey of ISO surface texture parameters. *CIRP J. Manuf. Sci. Technol.* **2017**, *19*, 84–92. [CrossRef]
22. Duo, Y.; Jinyuan, T.; Wei, Z.; Yuqin, W. Study on Roughness Parameters Screening and Characterizing Surface Contact Performance Based on Sensitivity Analysis. *J. Tribol.* **2022**, *144*, 1–9. [CrossRef]
23. Mironchenko, V. Correlation of roughness parameters Ra and Rq with an optical method of measuring them. *Meas. Tech.* **2009**, 52, 041502 . [CrossRef]
24. Nayak, P.R. Random process model of rough surfaces in plastic contact. *Wear* **1973**, *26*, 305–333. [CrossRef]
25. Li, L.; Tang, J.; Ding, H.; Liao, D.; Lei, D. On the linear transform technique for generating rough surfaces. *Tribol. Int.* **2021**, *163*, 1–13. [CrossRef]
26. Liao, D.R.; Shao, W.; Tang, J.Y.; Li, J.P. An Improved Rough Surface Modeling Method based on Linear Transformation Technique. *Tribol. Int.* **2018**, *119*, 786–794. [CrossRef]
27. Li, L.; Tang, J.; Wen, Y.; Shao, W. Characterization of ultrasonic-assisted grinding surface via the evaluation of the autocorrelation function. *Int. J. Adv. Manuf. Technol.* **2019**, *104*, 4219–4230. [CrossRef]
28. Nayak, P.R. Random Process Model of Rough Surfaces. *J. Tribol.* **1971**, *26*, 398–407. [CrossRef]
29. Li, L.; Tang, J.Y.; Wen, Y.Q.; Zhu, C.C. Numerical Simulation of Ultrasonic-Assisted Grinding Surfaces with Fast Fourier Transform. *J. Tribol.* **2020**, *142*, 1–10. [CrossRef]
30. Hauke, J.; Kossowski, T. Comparison of values of Pearson's and Spearman's correlation coefficients on the same sets of data. *Quaest. Geogr.* **2011**, *30*, 87–93. [CrossRef]

31. Yang, D.; Tang, J.; Zhou, W.; Wen, Y. Comprehensive study of relationships between surface morphology parameters and contact stress. *J. Mech. Sci. Technol.* **2021**, *35*, 4975–4985. [CrossRef]
32. Duo, Y.; Qibo, W.; Jinyuan, T.; Fujia, X.; Wei, Z.; Yuqin, W. Correlation analysis of roughness surface height distribution parameters and maximum mises stress. *Surf. Topogr. Metrol. Prop.* **2022**, *10*, 1–16. [CrossRef]
33. Shrestha, N. Detecting multicollinearity in regression analysis. *Am. J. Appl. Math. Stat.* **2020**, *8*, 39–42. [CrossRef]
34. Yang, D.; Tang, J.; Xia, F.; Zhou, W. Surface Roughness Characterization and Inversion of Ultrasonic Grinding Parameters based on Support Vector Machine. *J. Tribol.* **2022**, *144*, 1–9. [CrossRef]
35. *ISO Standard 25178-2:2012*; Geometrical Product Specifications (GPS)-Surface Texture: Area, Part 2: Terms, Definitions and Surface Texture Parameters. International Organization for Standardization: Geneva, Switzerland, 2012. Available online: https://www.iso.org/standard/42785.html (accessed on 25 August 2022).
36. Friendly, M.; Kwan, E. Where's Waldo? Visualizing collinearity diagnostics. *Am. Stat.* **2009**, *63*, 56–65. [CrossRef]
37. Wen, Y.; Tang, J.; Zhou, W.; Li, L.; Zhu, C. New analytical model of elastic-plastic contact for three-dimensional rough surfaces considering interaction of asperities. *Friction* **2022**, *10*, 217–231. [CrossRef]

materials

MDPI

Article

Hydrogen-Induced Dislocation Nucleation and Plastic Deformation of ⟨001⟩ and ⟨1$\bar{1}$0⟩ Grain Boundaries in Nickel

Jiaqing Li [1], Ziyue Wu [1], Lin Teng [1], Guanyu Deng [2], Rui Wang [2], Cheng Lu [2], Weidong Li [1], Xin Huang [1,*] and Yu Liu [3,*]

1 College of Chemical Engineering, Fuzhou University, Fuzhou 350116, China
2 School of Mechanical, Materials, Mechatronic and Biomedical Engineering, University of Wollongong, Wollongong, NSW 2522, Australia
3 School of Mechanical Engineering, Nantong University, Nantong 226019, China
* Correspondence: mmfhmt_hx@163.com (X.H.); liuyu_me@ntu.edu.cn (Y.L.)

Abstract: The grain boundary (GB) plays a crucial role in dominating hydrogen-induced plastic deformation and intergranular failure in polycrystal metals. In the present study, molecular dynamics simulations were employed to study the effects of hydrogen segregation on dislocation plasticity of a series of symmetrical tilt grain boundaries (STGBs) with various hydrogen concentrations. Our study shows that hydrogen both enhances and reduces dislocation nucleation events from STGBs, depending on different GB structures. Specifically, for ⟨001⟩ STGBs, hydrogen does not affect the mode of heterogeneous dislocation nucleation (HDN), but facilitates nucleation events as a consequence of hydrogen disordering the GB structure. Conversely, hydrogen retards dislocation nucleation due to the fact that hydrogen segregation disrupts the transformation of boundary structure such as Σ9 (2 2 $\bar{1}$) ⟨1$\bar{1}$0⟩ STGB. These results are helpful for deepening our understanding of GB-mediated hydrogen embrittlement (HE) mechanisms.

Keywords: molecular dynamics; hydrogen embrittlement; dislocation nucleation; plastic deformation

Citation: Li, J.; Wu, Z.; Teng, L.; Deng, G.; Wang, R.; Lu, C.; Li, W.; Huang, X.; Liu, Y. Hydrogen-Induced Dislocation Nucleation and Plastic Deformation of ⟨001⟩ and ⟨1$\bar{1}$0⟩ Grain Boundaries in Nickel. *Materials* **2022**, *15*, 6503. https://doi.org/10.3390/ma15186503

Academic Editor: Gueorgui Gueorguiev

Received: 23 August 2022
Accepted: 15 September 2022
Published: 19 September 2022

1. Introduction

As it is light, storable, energy-dense, and produces no direct emissions of pollutants or greenhouse gases, hydrogen has emerged as a critical pillar in any aspiring global net zero path. According to Goldman Sachs, hydrogen energy could develop into a market worth over $1 trillion a year [1]. However, the transportation and storage systems such as high-pressure pipelines, long-tube tailers, and vessels typically suffer from hydrogen embrittlement (HE) [2], severely obstructing hydrogen applications.

In the early 1870s, the H atom was firstly proposed to have damaging effects on iron and steel [3]. Since then, significant efforts have been made to characterize and comprehend the HE phenomenon of materials. Overall, the HE is a failure mechanism associated with the loss of plasticity caused by the presence of H in materials. The H atom may enter the material during production procedures such as welding or electroplating, namely called "internal hydrogen embrittlement" (IHE), or it can enter the material from the surrounding environment during service exposure, known as "environmental hydrogen embrittlement" (EHE) [3–5]. Solute H occupies and diffuses across interstitial lattice sites in metals and can be trapped to different degrees at defects [3,6]. So far, many HE mechanisms have been proposed. The most explained embrittlement mechanisms are hydrogen-enhanced decohesion (HEDE) [4,7–9] and hydrogen-enhanced localized plasticity (HELP) [10–14]. According to the HEDE mechanism, the decohesion occurs on account of metal charge transfer and weakening of interatomic bonds so that tensile separation of atoms occurs in preference to slip [5]. This mechanism can explain how the HE causes intergranular cracking along grain boundaries (GBs) [15]. In addition, the

HELP mechanism was proposed to explain the quasi-cleavage feature, which suggests that H atom enters and concentrates on the crack tip, and deformation occurs at the crack tip due to H-promoted dislocation motion [10,14,16]. Although these mechanisms are distinct in nature, they are not mutually exclusive and can frequently occur simultaneously, depending on materials, microstructures, and environmental factors.

GBs significantly affect the transport and segregation of H and fracture mechanisms in polycrystalline materials. The GB is more favorable for H adsorption than the interstitial lattice sites. The local critical H concentration in metals is the most crucial factor communed by all HE mechanisms. Therefore, to reach the critical H concentration, H atoms must be able to diffuse rapidly into the GBs and be efficiently trapped there. Numerous studies [17–21] have proven that the presence of geometrically required dislocations (GNDs) accelerates H diffusion along GBs. However, Curtin [22] showed that special GBs could trap H atoms and reduce H diffusion. H atoms trapped by GBs not only change the interaction between dislocations and GBs but also change the cleavage fracture along GBs [23]. Recent studies based on the observation of microstructures beneath the fracture features have shown that the presence of H changes the evolution of the dislocation structure. The fracture mode may develop from transgranular to intergranular [24]. Therefore, studies of HE mechanisms and damage have focused on H-GB interactions.

Due to the low concentrations and great diffusivity of H in engineering alloys, HE phenomena have historically been difficult to be analyzed using conventional experimental approaches. Recent advancements in experimental techniques such as atom probe tomography [25] and neutron tomography [26] have permitted direct microstructural investigations of H segregation. However, these experimental devices are complex to customize and difficult to implement. Consequently, numerous simulation-based studies have been developed to improve our understanding of HE in metals [27–33]. For example, Liang et al. [34] proposed that H accelerates dislocation microstructure evolution through a H-enhanced plasticity mechanism, and the interaction of dislocations with GBs can change the ultimate failure process. Li et al. [35] found that boundary disruption and the concentration of local stress states on the GB through plastic processes promote the embrittlement effect of H atoms in metallic materials. Therefore, simulation-based techniques play an important role in unveiling more details about HE in metals.

The present study aims to explore the dislocation nucleation and plastic deformation mechanisms of GBs tilted along $\langle 001 \rangle$ and $\langle 1\bar{1}0 \rangle$ axes in nickel with different H concentrations using molecular dynamics (MD) simulations. The content of the present study is organized as follows: the simulation method is introduced in Section 2, the GB structure, hydrogen segregation map and dislocation nucleation mechanisms are presented and discussed in Section 3, and our conclusions are summarized in Section 4.

2. Materials and Methods

The established bicrystal model with symmetric tilt grain boundary (STGB) is shown in Figure 1. The Z axis is defined as the tilt axis ($\langle 001 \rangle$ and $\langle 1\bar{1}0 \rangle$). The STGB was respectively constructed by rotating grain A and grain B with $\pm\frac{\theta}{2}$ along the tilt axes, forming a symmetrical tilt boundary. Table 1 presents the information of all chosen simulation models. The simulation model applied periodic boundary conditions along all directions, and the length of the model in the X direction and Y direction should be enough to minimize the image force.

In the present study, all MD simulations were performed using large-scale atomic/molecular massively parallel simulator (Lammps) software [36] with the embedded-atom method (EAM) interatomic potentials for Ni-H developed by Angelo et al. [37]. The Voronoi tessellation of atoms was constructed with the Voro++ code to find possible hydrogen trapping sites in each GB structure [38].

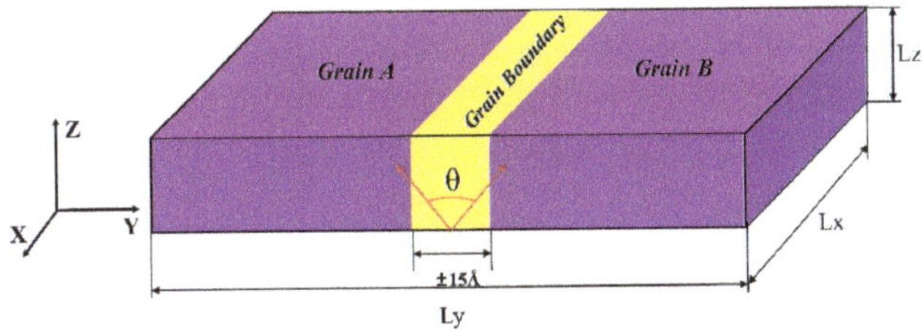

Figure 1. Schematic illustration of the STGB with tilt axis along the Z direction and the normal to the boundary plane along the Y direction.

Table 1. Characterization of all considered GBs, including the model size, misorientation angle (θ), GB structure unit, and GB energy γ (mJ·m^{-2}).

GB Type	Angle θ (°)	Model Size X × Y × Z (nm)	GB Structure	GB Energy γ (mJ·m^{-2})
Σ5 (3 1 0)⟨001⟩	36.87°	11.13 × 22.26 × 3.52	| *DBDB* |	1126.25
Σ29 (5 2 0)⟨001⟩	43.60°	15.16 × 22.75 × 3.52	| *BBD* |	1240.23
Σ5 (2 1 0)⟨001⟩	53.13°	15.74 × 22.04 × 3.52	| *BBB* |	1183.13
Σ17 (5 3 0)⟨001⟩	61.93°	16.42 × 24.63 × 3.52	| *BABA* |	1110.48
Σ37 (7 5 0)⟨001⟩	71.08°	15.14 × 24.24 × 3.52	| *BAAABAA* |	999.80
Σ11 (1 1 $\bar{3}$)⟨1$\bar{1}$0⟩	50.48°	13.21 × 23.35 × 4.98	| *CC* |	375.83
Σ57 (4 4 $\bar{5}$)⟨1$\bar{1}$0⟩	97.05°	18.79 × 26.58 × 4.98	| *CD* |	731.83
Σ3 (1 1 $\bar{1}$)⟨1$\bar{1}$0⟩	109.47°	16.38 × 23.17 × 4.98	| *DDDD* |	50.38
Σ123 (7 7 $\bar{5}$)⟨1$\bar{1}$0⟩	135.99°	11.04 × 23.42 × 4.98	| *EDDED* |	860.61
Σ9 (2 2 $\bar{1}$)⟨1$\bar{1}$0⟩	141.06°	11.95 × 23.23 × 4.98	| *EE* |	1097.45

The local *H* concentration at the GBs is represented by the following equation [39]:

$$C_H = \frac{N_H}{A} \quad (1)$$

where *A* is the cross-sectional area of the GB, and N_H is the number of *H* atoms in the region.

The GB energy is defined as:

$$E_{GB} = \frac{E_{region} - E_{atom} N_{atom,Ni}}{A} \quad (2)$$

where E_{region} is the GB energy of *Ni* atoms in the region ±15 Å above and below the GB, the cohesive energy of each *Ni* atom in bulk is designated as E_{atom}, $N_{atom,Ni}$ is the number of Ni atoms in the selected region.

MD simulations were performed under the isothermal–isobaric ensemble (NPT) for model with and without H, respectively. The Nose-Hoover method [40,41] was used to keep the system temperature at 10 K. The integration time step of MD was set to 0.001 ps, and the deformation was achieved by stretching the box dimensions and then averaging the displacement over each atom at a strain rate of 10^{-8} s^{-1}. Crystal structures were characterized by common neighbor analysis (CNA) [42] and centro-symmetry parameter (CSP) [43], and dislocation types were identified by the dislocation extraction algorithm (DXA) [44] in the Open Visualization Tool (OVITO) [45].

3. Results and Discussion

3.1. GB Energy and GB Structure

The GB structure described by the structural unit model (SUM) is shown in Figure 2. The Σ5 (2 1 0) STGB is composed of only B structural unit (SU), and the ratio of B SU decreases with an increase in the misorientation angle, as presented by Σ17 (5 3 0) and Σ37 (7 5 0) STGBs in Figure 2d,e. For the three ⟨1$\bar{1}$0⟩ tilted bicrystals with STGBs of Σ11 (1 1 $\bar{3}$), Σ3 (1 1 $\bar{1}$) and Σ9 (2 2 $\bar{1}$), the GB contains exactly *C* SU, *D* SU, *E* SU, respectively.

Each GB at the misorientation angle between θ = 50.48° and θ = 109.47° results from a combination of *C* SU and *D* SU. Likewise, the misorientation angle between θ = 109.47° and θ = 141.06° results from a combination of *D* SU and *E* SU. Taking the θ = 135.99° boundary as an example, there are two *E* SU and three *D* SU cycles to build the basic GB structure.

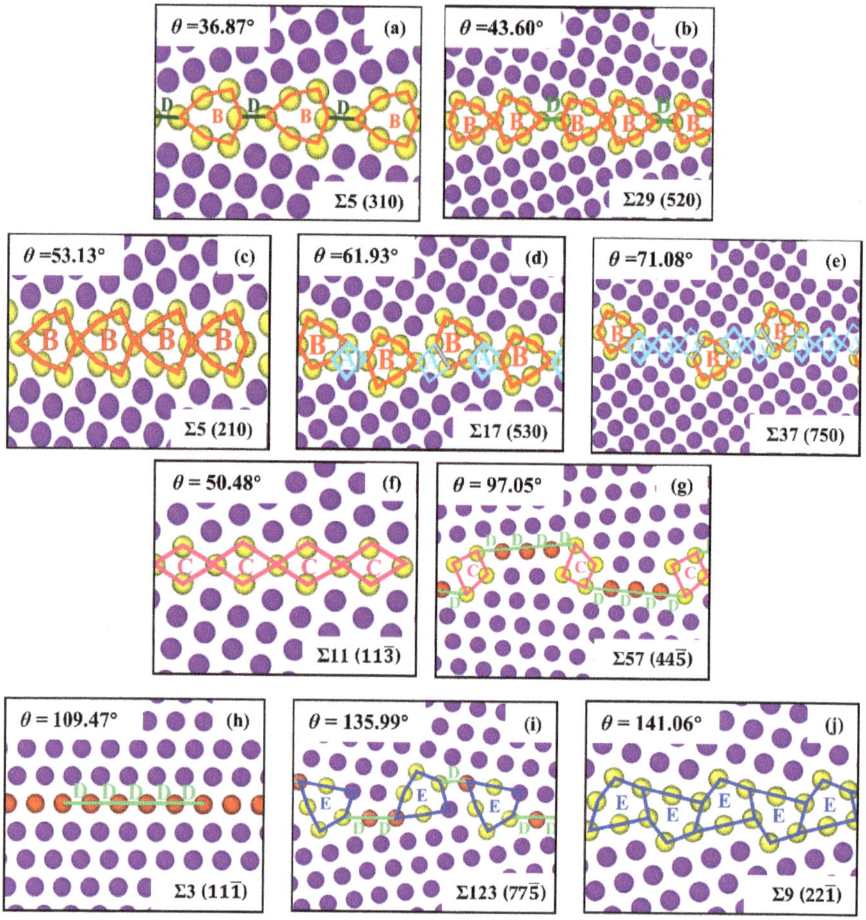

Figure 2. The atomistic configurations of several GB structure units along (**a**–**e**) ⟨001⟩ tilt axis and (**f**–**j**) $\langle 1\bar{1}0 \rangle$ tilt axis. Images are colored by CNA, where the atoms in FCC structure are colored in purple, the atoms in HCP structure are colored in red, the atoms in other structure are colored in yellow.

Figure 3 shows the GB energy as a function of misorientation angle for ⟨001⟩ and $\langle 1\bar{1}0 \rangle$ STGBs. Based on previous experiments [46] and simulation studies [47], it has been shown that there is a strong correlation between GB energy and GB structure. In Figure 3b, when the misorientation angle is θ = 109.47°, Σ3 $(1\ 1\ \bar{1})$ STGB has the lowest GB energy, corresponding to specific structural models. It is clear from Figure 2h that Σ3 $(1\ 1\ \bar{1})$ GB has a line defect structure, typically known as a coherent twin boundary. Such a GB formed along the dense plane has a lower energy, which is in accordance with a previous finding by Wolf and Philpot [48].

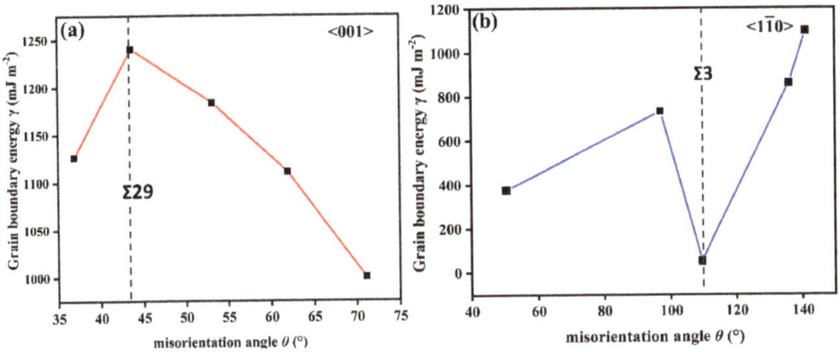

Figure 3. GB energy as a function of misorientation angle for (**a**) ⟨001⟩ and (**b**) ⟨1$\bar{1}$0⟩ STGBs. The GB energy for Ni is represented by the black square. The vertical dashed line indicates the GB of interest.

3.2. Hydrogen Segregation at GBs

A controversial area in comprehending HE occurrence is the role of GBs in the diffusion and trapping of H in FCC materials. Based on the GB structures identified in Figure 2, Figure 4 shows the H segregation energy diagrams for six typical GBs; the colored spheres stand for the possible sites where H atoms could be confined, while the black spheres represent Ni atoms. The H segregation energy diagrams at different GBs are variable because the structure of each GB is distinct. When comparing Figure 4d,f, it is clear that the Σ9 (2 2 $\bar{1}$) STGB traps more H atoms than the Σ11 (1 1 $\bar{3}$) STGB because the E SU has a larger free volume than the C SU. It is worth noting that for the Σ3 (11$\bar{1}$) STGB, essentially no H atoms are segregated at the boundary, and the H segregation energy is negligibly small. In contrast, the other STGBs possess multiple trapping sites and low value of H segregation energy.

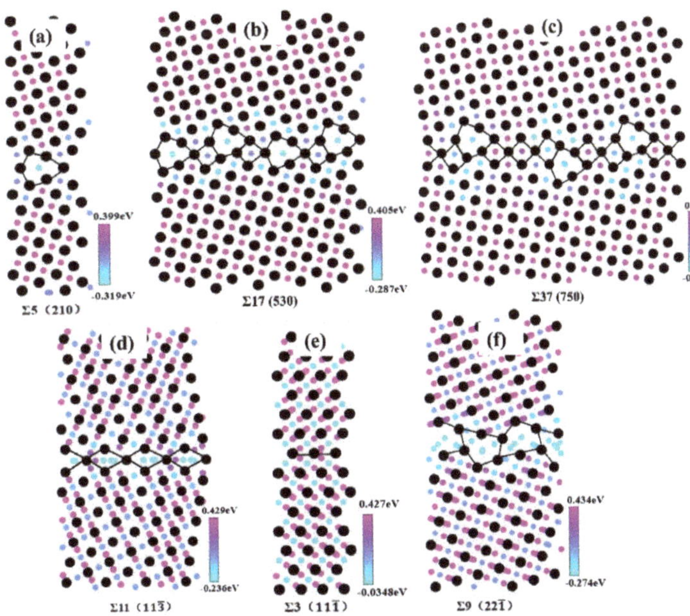

Figure 4. Hydrogen segregation energy diagrams for (**a**–**c**) ⟨001⟩ and (**d**–**f**) ⟨1$\bar{1}$0⟩ STGBs.

To find out the distribution of H segregation to STGBs, Figure 5 manifests that for the majority of GBs, possible H trapping sites with energies less than 0 eV are located within a region ±5 Å from the boundary planes. In the region away from the GBs, the segregation energy approaches 0 eV, being suggestive of a tendency for H diffusion from the bulk into the GBs. This observation agrees well with previous studies [49,50], which reported that some stable FCC structures such as Ni exhibited preferential trapping of H atoms at GBs. Recent experiments using atom probe tomography [25,51] also identified such H retention or trapping at GBs in metals.

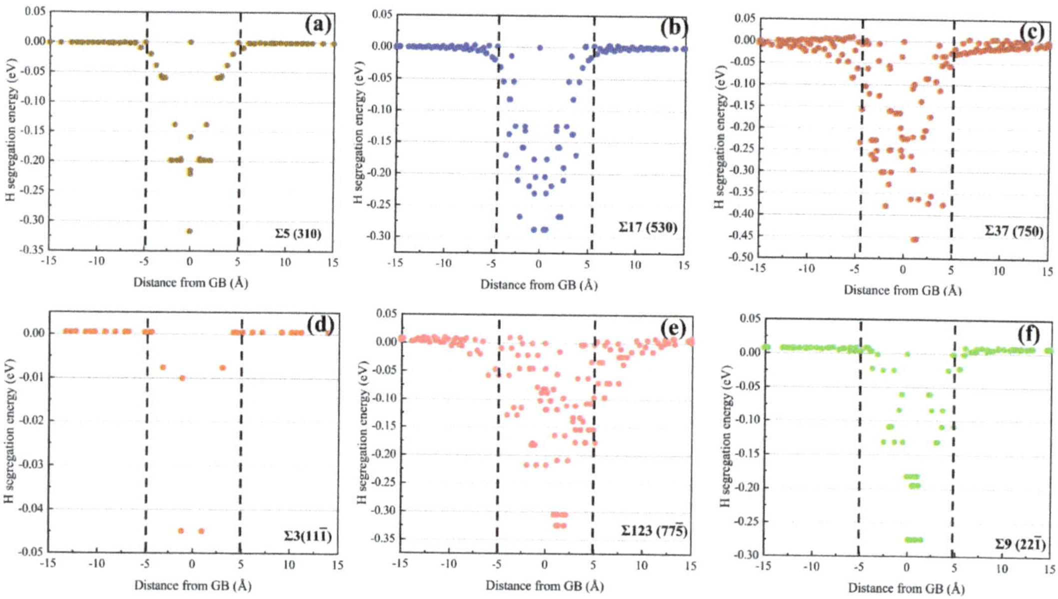

Figure 5. Hydrogen segregation energy distribution as a function of GB distance for **(a–c)** ⟨001⟩ and **(d–f)** ⟨1$\bar{1}$0⟩ STGBs.

3.3. Effects of Hydrogen on the Dislocation Nulceaiton and Plastic Deformation of GBs

3.3.1. Dislocation Nucleation Strength Influenced by Hydrogen in GB Model

The mechanical behavior and deformation mechanisms of polycrystalline metals largely depend on the characteristics of their defects or interfaces [52]. In the present study, each of the established bicrystal models was stretched along the Y-axis direction. All curves exhibit a series of stress decreases corresponding to dislocation nucleation. Specifically, the first stress drop is related with the onset of dislocation plasticity; therefore, the highest stress just prior to the first stress drop is referred to as the yield stress.

In order to reveal the connection between the misorientation angle and yield stress, Figure 6 shows the strain-stress curves of selected STGBs and the curves of misorientation angle of GBs with their corresponding yield stress. The yield stress of bicrystal models declines with increasing misorientation angle for ⟨001⟩ STGBs. On the other hand, for ⟨1$\bar{1}$0⟩ STGBs, two situations are presented: the yield stress increases with an increasing angle, while the stress in the bicrystal model abruptly reduces when the GB misorientation angle is beyond 109.47°. It can be observed that the yield stress of the Σ123 (7 7 $\bar{5}$) and Σ9 (2 2 $\bar{1}$) STGBs is lower than other GBs. Such a reduced yield stress can be linked with the E SUs [53].

To further investigate the effect of segregated H on the yield stress of different STGBs, the stress-strain curves of selected STGBs for two tilt axes at different H concentrations are shown in Figure 7. It is clear that, for ⟨001⟩ STGBs, the yield stress reduces with increasing

H concentration. For $\langle 1\bar{1}0 \rangle$ STGBs, the stress-strain curves vary in the Σ3 (1 1 $\bar{1}$) STGB and Σ123 (7 7 $\bar{5}$) STGB. The Σ3 (1 1 $\bar{1}$) STGB show a gradual decrease in the yield stress with increasing H concentration; however, Σ123 (7 7 $\bar{5}$) STGB manifests an increase in the yield stress. As mentioned above, such a difference in stress variation may be attributable to the aggregation of H atoms within the GB structure and the modification of the GB structure pattern during the stretching process.

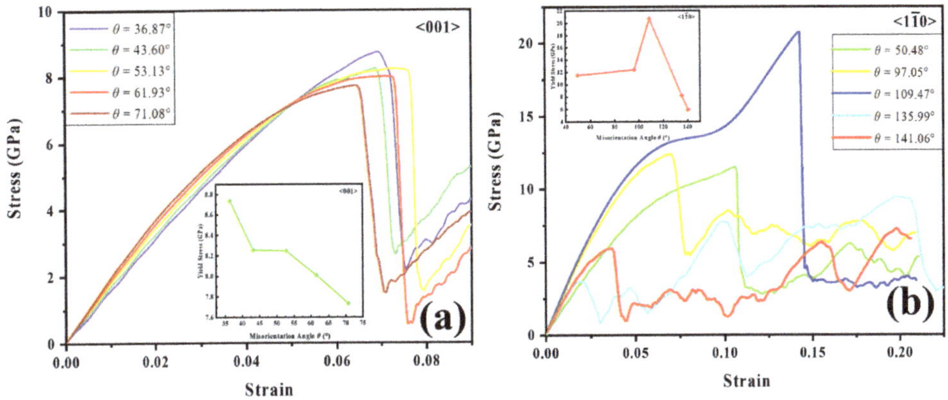

Figure 6. Stress-strain curves for (**a**) ⟨001⟩ and (**b**) $\langle 1\bar{1}0 \rangle$ STGBs. The insets show the dependency of yield stress on misorientation angle.

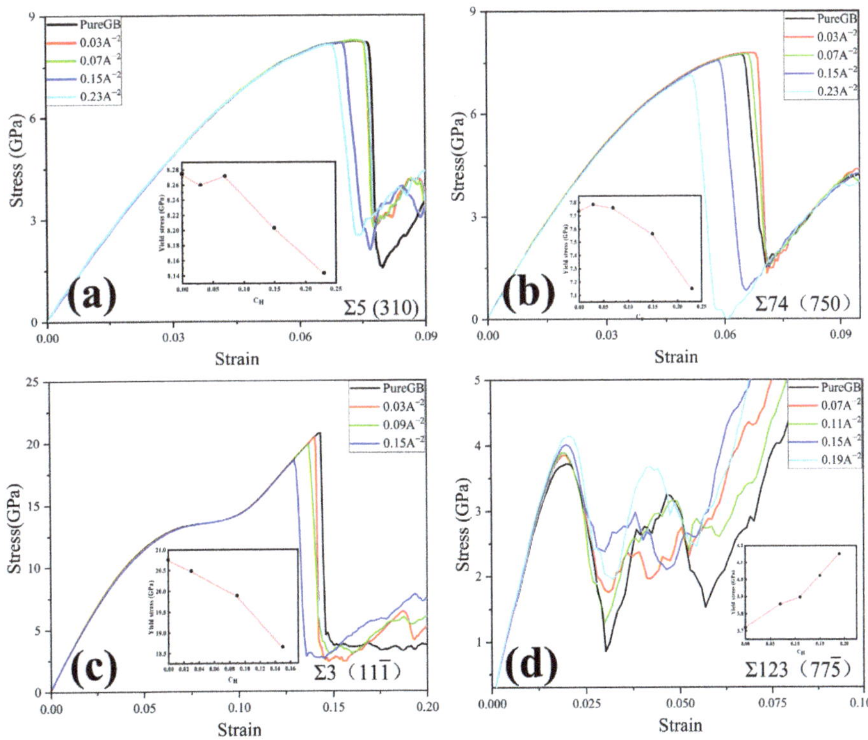

Figure 7. Stress-strain curves for (**a**,**b**) ⟨001⟩ and (**c**,**d**) $\langle 1\bar{1}0 \rangle$ STGBs with various hydrogen concentrations. The insets show the dependency of yield stress on hydrogen concentration.

3.3.2. The Influence of Hydrogen Segregation on Dislocation Nucleation of ⟨001⟩ STGB

In Figure 8, the exact heterogeneous dislocation nucleation (HDN) mechanism can be seen during tensile deformation of the Σ17 (5 3 0) STGB and Σ37 (7 5 0) STGB without and with H, respectively. The HDN mechanism operates via dislocation nucleation from the GBs, and dislocation loops made up of single Shockley dislocations subsequently slide into the grains.

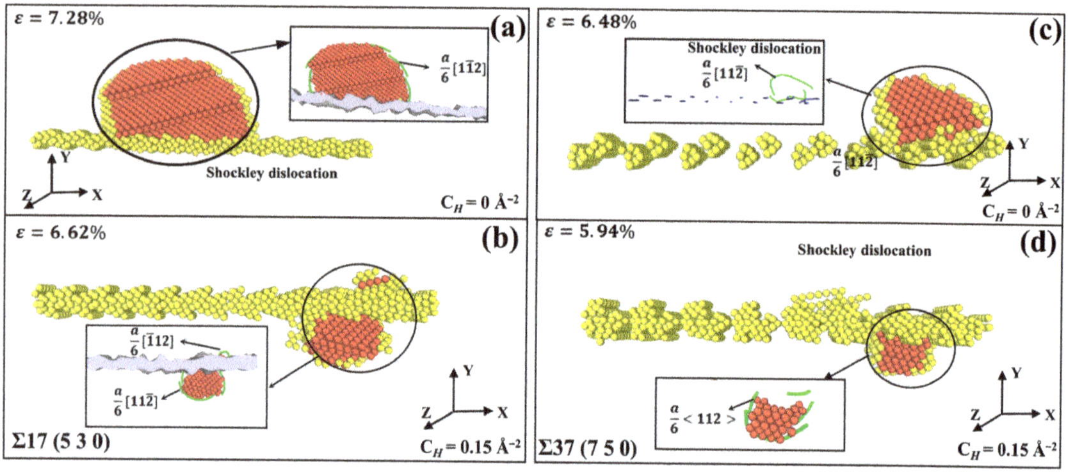

Figure 8. MD snapshots of dislocation nucleation from Ni bicrystal with (**a**,**b**) Σ17 (5 3 0) STGB and (**c**,**d**) Σ37 (7 5 0) STGB without and with H.

To further investigate the details of dislocation nucleation, the Σ5 (3 1 0) STGB at different deformation stages for various H concentrations is taken as an example. In Figure 9a, when the strain reaches ε= 7.24%, partial dislocation loops with edge and screw features concurrently form at the top and bottom of the GB. In light of DXA analysis, Shockley partial dislocations with the Burge vectors of $\frac{a}{6}[112]$ and $\frac{a}{6}[11\bar{2}]$ originate from the GB and propagate along the (111) and (11$\bar{1}$) planes, respectively. With increasing tensile stress, more dislocations gradually and continuously nucleate from the GB plane at a strain of ε = 7.34%.

Figure 9b shows the MD snapshots of Σ5 (3 1 0) STGB at C_H = 0.07Å^{-2}. In contrast to the H-free case, the boundary structure is significantly disordered, evidenced by a thicker boundary plane. Furthermore, it is clear that the dislocation plasticity occurs earlier; $\frac{a}{6}[11\bar{2}]$ dislocation nucleates at a strain of ε = 6.93%. This can be attributed to that H addition disorders the boundary structure and increases the strain energy, which promotes dislocation nucleation events [33]. As the tensile strain further increases, a series of dislocations slide continuously into the upper and lower grain regions, corresponding to the plastic deformation process.

In Figure 9c, when the concentration of H is increased to C_H = 0.15 Å^{-2}, dislocation nucleation becomes earlier, leading to a lower yield strain of ε = 6.70% and a reduced yield stress in Figure 7a. It is worth noting that dislocation events can only be observed in the lower grain, different from the H-free and C_H = 0.07Å^{-2} cases. It may be due to that the trapped H leading to an asymmetric GB structure, in which partial dislocations with Burger vectors of $\frac{a}{6}[11\bar{2}]$, $\frac{a}{6}[112]$ and $\frac{a}{6}[\overline{112}]$ are nucleated and emitted into the lower grain.

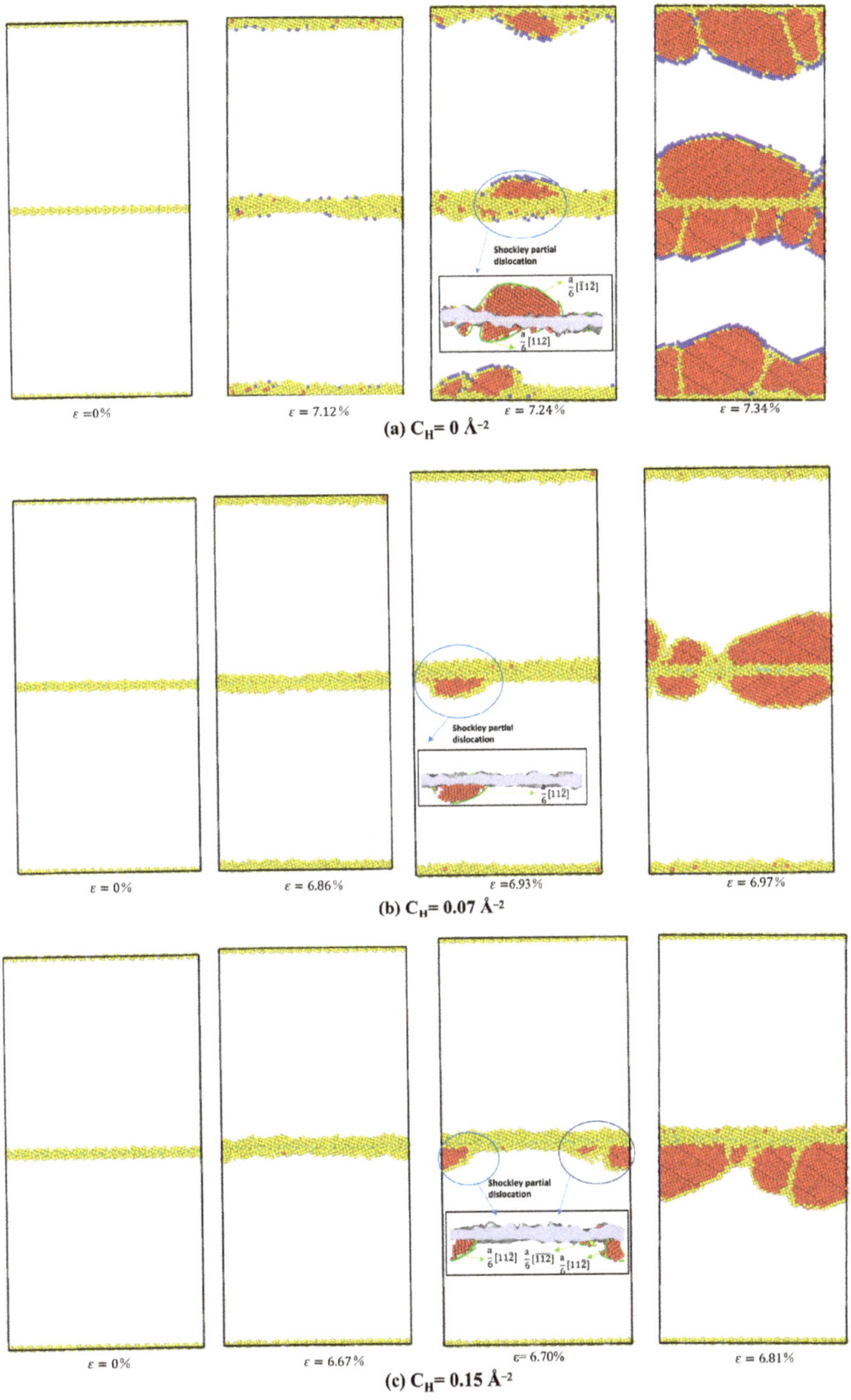

Figure 9. MD snapshots of dislocation nucleation from Ni bicrystal with Σ5 (3 1 0) STGB for various H concentrations. Only the defective arrangement of atoms is shown.

Overall, the results show that the prominent deformation mechanism of <0 0 1> STGBs is the HDN. H segregation disorders the GB structure and promotes the HDN event within the framework of HELP mechanism. This observation directly explains why the yield stress of <0 0 1> STGBs is reduced with increasing H concentration.

3.3.3. The Influence of Hydrogen Segregation on Dislocation Nucleation of $\langle 1\bar{1}0 \rangle$ STGB

The results of MD simulations for several $\langle 1\bar{1}0 \rangle$ STGBs are shown in Figures 10 and 11. Figure 10a,b depict the dislocation nucleation process of Σ11 (1 1 $\bar{3}$) STGB at 10K. The structure of Σ11(1 1 $\bar{3}$) STGB comprises a straightforward arrangement of C SUs with a relatively low GB energy value, making its nucleation mechanism quite special. The dislocations begin to form at the GB and slide along (111) and (11$\bar{1}$) planes into the upper and lower grains when the yield point is reached at ε = 10.06%, leaving behind a series of extrinsic stacking fault. In the presence of H, as depicted in Figure 10b, some dislocations with Burgers vectors of $\frac{a}{6}\left[\bar{1}12\right]$ and $\frac{a}{6}\left[1\bar{1}2\right]$ tend to appear, and both intrinsic and extrinsic stacking fault begin to form at the GB. The occurrence of intrinsic stacking fault can be associated with the fact that H atoms affect the structural alteration of the GB, which obstructs the conversion of intrinsic stacking fault to external stacking fault.

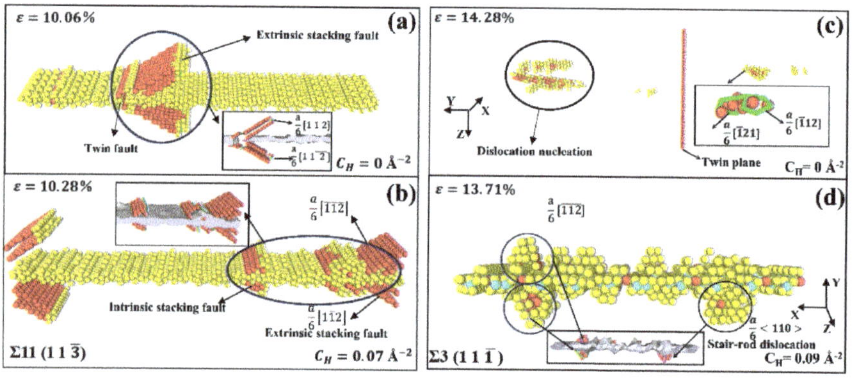

Figure 10. MD snapshots of dislocation nucleation from Ni bicrystal with (**a**,**b**) Σ11 (1 1 $\bar{3}$) STGB and (**c**,**d**) Σ3 (1 1 $\bar{1}$) STGB without and with H.

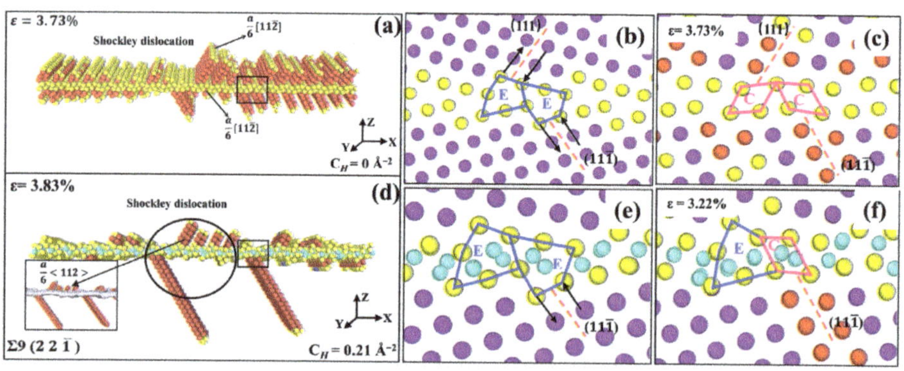

Figure 11. MD snapshots of dislocation nucleation from Ni bicrystal with Σ9 (2 2 $\bar{1}$) STGB without and with H. (**b**,**c**,**e**,**f**) Enlarged images for nucleation region of the boundary marked in (**a**,**d**), respectively.

For the Σ3 (1 1 $\bar{1}$) STGB, the GB energy is the lowest of all $\langle 1\bar{1}0 \rangle$ STGBs. As it has a simple GB structure, there is no extra free volume that can accommodate dislocations. As seen in Figure 10c, dislocations are nucleated from the grains in the form of dislocation loops, and there are no dislocations nucleating from the GB plane. Conversely, in the presence of H, as shown in Figure 10d, the local GB atoms change, and the arrangement of GB atom becomes increasingly disordered. The change in GB structure caused by the presence of H triggers dislocation nucleation from the boundary plane, with a series of Shockley dislocations and stair-rod dislocations nucleating at the GB when the yield stress is reached.

As mentioned previously, the $\Sigma 9$ $(2\,2\,\bar{1})$ STGB is comprised of E SUs. Because of the large free volume in E SUs, there is a specific structural deformation during dislocation nucleation. As shown in Figure 11a, the dislocations are nucleated from the GB, and a series of $\frac{a}{6}\langle 112\rangle$ Shockley partial dislocations propagate on the (111) and $(11\bar{1})$ planes. The nucleation event is accompanied by a transformation of E SUs into C SUs, as shown in Figure 11b,c. When considering the presence of H, Figure 11d shows that dislocation nucleation is retarded, and only several dislocations propagate from the GB. Atomic configurations reveal that H atoms segregated at the GB suppress the collapse of E SUs, thus hindering the nucleation event.

4. Summary

In the present study, H segregation and trapping at various $\langle 001\rangle$ and $\langle 1\bar{1}0\rangle$ bicrystal models with STGBs were identified using MD simulations. The effect of H on dislocation nucleation and plastic deformation of STGBs was subsequently investigated. Several conclusions are drawn as follows.

(1) The ability of the GB to trap H atoms is very sensitive to the GB structure and GB energy. The GBs with higher GB energy own a larger free volume, thus trapping more H atoms. There are essentially no H atoms segregating at the $\Sigma 3$ $(11\bar{1})$ STGB, while for other STGBs, H atoms are distributed within a region ± 5 Å from the boundary planes.

(2) For $\langle 001\rangle$ STGB, the plastic deformation mechanism is dominated by HDN. H segregation disorders the GB structure and promotes the HDN event, leading to a reduced yield stress with increasing H concentration.

(3) For $\langle 1\bar{1}0\rangle$ STGB, the yield stress is increased or decreased as the H concentration increases, depending on the GB structure. The reduced stress is associated with the H atoms disordering the boundary structure, while the increased stress is attributed to the fact that the presence of H inhibits the structural transformation and dislocation nucleation.

There are strong links between our simulation results and experimental observations. Analysis of stress–strain curves generated from uniaxial tension of $\langle 001\rangle$ STGBs indicates that the introduction of H causes a reduction in the yield stress required for dislocation nucleation from the boundary. This is in accordance with experimental observations of H-enhanced dislocation generation [54,55] as well as slip transfer across GBs [2,56,57]. In addition, it is found that H-segregated GBs tend to hold in their initial configurations and suppress structural evolution or predissociation, as shown in Figure 11. This observation was revealed by Ferreira et al. [58], who demonstrated that solute H could stabilize the dislocations and stop the evolution of structural configurations. These findings can provide a new perspective towards understanding the experimentally-observed HE in metals.

Author Contributions: J.L.: Methodology, Software, Writing—Original draft; C.L. and L.T.: Conceptualization, Supervision, Project administration; Z.W. and G.D.: Software, Writing—Review and editing; R.W. and W.L.: Visualization, Investigation; X.H. and Y.L.: Software, Validation. All authors have read and agreed to the published version of the manuscript.

Funding: This study was financially supported by the National Natural Science Foundation of China (No. 52205145), Research Program for Young and Middle-Aged Teachers of Fujian Province (No. JAT210038), Talent Program of Fuzhou University (No. XRC-22024).

Institutional Review Board Statement: Not applicable.

Informed Consent Statement: Not applicable.

Data Availability Statement: The data presented in this study are available on request from the corresponding author.

Conflicts of Interest: The authors declare no conflict of interest.

Nomenclature

GB	grain boundary
STGBs	symmetrical tilt grain boundaries
HDN	heterogeneous dislocation nucleation
HE	hydrogen embrittlement
IHE	internal hydrogen embrittlement
EHE	environmental hydrogen embrittlement
HEDE	hydrogen-enhanced decohesion
HELP	hydrogen enhanced localized plasticity
MD	molecular dynamics
EAM	embedded-atom method
LAMMPS	large-scale atomic/molecular massively parallel simulator
NPT	isothermal–isobaric ensemble
CNA	common neighbor analysis
CSP	centro-symmetry parameter
DXA	dislocation extraction algorithm
OVITO	open visualization tool

References

1. Khanna, M. COVID-19: A cloud with a silver Lining for renewable energy? *Appl. Econ. Perspect. Policy* **2021**, *43*, 73–85. [CrossRef] [PubMed]
2. Robertson, I.M.; Sofronis, P.; Nagao, A.; Martin, M.L.; Wang, S.; Gross, D.W.; Nygren, K.E. Hydrogen embrittlement understood. *Metall. Mater. Trans. A* **2015**, *46*, 2323–2341. [CrossRef]
3. Lynch, S. Hydrogen embrittlement phenomena and mechanisms. *Corros. Rev.* **2012**, *30*, 105–123. [CrossRef]
4. Troiano, A.R. The role of hydrogen and other interstitials in the mechanical behavior of metals. *Metallogr. Microstruct. Anal.* **2016**, *5*, 557–569. [CrossRef]
5. Oriani, R. A mechanistic theory of hydrogen embrittlement of steels. *Ber. Bunsenges. Phys. Chem.* **1972**, *76*, 848–857.
6. Lynch, S. Progress towards understanding mechanisms of hydrogen embrittlement and stress corrosion cracking. In Proceedings of the CORROSION 2007, Nashville, TN, USA, 11–15 March 2007; pp. 074931–0749355.
7. He, S.; Ecker, W.; Pippan, R.; Razumovskiy, V.I. Hydrogen-enhanced decohesion mechanism of the special Σ5 (0 1 2)[1 0 0] grain boundary in Ni with Mo and C solutes. *Comput. Mater. Sci.* **2019**, *167*, 100–110. [CrossRef]
8. Knott, J.F. Fracture toughness and hydrogen-assisted crack. *Hydrog. Eff. Mater.* **2013**, *387*, 385–408.
9. Li, X.; Ma, X.; Zhang, J.; Akiyama, E.; Wang, Y.; Song, X. Review of hydrogen embrittlement in metals: Hydrogen diffusion, hydrogen characterization, hydrogen mechanism and prevention. *Acta Metall. Sin. (Engl. Lett.)* **2020**, *33*, 759–773. [CrossRef]
10. Wasim, M.; Djukic, M.B.; Ngo, T.D. Influence of hydrogen-enhanced plasticity and decohesion mechanisms of hydrogen embrittlement on the fracture resistance of steel. *Eng. Fail. Anal.* **2021**, *123*, 105312. [CrossRef]
11. Li, J.; Lu, C.; Pei, L.; Zhang, C.; Tieu, K. Influence of solute hydrogen on the interaction of screw dislocations with vicinal twin boundaries in nickel. *Scripta Mater.* **2019**, *173*, 115–119. [CrossRef]
12. Huang, S.; Zhang, Y.; Yang, C.; Hu, H. Fracture strain model for hydrogen embrittlement based on hydrogen enhanced localized plasticity mechanism. *Int. J. Hydrogen Energy* **2020**, *45*, 25541–25554. [CrossRef]
13. Kim, S.K.; Lee, C.S.; Kim, M.H.; Lee, J.M. Numerical analysis of hydrogen transport using a hydrogen-enhanced localized plasticity mechanism. *Proc. World Acad. Sci. Eng. Technol.* **2012**, *58*, 398–401.
14. Nagao, A.; Dadfarnia, M.; Somerday, B.P.; Sofronis, P.; Ritchie, R.O. Hydrogen-enhanced-plasticity mediated decohesion for hydrogen-induced intergranular and "quasi-cleavage" fracture of lath martensitic steels. *J. Mech. Phys. Solids* **2018**, *112*, 403–430. [CrossRef]
15. Bechtle, S.; Kumar, M.; Somerday, B.P.; Launey, M.E.; Ritchie, R.O. Grain-boundary engineering markedly reduces susceptibility to intergranular hydrogen embrittlement in metallic materials. *Acta Mater.* **2009**, *57*, 4148–4157. [CrossRef]
16. Martin, M.L.; Dadfarnia, M.; Nagao, A.; Wang, S.; Sofronis, P. Enumeration of the hydrogen-enhanced localized plasticity mechanism for hydrogen embrittlement in structural materials. *Acta Mater.* **2019**, *165*, 734–750. [CrossRef]
17. Oudriss, A.; Creus, J.; Bouhattate, J.; Conforto, E.; Berziou, C.; Savall, C.; Feaugas, X. Grain size and grain-boundary effects on diffusion and trapping of hydrogen in pure nickel. *Acta Mater.* **2012**, *60*, 6814–6828. [CrossRef]

18. He, Y.; Su, Y.; Yu, H.; Chen, C. First-principles study of hydrogen trapping and diffusion at grain boundaries in γ-Fe. *Int. J. Hydrogen Energy* **2021**, *46*, 7589–7600. [CrossRef]
19. Muránsky, O.; Karatchevtseva, I.; Danon, A.E.; Holmes, R.; Huai, P.; Zhang, Z. Impact of dislocations and dislocation substructures on molten salt corrosion of alloys under plasticity-imparting conditions. *Corros. Sci.* **2020**, *176*, 108915. [CrossRef]
20. Ogawa, Y.; Takakuwa, O.; Okazaki, S.; Funakoshi, Y.; Matsuoka, S.; Matsunaga, H. Hydrogen-assisted fatigue crack-propagation in a Ni-based superalloy 718, revealed via crack-path crystallography and deformation microstructures. *Corros. Sci.* **2020**, *174*, 108814. [CrossRef]
21. Ogawa, Y.; Takakuwa, O.; Okazaki, S.; Okita, K.; Funakoshi, Y.; Matsunaga, H.; Matsuoka, S. Pronounced transition of crack initiation and propagation modes in the hydrogen-related failure of a Ni-based superalloy 718 under internal and external hydrogen conditions. *Corros. Sci.* **2019**, *161*, 108186. [CrossRef]
22. Song, J.; Curtin, W. A nanoscale mechanism of hydrogen embrittlement in metals. *Acta Mater.* **2011**, *59*, 1557–1569. [CrossRef]
23. Jothi, S.; Croft, T.; Brown, S. Influence of grain boundary misorientation on hydrogen embrittlement in bi-crystal nickel. *Int. J. Hydrogen Energy* **2014**, *39*, 20671–20688. [CrossRef]
24. Bertsch, K.; Wang, S.; Nagao, A.; Robertson, I. Hydrogen-induced compatibility constraints across grain boundaries drive intergranular failure of Ni. *Mater. Sci. Eng. A* **2019**, *760*, 58–67. [CrossRef]
25. Chen, Y.S.; Lu, H.; Liang, J.; Rosenthal, A.; Liu, H.; Sneddon, G.; Cairney, J.M. Observation of hydrogen trapping at dislocations, grain boundaries, and precipitates. *Science* **2020**, *367*, 171–175. [CrossRef] [PubMed]
26. Griesche, A.; Dabah, E.; Kannengiesser, T.; Kardjilov, N.; Hilger, A.; Manke, I. Three-dimensional imaging of hydrogen blister in iron with neutron tomography. *Acta Mater.* **2014**, *78*, 14–22. [CrossRef]
27. Tehranchi, A.; Curtin, W. Atomistic study of hydrogen embrittlement of grain boundaries in nickel: I. Fracture. *J. Mech. Phys. Solids* **2017**, *101*, 150–165. [CrossRef]
28. Yu, H.; Olsen, J.S.; Olden, V.; Alvaro, A.; He, J.; Zhang, Z. Cohesive zone simulation of grain size and misorientation effects on hydrogen embrittlement in nickel. *Eng. Fail. Anal.* **2017**, *81*, 79–93. [CrossRef]
29. Huang, S.; Chen, D.; Song, J.; McDowell, D.L.; Zhu, T. Hydrogen embrittlement of grain boundaries in nickel: An atomistic study. *NPJ Comput. Mater.* **2017**, *3*, 1–8. [CrossRef]
30. Alvaro, A.; Jensen, I.T.; Kheradmand, N.; Løvvik, O.; Olden, V. Hydrogen embrittlement in nickel, visited by first principles modeling, cohesive zone simulation and nanomechanical testing. *Int. J. Hydrogen Energy* **2015**, *40*, 16892–16900. [CrossRef]
31. Geng, W.; Freeman, A.J.; Wu, R.; Geller, C.; Raynolds, J. Embrittling and strengthening effects of hydrogen, boron, and phosphorus on a Σ 5 nickel grain boundary. *Phys. Rev. B* **1999**, *60*, 7149. [CrossRef]
32. Yamaguchi, M.; Shiga, M.; Kaburaki, H. First-principles study on segregation energy and embrittling potency of hydrogen in niσ5 (012) tilt grain boundary. *J. Phys. Soc. Jpn.* **2004**, *73*, 441–449. [CrossRef]
33. Li, J.; Lu, C.; Pei, L.; Zhang, C.; Wang, R.; Tieu, K. Atomistic simulations of hydrogen effects on tensile deformation behaviour of [0 0 1] twist grain boundaries in nickel. *Comput. Mater. Sci.* **2019**, *159*, 12–23. [CrossRef]
34. Liang, S.; Huang, M.; Zhao, L.; Zhu, Y.; Li, Z. Effect of multiple hydrogen embrittlement mechanisms on crack propagation behavior of FCC metals: Competition vs. synergy. *Int. J. Plast.* **2021**, *143*, 103023. [CrossRef]
35. Li, J.; Lu, C.; Pei, L.; Zhang, C.; Wang, R. Atomistic investigation of hydrogen induced decohesion of Ni grain boundaries. *Mech. Mater.* **2020**, *150*, 103586. [CrossRef]
36. Plimpton, S. Fast parallel algorithms for short-range molecular dynamics. *J. Comput. Phys.* **1995**, *117*, 1–19. [CrossRef]
37. Angelo, J.E.; Moody, N.R.; Baskes, M.I. Trapping of hydrogen to lattice defects in nickel. *Modell. Simul. Mater. Sci. Eng.* **1995**, 3, 289. [CrossRef]
38. Rycroft, C.H. VORO++: A three-dimensional Voronoi cell library in C++. *Chaos: Interdiscip. J. Nonlinear Sci.* **2009**, *19*, 041111. [CrossRef]
39. Li, J.; Lu, C.; Pei, L.; Zhang, C.; Wang, R.; Tieu, K. Influence of hydrogen environment on dislocation nucleation and fracture response of <1 1 0> grain boundaries in nickel. *Comput. Mater. Sci.* **2019**, *165*, 40–50. [CrossRef]
40. Hoover, W.G. Canonical dynamics: Equilibrium phase-space distributions. *Phys. Rev. A* **1985**, *31*, 1695. [CrossRef]
41. Nosé, S. A unified formulation of the constant temperature molecular dynamics methods. *J. Chem. Physics* **1984**, *81*, 511–519. [CrossRef]
42. Schiøtz, J.; Di Tolla, F.D.; Jacobsen, K.W. Softening of nanocrystalline metals at very small grain sizes. *Nature* **1998**, *391*, 561–563. [CrossRef]
43. Kelchner, C.L.; Plimpton, S.; Hamilton, J. Dislocation nucleation and defect structure during surface indentation. *Phys. Rev. B* **1998**, *58*, 11085. [CrossRef]
44. Stukowski, A.; Albe, K. Dislocation detection algorithm for atomistic simulations. *Model. Simul. Mater. Sci. Eng.* **2010**, *18*, 025016. [CrossRef]
45. Stukowski, A. Visualization and analysis of atomistic simulation data with OVITO–the Open Visualization Tool. *Model. Simul. Mater. Sci. Eng.* **2009**, *18*, 015012. [CrossRef]
46. Mori, T.; Miura, H.; Tokita, T.; Haji, J.; Kato, M. Determination of the energies of [001] twist boundaries in Cu with the shape of boundary SiO_2 particles. *Philos. Mag. Lett.* **1988**, *58*, 11–15. [CrossRef]
47. Spearot, D.E.; Sangid, M.D. Insights on slip transmission at grain boundaries from atomistic simulations. *Curr. Opin. Solid State Mater. Sci.* **2014**, *18*, 188–195. [CrossRef]

48. Wolf, D.; Phillpot, S. Role of the densest lattice planes in the stability of crystalline interfaces: A computer simulation study. *Mater. Sci. Eng. A* **1989**, *107*, 3–14. [CrossRef]
49. Harris, T.; Latanision, M. Grain boundary diffusion of hydrogen in nickel. *Metall. Trans. A* **1991**, *22*, 351–355. [CrossRef]
50. Ding, Y.; Yu, H.; Zhao, K.; Lin, M.; Xiao, S.; Ortiz, M.; Zhang, Z. Hydrogen-induced transgranular to intergranular fracture transition in bi-crystalline nickel. *Scr. Mater.* **2021**, *204*, 114122. [CrossRef]
51. Yan, F.; Mouton, I.; Stephenson, L.T.; Breen, A.J.; Chang, Y.; Ponge, D.; Gault, B. Atomic-scale investigation of hydrogen distribution in a TiMo alloy. *Scr. Mater.* **2019**, *162*, 321–325. [CrossRef]
52. Hahn, E.N.; Meyers, M.A. Grain-size dependent mechanical behavior of nanocrystalline metals. *Mater. Sci. Eng. A* **2015**, *646*, 101–134. [CrossRef]
53. Tschopp, M.A.; Tucker, G.J.; McDowell, D.L. Structure and free volume of <1 1 0> symmetric tilt grain boundaries with the E structural unit. *Acta Mater.* **2007**, *55*, 3959–3969. [CrossRef]
54. Robertson, I.M.; Birnbaum, H.K.; Sofronis, P. Hydrogen effects on plasticity. *Dislocations Solids* **2009**, *15*, 249–293.
55. Robertson, I.M. The effect of hydrogen on dislocation dynamics. *Eng. Fract. Mech.* **2001**, *68*, 671–692. [CrossRef]
56. Martin, M.L.; Somerday, B.P.; Ritchie, R.O.; Sofronis, P.; Robertson, I.M. Hydrogen-induced intergranular failure in nickel revisited. *Acta Mater.* **2012**, *60*, 2739–2745. [CrossRef]
57. Wang, S.; Martin, M.L.; Sofronis, P.; Ohnuki, S.; Hashimoto, N.; Robertson, I.M. Hydrogen-induced intergranular failure of iron. *Acta Mater.* **2014**, *69*, 275–282. [CrossRef]
58. Ferreira, P.J.; Robertson, I.M.; Birnbaum, H.K. Hydrogen effects on the character of dislocations in high-purity aluminum. *Acta Mater.* **1999**, *47*, 2991–2998. [CrossRef]

materials MDPI

Article

Study on the Hydrogen Embrittlement of Nanograined Materials with Different Grain Sizes by Atomistic Simulation

Jiaqing Li [1], Ziyue Wu [1], Fang Wang [2], Liang Zhang [3], Chilou Zhou [4], Cheng Lu [5], Lin Teng [1,*] and Qifeng Lin [6,*]

1 College of Chemical Engineering, Fuzhou University, Fuzhou 350116, China; jiaqing@fzu.edu.cn (J.L.); 210427154@fzu.edu.cn (Z.W.)
2 Fujian Special Equipment Inspection and Research Institute, Fuzhou 350008, China; wangfang2080@163.com
3 International Joint Laboratory for Light Alloys (MOE), College of Materials Science and Engineering, Chongqing University, Chongqing 400044, China; liangz@cqu.edu.cn
4 School of Mechanical and Automotive Engineering, South China University of Technology, Guangzhou 510641, China; mezcl@scut.edu.cn
5 School of Mechanical, Materials, Mechatronic and Biomedical Engineering, University of Wollongong, Wollongong, NSW 2522, Australia; chenglu@uow.edu.au
6 College of Computer and Data Science, Fuzhou University, Fuzhou 350116, China
* Correspondence: tenglin@fzu.edu.cn (L.T.); linqf@fzu.edu.cn (Q.L.)

Abstract: Although hydrogen embrittlement (HE) behavior has been extensively studied in bulk materials, little is known about H-related deformation and the fracture of nanograined materials. In this study, H segregation and HE mechanisms of nanograined Fe with different grain sizes are unveiled, following the employment of classical molecular dynamics simulations. The H segregation ratio increased, but the local H concentration at the grain boundaries (GBs) decreased with decreases in the grain size at a given bulk H concentration. The results demonstrate that H atoms increased the yield stress of nanograined models irrespective of the grain size. Furthermore, it is revealed that brittle fractures were inhibited, and the resistance to HE increased as the grain size decreased, due to the fact that the small-grain models had a lower local H concentration at the GBs and an enhanced GB-mediated intergranular deformation. These results are a clear indication of the utility of grain refinement to resist H-induced brittle failure.

Keywords: hydrogen embrittlement; nanograined materials; intergranular deformation; brittle fracture; grain refinement

Citation: Li, J.; Wu, Z.; Wang, F.; Zhang, L.; Zhou, C.; Lu, C.; Teng, L.; Lin, Q. Study on the Hydrogen Embrittlement of Nanograined Materials with Different Grain Sizes by Atomistic Simulation. *Materials* **2022**, *15*, 4589. https://doi.org/10.3390/ma15134589

Academic Editor: Boris B. Straumal

Received: 31 May 2022
Accepted: 27 June 2022
Published: 29 June 2022

1. Introduction

As a renewable energy source, hydrogen gas will be widely used in transportation, energy storage, industry, construction and other fields in the near future [1,2]. The global market of hydrogen energy is now about 125 billion US dollars and has the potential to double to 250 billion US dollars by 2030 and exceed 1 trillion US dollars [3,4]. Safe and economical hydrogen storage and transportation is the key to the practical and industrial utilization of hydrogen energy. However, hydrogen storage and transportation systems such as pipelines, long-tube trailers and hydrogen refueling stations operate under high-pressure and high-purity hydrogen environments for a long time, which usually leads to the 'hydrogen embrittlement' (HE) phenomenon [5,6]. As the name suggests, HE causes a loss in ductility, an increase in the fatigue crack growth rate and often includes sudden brittle failure and fractures [7–9]. Over the years, safety accidents related to HE problems have been widely reported and have led to serious casualties and property losses.

Under various service environments, H has the ability to absorb into metallic materials and diffuse through materials due to its small size. These diffusible H atoms are more prone to be trapped by microstructural features such as vacancies [10,11], dislocations [12,13], grain boundaries (GBs) [14,15], crack tips [16,17] and precipitates [15,18]. Extensive experimental studies have revealed that H-related failure is often cleavage-like and intergranular, underlining the importance of GBs in the HE of polycrystalline metals [19–22]. Various GB-related HE theories have been proposed and developed. One is H-enhanced decohesion (HEDE), where interstitial H weakens the cohesive strength of GBs by the dilatation of the atomic lattice [23–26]. However, HEDE alone cannot explain the observed strong plasticity activities beneath H-induced quasi-cleavage and intergranular facets. H-enhanced localized plasticity (HELP) posits that H promotes dislocation plasticity and dislocation–GB interactions [27–30], which leads to material softening. This mechanism has been proposed as crucial for the creation of the environment for the activation of intergranular failure [20,31].

With the importance of GBs in H-related failure, increasing the GB density via grain refinement is anticipated to affect the susceptibility of materials to intergranular HE. Severe plastic deformation (SPD) techniques, such as high-pressure torsion (HPT) [32–37], accumulative roll bonding (ARB) [38–40], equal channel angular pressing (ECAP) [41–43] and dynamic plastic deformation (DPD) [44] have been widely applied to achieve grain refinement, modify the texture and improve the mechanical properties of various metallic materials. In comparison with other SPD techniques, HPT is much more efficient in refining the microstructure and improving the mechanical strength. Besides, the HE performance of SPD-processed materials has attracted much research interest. For example, recently, Mine et al. [32] conducted HPT and subsequent annealing to produce ultrafine-grained-type 304 steel and found that ultrafine-grained specimens reduced the HE susceptibility and showed a good combination of strength and ductility. Fang et al. [44] developed a nanotwinned 304 austenite stainless steel through the DPD technique, pointing out that the DPD-annealed sample with 41% nanotwins showed a significantly high HE resistance. Although these experiments elucidated the effect of grain size on HE sensitivity, samples were usually on a micrometer scale. When grain size is decreased to a nanometer scale and a critical grain size where the GB role is more prominent, the GB-related HE fracture of nanograined materials should be alien from that of the ultrafine and coarse-grain counterparts. Unfortunately, due to the limitations of current experimental techniques on determining the nanoscale distribution of H atoms, little is known about the effect of H on nanograined materials.

Considering that atomistic simulation is capable of identifying the nanoscale H trapping at GBs [14,19,26,29,45–47] and quantifying the plasticity activities of nanograined materials [48–51], it is employed here to probe the HE of nanograined materials, with the emphasis on the influence of grain size on H segregation, H-related deformation and H-induced fracture. As steels are more susceptible to HE, α-Fe was selected as the model material. The present study is organized as follows: The details of the simulation setup are described in Section 2. H segregation and deformation mechanisms of nanograined Fe models without and with solute H are given in Section 3. The simulation results are discussed in Section 4, followed by the main conclusions in Section 5.

2. Materials and Methods

All molecular dynamics (MD) simulations were carried out by the open-source Large-scale Atomic/Molecular Massively Parallel Simulator (LAMMPS) [52] with the Finnis–Sinclair-type embedded atom method potential for Fe-H [28]. The initial samples were geometrically established via the Voronoi program. Periodic boundary conditions were prescribed along all three directions. A total of 8 nanograined Fe models with different in-plane (x-y plane) grain diameters of 4 nm, 6 nm, 8 nm, 10 nm, 12 nm, 14 nm, 16 nm and 18 nm were considered, and denoted as NG_d with d being the grain size. The initial configurations were optimized by performing the conjugate gradient energy minimization technique. Crystal structures were calculated by common neighbor analysis (CNA) and centro-symmetry parameter (CSP), and dislocation plasticity was analyzed by the dislocation extraction algorithm (DXA) using the Open Visualization Tool (OVITO) [53], in which output data generated in MD simulations, atomistic Monte-Carlo and other particle-based simulations can be visualized and analyzed. For an investigation of the HE of nanograined

Fe, models filled with H atoms, which corresponded to bulk H concentrations (C_0) of 0.25%, 0.5%, 0.75% and 1%, were prepared. After the insertion of H atoms, the created models were first relaxed at 700 K for 1 ns and were then cooled down at a temperature of 300 K for 1 ns, followed by a further relaxation at 300 K for 3 ns [54]. It was assumed that the heating process can bring boundary structure and H segregation into a state of equilibrium. Uniaxial tension was conducted by stretching the simulation boxes along the x direction while keeping the y and z directions stress free under the isothermal–isobaric ensemble (NPT). The temperature of the whole system was maintained at 300 K using the Nose–Hoover thermostat [55]. Two different timesteps of 0.5 fs and 1 fs were used for the models in the presence and absence of H, respectively [16]. It is worth noting that the applied tensile strain rate was set at 5×10^8/s, which is much higher by several orders of magnitude than that in real experiments. Under such a high strain rate, the H diffusion process over the time scales of loading is typically not captured. However, this is an inherent limitation of the MD method, and it will be presented that the influence of H atoms on the response of nanograined Fe with different sizes is in line with experimental observations of the H-induced failure.

3. Results

H segregation in metals has a profound effect on mechanical properties and ultimate failure. In the following, H segregation at the GBs of nanograined Fe models is presented in Section 3.1, and its influence on the mechanical behavior, deformation mechanisms and fracture response of nanograined Fe models with different grain sizes is given in Sections 3.2 and 3.3.

3.1. GB Structures and H Segregation in Nanograined Fe Models

The equilibrated structures of nanograined Fe are shown in Figure 1a. The considered GBs in nanograined models are characterized by the GB atom ratio and the GB free volume. The GB atom ratio can be obtained by the number of GB atoms against the total number of Fe atoms, while the GB free volume is defined as:

$$\Delta V_{GB} = \frac{V_{nano} - N_{Fe}V_{atom}}{N_{Fe}V_{atom}} \quad (1)$$

where V_{nano} is the volume of the nanograined model, N_{Fe} means the total number of Fe atoms and V_{atom} is the volume of one perfect Fe atom.

To investigate H segregation in nanograined Fe models, H atoms were added into models and diffused into GBs. The H distribution in four models is depicted in Figure 1b. It is clear that most of the H atoms were trapped at GBs and dislocation cores, while other H atoms occupied bulk sites. During the H diffusion process, the H diffusion coefficient can be calculated as [56]:

$$D_{eff} = lim_{t\to\infty} \frac{1}{6N} \sum_{i=1}^{N} \left(\vec{r_i}(t) - \vec{r_i}(t_0)\right)^2 \quad (2)$$

where $\frac{1}{N}\sum_{i=1}^{N}\left(\vec{r_i}(t) - \vec{r_i}(t_0)\right)^2$ defines the mean-squared displacement (MSD) of H atoms. N is the total number of H atoms, $\vec{r_i}(t_0)$ is the original position of the ith H atom and $\vec{r_i}(t)$ is the position of the ith H atom at time t. The MSD curves vs. simulation time are shown in Figure 2a, in which H diffusion velocity (the slope of plots) was initially high, and then decreased as the H atoms arrived at the GBs. Taking the NG_6 as an example, after 0.2 ns H atoms were trapped by the GBs, and the H diffusion became relatively low due to the GB trapping effect [54]. With the increase in grain size the GB density decreased and the GB trapping effect consequently weakened, rendering fast H diffusion. The H diffusion coefficients calculated from our simulations are shown in Figure 2b and are compared with the available experimental values [57,58]. It is interesting to note that the H diffusion at a lower concentration (C_0 = 0.5%) was found to be higher than that at a higher concentration (C_0 = 1%). This confirms that the H diffusion process can be influenced by other H interstitials.

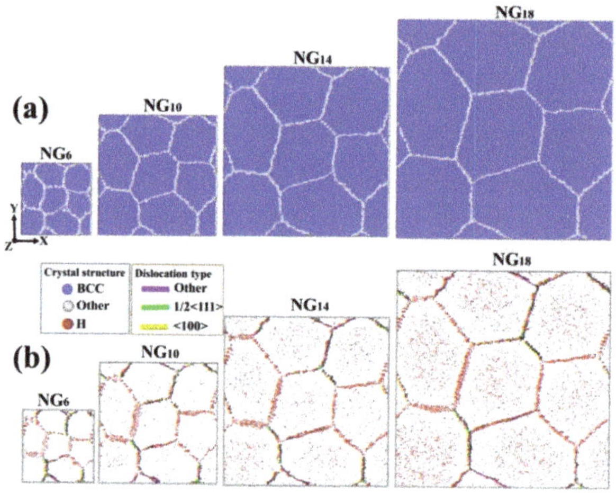

Figure 1. (**a**) Atomic configurations of nanograined Fe models with various grain sizes; (**b**) H distribution in nanograined models. Atoms are colored according to CNA, where Fe atoms with a perfect structure are colored blue, atoms at GBs are colored white and H atoms are colored red. Possible dislocations are indicated by their types.

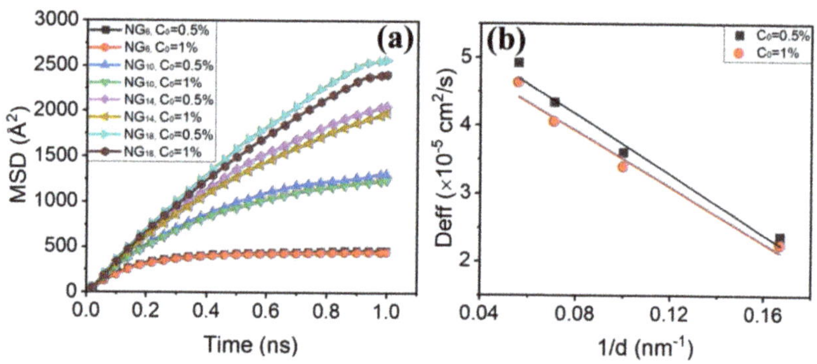

Figure 2. (**a**) MSD of H atoms in models with different grain sizes vs. time; (**b**) D_{eff} as a function of 1/d.

After the H diffusion process, the H atoms segregating at the GBs were counted and divided by total number of H atoms to obtain the segregation ratio (f_{seg}). The corresponding values are tabulated in Table 1. For the nanograined models, f_{seg} was almost insensitive to C_0. The reason can be attributed to the fact that in an equilibrated state, f_{seg} depends on the number of trapping sites that GBs provide; for the same model, the trapping sites were constant and f_{seg} was therefore unchanged. Furthermore, it can be seen that f_{seg} decreased with increases in the grain size, which is indicative of the fact that small-grain models have a stronger ability to trap H atoms. This observation can be linked with the GB atom ratio and ΔV_{GB}; small-grain models have higher values of the GB atom ratio and ΔV_{GB}, suggesting that more trapping sites can accommodate H atoms.

Table 1. GB characters and H segregation properties in NG_6, NG_{10}, NG_{14} and NG_{18}.

Grain Size (nm)	ΔV_{GB} (%)	GB Atom Ratio (%)	f_{seg}, $C_0 = 0.5\%$	f_{seg}, $C_0 = 1\%$	$C_{H\text{-}GB}$, $C_0 = 0.5\%$	$C_{H\text{-}GB}$, $C_0 = 1\%$
6	0.63	14.71	0.862	0.860	0.012	0.023
10	0.39	8.60	0.841	0.837	0.017	0.035
14	0.28	6.42	0.835	0.821	0.023	0.048
18	0.23	5.51	0.785	0.780	0.025	0.052

The local H concentration at GBs ($C_{H\text{-}GB}$) is calculated as the number of H atoms at GBs divided by the boundary area. Table 1 shows that the local H concentration increased with increases in the grain size. Due to the grain refinement, small-grain models have a higher segregation ratio, but a higher GB density at the same time. As a result of the 'dilution effect', the $C_{H\text{-}GB}$ of small-grain models is lower than that of large-grain ones, evidenced by that the $C_{H\text{-}GB}$ of NG_6 was only half of that of NG_{18}.

3.2. Mechanical Behavior and Deformation Mechanisms of Nanograined Fe Models without H

The stress–strain relationships of nanograined Fe cases with varying grain sizes are plotted in Figure 3. The tensile stress is calculated by dividing the stress tensor of the entire system over the volume of the system, and tensile strain is obtained from the applied strain rate multiplied by the deformation time. It is clear from the slope of stress–strain relationships that Young's modulus was reduced with decreases in the grain size. This phenomenon may be associated with the fact that models with smaller grain sizes have a higher GB fraction, which promotes the intergranular deformation rather than intragranular deformation, leading to a lower Young's modulus. On the other hand, Figure 3b clearly shows that the peak stress increased as the grain size decreased to 10 nm, but a reverse relation occurred with a further reduction in the grain size, namely 'inverse' Hall–Petch behavior.

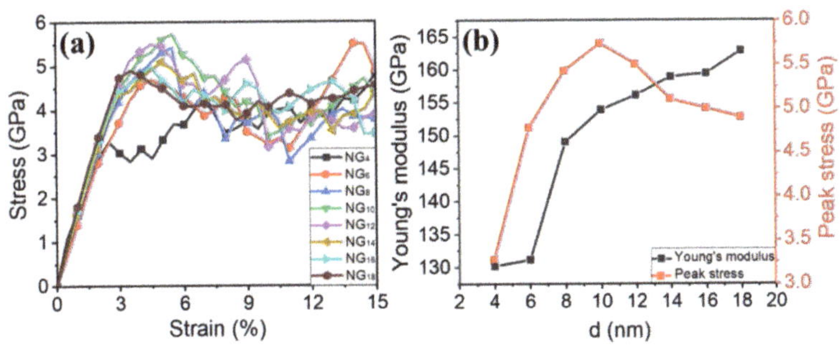

Figure 3. (**a**) Tensile stress–strain relationships of nanograined Fe cases with various grain sizes; (**b**) Young's modulus and peak stress as a function of d.

Figure 4 presents the atomic configurations of the models with grain sizes of 18 nm, 10 nm and 6 nm, respectively. In the case of NG_{18}, dislocation-dominated plasticity was the main deformation mechanism, as seen in Figure 4a. Heterogeneous dislocation nucleation and emission from the GB occurred at a strain of $\varepsilon = 8.05\%$, which is in accordance with our previous findings that GBs are the dislocation source [23,29]. Besides, dislocations were also generated from the grain interior in several locations. These intragranular dislocation mechanisms dominated the plastic deformation processes. Deformation twins only occurred at some certain grains and remained almost unchanged with further applied tensile strain, indicating that these twins play a secondary role in the deformation process of the NG_{18} model.

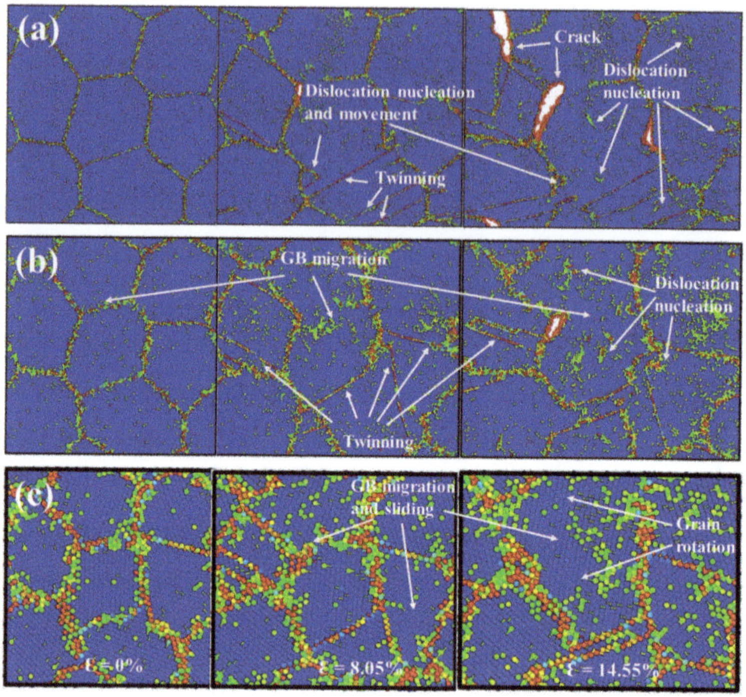

Figure 4. (**a**–**c**) The atomic snapshots of NG_{18}, NG_{10} and NG_6 at different tensile strains under tension, respectively. Images are colored by CSP, and various deformation mechanisms are indicated by arrows.

Figure 4b shows the atomic snapshots of the NG_{10} model in the absence of H. It is clear that the GB-mediated process was increasingly more important, as the GB structures became disordered during the deformation process. GB migration and grain growth occurred at a strain of $\varepsilon = 8.05\%$ with two separated grains growing into one large grain. On the other hand, there were more deformation twins in the NG_{10} model than NG_{18} model. These twins nucleated from the GBs by a continuous emission of partials and evolved in the grain interior. Previous reports of nanocrystalline Mo and gold films pointed out that nanotwins not only produced plastic strain and intergranular failure [59], but also assisted grain coarsening by changing the local grain misorientation and mobilizing the GB during the plastic deformation process [60]. Our results further confirm that these twins are important deformation carriers in nanograined Fe materials and induce the grain rotation and growth. Next, the intergranular deformation of the NG_6 model is examined and shown in Figure 4c. Under tension, most of grains evolved from their initial shapes, and a majority of GBs migrated from their original positions. The occurrence of GB sliding, and migration accompanied by grain rotation and growth suggests that GB-mediated deformation is the dominating deformation mechanism of the NG_6 model. It is worth noting that there were no dislocation nucleation or dislocation pile-up events, unlike the cases of NG_{18} and NG_{10}.

The aforementioned deformation process indicates a change in the deformation mechanisms from intragranular mode to intergranular mode. When the grain size was above the 10 nm, nanograined models deformed primarily by the intragranular mode. Specifically, full dislocations nucleated from the grain interior and GBs, and further dislocation emission and movement occurred. However, when the grain size was less than 10 nm, the dominant deformation mechanism of nanograined Fe was intergranular deformation. The transition can be associated with the fact that the decreased grain size resulted in the increase in the GB atom ratio, as tabulated in Table 1. The GB atom ratio of NG_{18} was 5.51%, while the value increased to 14.71% in the case of NG_6. The higher GB atom ratio and GB density encouraged the GB-mediated deformation processes such as GB sliding and migration. The discrepancy of deformation mechanisms of nanograined Fe models suggests that HE mechanisms should be dependent on the grain size, which will be discussed in the following paragraphs.

3.3. Deformation Mechanisms and the Fracture Response of Nanograined Fe Models with H

Uniaxial tension was carried out for each nanograined Fe model along the x direction with various bulk H concentrations considered for compassion. The tensile stress–strain curves of nanograined Fe models with five bulk H concentrations are plotted in Figure 5. It is obvious that the yield stress was higher when the bulk H concentration increased, which is indicative of the fact that H atoms impede the onset of plastic deformation. The embrittling effect that stems from H atoms can be observed, as the fracture strain was reduced with increases in the bulk H concentration. Furthermore, it was found that the susceptibility of nanograined Fe to intergranular embrittlement due to H was dependent on the grain size. In order to reveal these H-related deformations and fractures, the atomic configurations of the corresponding models with various bulk H concentrations are elaborated in Figures 6–8.

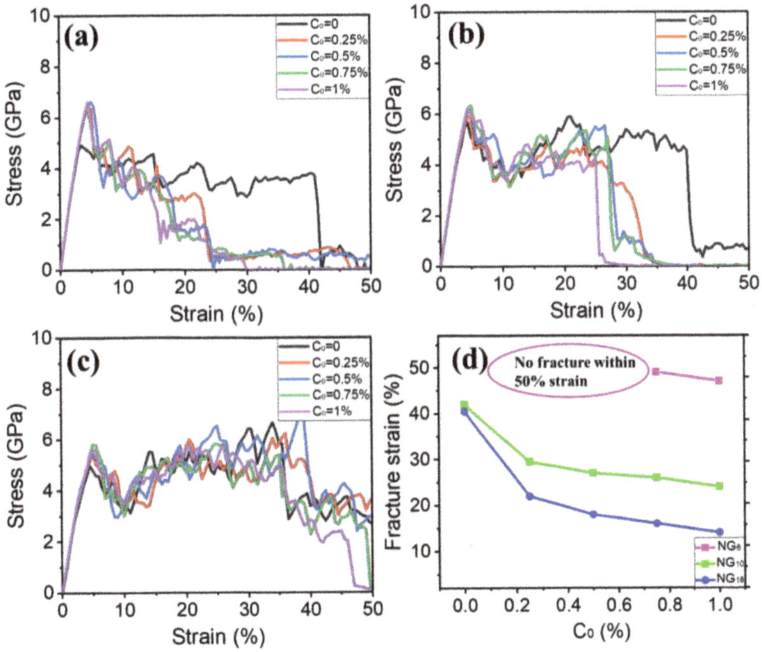

Figure 5. (**a**–**c**) Tensile stress–strain curves of NG_{18}, NG_{10} and NG_6 models, respectively, with different bulk H concentrations; (**d**) fracture strain as a function of C_0.

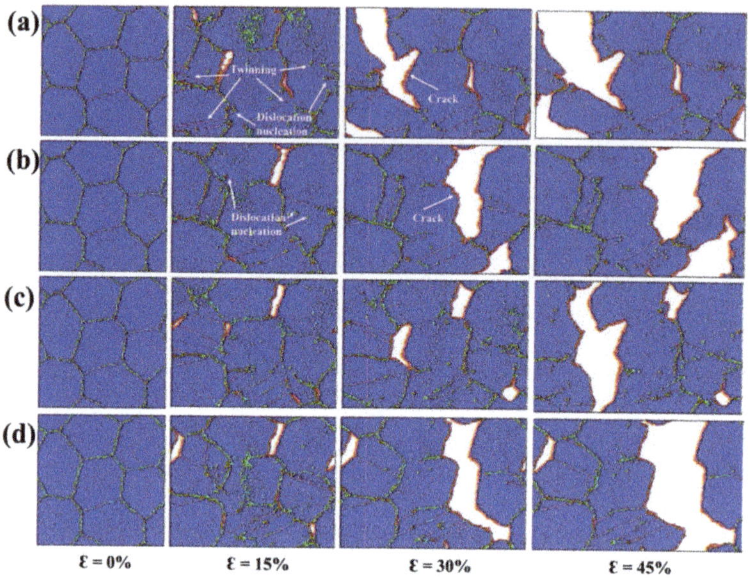

Figure 6. (**a–d**) The atomic snapshots of NG_{18} at 0%, 15%, 30% and 45% tensile strains with bulk H concentrations of 0%, 0.25%, 0.5%, 0.75% and 1%, respectively.

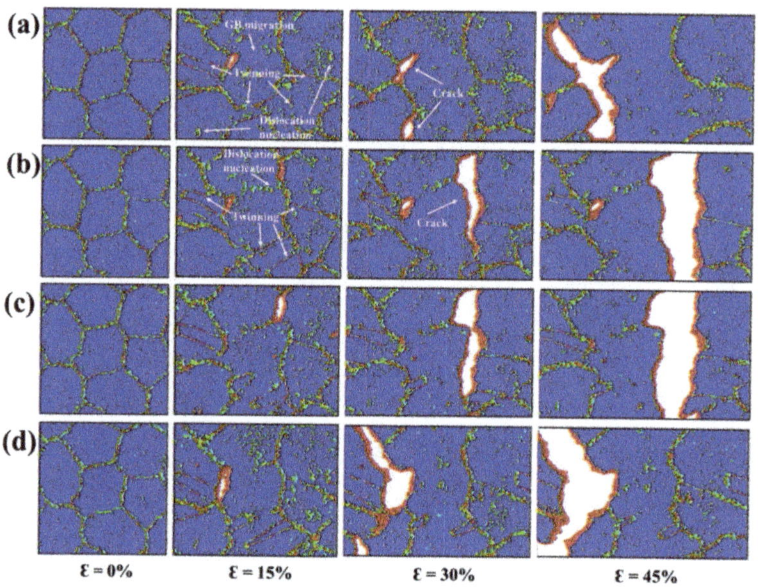

Figure 7. (**a–d**) The atomic snapshots of NG_{10} at 0%, 15%, 30% and 45% tensile strains with bulk H concentrations of 0%, 0.25%, 0.5%, 0.75% and 1%, respectively.

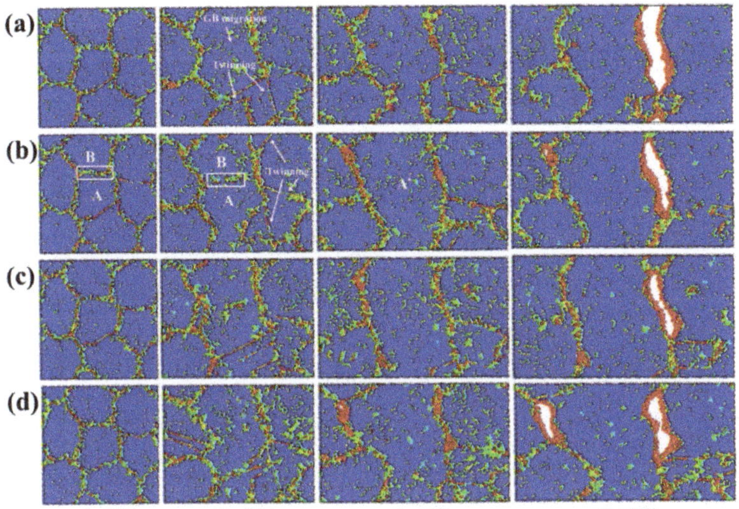

Figure 8. (**a**–**d**) The atomic snapshots of NG_6 at 0%, 15%, 30% and 45% tensile strains with bulk H concentrations of 0%, 0.25%, 0.5%, 0.75% and 1%, respectively.

The atomic configurations of NG_{18} at 0%, 15%, 30% and 45% tensile strains with bulk H concentrations of 0%, 0.25%, 0.5%, 0.75% and 1% are presented in Figure 6. At low H concentrations, the plastic process occurred by means of the nucleation of dislocations from the grain interior and GBs. Some twins were also observed across the grain interior, which suggests an intragranular mode rather than intergranular mode. The failure feature was ductile as the fracture surfaces were curved. With the increase in the H concentration, the nucleation of dislocations and twins was suppressed, accompanied by a remarkable loss in ductility. This is direct evidence that H atoms increase the yield stress of NG_{18}. At a strain of 15%, there were more activated cracks with 0.5% and 1% H atoms than that with 0% and 0.25% H atoms. These cracks advanced along the boundary plane in a brittle manner with further applied strain, leading to an ultimate failure. Such embrittlement is in accordance with the HEDE mechanism [23–26], where the cohesive strength of GBs is significantly weakened, and GB fracture is promoted.

The atomic configurations of NG_{10} at 0%, 15%, 30% and 45% tensile strains with bulk H concentrations of 0%, 0.25%, 0.5%, 0.75% and 1% are presented in Figure 7. It can be observed that dislocation nucleation events occurred during the deformation process. Moreover, GB-mediated deformation involving GB migration, GB sliding, grain rotation and growth was more prominent in NG_{10} than that in NG_{18}. With the increase in the H concentration, the nucleation of dislocations and twins was suppressed, and GB-mediated deformation was also hindered. It is clear that the GB migration event was reduced in the presence of H. The presence of H severely destroyed the local boundary structures, a phenomenon discussed in our previous study. In terms of ⟨100⟩ mode GB, H atoms enhanced the dislocation interactions by changing the GB dislocation configurations, thereby suppressing the collective gliding of GB dislocations and consequent GB migration. Regarding ⟨111⟩ mode GB, H atoms disordered local boundary structures, and thus inhibited the GB structural transformation and changed the coupling mode [61]. Such H-impeded coupled GB motion increases the yield stress, as shown in Figure 5. In terms of failure mode, the fracture surfaces were composed of more disordered atoms, and were more curved than those in NG_{18}, suggesting that small-grain models are more resistant to HE.

The atomic configurations of NG_6 at 0%, 15%, 30% and 45% tensile strains with bulk H concentrations of 0%, 0.25%, 0.5%, 0.75% and 1% are presented in Figure 8. The segregated H atoms prevented the GB-related intergranular deformation. With increases in the H concentration, grains kept their initial shapes, while GB migration, GB sliding and grain rotation decreased. The nucleation of dislocations was also inhibited by the segregated H atoms, as few dislocation plasticity events were found at the stage of deformation. Note that despite the dragging effect of solute H on GB mobility, the normal GB displacement can still occur at a higher tensile strain. For example, the boundary between grain A and B was pinned by segregated H atoms at a strain of 15%, but migrated at the strain of 30%, leading to grain A evolving into grain A′ at the expense of grain B, as shown in Figure 8b.

By comparing with NG_{18} and NG_{10} models, it is clear that NG_6 was the most resistant to the HE. The NG_6 model featured the fracture surfaces that were composed of very disordered atoms with higher CSP values. Besides, the formed cracks could evolve into the ultimate failure, contrary to the flat fracture surfaces of the NG_{18} and NG_{10} models. This phenomenon can be attributed to two aspects. One is that the NG_6 model had a higher GB density, which contributed to the GB-related intergranular deformation. Such intergranular deformation relieved the accumulated stress around GBs and thus inhibited fracture process. The other is associated with local H concentration at GBs. As tabulated in Table 1, the $C_{H\text{-}GB}$ of small-grain models was lower than that of large-grain ones, evidenced by that the $C_{H\text{-}GB}$ of NG_6 was only half of that of NG_{18}. With a lower $C_{H\text{-}GB}$, the embrittling effect of H atoms on NG_6 model was mitigated.

4. Discussion

The simulations show that there was a strong dependence of H segregation, H-related deformation and fracture on the grain size of nanograined Fe models. The local H concentration at GBs was reduced with the decrease in the grain size at a given bulk H concentration. H segregation increased the yield stress of nanograined models and impeded the onset of plastic deformation. Furthermore, the susceptibility of nanograined Fe to HE decreased as the grain size decreased.

Previous studies have measured diffusible H content by the thermal desorption spectrum, finding that the ultrafine-grained specimen introduced the largest amount of diffusible H, but the local H content per unit surface area of GB was reduced with the decrease in the mean grain size [62,63]. These studies investigated H segregation properties on a micrometer scale but were not able to provide nanoscale H diffusion and distribution around the defects. The present study explains H segregation from the atomistic perspective. The trapping ability is dependent on the number of trapping sites at GBs; therefore, the H segregation ratio is higher in the small-grain models that have more possible trapping sites. However, due to the 'dilution effect' of the high GB density, the local H concentration at GBs of small-grain models is still lower than that of large-grain models.

Figure 5 shows the H-induced hardening effect, where H segregation increased the yield stress and impeded the onset of plasticity. When the grain size was above 10 nm, the normal Hall–Patch relationship operated. Dislocations were nucleated from the grain interior and GBs, and the intragranular dislocation mechanism was the dominant strength-controlling process. H atoms can produce hardness as a result of possible H-dislocation interactions. H concentrations generate the Cottrell atmospheres, leading to the decrease in dislocation movement and a dislocation pinning effect [64–66], or an increased slip planarity [16,67]. Our results confirm that H atoms suppress the nucleation of dislocations and twins, accompanied by a large amount of ductility loss. This is direct evidence that H atoms increase the yield stress. As the grain size was decreased below 10 nm, there was a transition of deformation mode into GB-mediated intergranular deformation involving GB migration, GB sliding, grain rotation and growth. Previous MD simulations have pointed out that H atoms hinder GB mobility by disordering local structures of GBs [61]. Keeping this in mind, H atoms increase the yield stress corresponding to GB mobility.

Besides the change in the mechanical behavior of nanograined Fe models after H segregation, it is important to consider the ultimate fracture in the presence of H. Figure 5d compares the fracture strain of different nanograined models with various H concentrations. The small-grain models were less susceptible to HE, e.g., there was no observed failure of

NG_6 with bulk H concentrations of 0%, 0.25% and 0.5%. Even though the H concentration reached 1%, the fracture strain was at a very high value (about 46%). Conversely, the NG_{18} model fractured at a low strain of 22% when the bulk H concentration was 0.25%, and fracture strain was further decreased with increases in the bulk H concentration. Bai et al. [62] showed that due to a high local H concentration, secondary cracks formed on the fracture surfaces of the H-charged specimens with coarser grain sizes, which resulted in the earlier fracture. Figures 6–8 reveal this point as there were more formed cracks along GBs of NG_{18} due to a high local H concentration, and an easier brittle fracture compared to NG_6. Apart from this, the GB-mediated intergranular deformation of small-grain models might also be resistant to HE. Solute H atoms are detached and left behind from GBs during GB migration process, consequently the cohesive strength of GBs cannot be weakened by solute H, and HEDE cannot operate [61]. These results clearly indicate one possible pathway by utilizing grain refinement to resist H-induced brittle failure, which is expected to prevent the HE problem of H storage and transportation systems and promote H energy used in transportation, energy storage and other fields.

5. Conclusions

In the present study, the effect of grain size on the HE behavior of nanograined Fe materials was investigated by MD simulations. It was found that the H segregation ratio increased, but local H concentration at GBs decreased with decreases in the grain size. When the grain size was above 10 nm, the intragranular dislocation mechanism was the dominant strength-controlling process, while GB-mediated intergranular deformation operated with the grain size below 10 nm.

H segregation increased the yield stress of nanograined models irrespective of the deformation mode. Furthermore, the simulations showed that small-grain models had less local H concertation at GBs and formed cracks, thus inhibiting ultimate fracture. These results are a clear indication of the utility of grain refinement to resist H-induced brittle failure.

Author Contributions: J.L.: Methodology, Software, Writing—original draft; C.L. and L.T.: Conceptualization, Supervision, Project administration; Z.W. and L.Z.: Software, Writing—review and editing; F.W. and C.Z.: Visualization, Investigation; Q.L.: Software, Validation. All authors have read and agreed to the published version of the manuscript.

Funding: This study was financially supported by the Research Program for Young and Middle-Aged Teachers of Fujian Province (No. JAT210038), Talent Program of Fuzhou University (No. XRC-22024), National Natural Science Foundation of China (No. 52071034; No. 52130107), Guangdong Basic and Applied Basic Research Foundation (No. 2019A1515011157) and Science and Technology Program of Guangzhou (No. 202002030275).

Institutional Review Board Statement: Not applicable.

Informed Consent Statement: Not applicable.

Data Availability Statement: The data presented in this study are available on request from the corresponding author.

Conflicts of Interest: The authors declare no conflict of interest.

Nomenclature

HE	hydrogen embrittlement
GBs	grain boundaries
HEDE	hydrogen-enhanced decohesion
HELP	hydrogen enhanced localized plasticity
SPD	severe plastic deformation
HPT	high pressure torsion
ARB	accumulative roll bonding

ECAP	equal channel angular pressing
DPD	dynamic plastic deformation
MD	molecular dynamics
LAMMPS	large-scale atomic/molecular massively parallel simulator
OVITO	open visualization tool
CNA	common neighbor analysis
CSP	centro-symmetry parameter
DXA	dislocation extraction algorithm
NPT	Isothermal–isobaric ensemble
MSD	mean-squared displacement

References

1. Andersson, J.; Grönkvist, S. Large-scale storage of hydrogen. *Int. J. Hydrogen Energy* **2019**, *44*, 11901–11919. [CrossRef]
2. Dawood, F.; Anda, M.; Shafiullah, G.M. Hydrogen production for energy: An overview. *Int. J. Hydrogen Energy* **2020**, *45*, 3847–3869. [CrossRef]
3. Parra, D.; Valverde, L.; Pino, F.J.; Patel, M.K. A review on the role, cost and value of hydrogen energy systems for deep decarbonisation. *Renew. Sustain. Energy Rev.* **2019**, *101*, 279–294. [CrossRef]
4. Ghaib, K.; Ben-Fares, F.-Z. Power-to-methane: A state-of-the-art review. *Renew. Sustain. Energy Rev.* **2018**, *81*, 433–446. [CrossRef]
5. Bechtle, S.; Kumar, M.; Somerday, B.; Launey, M.; Ritchie, R. Grain-boundary engineering markedly reduces susceptibility to intergranular hydrogen embrittlement in metallic materials. *Acta Mater.* **2009**, *57*, 4148–4157. [CrossRef]
6. Koyama, M.; Akiyama, E.; Lee, Y.-K.; Raabe, D.; Tsuzaki, K. Overview of hydrogen embrittlement in high-Mn steels. *Int. J. Hydrogen Energy* **2017**, *42*, 12706–12723. [CrossRef]
7. Zhou, C.; Fang, B.; Wang, J.; Tang, D.; Tao, H.; He, Y.; Zhou, Z.; Chen, C.; Zhang, L. Hydrogen embrittlement resistance of TWIP (twinning-induced plasticity) steel in high pressure hydrogen environment. *Int. J. Fatigue* **2021**, *151*, 106362. [CrossRef]
8. Wasim, M.; Djukic, M.B.; Ngo, T.D. Influence of hydrogen-enhanced plasticity and decohesion mechanisms of hydrogen embrittlement on the fracture resistance of steel. *Eng. Fail. Anal.* **2021**, *123*, 105312. [CrossRef]
9. Robertson, I.M.; Sofronis, P.; Nagao, A.; Martin, M.; Wang, S.; Gross, D.W.; Nygren, K.E. Hydrogen embrittlement understood. *Metall. Mater. Trans. B* **2015**, *46*, 1085–1103. [CrossRef]
10. Nagumo, M.; Takai, K. The predominant role of strain-induced vacancies in hydrogen embrittlement of steels: Overview. *Acta Mater.* **2019**, *165*, 722–733. [CrossRef]
11. Li, S.; Li, Y.; Lo, Y.C.; Neeraj, T.; Srinivasan, R.; Ding, X.; Sun, J.; Qi, L.; Gumbsch, P.; Li, J. The interaction of dislocations and hydrogen-vacancy complexes and its importance for deformation-induced proto nano-voids formation in α-Fe. *Int. J. Plast.* **2015**, *74*, 175–191. [CrossRef]
12. Kapci, M.F.; Schön, J.C.; Bal, B. The role of hydrogen in the edge dislocation mobility and grain boundary-dislocation interaction in α-Fe. *Int. J. Hydrogen Energy* **2021**, *46*, 32695–32709. [CrossRef]
13. Taketomim, S.; Matsumoto, R.; Miyazaki, N. Atomistic simulation of the effects of hydrogen on the mobility of edge dislocation in alpha iron. *J. Mater. Sci.* **2008**, *43*, 1166–1169. [CrossRef]
14. Xing, X.; Deng, G.; Zhang, H.; Cui, G.; Liu, J.; Li, Z.; Wang, B.; Li, S.; Qi, C. Molecular dynamics studies of hydrogen effect on intergranular fracture in α-iron. *Materials* **2020**, *13*, 4949. [CrossRef]
15. Chen, Y.-S.; Lu, H.; Liang, J.; Rosenthal, A.; Liu, H.; Sneddon, G.; McCarroll, I.; Zhao, Z.; Li, W.; Guo, A.; et al. Observation of hydrogen trapping at dislocations, grain boundaries, and precipitates. *Science* **2020**, *367*, 171–175. [CrossRef]
16. Song, J.; Curtin, W.A. Atomic mechanism and prediction of hydrogen embrittlement in iron. *Nat. Mater.* **2012**, *12*, 145–151. [CrossRef]
17. Song, J.; Curtin, W.A. A nanoscale mechanism of hydrogen embrittlement in metals. *Acta Mater.* **2011**, *59*, 1557–1569. [CrossRef]
18. Peral, L.; Zafra, A.; Belzunce, J.; Rodríguez, C. Effects of hydrogen on the fracture toughness of CrMo and CrMoV steels quenched and tempered at different temperatures. *Int. J. Hydrogen Energy* **2019**, *44*, 3953–3965. [CrossRef]
19. Tehranchi, A.; Curtin, W.A. Atomistic study of hydrogen embrittlement of grain boundaries in nickel: I. Fracture. *J. Mech. Phys. Solids* **2017**, *101*, 150–165. [CrossRef]
20. Wang, S.; Martin, M.L.; Sofronis, P.; Ohnuki, S.; Hashimoto, N.; Robertson, I.M. Hydrogen-induced intergranular failure of iron. *Acta Mater.* **2014**, *69*, 275–282. [CrossRef]
21. Ogawa, Y.; Birenis, D.; Matsunaga, H.; Takakuwa, O.; Yamabe, J.; Prytz, Ø.; Thøgersen, A. The role of intergranular fracture on hydrogen-assisted fatigue crack propagation in pure iron at a low stress intensity range. *Mater. Sci. Eng. A* **2018**, *733*, 316–328. [CrossRef]
22. Shibata, A.; Murata, T.; Takahashi, H.; Matsuoka, T.; Tsuji, N. Characterization of hydrogen-related fracture behavior in as-quenched low-carbon martensitic steel and tempered medium-carbon martensitic steel. *Metall. Mater. Trans. A* **2015**, *46*, 5685–5696. [CrossRef]
23. Wang, S.; Martin, M.; Robertson, I.M.; Sofronis, P. Effect of hydrogen environment on the separation of Fe grain boundaries. *Acta Mater.* **2016**, *107*, 279–288. [CrossRef]

24. Huang, S.; Chen, D.; Song, J.; McDowell, D.L.; Zhu, T. Hydrogen embrittlement of grain boundaries in nickel: An atomistic study. *npj Comput. Mater.* **2017**, *3*, 28. [CrossRef]
25. Tehranchi, A.; Curtin, W.A. Atomistic study of hydrogen embrittlement of grain boundaries in nickel: II. Decohesion. *Model. Simul. Mater. Sci. Eng.* **2017**, *25*, 075013. [CrossRef]
26. Li, J.; Lu, C.; Pei, L.; Zhang, C.; Wang, R. Atomistic investigation of hydrogen induced decohesion of Ni grain boundaries. *Mech. Mater.* **2020**, *150*, 103586. [CrossRef]
27. Beachem, C.D. A new model for hydrogen-assisted cracking (hydrogen "embrittlement"). *Metall. Mater. Trans. B* **1972**, *3*, 441–455. [CrossRef]
28. Birnbaum, H.K.; Sofronis, P. Hydrogen-enhanced localized plasticity—A mechanism for hydrogen-related fracture. *Mater. Sci. Eng. A* **1994**, *176*, 191–202. [CrossRef]
29. Li, J.; Lu, C.; Pei, L.; Zhang, C.; Tieu, K. Influence of solute hydrogen on the interaction of screw dislocations with vicinal twin boundaries in nickel. *Scrip. Mater.* **2019**, *173*, 115–119. [CrossRef]
30. Zhu, Y.; Li, Z.; Huang, M. Solute hydrogen effects on plastic deformation mechanisms of α-Fe with twist grain boundary. *Int. J. Hydrogen Energy* **2018**, *43*, 10481–10495. [CrossRef]
31. Martin, M.L.; Dadfarnia, M.; Nagao, A.; Wang, S.; Sofronis, P. Enumeration of the hydrogen-enhanced localized plasticity mechanism for hydrogen embrittlement in structural materials. *Acta Mater.* **2019**, *165*, 734–750. [CrossRef]
32. Mine, Y.; Matsumoto, S.; Horita, Z. Strengthening and hydrogen embrittlement of ultrafine-grained Fe–0.01mass% C alloy processed by high-pressure torsion. *Corros. Sci.* **2011**, *53*, 2969–2977. [CrossRef]
33. Mine, Y.; Horita, N.; Horita, Z.; Takashima, K. Effect of ultrafine grain refinement on hydrogen embrittlement of metastable austenitic stainless steel. *Int. J. Hydrogen Energy* **2017**, *42*, 15415–15425. [CrossRef]
34. Deng, G.; Zhao, X.; Su, L.; Wei, P.; Zhang, L.; Zhan, L.; Chong, Y.; Zhu, H.; Tsuji, N. Effect of high pressure torsion process on the microhardness, microstructure and tribological property of Ti6Al4V alloy. *J. Mater. Sci. Technol.* **2021**, *94*, 183–195. [CrossRef]
35. Wei, P.; Lu, C.; Tieu, K.; Su, L.; Deng, G.; Huang, W. A study on the texture evolution mechanism of nickel single crystal deformed by high pressure torsion. *Mater. Sci. Eng. A* **2017**, *684*, 239–248. [CrossRef]
36. Deng, G.; Chong, Y.; Su, L.; Zhan, L.; Wei, P.; Zhao, X.; Zhang, L.; Tian, Y.; Zhu, H.; Tsuji, N. Mechanisms of remarkable wear reduction and evolutions of subsurface microstructure and nano-mechanical properties during dry sliding of nano-grained Ti6Al4V alloy: A comparative study. *Tribol. Int.* **2022**, *169*, 107464. [CrossRef]
37. Deng, G.; Bhattacharjee, T.; Chong, Y.; Zheng, R.; Bai, Y.; Shibata, A.; Tsuji, N. Influence of Fe addition in CP titanium on phase transformation, microstructure and mechanical properties during high pressure torsion. *J. Alloys Compd.* **2020**, *822*, 153604. [CrossRef]
38. Su, L.; Lu, C.; Deng, G.; Tieu, K. Microstructure and mechanical properties of AA5005/AA6061 laminated composite processed by accumulative roll bonding. *Metall. Mater. Trans. B* **2014**, *45*, 515–522. [CrossRef]
39. Nie, N.; Su, L.; Deng, G.; Li, H.; Yu, H.; Tieu, A.K. A review on plastic deformation induced surface/interface roughening of sheet metallic materials. *J. Mater. Res. Technol.* **2021**, *15*, 6574–6607. [CrossRef]
40. Su, L.; Lu, C.; Deng, G.; Tieu, A.K.; Sun, X. Microstructure and mechanical properties of 1050/6061 laminated composite processed by accumulative roll bonding. *Rev. Adv. Mater. Sci.* **2013**, *33*, 33–37.
41. Deng, G.; Lu, C.; Su, L.; Tieu, A.; Yu, H.; Liu, X. Investigation of sample size effect on the deformation heterogeneity and texture development during equal channel angular pressing. *Comput. Mater. Sci.* **2013**, *74*, 75–85. [CrossRef]
42. Deng, G.; Lu, C.; Su, L.; Tieu, A.K.; Li, J.; Liu, M.; Zhu, H.; Liu, X. Influence of outer corner angle (OCA) on the plastic deformation and texture evolution in equal channel angular pressing. *Comput. Mater. Sci.* **2014**, *81*, 79–88. [CrossRef]
43. Su, L.; Deng, G.; Luzin, V.; Wang, H.; Wang, Z.; Yu, H.; Li, H.; Tieu, A.K. Effect of cryogenic temperature equal channel angular pressing on microstructure, bulk texture and tensile properties of AA1050. *Mater. Sci. Eng. A* **2020**, *780*, 139190. [CrossRef]
44. Fan, Y.H.; Cui, F.; Lu, L. A nanotwinned austenite stainless steel with high hydrogen embrittlement resistance. *J. Alloy. Comp.* **2019**, *788*, 1066–1075. [CrossRef]
45. Ding, Y.; Yu, H.; Zhao, K.; Lin, M.; Xiao, S.; Ortiz, M.; He, J.; Zhang, Z. Hydrogen-induced transgranular to intergranular fracture transition in bi-crystalline nickel. *Scrip. Mater.* **2021**, *204*, 114122. [CrossRef]
46. Li, J.; Pei, L.; Lu, C.; Godbole, A.; Michal, G. Hydrogen effects on the mechanical behaviour and deformation mechanisms of inclined twin boundaries. *Int. J. Hydrogen Energy* **2021**, *46*, 16127–16140. [CrossRef]
47. Hüter, C.; Shanthraj, P.; McEniry, E.; Spatschek, R.; Hickel, T.; Tehranchi, A.; Guo, X.; Roters, F. Multiscale modelling of hydrogen transport and segregation in polycrystalline steels. *Metals* **2018**, *8*, 430. [CrossRef]
48. Zhang, L.; Shibuta, Y.; Lu, C.; Huang, X. Interaction between nano-voids and migrating grain boundary by molecular dynamics simulation. *Acta Mater.* **2019**, *173*, 206–224. [CrossRef]
49. Zhang, L.; Lu, C.; Tieu, K.; Shibuta, Y. Dynamic interaction between grain boundary and stacking fault tetrahedron. *Scripta Mater.* **2018**, *144*, 78–83. [CrossRef]
50. Dupraz, M.; Sun, Z.; Brandl, C.; Van Swygenhoven, H. Dislocation interactions at reduced strain rates in atomistic simulations of nanocrystalline Al. *Acta Mater.* **2018**, *144*, 68–79. [CrossRef]
51. Zhu, Y.; Li, Z.; Huang, M.; Liu, Y. Strengthening mechanisms of the nanolayered polycrystalline metallic multilayers assisted by twins. *Int. J. Plast.* **2015**, *72*, 168–184. [CrossRef]
52. Plimpton, S. Fast parallel algorithms for short-range molecular dynamics. *J. Comput. Phys.* **1995**, *117*, 1–19. [CrossRef]

53. Stukowski, A. Visualization and analysis of atomistic simulation data with OVITO—The open visualization tool. *Model. Simul. Mater. Sci. Eng.* **2010**, *18*, 015012. [CrossRef]
54. Li, J.; Lu, C.; Wang, L.; Pei, L.; Godbole, A.; Michal, G. The role of hydrogen on the behavior of intergranular cracks in bicrystalline α-Fe nanowires. *Nanomaterials* **2021**, *11*, 294. [CrossRef]
55. Li, X.; Wei, Y.; Lu, L.; Lu, K.; Gao, H. Dislocation nucleation governed softening and maximum strength in nano-twinned metals. *Nature* **2010**, *464*, 877–880. [CrossRef]
56. Zhou, X.-Y.; Zhu, J.-H.; Wu, H.-H. Molecular dynamics studies of the grain-size dependent hydrogen diffusion coefficient of nanograined Fe. *Int. J. Hydrogen Energy* **2021**, *46*, 5842–5851. [CrossRef]
57. Nagano, M.; Hayashi, Y.; Ohtani, N.; Isshiki, M.; Igaki, K. Hydrogen diffusivity in high-purity alpha-iron. *Scr. Metall.* **1982**, *16*, 973–976. [CrossRef]
58. Grabke, H.J.; Riecke, E. Absorption and diffusion of hydrogen in steels. *Mater. Tehnol.* **2000**, *34*, 331.
59. Zhang, Y.; Millett, P.C.; Tonks, M.; Biner, S.B. Deformation twins in nanocrystalline body-centered cubic mo as predicted by molecular dynamics simulations. *Acta Mater.* **2012**, *60*, 6421–6428. [CrossRef]
60. Luo, X.-M.; Zhu, X.-F.; Zhang, G.-P. Nanotwin-assisted grain growth in nanocrystalline gold films under cyclic loading. *Nat. Commun.* **2014**, *5*, 3021. [CrossRef]
61. Li, J.; Lu, C.; Pei, L.; Zhang, C.; Wang, R.; Tieu, K. Effects of H segregation on shear-coupled motion of ⟨110⟩ grain boundaries in α-Fe. *Int. J. Hydrogen Energy* **2019**, *44*, 18616–18627. [CrossRef]
62. Bai, Y.; Momotani, Y.; Chen, M.; Shibata, A.; Tsuji, N. Effect of grain refinement on hydrogen embrittlement behaviors of high-Mn TWIP steel. *Mater. Sci. Eng. A* **2016**, *651*, 935–944. [CrossRef]
63. Zan, N.; Ding, H.; Guo, X.; Tang, Z.; Bleck, W. Effects of grain size on hydrogen embrittlement in a Fe-22Mn-0.6C TWIP steel. *Int. J. Hydrogen Energy* **2015**, *40*, 10687–10696. [CrossRef]
64. Takakuwa, O.; Mano, Y.; Soyama, H. Increase in the local yield stress near surface of austenitic stainless steel due to invasion by hydrogen. *Int. J. Hydrogen Energy* **2014**, *39*, 6095–6103. [CrossRef]
65. Murakami, Y.; Kanezaki, T.; Mine, Y. Hydrogen effect against hydrogen embrittlement. *Metall. Mater. Trans. A* **2010**, *41*, 2548–2562. [CrossRef]
66. Song, J.; Curtin, W.A. Mechanisms of hydrogen-enhanced localized plasticity: An atomistic study using α-Fe as a model system. *Acta Mater.* **2014**, *68*, 61–69. [CrossRef]
67. Nibur, K.A.; Bahr, D.F.; Somerday, B.P. Hydrogen effects on dislocation activity in austenitic stainless steel. *Acta Mater.* **2006**, *54*, 2677–2684. [CrossRef]

MDPI
St. Alban-Anlage 66
4052 Basel
Switzerland
Tel. +41 61 683 77 34
Fax +41 61 302 89 18
www.mdpi.com

Materials Editorial Office
E-mail: materials@mdpi.com
www.mdpi.com/journal/materials

www.ingramcontent.com/pod-product-compliance
Lightning Source LLC
Chambersburg PA
CBHW061813210926
51850CB00006B/696
9783036573892